Lecture Notes in Computer Science 6262

Commenced Publication in 1973
Founding and Former Series Editors:
Gerhard Goos, Juris Hartmanis, and Jan van Leeuwen

W0246096

Editorial Board

Pablo García Bringas
Abdelkader Hameurlain
Gerald Quirchmayr (Eds.)

Database and Expert Systems Applications

21st International Conference, DEXA 2010
Bilbao, Spain, August 30 - September 3, 2010
Proceedings, Part II

 Springer

Volume Editors

Pablo García Bringas
University of Deusto, DeustoTech Computing
Avda. Universidades, 24, 48007 Bilbao, Spain
E-mail: pablo.garcia.bringas@deusto.es

Abdelkader Hameurlain
Paul Sabatier University, Institut de Recherche
en Informatique de Toulouse (IRIT)
118, route de Narbonne, 31062 Toulouse Cedex, France
E-mail: hameur@irit.fr

Gerald Quirchmayr
University of Vienna, Faculty of Computer Science
Department of Distributed Systems and Multimedia Systems
Liebiggasse 4/3-4, 1010 Vienna, Austria
E-mail: gerald.quirchmayr@univie.ac.at

Library of Congress Control Number: 2010932618

CR Subject Classification (1998): H.4, I.2, H.3, C.2, H.5, J.1

LNCS Sublibrary: SL 3 – Information Systems and Application,
incl. Internet/Web and HCI

ISSN	0302-9743
ISBN-10	3-642-15250-3 Springer Berlin Heidelberg New York
ISBN-13	978-3-642-15250-4 Springer Berlin Heidelberg New York

springer.com

© Springer-Verlag Berlin Heidelberg 2010
Printed in Germany

Typesetting: Camera-ready by author, data conversion by Scientific Publishing Services, Chennai, India
Printed on acid-free paper   06/3180

Preface

We welcome you to the proceedings of the 21st International Conference on Database and Expert Systems Applications held in Bilbao. With information and database systems being a central topic of computer science, it was to be expected that the integration of knowledge, information and data is today contributing to the again rapidly increasing attractiveness of this field for researchers and practitioners.

Since its foundation in 1990, DEXA has been an annual international conference, located in Europe, which showcases state-of-the-art research activities in these areas. DEXA 2010 continued this tradition and provided a forum for presenting and discussing research results in the area of database and intelligent systems and advanced research topics, applications and practically relevant issues related to these areas. It offered attendees the opportunity to extensively discuss requirements, problems, and solutions in the field in the pleasant atmosphere of the city of Bilbao, which is known for its driving industriousness, its top cultural venues and its rich and inspiring heritage and lifestyle. The University of Deusto with its great educational and research traditions, and the hospitality which the university and the city are so famous for, set the stage for this year's DEXA conference.

This volume contains the papers selected for presentation at the DEXA conference. DEXA 2010 attracted 197 submissions, and from these the Program Committee, based on the reviews, accepted two categories of papers: 45 regular papers and 36 short papers. Regular papers were given a maximum of 15 pages in the proceedings to report their results. Short papers were given an 8-page limit. Decisions made by members of the Program Committee were not always easy, and due to limited space a number of submissions had to be left out.

We would like to thank all those who contributed to the success of DEXA 2010: the hard work of the authors, the Program Committee, the external reviewers, and all the institutions (University of Deusto and University of Linz/FAW) that actively supported this conference and made it possible. Our special thanks go to Gabriela Wagner, manager of the DEXA organization, for her valuable help and efficiency in the realization of this conference.

We thank the DEXA Association and the University of Deusto for making DEXA 2010 a successful event. Without the continuous efforts of the General Chair, Pablo Garcia Bringas and his team, and the active support of José Luis del Val from Deusto University, this conference would not have been able to take place in the charming city of Bilbao.

June 2009   Abdelkader Hameurlain
                                                             Gerald Quirchmayr

Organization

Honorary Chairs

Makoto Takizawa   Seikei University, Japan

General Chair

Pablo Garcia Bringas   University of Deusto, Spain

Conference Program Chairs

Abdelkader Hameurlain   IRIT, Paul Sabatier University, Toulouse, France
Gerald Quirchmayr   University of Vienna, University of South
                                   Australia, Austria, Australia

Program Committee

Witold Abramowicz	Poznan University of Economics, Poland
Osman Abul	TOBB University, Turkey
Rafael Accorsi	University of Freiburg, Germany
Hamideh Afsarmanesh	University of Amsterdam, The Netherlands
Riccardo Albertoni	CNR-IMATI-GE, Italy
Toshiyuki Amagasa	University of Tsukuba, Japan
Sarabjot Singh Anand	University of Warwick, UK
Rachid Anane	Coventry University, UK
Annalisa Appice	Università degli Studi di Bari, Italy
José Enrique Armendáriz-Iñigo	Universidad Pública de Navarra, Spain
Mustafa Atay	Winston-Salem State University, USA
Ramazan S. Aygun	University of Alabama in Huntsville, USA
James Bailey	University of Melbourne, Australia
Spiridon Bakiras	City University of New York, USA
Ladjel Bellatreche	ENSMA-Poitiers University, France
Morad Benyoucef	University of Ottawa, Canada
Catherine Berrut	Grenoble University, France
Leopoldo Bertossi	Carleton University, Canada
Bishwaranjan Bhattacharjee	IBM Thomas J. Watson Research Center, USA
Debmalya Biswas	SAP Research, Germany
Agustinus Borgy Waluyo	Institute for Infocomm Research, Singapore
Patrick Bosc	IRISA/ENSSAT, France
Athman Bouguettaya	CSIRO, Australia

Danielle Boulanger   University of Lyon, France
Omar Boussaid   University of Lyon, France
Stephane Bressan   National University of Singapore, Singapore
Patrick Brezillon   University of Paris VI, France
Pablo Garcia Bringas   Deusto University Bilbao, Spain
Yingyi Bu   Microsoft, China
Luis M. Camarinha-Matos   Universidade Nova de Lisboa + Uninova,
                                    Portugal
Yiwei Cao   RWTH Aachen University, Germany
Barbara Carminati   Università degli Studi dell'Insubria, Italy
Silvana Castano   Università degli Studi di Milano, Italy
Barbara Catania   Università di Genova, Italy
Michelangelo Ceci   University of Bari, Italy
Wojciech Cellary   University of Economics at Poznan, Poland
Sharma Chakravarthy   The University of Texas at Arlington, USA
Badrish Chandramouli   Microsoft Research , USA
Cindy Chen   University of Massachusetts Lowell, USA
Phoebe Chen   Deakin University, Australia
Shu-Ching Chen   Florida International University, USA
Hao Cheng   University of Central Florida, USA
James Cheng   Nanyang Technological University, Singapore
Reynold Cheng   The University of Hong Kong, China
Max Chevalier   IRIT - SIG, Université de Toulouse, France
Byron Choi   Hong Kong Baptist University, Hong Kong
Henning Christiansen   Roskilde University, Denmark
Soon Ae Chun   City University of New York, USA
Eliseo Clementini   University of L'Aquila, Italy
Martine Collard   University of Nice, France
Gao Cong   Microsoft Research Asia, China
Oscar Corcho   Universidad Politécnica de Madrid, Spain
Bin Cui   Peking University, China
Carlo A. Curino   Politecnico di Milano, Italy
Emiran Curtmola   University of California, San Diego, USA
Alfredo Cuzzocrea   University of Calabria, Italy
Deborah Dahl   Conversational Technologies
Violeta Damjanovic   Salzburg Research Forschungsgesellschaft
                                    m.b.H., Austria
Jérôme Darmont   Université Lumière Lyon 2, France
Valeria De Antonellis   Università di Brescia, Italy
Andre de Carvalho   University of Sao Paulo, Brazil
Vincenzo De Florio   University of Antwerp, Belgium
Guy De Tré   Ghent University, Belgium
Olga De Troyer   Vrije Universiteit Brussel, Belgium
Roberto De Virgilio   Università Roma Tre, Italy
John Debenham   University of Technology, Sydney, Australia
Hendrik Decker   Universidad Politécnica de Valencia, Spain

Zhi-Hong Deng   Peking University, China
Vincenzo Deufemia   Università degli Studi di Salerno, Italy
Claudia Diamantini   Università Politecnica delle Marche, Italy
Juliette Dibie-Barthélemy   AgroParisTech, France
Ying Ding   Indiana University, USA
Zhiming Ding   Chinese Academy of Sciences, China
Gillian Dobbie   University of Auckland, New Zealand
Peter Dolog   Aalborg University, Denmark
Dejing Dou   University of Oregon, USA
Cedric du Mouza   CNAM, France
Johann Eder   University of Vienna, Austria
Suzanne M. Embury   The University of Manchester, UK
Christian Engelmann   Oak Ridge National Laboratory, USA
Jianping Fan   University of North Carolina at Charlotte, USA
Cécile Favre   University of Lyon, France
Bettina Fazzinga   University of Calabria, Italy
Leonidas Fegaras   The University of Texas at Arlington, USA
Yaokai Feng   Kyushu University, Japan
Stefano Ferilli   University of Bari, Italy
Eduardo Fernandez   Florida Atlantic University, USA
Filomena Ferrucci   Università di Salerno, Italy
Flavius Frasincar   Erasmus University Rotterdam, The Netherlands
Hiroaki Fukuda   Keio University, Japan
Benjamin Fung   Concordia University, Canada
Steven Furnell   University of Plymouth, UK
Aryya Gangopadhyay   University of Maryland Baltimore County, USA
Sumit Ganguly   Indian Institute of Technology, Kanpur, India
Yunjun Gao   Zhejiang University, China
Dragan Gasevic   Athabasca University, Canada
Manolis Gergatsoulis   Ionian University, Greece
Bernard Grabot   LGP-ENIT, France
Fabio Grandi   University of Bologna, Italy
Carmine Gravino   University of Salerno, Italy
Nathan Griffiths   University of Warwick, UK
Sven Groppe   Lübeck University, Germany
Crina Grosan   Babes-Bolyai University Cluj-Napoca, Romania
William Grosky   University of Michigan, USA
Volker Gruhn   Leipzig University, Germany
Stephane Grumbach   INRIA, France
Jerzy Grzymala-Busse   University of Kansas, USA
Francesco Guerra   Università degli Studi Di Modena e Reggio
                                 Emilia, Italy
Giovanna Guerrini   University of Genova, Italy
Levent Gurgen   CEA-LETI Minatec, France
Antonella Guzzo   University of Calabria, Italy
Abdelkader Hameurlain   Paul Sabatier University, Toulouse, France

Ibrahim Hamidah	Universiti Putra Malaysia, Malaysia
Wook-Shin Han	Kyungpook National University, Korea
Takahiro Hara	Osaka University, Japan
Theo Härder	TU Kaiserslautern, Germany
Saven He	Microsoft Research at Asia, China
Francisco Herrera	University of Granada, Spain
Birgit Hofreiter	University of Liechtenstein, Liechtenstein
Steven Hoi	Nanyang Technological University, Singapore
Estevam Rafael Hruschka Jr	Carnegie Mellon University, USA
Wynne Hsu	National University of Singapore, Singapore
Yu Hua	Huazhong University of Science and Technology, China
Jimmy Huang	York University, Canada
Xiaoyu Huang	South China University of Technology, China
San-Yih Hwang	National Sun Yat-Sen University, Taiwan
Ionut Emil Iacob	Georgia Southern University, USA
Renato Iannella	National ICT Australia (NICTA), Australia
Sergio Ilarri	University of Zaragoza, Spain
Abdessamad Imine	University of Nancy, France
Yoshiharu Ishikawa	Nagoya University, Japan
Mizuho Iwaihara	Waseda University, Japan
Adam Jatowt	Kyoto University, Japan
Peiquan Jin	University of Science and Technology, China
Jan Jurjens	Open University and Microsoft Research, UK
Ejub Kajan	State University of Novi Pazar, Serbia
Ken Kaneiwa	National Institute of Information and Communications Technology(NICT), Japan
Anne Kao	Boeing Phantom Works, USA
Dimitris Karagiannis	University of Vienna, Austria
Stefan Katzenbeisser	Technical University of Darmstadt, Germany
Yiping Ke	Chinese University of Hong Kong, Hong Kong
Myoung Ho Kim	KAIST, Korea
Sang-Wook Kim	Hanyang University, Korea
Markus Kirchberg	Institute for Infocomm Research, A*STAR, Singapore
Hiroyuki Kitagawa	University of Tsukuba, Japan
Carsten Kleiner	University of Applied Sciences&Arts Hannover, Germany
Ibrahim Korpeoglu	Bilkent University, Turkey
Harald Kosch	University of Passau, Germany
Michal Krátký	VSB-Technical University of Ostrava, Czech Republic
Petr Kroha	Technische Universität Chemnitz-Zwickau, Germany
Arun Kumar	IBM India Research Lab., India
Ashish Kundu	Purdue University, USA

Josef Küng	University of Linz, Austria
Lotfi Lakhal	University of Marseille, France
Kwok-Wa Lam	University of Hong Kong, Hong Kong
Nadira Lammari	CNAM, France
Gianfranco Lamperti	University of Brescia, Italy
Andreas Langegger	University of Linz, Austria
Anne Laurent	LIRMM, University of Montpellier 2, France
Mong Li Lee	National University of Singapore, Singapore
Young-Koo Lee	Kyung Hee University, Korea
Alain Toinon Leger	Orange - France Telecom R&D, France
Daniel Lemire	Université du Québec à Montréal, Canada
Scott Leutenegger	University of Denver, USA
Pierre Lévy	Public Health Department, France
Lenka Lhotska	Czech Technical University, Czech Republic
Wenxin Liang	Dalian University of Technology, China
Lipyeow Lim	IBM T. J. Watson Research Center, USA
Hong Lin	University of Houston-Downtown, USA
Tok Wang Ling	National University of Singapore, Singapore
Sebastian Link	Victoria University of Wellington, New Zealand
Volker Linnemann	University of Lübeck, Germany
Chengfei Liu	Swinburne University of Technology, Australia
Chuan-Ming Liu	National Taipei University of Technology, Taiwan
Fuyu Liu	University of Central Florida, USA
Hong-Cheu Liu	Diwan University, Taiwan
Hua Liu	Xerox Research Center at Webster, USA
Jorge Lloret Gazo	University of Zaragoza, Spain
Miguel Ángel López Carmona	University of Alcalá de Henares, Spain
Peri Loucopoulos	Loughborough University, UK
Chang-Tien Lu	Virginia Tech, USA
James J. Lu	Emory University, Atlanta, USA
Jianguo Lu	University of Windsor, Canada
Alessandra Lumini	University of Bologna, Italy
Qiang Ma	Kyoto University, Japan
Stéphane Maag	TELECOM & Management SudParis, France
Nikos Mamoulis	University of Hong Kong, Hong Kong
Vladimir Marik	Czech Technical University, Czech Republic
Elio Masciari	ICAR-CNR, Italy
Norman May	SAP Research Center, Germany
Jose-Norberto Mazon	University of Alicante in Spain, Spain
Dennis McLeod	University of Southern California, USA
Brahim Medjahed	University of Michigan - Dearborn, USA
Carlo Meghini	ISTI-CNR, Italy
Rosa Meo	University of Turin, Italy
Farid Meziane	Salford University, UK
Harekrishna Misra	Institute of Rural Management Anand, India

Jose Mocito	INESC-ID/FCUL, Portugal
Lars Mönch	FernUniversität in Hagen, Germany
Anirban Mondal	University of Tokyo, Japan
Hyun Jin Moon	UCLA Computer Science, USA
Yang-Sae Moon	Kangwon National University, Korea
Reagan Moore	San Diego Supercomputer Center, USA
Franck Morvan	IRIT, Paul Sabatier University, Toulouse, France
Yi Mu	University of Wollongong, Australia
Mirco Musolesi	University of Cambridge, UK
Tadashi Nakano	University of California, Irvine, USA
Ullas Nambiar	IBM India Research Lab, India
Ismael Navas-Delgado	University of Málaga, Spain
Wilfred Ng	University of Science & Technology, Hong Kong
Christophe Nicolle	University of Burgundy, France
Javier Nieves Acedo	Deusto University, Spain
Selim Nurcan	University of Paris 1 Pantheon Sorbonne, France
Joann J. Ordille	Avaya Labs Research, USA
Mehmet Orgun	Macquarie University, Australia
Luís Fernando Orleans	Federal University of Rio de Janeiro, Brazil
Mourad Oussalah	University of Nantes, France
Gultekin Ozsoyoglu	Case Western Reserve University, USA
George Pallis	University of Cyprus, Cyprus
Christos Papatheodorou	Ionian University, Corfu, Greece
Paolo Papotti	Università Roma Tre, Italy
Marcin Paprzycki	Polish Academy of Sciences, Warsaw Management Academy, Poland
Oscar Pastor	Universidad Politecnica de Valencia, Spain
Jovan Pehcevski	MIT University, Skopje, Macedonia
David Pinto	BUAP University, Mexico
Clara Pizzuti	CNR, ICAR, Italy
Jaroslav Pokorny	Charles University in Prague, Czech Republic
Giuseppe Polese	University of Salerno, Italy
Pascal Poncelet	LIRMM, France
Elaheh Pourabbas	National Research Council, Italy
Xiaojun Qi	Utah State University, USA
Fausto Rabitti	ISTI, CNR Pisa, Italy
Claudia Raibulet	Università degli Studi di Milano-Bicocca, Italy
Isidro Ramos	Technical University of Valencia, Spain
Praveen Rao	University of Missouri Kansas City, USA
Jan Recker	Queensland University of Technology, Australia
Manjeet Rege	Rochester Institute of Technology, USA
Rodolfo F. Resende	Federal University of Minas Gerais, Brazil
Claudia Roncancio	Grenoble University / LIG, France
Kamel Rouibah	College of Business Administration, Kuwait
Edna Ruckhaus	Universidad Simon Bolivar, Venezuela
Massimo Ruffolo	University of Calabria, Italy

Igor Ruiz Agúndez	Deusto University, Spain
Giovanni Maria Sacco	University of Turin, Italy
Fereidoon (Fred) Sadri	University of North Carolina at Greensboro, USA
Simonas Saltenis	Aalborg University, Denmark
Jose Samos	Universidad de Granada, Spain
Demetrios G. Sampson	University of Piraeus, Greece
Carlo Sansone	Università di Napoli "Federico II", Italy
Paolo Santi	Istituto di Informatica e Telematica, Italy
Igor Santos Grueiro	Deusto University, Spain
Ismael Sanz	Universitat Jaume I, Spain
N.L. Sarda	I.I.T. Bombay, India
Marinette Savonnet	University of Burgundy, France
Raimondo Schettini	Università degli Studi di Milano-Bicocca, Italy
Erich Schweighofer	University of Vienna, Austria
Florence Sedes	IRIT Toulouse, France
Nazha Selmaoui	University of New Caledonia, France
Patrick Siarry	Université Paris 12 (LiSSi), France
Gheorghe Cosmin Silaghi	Babes-Bolyai University of Cluj-Napoca, Romania
Hala Skaff-Molli	Université Henri Poincaré, France
Giovanni Soda	University of Florence, Italy
Leonid Sokolinsky	South Ural State University, Russia
MoonBae Song	Samsung Electronics, Korea
Adrian Spalka	CompuGROUP Holding AG, Germany
Bala Srinivasan	Monash University, Australia
Umberto Straccia	Italian National Research Council, Italy
Darijus Strasunskas	Norwegian University of Science and Technology (NTNU), Norway
Martin J. Strauss	Michigan University, USA
Lena Stromback	Linköpings Universitet, Sweden
Aixin Sun	Nanyang Technological University, Singapore
Raj Sunderraman	Georgia State University, USA
Ashish Sureka	Infosys Technologies Limited, India
Jun Suzuki	University of Massachusetts, Boston, USA
Jie Tang	Tsinghua University, China
David Taniar	Monash University, Australia
Cui Tao	Brigham Young University, USA
Maguelonne Teisseire	LIRMM, University of Montpellier 2, France
Sergio Tessaris	Free University of Bozen-Bolzano, Italy
Olivier Teste	IRIT, University of Toulouse, France
Stephanie Teufel	University of Fribourg, Switzerland
Jukka Teuhola	University of Turku, Finland
Taro Tezuka	Ritsumeikan University, Japan
J.M. Thevenin	University of Toulouse, France
Philippe Thiran	University of Namur, Belgium
Helmut Thoma	University of Basel, Switzerland

External Reviewers

Shafiq Alam
Amin Anjomshoaa
Yuki Arase
Zahoua Aoussat
M. Asif Naeem
Naser Ayat
Zhifeng Bao
Nicolas Béchet
Tom Heydt-Benjamin
Jorge Bernardino
Vineetha Bettaiah
Paulo Alves Braz
Souad Boukheddouma
Laurynas Biveinis
Guillaume Cabanac
Marc Chastand
Himanshu Chauhan
Yi Chen
Anna Ciampi
Camelia Constantin
Lucie Copin
Nadine Cullot
Theodore Dalamagas
Claus Dabringer
Enrique de la Hoz
Kuntal Dey
Raimundo F. Dos Santos
Fabrizio Falchi

Ming Fang
Nikolaos Fousteris
Filippo Furfaro
George Giannopoulos
Sergio A. Gómez
Shen Ge
Alberto Gemelli
Massimiliano Giordano
Wang Hao
Ben He
Yukai He
Tok Wee Hyong
Vivian Hu
Dino Ienco
Evan Jones
Christine Julien
Hideyuki Kawashima
Nikos Kiourtis
Cyril Labbé
Maria Laura Maag
Ki Yong Lee
Jianxin Li
Jing Li
Wei Li
Weiling Li
Haishan Liu
Xutong Liu
Ivan Marsa-Maestre

Michele Melchiori
Sumit Mittal
Riad Mokadem
Sébastien Nedjar
Afonso Araujo Neto
Ermelinda Oro
Ruggero G. Pensa
Yoann Pitarch
Domenico Potena
Han Qin
Julien Rabatel
Hassan Sanaifar
Federica Sarro
Pasquale Savino
Wei Shen
Chad M.S. Steel
Umberto Straccia
Jafar Tanha
Manolis Terrovitis
Ronan Tournier
Antoine Veillard
Daya Wimalasuriya
Huayu Wu
Kefeng Xuan
Jun Zhang
Geng Zhao
Rui Zhou

Table of Contents – Part II

Data Mining Systems

Parallelism and Query Planning

Data Warehousing and Decision Support Systems

Temporal, Spatial and High Dimensional Databases (Short Papers)

Data Warehousing and Data Mining Algorithms (Short Papers)

Data Mining Algorithms (I)

Information Retrieval and Database Systems (Short Papers)

Query Processing and Optimization

Application of DB Systems, Similarity Search and XML

Data Mining Algorithms (II)

Pervasive Data and Sensor Data Management (Short Papers)

Data Mining Algorithms (III)

Table of Contents – Part I

Data Management Algorithms and Performance

Decision Support Systems and Performance (Short Papers)

Data Streams (Short Papers)

XML Databases, Programming Language and Cooperative Work (Short Papers)

Query Processing and Optimization (Short Papers)

Data Privacy and Security

Temporal, Spatial, and High Dimensional Databases

Semantic Web and Ontologies (Short Papers)

Mining and Explaining Relationships in Wikipedia

Xinpeng Zhang, Yasuhito Asano, and Masatoshi Yoshikawa

Kyoto University, Kyoto, Japan 606-8501
{xinpeng.zhang@db.soc.,asano@,yoshikawa@}i.kyoto-u.ac.jp

Abstract. Mining and explaining relationships between objects are challenging tasks in the field of knowledge search. We propose a new approach for the tasks using disjoint paths formed by links in Wikipedia. To realizing this approach, we propose a naive and a generalized flow based method, and a technique of avoiding flow confluences for forcing a generalized flow to be disjoint as possible. We also apply the approach to classification of relationships. Our experiments reveal that the generalized flow based method can mine many disjoint paths important for a relationship, and the classification is effective for explaining relationships.

Keywords: link analysis, generalized max-flow, Wikipedia mining, relationship.

1 Introduction

Knowledge search has recently been researched to obtain knowledge of a single object and relations between multiple objects, such as humans, places or events. Wikipedia is widely used for searching knowledge of objects. In Wikipedia, the knowledge of an object is gathered in a single page updated constantly by a number of volunteers. Wikipedia covers objects in numerous categories, such as people, science, geography, politic, and history. Therefore, Wikipedia is usually a better choice than typical keyword search engines for searching knowledge of a single object.

A user might desire to search not only knowledge about a single object, but also knowledge about a relationship between two objects. For example, a user would desire to know the relationship between petroleum and a certain country, or to know the financial relationships between the USA and other countries. Typical keyword search engines are inadequate for discovering knowledge about a relationship; it is difficult for a user to find and organize the information about a relationship from numerous search result web pages. The main issue for analyzing relationships arises from the fact that two kinds of relationships exist: "explicit relationships" and "implicit relationships." In Wikipedia, an explicit relationship is represented as a link. A user could understand an explicit relationship easily by reading text surrounding the anchor text of the link. For example, an explicit relationship between petroleum and plastic might be represented by a link from page "Plastic" to page "Petroleum." A user could understand its meaning by reading the text "plastic is mainly produced from *petroleum*" surrounding the anchor text "petroleum" on page "Plastic." An implicit relationship is represented by multiple links and pages in Wikipedia. For example, the Gulf of Mexico is a major oil producer in the USA. This fact could be an implicit relationship represented by two links in Wikipedia: one between "Petroleum" and "Gulf of Mexico" and the other one between

P. García Bringas et al. (Eds.): DEXA 2010, Part II, LNCS 6262, pp. 1–16, 2010.

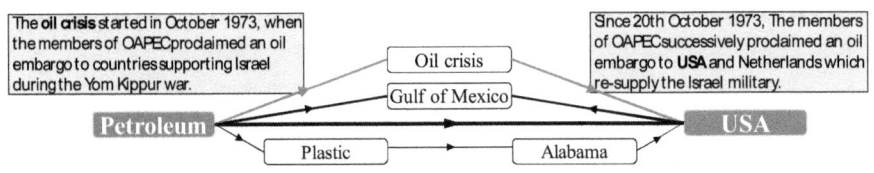

Fig. 1. Explaining the relationship between Petroleum and the USA

"Gulf of Mexico" and the "USA." It is difficult for a user to discover or understand an implicit relationship without investigating a number of pages and links. Therefore, it is an interesting problem to mine and explain an implicit relationship in Wikipedia.

Several methods [1,2,3] have been proposed for analysing relationships on an *information network* (V, E), a directed graph where V is a set of objects; edges in E represents explicit relationships between objects. A *Wikipedia information network* can be defined, whose vertices are pages of Wikipedia and edges are links between pages. In this paper, we propose a new approach for explaining a relationship from a source object s to a destination object t on a Wikipedia information network by mining disjoint s-t paths, that is, paths connecting s and t sharing no vertices except s and t with each other. For example, four disjoint paths linking "Petroleum" to "USA" depicted in Fig. 1, explain the implicit relationship between petroleum and the USA. We mine s-t paths in a network, because a user could understand the meaning of a path easily by tracing the links in the path from s to t. Tracing each link can be done by understanding the meaning of an explicit relationship represented by the link. For example, if users read the snippets shown in Fig. 1 from left to right, then they can understand the top path containing "Oil crisis". They can understand why the oil crisis is important to the relationship between petroleum and the USA. We will explain our motivation for mining "disjoint" paths for explaining a relationship in Section 2. Our motivation is mainly based on an idea that the same or similar paths should not appear multiple times in the mined paths. A similar idea is widely accepted in the field of document retrieval: a search result should not contain same or similar documents [4,5,6].

To mine paths important for a relationship in Wikipedia, we first propose a naive method based on CFEC [2], which is a method for measuring the strength of a relationship. The naive method adopts the scheme for computing the weight of a path of CFEC, although it cannot mine disjoint paths. We then propose a method to mine disjoint paths important for a relationship, based on the generalized max-flow model proposed by Zhang et al. [1]. For a relationship between two objects s and t, we compute a generalized max-flow emanating from s to t. We then output paths along which a large amount of flow is sent as paths important for the relationship. To force a generalized flow to be sent along disjoint paths as much as possible, we propose a new technique using vertex capacities. We also construct an interface for understanding a relationship by visualizing the top-k important mined paths and snippets for explaining the paths. We obtain snippets for every edge (u, v) in the paths by extracting text surrounding the anchor text of link v on page u in Wikipedia.

As an application of our approach, we propose a method for classifying relationships between a common source object and different destination objects, e.g. relationships

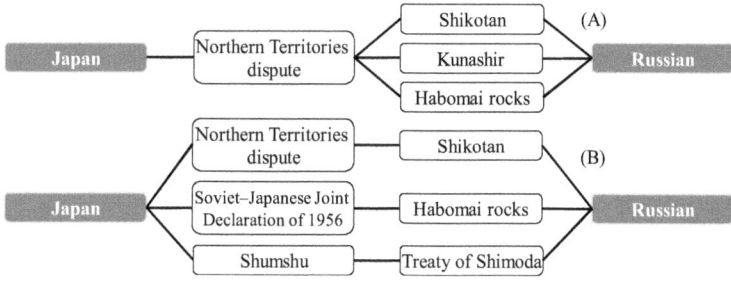

Fig. 2. Dependent paths (A) and disjoint paths (B)

between petroleum and countries, by analyzing mined paths for the relationships. For this example, the method classifies the countries into two groups which could correspond to "petroleum exporting countries" and "petroleum consuming countries."

Our experiment results reveal that the generalized flow based method mines important paths more than the method based on CFEC does, and that the proposed technique using vertex capacities is useful for mining more disjoint paths. We also confirm that the classification method is helpful for understanding relationships through case studies.

The rest of this paper is organized as follows. Section 3 reviews related work. Section 4 presents the methods for mining paths important for a relationship in Wikipedia, and the method for classifying relationships. Section 5 reports the experimental results. Section 6 concludes the paper.

2 Mining Disjoint Paths for Explaining Relationships

Users prefer not to read similar documents repeatedly, and they might desire to obtain various kinds of knowledge by reading small number of documents. Therefore, recent document information retrieval methods [4,5,6] adopt an idea that redundant information should be minimized in the top-ranked documents by removing documents similar to a higher ranked documents. For example, given a query "foreign relations of the USA," a set of the top-ranked documents should cover relations between the USA and various countries, the set should not contain a number of similar documents explaining the relation between the USA and a certain country.

Applying the idea to the problem of mining paths important for a relationship on an information network, we should avoid outputting redundant objects in the mined paths. Disjoint paths connecting two object s and t are paths sharing no vertices except s and t with each other. If we could mine disjoint paths connecting s and t, we then could prevent an object except s and t from appearing multiple times in the mined paths.

Fig. 2 (A) and (B) depict graphs constituted by three dependent paths and three disjoint paths, respectively. Both graphs explain the relationship about the territorial problem between Japan and Russian. All the three dependent paths in Fig. 2 (A) contain the same object "Northern Territories dispute" which represents the dispute between Japan and Russia over sovereignty over the South Kuril Islands, including Shikotan, Kunashir and Habomai rocks. If a user knows about the dispute, then the user could not get any

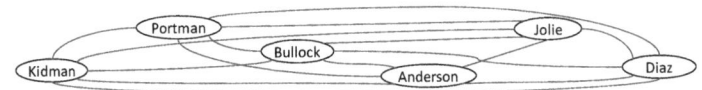

Fig. 3. A snapshot of a connection subgraph for the relationship between Kidman and Diaz

new knowledge from the dependent paths. On the other hand, the three disjoint paths depicted in Fig. 2 (B) contains no redundant object. A user could obtain knowledge other than the Northern Territories dispute such as knowledge about the "Soviet-Japanese Joint Decalaration of 1956" and the "Treaty of Shimoda." Furthermore, it is easy to understand the meaning of each disjoint path by tracing the links in the path from left to right, as discussed in Section 1. In contrast, dependent paths make a graph be too complicated to find out the order of tracing links in some cases, such as those in the graph depicted in Fig. 3. Therefore, in this paper, we aim to mine disjoint paths important for a relationship on an Wikipedia information network.

3   Related Work

Measuring the strength of an implicit relationship is one approach for explaining the relationship. Zhang et al. [1] model a relationship between two objects in a Wikipedia information network using a generalized max-flow. They ascertained a method using the model that can measure the strength of an implicit relationship more correctly than previous methods can. Several kinds of questions about relationships can be answered by measuring relationships. For example, a user could know which one of two specified countries has a stronger relationship to petroleum. However, measuring strength alone is insufficient for understanding relationships. A user would desire to know what objects constitute a relationship or what roles they play in the relationship.

Another approach to explain relationships might be extracting a "connection subgraph" [2,3,7,8]. Faloutsos et al. [3] model an information network as an electric network [9], and model the weight of a path as the current delivered by the path. Given two query vertices s and t and an undirected graph G, they extract a connected subgraph H containing s and t and limited number of other vertices that maximize the weight of H, the sum of the weights of all the paths in H. Extending the problem into more than two query vertices, Tong and Faloutsos [7] proposed *CEPS* problem. Koren et al. [2] proposed CFEC to outputs a small subgraph on which the strength of the relationship measured approximates to that measured on the original graph.

Fig. 3 is an example of a connection subgraph presented by Faloutsos et al. [3]. Vertices in a connection subgraph represent objects that are considered important for a relationship. Therefore, a user could know what objects constitute a relationship. However, it is still difficult to know what roles the objects play in the relationship using the connection subgraph. A connection subgraph usually contains several dependent paths, so that it can not present the order of tracing links. For the above example, a user would wonder which order is proper: the order from "Kidman," "Bullock," "Anderson," to "Diaz?"; or exchanging "Bullock" and "Anderson" in the order? Without the order of tracing links, a user could not understand the roles of objects correctly by reading Wikipedia pages. Therefore, a connection subgraph is inadequate for understanding

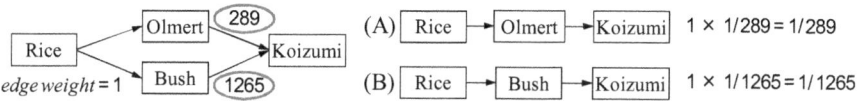

Fig. 4. An example for CFEC

a relationship. We ascertain through experiments in Section 5.2 that methods [2] extracting connection subgraph are inadequate for mining disjoint paths important for a relationship.

To create a connection subgraph for a relationship, the methods [2,3,8] discussed above first compute the weights of paths using random walk [9]. They define the weight of a path fundamentally as the product of the weights of the edges composing the path divided by the product of the weights of the edges incident to every vertex in the path. Therefore, random walk based methods have a property that they compute the weight of a path extremely small if a popular object—an object linked by or from many other objects—exists in the path. We claim that this property is unsuitable for mining paths important for a relationship through experiments discussed in Section 5.2.

Zhu et al. [10] extract explicit relationships between pairs of people from the Web. Some semantic based methods extract good paths between two entities in an RDF graph [11,12,13]. Assuming semantics in an information network is beyond the scope of this paper. Therefore, we do not adopt the ideas used by these methods.

4 Methods for Mining Disjoint Paths in Wikipedia

We now present our methods for mining disjoint paths important for a relationship. To the best of our knowledge, no such method was proposed. We first propose a naive method based on CFEC [2] in section 4.1.

4.1 Naive Method Based on CFEC

Given a graph G, a source vertex s and a destination vertex t, CFEC first finds the n shortest paths between s and t. It then computes the strength of the relationship between s and t using random walk on the paths [9]. In CFEC, the weight of a path $p = (s = v_1, v_2, ..., v_\ell = t)$ from s to t is defined as $w_{sum}(v_1) \cdot \prod_{i=1}^{\ell-1} \frac{w(v_i, v_{i+1})}{w_{sum}(v_i)}$, where $w(u, v)$ is the weight of edge (u, v) and $w_{sum}(v)$ is the sum of the weights of the edges going from vertex v. For example, Fig. 4 depicts two paths between "Rice" and "Koizumi." The number shown beside a vertex o_i is the number of links going from the vertex, which equals to $w_{sum}(o_i)$ if the weight of every edge is 1. The weights of path (A) and path (B) become 1/289 and 1/1265, respectively.

We now propose a naive method for mining paths important for a relationship between a source object s and a destination object t in Wikipedia based on CFEC [2].

(1) Construct a network $G = (V, E)$ using pages and links within at most m hop links from s or t in Wikipedia. (2) Set the weight of every edge $e \in E$ to 1, compute the top-k paths in decreasing order of the path weight. (3) For each edge (u, v) in the

top-k paths, extract an explanatory snippet, i.e., text surrounding the anchor text of link v on page u, using a KWIC concordance tool [14].

The top-k paths mined by the method probably contain some dependent paths. Although we can select some disjoint paths among the mined paths, we cannot determine in advance how many paths should be mined to obtain a specified number of disjoint paths. Moreover, this method has a problem because of popular objects, as discussed in Section 3. For example, in Fig. 4, the weight of path (B) is significantly smaller than that of path (A), because "Bush" is more popular, i.e., linked from or to more objects, than "Olmert." Consequently, important paths containing a popular object seldom appear in the top-k paths mined by the method.

4.2   Improvements Using Doubled Network and Domain-Based Weight

We now discuss two improvements for mining important paths: a doubled network and an edge weight function using the category information on Wikipedia. Both improvements were originally proposed for measuring a relationship [1].

A path constituted by links of different directions could be important for a relationship in Wikipedia. For example, the path *(Petroleum, Plastic, Alabama, USA)* in Fig. 1 is formed by links of different directions. Tha path would correspond to an important fact between "Petroleum" and the "USA" that the Alabama State of the *USA* produces a large quantity of plastic from *petroleum*. To mine such paths, we construct a doubled network [1] by adding to every original edge a reversed edge whose direction is opposite to the original one. For example, Fig. 6(B) depicts the doubled network G' for G in Fig. 6(A).

For mining important paths, it is desired to assign a larger weight to an important edge. The importance of an edge depends on its roles for the relationship. Let consider the relationship between the American politician "Rice" and the Japanese politician "Koizumi" depicted in Fig. 4. In the example, we should assign a larger weight to the primarily important edges connecting American and Japanese politicians, than probably unimportant edges connecting American and Israeli Politicians. Edges connecting American politicians or Japanese politicians would be secondarily important. Therefore, the weight of an edge is best determined according to the kinds of objects that the source and the destinations are. To realize such a weight assignment, we must construct groups of objects in Wikipedia, such as "Japanese politicians" and "baseball players". In Wikipedia, a page corresponding to an object belongs to at least one category. For example, "George W. Bush" belongs to the category "Presidents of the USA." However, categories cannot be used as groups directly because the category structure of Wikipedia is too fractionalized. Given a category c_i, Zhang et al. [1] construct the group for c_i by grouping c_i and its descendant categories which represent sub concepts of c_i together. Let S be the set of objects belonging to a category in group for a category of the source. Similarly, let T be the set of objects for the destination. Zhang et al. [1] then proposed an edge weight function to assign a high weight to edges connecting an object in S with an object in T; assign a medium weight to edges connecting objects in s or edges connecting objects in t; and assign low weights to other kinds of edges. Zhang et al. [1] assign a smaller weight to a reversed edge in the doubled network than that of its original

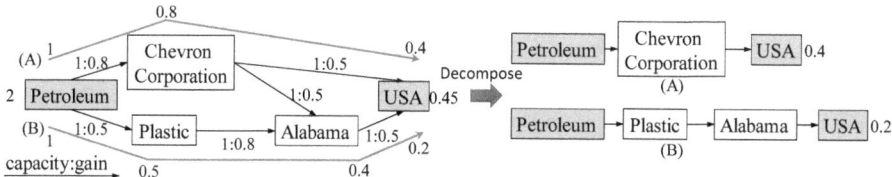

Fig. 5. A generalized max-flow and its decomposition

one. We omit the details because of space limitations. The groups and the weight function is useful for mining important paths directly.

4.3 Generalized Max-Flow Model

The naive method was unable to mine disjoint paths. We propose a generalized flow based method that could mine disjoint paths in Section 4.4. Before introducing the method, we explain its basis: the generalized max-flow model proposed by Zhang et al. [1] for computing the strength of a relationship.

The generalized max-flow problem [15][16] is identical to the classical max-flow problem except that every edge e has a gain $\gamma(e) > 0$; the value of a flow sent along edge e is multiplied by $\gamma(e)$. Let $f(e) \geq 0$ be the amount of flow f on edge e, and $\mu(e) \geq 0$ be the capacity of edge e. The capacity constraint $f(e) \leq \mu(e)$ must hold for every edge e. The goal of the problem is to send a flow emanating from the source into the destination to the greatest extent possible, subject to the capacity constraints. Let *generalized network* $G = (V, E, s, t, \mu, \gamma)$ be information network (V, E) with the source $s \in V$, the destination $t \in V$, the capacity μ, and the gain γ. Fig. 5 depicts an example of a generalized max-flow. 0.4 units and 0.2 units of the flow arrive at "USA" along path (A) and path (B), respectively.

To use edges of both directions, Zhang et al. [1] construct a *doubled network*, as discussed in Section 4.2. The reversed edge e_{rev} for every edge e in G is assigned with $\mu(e_{rev}) = \mu(e)$ and $\gamma(e_{rev}) = rev(e) = \lambda \times \gamma(e), 0 \leq \lambda \leq 1$, as depicted in Fig. 6 (B). Also, a new constraint $f(e)f(e_{rev}) = 0$ for every edge e is introduced to satisfy the capacity constraint on the doubled network. To assign gain for edges, Zhang et al. [1] use the edge weight function introduced in Section 4.2.

4.4 A Generalized Flow Based Method

We propose a generalized flow based method to mine disjoint paths important for a relationship from object s to object t in Wikipedia. We use a new technique for avoiding "confluences" of a generalized max-flow by setting vertex capacities.

We first present the method as follows.

(1) Construct a generalized network $G = (V, E, s, t, \mu, \gamma)$ using pages and links within at most m hop links from s or t in Wikipedia.
(2) Construct the doubled network G' for G, determine edge gain γ using the edge weight function discussed in Section 4.2, and set vertex capacities discussed later.

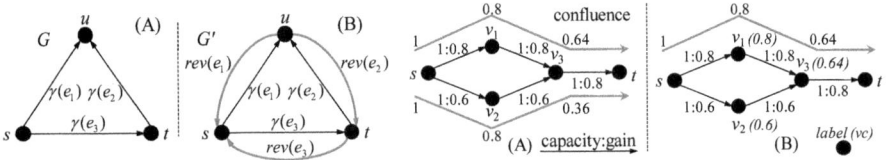

Fig. 6. A doubled network   Fig. 7. The flow confluence problem

(3) Compute a generalized max-flow f emanating from s into t on G'.
(4) Decompose the flow f into flows on a set P of paths [16]. Let $df(p_i)$ denote the value of flow on a path p_i, for $i = 1, 2, ..., |P|$. For example, the flow on the network depicted in Fig. 5 is decomposed into flows on two paths (A) and (B). The value of the decomposed flow on path (A) is 0.4; that on path (B) is 0.2.
(5) Output the top-k paths in decreasing order of $df(p_i)$.
(6) For each edge (u, v) in the top-k paths, extracts an explanatory snippet, i.e., text surrounding the anchor text of link v on page u, using a KWIC concordance tool [14].

We next discuss the new technique of setting vertex capacity. The generalized max-flow problem is a natural extension of the classical max-flow problem whose flow is always sent along disjoint paths. A problem arises, however, which is attributable to the gain: a flow can be confluent at a vertex except s and t. For example, Fig. 7 (A) depicts a confluence of flow at vertex v_3; the amount of the flow sent along (v_1, v_3) becomes 0.64 at v_3. That along (v_2, v_3) becomes 0.36. The flow can be confluent at v_3 and can be sent along (v_3, t). If a generalized max-flow is confluent at many vertices, then the paths composing the flow become dependent paths. Consequently, the top-k paths obtained in (5) might contain some dependent paths. One idea to solve the problem is to introduce a constraint that a flow must be sent along vertex disjoint paths. Unfortunately, no polynomial-time algorithm exists, to the best of our knowledge, to solve the generalized max-flow with the constraint.

We propose an approach to prevent a flow from being confluent to the greatest extent possible. Concretely, we set the capacity of every edge to one and set the capacity of every vertex v, except s and t, to $\mu'(v) = \max_{p \in P} \prod_{e \in p} \gamma(e)$, the maximum production of the gains of the edges in a path, where P is the set of all paths from s to v. The vertex capacity of s or t is set to ∞. The capacities of all the vertices can be computed easily by solving a single source shortest path problem setting the length of edge e to $-\log(\gamma(e))$. Because the capacity of every edge is setting to 1, the largest value of the flow could be sent to node v along a single path going from s to v is $\mu'(v)$. Therefore, by setting the capacity of vertex v to $\mu'(v)$, most of the time, we could prevent a flow to be sent to v along more than one paths. For example, Fig. 7 (B) depicts the vertex capacity function for the generalized network depicted in Fig. 7 (A). Although a flow is confluent at vertex v_3 in Fig. 7 (A), the confluence does not happen in Fig. 7 (B) because of the vertex capacity. We examine how effective the vertex capacity function is using experiments discussed in Section 5.2.

Country	Elucidatory Objects
Japan	Oil crisis, Niigata, Nihon Shoki, Kyushu Oil Co., Ltd., added-profit trade, Crude oil, Nippon Oil Corp., Japanese post-war economic miracle, ...
Saudi Arabia	Ghawar Field, OAPEC, Crude oil, Oil field, Price of petroleum, Oil-producing Country, Rub' al Khali, Arabian Oil Company, OPEC, ...
Kuwait	Burgan Field, OAPEC, Crude oil, Oil field, Oil-producing Country, OPEC, Asphalt, Gulf War, Middle East War, ...

Fig. 8. Elucidatory objects for relationships from petroleum to each country

Group	Countries	Label
0	Saudi Arabia, Kuwait, Iran, Bahrain, Libya	Oil crisis, OAPEC, Oil-producing country, Middle East, Oil field, Price of petroleum, Saudi Aramco, Iran–Iraq War, Asphalt
1	Japan, USA, Russia, China, UK	Crude oil, Middle East, Asphalt, Oil field, Iraq, Iran, Price of petroleum, North Sea oil, Sudan

Fig. 9. Classification for relationships between petroleum and countries

4.5 Classification for Relationships

In this section, given a set of relationships between a common source object and different destination objects, we apply our method for mining paths to classify the destination objects in the relationships. For example, given a set of relationships between petroleum and countries, we classify the countries into groups. We first mine the top-k paths for each relationship, say $k = 50$. We define *elucidatory objects* for a relationship as the objects in the paths, except the source and destination. Intuitively, similar relationships could share many common elucidatory objects. For example, Fig. 8 presents some elucidatory objects for Japan, Saudi Arabia, and Kuwait. Saudi Arabia and Kuwait, which are both oil-producing countries in the Middle East, share many common elucidatory objects. On the other hand, Japan shares almost no elucidatory objects with them.

We apply a frequent itemsets based clustering method named FIHC [17] to our classification. In fact, FIHC is used to classify documents using sets of words appearing together in many documents. Using elucidatory objects instead of words, we can obtain clusters of destinations. Every cluster is also assigned a label which is a set of elucidatory objects shared by every relationship in the cluster. In some cases, the clusters obtained by FIHC could be too numerous for a user to understand. Our classification method unifies them into fewer groups in response to a user's request, by computing similarities between clusters according to frequent elucidatory objects of every destination in the each cluster. We omit details of the classification method because of space limitations. As an example, our method classifies the relationships from petroleum to the top-10 countries strongly related to petroleum into two groups. Fig. 9 presents the groups of the countries. By investigating the labels of groups, a user could understand that group 0 and group 1 respectively correspond to "petroleum exporting countries" and "petroleum consuming countries."

5   Experiments and Evaluation

5.1   Dataset and Environment

We perform experiments on a Japanese Wikipedia dataset (2009/05/13 snapshot). We first extract 27,380,916 links that appeared in all pages. We then remove pages that are not corresponding to objects, such as each day, month, category, person list, and portal. We also remove links to such pages, and obtain 11,504,720 remaining links.

We implemented our program in Java and performed experiments on a PC with four 3.0 GHz CPUs (Xeon), 64 GB of RAM, and a 64-bit operating system (Windows Vista).

5.2   Evaluation of Mined Paths

In this section, we first investigate whether paths mined by our methods are actually important for a relationship. We then examine how many of the mined paths are disjoint.

Let the following five symbols represent our methods below. (o) is the naive method explained in Section 4.1. (e), (d), (de) are the naive methods using improvements described in Section 4.2: (e) the edge weight function, (d) the doubled network, and (de) the both ones. (g) is the generalized flow based method proposed in Section 4.4. We select 105 relationships between two objects of the following six types: (1) two politicians, (2) two countries, (3) a politician and a country, (4) petroleum and a country, (5) Buddhism and a country, and (6) two countries' cuisines. To mine paths important for a relationship between s and t, we construct a network G using pages and links within at most three hop links from s or t in Wikipedia, for all methods. Careful observation of Wikipedia pages revealed that several paths formed by three links are important for a relation, although we were able to find few important paths formed by four links. In preliminary experiments, we also find that paths formed by four links seldom appear in the top-k paths mined on G constructed using four hop links.

Path Importance.   It is desired to consider the following two questions to evaluate our methods: (Q1) How many mined paths are important for a relationship; and (Q2) Do most paths important for a relationship could be found by our methods? To answer these two questions, we evaluate the importance of the mined paths by human subjects.

We first randomly select 10 from the 105 relationships (one or two from each of the six types) explained above. For each relationship, we mine a set of the top-20 paths by each of our five methods. Let P be the union of these five sets. On average, P contains 50-60 paths, because the sets mined by different methods usually overlap. We then ask 10 testers to evaluate every path p in P and every edge $e(u, v)$ in p. To each edge $e(u, v)$, every tester assigns an integer score 0, 1, or 2 representing the strength of the explicit relationship between u and v, by reading the explanatory snippet of $e(u, v)$. A higher score was assigned to a stronger relationship. To each path p, every tester assigns an integer score 0, 1, or 2 representing how important p is for the relationship between the source s and the destination t, by reading the snippets of the edges in p along the direction from s to t. A higher score was assigned to a more important path. We then compute the average score of every edge and every path for each relationship.

We present some examples of the assignments here. All the testers assigned score 2 to the top path and the two edges in the path depicted in Fig. 1. For the relationship from petroleum to Saudi Arabia, path *(Petroleum, Burgan Field, Saudi Arabia)* is mined. The snippet of edge *(Burgan Field, Saudi Arabia)* is "Burgan Field in Kuwait is still one of the world's easiest production sites now, which differs from the Ghawar Field in Saudi Arabia." Most testers assigned score 0 to the path and the edge. Let us consider another path *(George W. Bush, Yasuo Fukuda, Junichiro Koizumi)*. Each of "Yasuo Fukuda" and "Junichiro Koizumi" was a prime minister of Japan during the tenure of "George W. Bush" as the president of the USA. Most testers think that both the two edges in the path represent strong explicit relationships. However, they think that "Yasuo Fukuda" is unimportant in the relationship. Consequently, they assigned score 0 to the path.

It is difficult to find all important paths for a relationship in Wikipedia. Therefore, we could not adopt the conventional precision and recall based evaluation to answer questions Q1 and Q2 presented above. Instead, we introduce two measures : "Important Path Ratio $(PRatio)$" for Q1, "Retrieved Important Path Ratio $(RPRatio)$." Let $TP@n$ be the set of the top-n paths mined by a method, let $AP@n$ be the union of the top-n paths mined by every method, and let $s(p)$ be the average score of path p, then

$$PRatio@n = \frac{\sum_{p \in TP@n} s(p)}{2 \times n}, \qquad RPRatio@20 = \frac{\sum_{p \in TP@20} s(p)}{\sum_{p \in AP@20} s(p)}.$$

These ratios vary from 0 to 1. For a method, if all the average scores of every path in $TP@20$ are 2, then $PRatio@20$ is 1; if the other methods yielded no path with an average score greater than 0 except the ones in $TP@20$, then its $RPRation@20$ becomes 1. Therefore, $PRatio$ corresponds to an absolute evaluation; $RPRatio$ corresponds to a relative evaluation. Similarly, we define "Important Edge Ratio $(ERatio)$" and "Retrieved Important Edge Ratio $(RERatio)$." Let $TE@n$ be the set of edges in $TP@n$, let $AE@n$ be the set of edges in $AP@n$, and let $s(e)$ be the average score of edge e, then

$$ERatio@n = \frac{\sum_{e \in TE@n} s(e)}{2 \times |TE@n|}, \qquad RERatio@20 = \frac{\sum_{e \in TE@20} s(e)}{\sum_{e \in AE@20} s(e)}.$$

Table 1 presents the $PRatios$ and $RPRatios$ of our five methods (g), (d), (de), (o), and (e) for only five relationships because of space limitations. Similar results are obtained for the remaining five relationships. The shaded cells emphasize the maximum ratios of each relationship. The generalized flow based method (g) yields most of the highest ratios. The ratios obtained by the methods without the double network, (o) and (e), are significantly low for some relationships, such as the relationship between petroleum and the USA. Using only the original directions of edges, few directed paths exist from the source to the destination in the network. However, several important paths formed by edges of different directions are mined by using the doubled network.

Fig. 10 presents the average $ERatios$ and $RERatio$ of the edges of all the 10 relationships obtained by each method. Similarly, Fig. 11 presents the average $PRatios$ and $RPRatio$ of the paths of the 10 relationships. Method (g) produces the highest average ratios for both paths and edges. All methods yield high average $ERatios$; they can mine many edges representing strong explicit relationships. However, such edges do not necessarily constitute a path important for a relationship. For example, the path *(George W. Bush, Yasuo Fukuda, Junichiro Koizumi)* discussed above is constituted by

12   X. Zhang, Y. Asano, and M. Yoshikawa

Table 1. $PRatio$s and $RPRatio$s of paths

Relationship	Method	$PRatio$ @5	$PRatio$ @10	$PRatio$ @20	$RPRatio$ @20
Japan – Russia	g	0.88	0.87	0.71	0.64
	d	0.76	0.56	0.54	0.49
	de	0.84	0.69	0.61	0.55
	o	0.52	0.49	0.47	0.43
	e	0.68	0.57	0.50	0.45
Petroleum – USA	g	0.92	0.75	0.68	0.85
	d	0.44	0.38	0.47	0.59
	de	0.76	0.57	0.53	0.66
	o	0.10	0.10	0.10	0.10
	e	0.10	0.10	0.10	0.10
Buddhism – Sri Lanka	g	0.70	0.54	0.50	0.45
	d	0.58	0.5	0.43	0.39
	de	0.58	0.53	0.46	0.41
	o	0.64	0.52	0.48	0.43
	e	0.78	0.52	0.48	0.43
Yoshiro Mori – China	g	0.40	0.46	0.45	0.47
	d	0.52	0.5	0.35	0.37
	de	0.52	0.48	0.38	0.39
	o	0.42	0.3	0.26	0.27
	e	0.44	0.35	0.27	0.28
George W. Bush – Junichiro Koizumi	g	0.98	0.8	0.76	0.49
	d	0.80	0.82	0.74	0.47
	de	0.8	0.82	0.73	0.47
	o	0.96	0.86	0.79	0.50
	e	0.96	0.88	0.79	0.50

such edges, but is not important. With respect to paths, the method (g) produces significantly higher average ratios than the other methods. The methods without the doubled network, (o) and (e), produce the lowest average ratios for paths, because paths formed by edges of different directions cannot be mined by them. The methods without the edge weight function, (o) and (d), produce lower ratios for paths than those using the edge weight, (e) and (de), respectively. Therefore, we conclude that the generalized flow based method is the best for mining many paths important for relationships, and that the doubled network and the edge weight function are effective.

Evaluation of Disjoint Paths. We have proposed a technique for avoiding confluences using the vertex capacity function in section 4.4. We examine how many disjoint paths were mined by each method, and how effective the technique is. For the selected 105 relationships, we first mine the top 20 and the top 50 paths by each of our five methods and the generalized flow based method without the technique. We then count the disjoint paths in the mined paths for each relationship. Fig. 12 (A) and Fig. 12 (B) depict the average number of disjoint paths in the top-20 and that in the top-50 paths, respectively. The symbol (g, w/ vc) denotes the generalized flow based method with the technique, and (g, w/o vc) denotes that without the technique. Method (g, w/ vc) produced the highest average number for both the top-20 and top-50 paths; especially, all the top-20

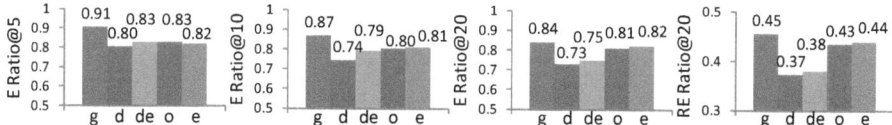

Fig. 10. Average *ERatio* and *RERatio* of edges of all the 10 relationships

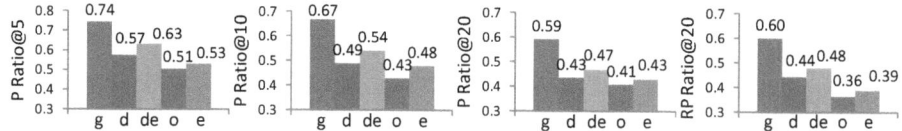

Fig. 11. Average *PRatio* and *RPRatio* of paths of all the 10 relationships

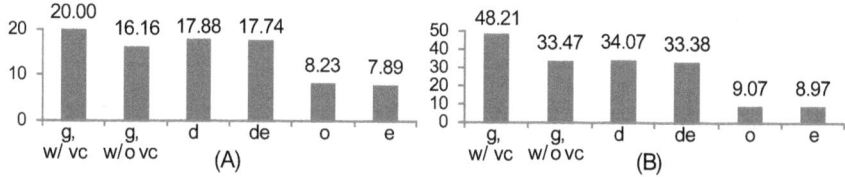

Fig. 12. Number of disjoint paths mined by each method

paths are disjoint for all the 105 relationships. The naive methods without the doubled network, (o) and (e), produced the lowest average numbers. Consequently, we observed the following three facts: (1) Our technique is effective in mining disjoint paths. (2) The naive method is inadequate for mining disjoint paths. (3) The doubled network is effective in mining disjoint paths formed by edges of different directions.

As discussed in Section 2, we mine disjoint path to prevent an object appearing multiple times in the mined paths. Therefore, we also evaluate how frequent an object appears in the top-k paths. We compute the average object frequency $f@k$ in the top-k paths important for a relationship between s and t mined by each method. Let O_k be the set of objects in the top-k paths, except s and t, and let $n_{k,j}$ denote how many times the j-th object $o_{k,j} \in O_k$ appears in the top-k paths. Then, the average object frequency is defined as $f@k = \frac{\sum_j n_{k,j}}{|O_k|}$. Note that $f@k$ is at least 1. If $f@k = 1$, then every object appears only once in the top-k paths; if a number of objects appear many times in the top-k paths, then $f@k$ becomes larger than 1. Fig. 13 illustrates the average value of $f@k$ in the top-k paths mined by each method, for the selected 105 relationships. The method (g, w/ vc) has the lowest $f@k$ among all methods; especially, $f@100 = 1.04$ is almost equal to 1. That is, almost all objects appear only once in the top-100 paths mined by the method (g, w/ vc). The method without setting the vertex capacity (g, w/o vc) has higher $f@k$ than the method (g, w/ vc). The naive methods without the doubled network, (o) and (e), produced the highest $f@k$; the values of $f@k$ increase dramatically as k increases. Consequently, we conclude that our technique of using the

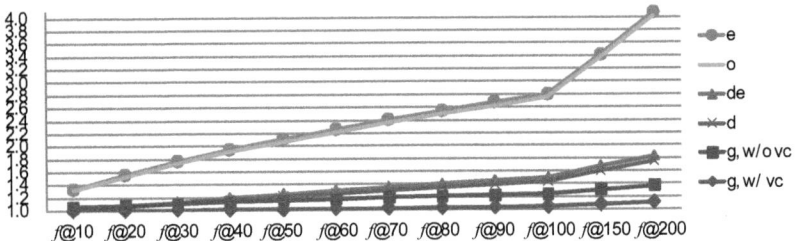

Fig. 13. Average object frequency in the top-k paths mined by each method

Table 2. Objects in the paths for relationship between "Japanese cuisine" and "Chinese cuisine"

The generalized flow base method (g)
Karaage (269), Chili pepper (402), Soy milk (103), Sesame oil (95), Mochi (345), Dashi (305), Ginger (344), Donburi (119), Tonkatsu (215), Sashimi (477), Fried vegetables (87), Jiaozi (341), Jellied fish (45), Yatai (412), Chazuke (164), Kenchin soup (47), Western Cuisine (77), Crab stick (58), Japanese noodle (1038)
The naive method (de)
Nouvelle Chinois (12), Three major world cuisine (10), Seafood (12), Wynn Macau (13), Cooking School of West Japan (15), The Family Restaurant (15), Kazuhiko Cheng (31), Radisson Hotel Bangkok (21), Hotel Laforet Tokyo (17), Banyan Tree Bangkok (18), Jellied fish (45), West (Japanese restaurant chain) (19), Soup spoon (38), Grand Hyatt Fukuoka (20), Grand Hyatt Singapore (16), Ship dish person (20), Resort Okinawa Marriott & Spa (21), Hyatt Regency Osaka (22)

vertex capacity function is effective for avoiding redundant objects in the mined paths. Many objects appear frequently in the paths mined by the naive methods, although the doubled network is helpful for alleviating the redundancy issue.

Case Studies for Understanding Relationships. Table 2 presents the elucidatory objects in the top-20 paths important for the relationship between "Japanese cuisine" and "Chinese cuisine," mined by methods (g) and (de), respectively. The number in the parentheses behind each object is the number of links going from or to the page representing the object in Wikipedia. Each object shown in Table 2 constitutes a mined path, e.g. "Karaage" constitutes *(Japanese cuisine, Karaage, Chinese cuisine)*. Method (g) mines many Japanese foods originated in China, such as karaage, mochi, fried vegetables, jiaozi, Japanese noodle, and soy milk. Method (g) also mines some cooking ingredients used in both cuisines, such as chili pepper, sesame oil, and ginger. On the other hand, most elucidatory objects mined by method (de) are hotels or restaurants purveying both cuisines. As discussed in Section 3, random walk based methods always underestimate popular objects; inversely, the weights of paths constituted by objects having few links are always overestimated. As shown in Table 2, these hotels and restaurants have few links in Wikipedia. Therefore, the naive methods based on CFEC overestimate paths constituted by objects corresponding to these pages. However, method (g) mined many important objects regardless of how many links the objects have. Therefore,

Group	Industries	Label
0	Retailing, information and communication industry, Service, Traffic, Finance industry	Convenience store, Vending machine, Emissions Trading, Transportation, Niigata, Industry, Senko Co.,Ltd., Shima Spain Village
1	Construction industry, Manufacturing	Fuel cell, Biofuel, Carbon footprint, Convenience store, Emissions Trading, Industrial evolution, USA
2	Agriculture, Forestry, Fishing industry	Afforestation, Local production for local consumption, Biodiesel, Biomass, Biofuel

Fig. 14. Classification for relationships between CO_2 and industries

we confirm that the generalized flow based method is more appropriate than the naive methods for mining paths important for a relationship in Wikipedia.

5.3 Case Study: Classification for Relationships

We present an example of our classification for relationships from carbon dioxide, CO_2, to the top-10 industries strongly related to CO_2. Our method discussed in Section 4.5 then classifies the 10 industries into three groups. Fig. 14 presents the groups and the label for each group. By investigating the groups and the labels, a user could understand the classification. In fact, the groups 0, 1, and 2 respectively correspond to the tertiary sector, the secondary sector, and the primary sector of the economy. The label for group 2 includes "Afforestation," "Local production for local consumption," "Biodiesel," "Biomass," and "Biofuel," which are approaches performed in "Agriculture," "Forestry," or "Fishing industry," for decreasing CO_2 emissions. The label for group 0 also contains objects related to CO_2 emitted by the industries in the group. For example, "Shima Spain Village" is a famous amusement park in Japan, and "Senko Co. Ltd." is a Japanese Logistics company, both of which use renewable energy; "Niigata" is one of the top three cities having high CO_2 emissions per capita in the transportation industry of Japan. Similarly, the label for group 1 is helpful for understanding the relationships between CO_2 and the industries in group 1. Consequently, we confirmed that our classification method could give a user a better understanding of relationships.

6   Conclusion

We proposed a new approach for explaining a relationship between two objects by mining disjoint paths connecting the objects on Wikipedia. We realized the approach by proposing the naive method and the generalized flow based method. Our experiments revealed that the generalized flow based method can mine many disjoint paths important for relationships, and that the proposed technique for avoiding flow confluences is very effective in improving the method. We ascertained that our classification, proposed as an application of our approach, is also helpful for understanding relationships.

We plan to apply the mined paths for a relationship to Web search of images and texts explaining the relationship.

Acknowledgment

This work was supported in part by the National Institute of Information and Communications Technology, Japan.

References

1. Zhang, X., Asano, Y., Yoshikawa, M.: Analysis of implicit relations on Wikipedia: Measuring strength through mining elucidatory objects. In: Kitagawa, H., Ishikawa, Y., Li, Q., Watanabe, C. (eds.) DASFAA 2010. LNCS, vol. 5981, pp. 460–475. Springer, Heidelberg (2010)
2. Koren, Y., North, S.C., Volinsky, C.: Measuring and extracting proximity in networks. In: Proc. of 12th ACM SIGKDD Conference, pp. 245–255 (2006)
3. Faloutsos, C., McCurley, K.S., Tomkins, A.: Fast discovery of connection subgraphs. In: Proc. of 10th ACM SIGKDD Conference, pp. 118–127 (2004)
4. Zhang, B., Li, H., Liu, Y., Ji, L., Xi, W., Fan, W., Chen, Z., Ma, W.Y.: Improving web search results using affinity graph. In: Proc. of 28th SIGIR, pp. 504–511 (2005)
5. Chen, H., Karger, D.R.: Less is more: probabilistic models for retrieving fewer relevant documents. In: Proc. of 29th SIGIR, pp. 429–436 (2006)
6. Clarke, C.L., Kolla, M., Cormack, G.V., Vechtomova, O., Ashkan, A., Büttcher, S., MacKinnon, I.: Novelty and diversity in information retrieval evaluation. In: Proc. of 31th SIGIR, pp. 659–666 (2008)
7. Tong, H., Faloutsos, C.: Center-piece subgraphs: Problem definition and fast solutions. In: Proc. of 12th ACM SIGKDD Conference, pp. 404–413 (2006)
8. Cheng, J., Ke, Y., Ng, W., Yu, J.X.: Context-aware object connection discovery in large graphs. In: Proc. of 25th ICDE, pp. 856–867 (2009)
9. Doyle, P.G., Snell, J.L.: Random Walks and Electric Networks, vol. 22. Mathematical Association America, New York (1984)
10. Zhu, J., Nie, Z., Liu, X., Zhang, B., Wen, J.R.: Statsnowball: a statistical approach to extracting entity relationships. In: Proc. of 18th WWW, pp. 101–110 (2009)
11. Anyanwu, K., Maduko, A., Sheth, A.P.: Semrank: ranking complex relationship search results on the semantic web. In: Proc. of 14th WWW, pp. 117–127 (2005)
12. Kasneci, G., Suchanek, F.M., Ifrim, G., Ramanath, M., Weikum, G.: Naga: Searching and ranking knowledge. In: Proc. of 24th ICDE, pp. 953–962 (2008)
13. Aleman-Meza, B., Halaschek-Wiener, C., Arpinar, I.B., Sheth, A.P.: Context-aware semantic association ranking. In: Proc. of 1st SWDB, pp. 33–50 (2003)
14. Manning, C., Schutze, H.: Foundations of Statistical Natural Language Processing. MIT Press, Cambridge (1999)
15. Wayne, K.D.: Generalized Maximum Flow Algorithm. PhD thesis, Cornell University, New York, U.S (January 1999)
16. Ahuja, R.K., Magnanti, T.L., Orlin, J.B.: Network Flows: Theory, Algorithms, and Applications. Prentice-Hall, New Jersey (1993)
17. Fung, B.C.M., Wang, K., Ester, M.: Hierarchical document clustering using frequent itemsets. In: Proc. of 3rd SDM (2003)

Publishing Time-Series Data under Preservation of Privacy and Distance Orders

Yang-Sae Moon[1], Hea-Suk Kim[1], Sang-Pil Kim[1], and Elisa Bertino[2]

[1] Department of Computer Science, Kangwon National University, Korea
{ysmoon,hskim,spkim}@kangwon.ac.kr
[2] Department of Computer Science, Purdue University, USA
bertino@cs.purdue.edu

Abstract. In this paper we address the problem of preserving mining accuracy as well as privacy in publishing sensitive time-series data. For example, people with heart disease do not want to disclose their electrocardiogram time-series, but they still allow mining of some accurate patterns from their time-series. Based on this observation, we introduce the related assumptions and requirements. We show that only randomization methods satisfy all assumptions, but even those methods do not satisfy the requirements. Thus, we discuss the randomization-based solutions that satisfy all assumptions and requirements. For this purpose, we use the *noise averaging effect* of piecewise aggregate approximation (PAA), which may alleviate the problem of destroying distance orders in randomly perturbed time-series. Based on the noise averaging effect, we first propose two naive solutions that use the random data perturbation in publishing time-series while exploiting the *PAA distance* in computing distances. There is, however, a tradeoff between these two solutions with respect to uncertainty and distance orders. We thus propose two more advanced solutions that take advantages of both naive solutions. Experimental results show that our advanced solutions are superior to the naive solutions.

Keywords: data mining, time-series data, privacy preservation, similarity search, data perturbation.

1 Introduction

In recent years privacy preserving data mining (PPDM) [2,4] has been investigated extensively motivated by the current practice by private and public organizations of collecting large amounts of often sensitive data. The aim of PPDM algorithms is to extract relevant knowledge from a large amount of data while protecting at the same time sensitive information [4]. PPDM algorithms can be classified into four categories [1]: random data perturbation, k-anonymization, distributed privacy preservation, and privacy preservation of mining results.

In this paper we address the problem of preserving both privacy and mining accuracy in publishing sensitive time-series data. Time-series data have been widely used in many applications, and data mining on time-series data has been

P. García Bringas et al. (Eds.): DEXA 2010, Part II, LNCS 6262, pp. 17–31, 2010.
© Springer-Verlag Berlin Heidelberg 2010

Fig. 1. A data flow model of independent data sources and third parties

actively studied [8,10,11]. Fig. 1 shows the data flow model that we assume. We use the centralized model [12,13] where multiple independent data sources provide their time-series to third parties. In this model, however, data sources do not trust third parties, so they do not wish to provide their original time-series, but they could still provide appropriately distorted time-series to get meaningful mining results. Thus, we can say that accuracy preservation of mining results as important as privacy preservation of sensitive time-series data.

To address both privacy and mining accuracy, we first setup a privacy model that addresses the underlying assumptions and requirements. Our model has three assumptions: *full disclosure, equi-uncertainty,* and *independency.* Full disclosure means that all information used in distorting and publishing time-series can be revealed to third parties or attackers. Equi-uncertainty means that each of distorted time-series has the same amount of uncertainty, which represents the degree of difference between original and distorted time-series [13]. Independency means that each time-series can be independently distorted without considering other time-series. To meet our main goal of preserving privacy and mining accuracy, we also derive two preservation requirements: *uncertainty* and *distance order.* Uncertainty preservation means that original time-series cannot be reconstructed from the published, distorted time-series. Distance order preservation means that relative distance orders among time-series must be preserved after the distortion. According to our analysis, only the random perturbation methods [1,2,13] satisfy all three assumptions, but even these methods do not satisfy both the requirements. Therefore, we discuss the randomization-based solutions that satisfy all assumptions and requirements of the privacy model.

For the purpose of preserving distance orders, we use the *noise averaging effect* of piecewise aggregate approximation (PAA) [8]. This notion is derived from a simple intuition that the summation of random noise eventually converges to 0. The noise averaging effect can alleviate the problem of distorting distance orders in randomly perturbed time-series. PAA extracts a fixed number of averages from a long time-series and uses those averages to compute the distance [8]. Since PAA uses the averages in computing distances, it naturally exploits the noise averaging effect on the distorted/published time-series. To exploit this noise averaging effect, we use *PAA distances* in computing distances of the distorted time-series.

In this paper we propose naive and advanced solutions based on the random perturbation and the noise averaging effect. Our first solution simply adopts the random perturbation in publishing time-series, but it uses PAA distances to preserve distance orders by exploiting the noise averaging effect. The simple random perturbation, however, can be attacked by the wavelet filter [13]. We thus propose another solution that uses the recent wavelet-based perturbation [13] which is secure against the wavelet filtering attack. These two solutions, however, are in a tradeoff relationship with respect to uncertainty and distance orders. We thus propose two more advanced solutions that take advantages of both naive solutions. Our advanced solutions can be seen as an engineering approach that preserves uncertainty and distance orders as much as possible through the recent wavelet-based perturbation and the noise averaging effect of PAA.

Our solutions provide a very practical approach to publish time-series data which well preserves mining accuracy as well as privacy. We do not use any complicated cryptography techniques or SMC protocols [5], but simply adopt an intuitive notion of the noise averaging effect and the wavelet-based secure perturbation method. The contributions of the paper can be summarized as follows. (1) We present a privacy model characterized by assumptions required in a centralized data flow model and requirements for preserving privacy and mining accuracy. (2) We discuss the notion of the noise averaging effect and show its effectiveness in computing distances of the perturbed time-series. (3) We propose two naive solutions that exploit the noise averaging effect and introduce two more advanced solutions that represent a compromise in the tradeoff relationships between those naive solutions. (4) Through extensive experiments we showcase the superiority of our advanced solutions.

2   Proposed Privacy Model

Our privacy model uses the Euclidean distance [10,11,13] as the metric of (dis)similarity between time-series since it has been widely used in many clustering or classification algorithms [11]. Given two time-series $X = \{x_1, \ldots, x_n\}$ and $Y = \{y_1, \ldots, y_n\}$, the Euclidean distance $D(X, Y)$ is defined as $\sqrt{\sum_{i=1}^{n}(x_i - y_i)^2}$. Our model assumes that each data source first distorts its time-series X to $X^d = \{x_1^d, \ldots, x_n^d\}$ independently, and then publishes the distorted time-series X^d. Attackers may try to recover the original time-series X from the published time-series X^d for the malicious purpose of obtaining privacy-sensitive data. We denote by $X^r = \{x_1^r, \ldots, x_n^r\}$ the recovered time-series recovered from X^d.

Under the data flows in Fig. 1, our privacy model has three assumptions.

- *Full disclosure*: We assume that all information used in distorting time-series can be revealed to third parties or attackers. It means that distortion techniques and related parameters can be published.
- *Equi-uncertainty*: We assume that every published time-series has the same amount of distorted information. In other words, all distorted time-series have the same amount of uncertainty. Here, the uncertainty represents the degree of difference between original and distorted time-series. (We will formally define it below.)
- *Independency*: We assume that each time-series can be independently distorted without considering other time-series. This is because, as shown in Fig. 1, time-series are scattered in multiple independent data sources, and

those sources do not interact with each other. Thus, each data source independently distorts its time-series without interacting with other data sources or third parties.

The major goal of our privacy model is to preserve privacy and at the same time assure mining accuracy. To discuss about the privacy preservation first, we use the *uncertainty* [13], also known as the mean square error or discrepancy [6], as the metric of privacy. Uncertainty between an original time-series X and its distorted time-series X^d is defined as $u(X, X^d) = \sum_{i=1}^{n} |x_i - x_i^d|^2$; uncertainty between X and its recovered time-series X^r is defined as $u(X, X^r) = \sum_{i=1}^{n} |x_i - x_i^r|^2$. The former uncertainty $u(X, X^d)$ can be seen as the noise amount enforced by the data source of X; the latter uncertainty $u(X, X^r)$ the noise amount remaining after the attack. Thus, the smaller difference between $u(X, X^d)$ and $u(X, X^r)$ the better privacy preservation is [13]. Based on this observation, we formally define the uncertainty preservation and derive the privacy requirement.

Definition 1. *Given an original time-series X, its distorted time-series X^d, and its recovered time-series X^r, we say that the uncertainty of X^d is preserved if $|u(X, X^d) - u(X, X^r)|$ is less than the user-specified threshold.* □

Requirement 1. Uncertainty of the published time-series needs to be preserved to assure that original time-series cannot be reconstructed from the published ones. □

We next discuss about the mining accuracy preservation. Different mining techniques use different accuracy measures, and we thus introduce a notion of *distance orders* as a general measure of mining accuracy. Distance orders represent the relative orders among distances between time-series. In general, preserving both the absolute distances between time-series and their privacy is difficult. However, preserving the relative orders among distances is enough for providing higher accuracy in most mining algorithms [7]. Based on this observation, we use the notion of distance order preservation for assuring mining accuracy.

Definition 2. *Let O, A, and B be time-series, and O^d, A^d, and B^d be the corresponding distorted time-series, respectively. We say that the distance order among O, A, and B is preserved if one of the following implications holds (i.e., if their relative order of distances is not changed).*

$$D(O, A) \leq D(O, B) \implies D(O^d, A^d) \leq D(O^d, B^d),$$
$$D(O, A) \geq D(O, B) \implies D(O^d, A^d) \geq D(O^d, B^d)$$

Using Definition 2 we now derive the mining accuracy requirement.

Requirement 2. Distance orders among time-series need to be preserved for the purpose of providing high quality of mining results. □

We analyzed existing solutions with respect to the assumptions and requirements in our privacy model. In our comparison, we only considered approaches that can be applied to time-series data. (For the detailed analysis of existing solutions, refer to Section 6.) The analysis result shows that, except for random data perturbation solutions in [2,13], all privacy preserving solutions do not satisfy one or more assumptions. This is because most solutions assume some constraints on distortion techniques or underlined environments. The randomization methods, which distort time-series by adding white noise locally and independently, satisfy all three assumptions, but even those methods do not satisfy both the requirements. Therefore, in this paper we aim at finding the best solution that solves the following problem.

Problem Statement. In publishing time-series, find a solution that satisfies the three assumptions of full disclosure, equi-uncertainty, and independency and the two requirements of privacy preservation and distance order preservation. □

3   PAA-Based Intuitive Solutions

3.1   Noise Averaging Effect of PAA

Random data perturbation (*randomization* in short) generates white noise based on uniform or Gaussian distributions and adds that noise to original time-series. More formally, for a time-series X, the randomization generates a noise time-series $N = \{n_1, \ldots, n_n\}$ with mean 0 and standard deviation σ and constructs a distorted time-series as $X^d = \{x_1 + n_1, \ldots, x_n + n_n\}$. Obviously, the standard deviation σ equals to $u(X, X^d)$, the uncertainty between original and published time-series. Full disclosure, the first assumption, is satisfied since we do not hide any information including the mean and standard deviation. Equi-uncertainty, the second assumption, is also satisfied since we use the same standard deviation for all time-series. Independency, the third assumption, is trivially satisfied since each time-series reflects its own noise time-series. Thus, the only thing we need to consider in randomization is whether it satisfies two requirements or not.

Randomization is known to well preserve privacy, but not mining accuracy [11]. White noise makes it difficult to disclose the exact value of each entry, but at the same time it destroys distance orders among time-series. As we increase the amount of noise for better privacy, mining accuracy decreases rapidly [11]. To solve this problem, we use the *noise averaging effect*, which is derived from a simple intuition that the summation of white noise eventually converges to 0 since the mean of noise is 0. To exploit the noise averaging effect in computing the distance between two distorted time-series, we simply use their averages of multiple entries instead of individual entries.

In this paper we use the noise averaging effect of PAA [8]. PAA transforms a time-series $X (= \{x_1, \ldots, x_n\})$ and its distorted time-series $X^d (= \{x_1^d, \ldots, x_n^d\})$ to their averaged sequences $\bar{X} (= \{\bar{x}_1, \ldots, \bar{x}_f\})$ and $\bar{X}^d (= \{\bar{x}_1^d, \ldots, \bar{x}_f^d\})$:

$$\bar{x}_i = \frac{f}{n} \sum_{j=\frac{n}{f}(i-1)+1}^{\frac{n}{f}i} x_j, \quad \bar{x}_i^d = \frac{f}{n} \sum_{j=\frac{n}{f}(i-1)+1}^{\frac{n}{f}i} x_j^d \tag{1}$$

As shown in Eq. (1), PAA gets an average from each interval, and we thus naturally exploit the noise averaging effect if we use PAA in computing the distance. For this purpose, we define the *PAA distance* between two distorted time-series as follows.

Definition 3. Given two distorted time-series X^d and Y^d, their *PAA distance*, denoted as $PD(X^d, Y^d)$, is defined as follows:

$$PD(X^d, Y^d) = D(\bar{X}^d, \bar{Y}^d) = \sqrt{\sum_{i=1}^{f} (\bar{x}_i^d - \bar{y}_i^d)^2}. \qquad (2)$$

Using PAA distances instead of original distances we may alleviate the problem of destroying distance orders.

3.2 RAND: Random Data Perturbation and PAA Distances

Our first solution, called *RAND*, uses the randomization without any modification in distorting time-series, but it uses the PAA distance in comparing distance orders. Algorithm 1 shows the distortion procedure of RAND, which is simple and self-explained. In Line 1, GaussRand$(0, \sigma)$ generates a white noise based on the Gaussian distribution. According to the data flow model of Fig. 1, each data source publishes its time-series using RAND, and third parties mine the meaningful patterns using the PAA distance to get the higher mining accuracy.

Algorithm 1. RAND($X = \{x_1, \ldots, x_n\}, \sigma$)

1: Generate a noise time-series N where $n_i :=$ GaussRand$(0, \sigma)$;
2: Make a distorted time-series X^d from X and N;   // $x_i^d := x_i + n_i$
3: Publish the distorted time-series X^d to third parties;

According to our preliminary experiment, the PAA distance in RAND closely preserves distance orders, and it thus improves the mining accuracy. RAND, however, has a critical problem in preserving privacy. The problem is that white noise can be easily removed by the wavelet filter [13]. Example 1 shows how we can remove the white noise from the distorted time-series.

Example 1. In Fig. 2, we first distort an original time-series X to X^d by adding 20% of noise. We then perform the discrete wavelet transform (DWT) on X^d and get wavelet coefficients from X^d. Through DWT, most energy is concentrated on the first few coefficients. We next filter the less energy coefficients whose absolute values are less than σ [13]. We finally recover the time-series through the inverse DWT. As a result, the recovered time-series X^r has only 4.8% of white noise. It means that the uncertainty is significantly reduced from 20% $(= u(X, X^d))$ to 4.8% $(= u(X, X^r))$ by the wavelet filter.   □

Likewise, the uncertainty, i.e., the white noise can be removed by the wavelet filter, and we can say that privacy is not well preserved in RAND.

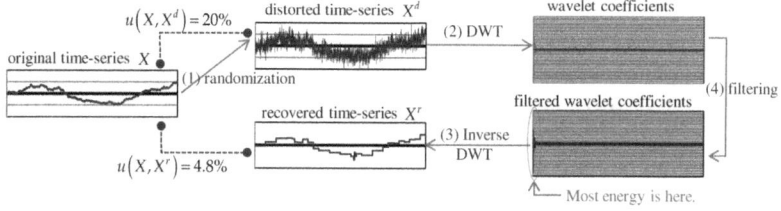

Fig. 2. Recovery of similar time-series by the wavelet filter

3.3 WAVE: Wavelet-Based Noise and PAA Distances

Papadimitriou et al.[13] pointed out the privacy problem of randomization and proposed a novel solution to avoid the filtering attack. Their solution generates a noise time-series by considering wavelet coefficients of an original time-series; more specifically, less energy coefficients have no contribution to making a noise time-series, but higher energy coefficients have much contribution to making it. Fig. 3 shows how their solution makes a distorted time-series. As shown in the figure, we first get a sequence of wavelet coefficients, $X^w = \{x_1^w, \ldots, x_n^w\}$, from an original time-series X. We then construct a sequence of noise coefficients, $N^w = \{n_1^w, \ldots, n_n^w\}$, based on X^w. A noise coefficient n_i^w is set to 0 if its corresponding wavelet coefficient x_i^w is less than the given uncertainty σ; in contrast, n_i^w is set to GaussRand$(0, c \cdot \sigma)$ if x_i^w is not less than σ, where $c = \sqrt{n/|\{x_i^w | x_i^w \geq \sigma\}|}$ (refer to [13] for details). This process explains that less energy coefficients are ignored, but higher energy coefficients have much noise. We next make a noise time-series N from N^w through the inverse DWT. We finally obtain a distorted time-series X^d by adding N to X and publish it to third parties. Papadimitriou et al.[13] showed that the noise of the resulting time-series was not removed by the wavelet filter, and the uncertainty was preserved well.

We now propose another randomization method, called *WAVE*, which uses the wavelet-based noise [13] in distorting time-series. WAVE solves the recovering problem of RAND by using the wavelet-based noise, but it still uses the PAA distance in comparing distance orders. The formal algorithm of WAVE is given

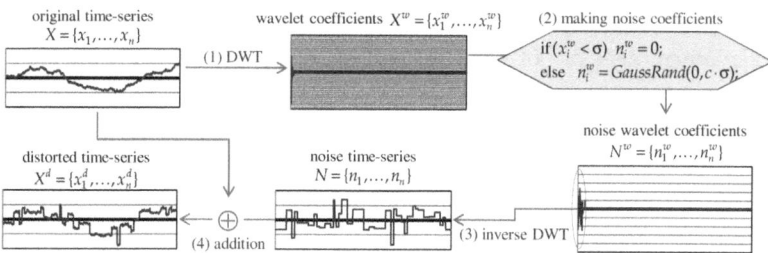

Fig. 3. WAVE-based noise time-series and its distorted time-series

in Algorithm 2. In Line 1, we get wavelet coefficients from the given time-series. In Lines 2-6, we obtain noise coefficients by considering energy of original coefficients. In Line 7, we make a noise time-series from the noise coefficients through the inverse DWT. In Lines 8-9, we construct a distorted time-series and publish it to third parties. Like RAND, each data source publishes its time-series using WAVE, and third parties use the PAA distance to mine the meaningful patterns.

Algorithm 2. WAVE($X = \{x_1, \ldots, x_n\}, \sigma$)

1: Get a sequence X^w of wavelet coefficients from X;
2: $c := \sqrt{n / |\{x_i^w | x_i^w \geq \sigma\}|}$;
3: **for** $i := 1$ **to** n **do** // get a seq. N^w of noise wavelet coefficients
4:    **if** $x_i^w < \sigma$ **then** $n_i^w := 0$;
5:    **else** $n_i^w :=$ GaussRand$(0, c \cdot \sigma)$;
6: **end-for**
7: Make a noise time-series N from N^w through the inverse DWT;
8: Get a distorted time-series X^d from X and N; // $x_i^d = x_i + n_i$
9: Publish the distorted time-series X^d to third parties;

WAVE, however, incurs another problem of destroying distance orders. According to our experiment, distance orders are severely destroyed in WAVE even though we use the PAA distance. The reason why WAVE destroys distance orders is that only a few high levels of wavelet coefficients are considered to make a noise time-series. That is, most noise is concentrated on a very small part of noise wavelet coefficients. This concentration simply makes WAVE stronger against the wavelet filtering attack and preserves the uncertainty well, but at the same time it significantly destroys distance orders.

4   Advanced Solutions

As we explained in Section 3, RAND has a problem in preserving privacy; WAVE has another problem in preserving distance orders. In other words, RAND is an extreme example of focusing on distance orders; WAVE is an extreme example of focusing on privacy. It means that there can be some intermediate solutions in between RAND and WAVE. In this section we discuss those advanced solutions that take advantages of both RAND and WAVE.

4.1   SNIL: Spread Noise to Intermediate Wavelet Levels

Our first advanced solution, called *SNIL*, spreads the noise to intermediate levels of wavelet coefficients when making a noise time-series. SNIL comes from an observation that WAVE concentrates most noise to only a few high levels of wavelet coefficients, and this concentration destroys distance orders. Based on this observation, SNIL spreads the noise to several intermediate levels instead of a few high levels. By not using a few high levels, some of noise can be removed by the wavelet filter; in contrast, by using intermediate levels rather than low levels, much noise can be preserved as the uncertainty. Moreover, using the PAA distance SNIL preserves distance orders relatively well since it spreads the noise to several levels of wavelet coefficients.

Algorithm 3 shows the distortion procedure of SNIL. As inputs to the algorithm, start and end levels, l_s and l_e, are given together with an original time-series X and the uncertainty σ. We let l_s be lower than l_e in wavelet levels. Thus, if the highest level is $L\,(=\log_2 n)$, the start and end levels, l_s and l_e, have 2^{L-l_s} and $2^{L-l_e}\,(<2^{L-l_s})$ coefficients, respectively. Like Algorithm 2, in Line 2 we compute how many coefficients will have the noise and obtain the constant factor c of that noise. In Lines 3-6 we assign the noise to the coefficients whose levels are in between start and end levels. This noise assignment spreads the given uncertainty to intermediate levels of l_s to l_e. The rest of Algorithm 3 is the same as Algorithm 2. Compared to WAVE, SNIL spreads the noise to more wavelet levels. This spread leads a well distribution of noise compared with WAVE, and through this well distribution we get the noise time-series that is strong to the wavelet filter and adequate to the PAA distance.

Algorithm 3. SNIL$(X = \{x_1, \ldots, x_n\}, \sigma, \text{start } l_s, \text{end } l_e)$

1: Get a sequence X^w of wavelet coefficients from X;
2: $c := \sqrt{n/\sum_{l=l_s}^{l_e} 2^{L-l}}$;   // $\sum_{l=l_s}^{l_e} 2^{L-l}$: the number of coefficients in l_s to l_e levels
3: **for** $i := 1$ **to** n **do**   // get a seq. N^w of noise wavelet coefficients
4:    **if** x_i^w's level is in $[l_s, l_e]$ **then** $n_i^w :=$ GaussRand$(0, c \cdot \sigma)$;
5:    **else** $n_i^w := 0$;
6: **end-for**
7: Make a noise time-series N from N^w through the inverse DWT;
8: Get a distorted time-series X^d from X and N;   // $x_i^d = x_i + n_i$
9: Publish the distorted time-series X^d to third parties;

Start and end levels of l_s and l_e represent on which levels we concentrate the noise. If those levels are close to the highest level $(= L)$, the resulting time-series becomes similar to that of WAVE; in contrast, if those levels are close to the lowest level $(= 1)$, the resulting time-series becomes similar to that of RAND. In this paper we use $\lceil \frac{1}{2} \log_2 n \rceil$ as l_s and $\lfloor \frac{3}{4} \log_2 n \rfloor$ as l_e. This is based on the real experimental result on random walk time-series such that if l_s is lower than $\lceil \frac{1}{2} \log_2 n \rceil$, too much noise is removed by the wavelet filter, and if l_e is higher than $\lfloor \frac{3}{4} \log_2 n \rfloor$, too many distance orders are destroyed. We also note that optimal l_s and l_e vary slightly according to the data set. For simplicity, however, we use $\lceil \frac{1}{2} \log_2 n \rceil$ and $\lfloor \frac{3}{4} \log_2 n \rfloor$ only in the experiment of Section 5.

4.2 DAPI: Divide And Perturb Independently

The second advanced solution, called *DAPI*, divides a time-series into several pieces and perturbs those pieces independently. DAPI exploits the noise averaging effect on each piece of a time-series by adding the noise to that piece independently. Regarding each piece as a unit of computing the PAA distance, we can easily exploit the noise averaging effect in DAPI. On the other hand, considering the wavelet filter we use the same distortion procedure of WAVE in distorting each piece, i.e., it computes a wavelet-based noise for each piece and adds the noise to that original piece.

Fig. 4 depicts the distortion procedure of DAPI. (We omit the detailed algorithm due to space limitation.) As shown in the figure, before adding the noise, we divide a time-series of length n into p pieces of length $\frac{n}{p}$ and localize the noise to individual pieces; after dividing the time-series, we add the same amount of noise to those pieces independently to enforce the given uncertainty σ.

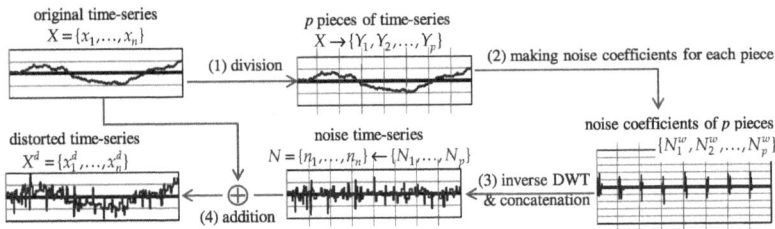

Fig. 4. DAPI-based noise time-series and its distorted time-series

To use DAPI in distorting time-series, we need to determine p, the number of pieces. We note that, as p increases, the number of entries $(= \frac{n}{p})$ contained in a piece decreases. The smaller number of entries (i.e., the larger number of pieces) makes it difficult to add the given uncertainty correctly, but it well preserves distance orders. This is because a small length time-series cannot have a large amount of noise while its average relatively well reflects all entries. In contrast, the larger number of entries (i.e., the smaller number of pieces) well preserves the uncertainty, but it does not preserve distance orders well. In other words, as we increase the number of pieces, DAPI shows a similar trend with RAND; in contrast, as we decrease the number, it shows a similar trend with WAVE. We use $\frac{1}{2}$ and $\frac{3}{4}$ of wavelet levels in SNIL, and thus, to be consistent with SNIL, we choose their average $\frac{7}{8}$ $\left(= \left(\frac{1}{2} + \frac{3}{4}\right)/2\right)$ as the number of pieces in DAPI. More precisely, we set p to a factor of n that is closest to $\frac{7}{8}\log_2 n$.

5   Experimental Evaluation

From the UCR data [9], we selected three data sets, CBF (143 time-series of length 60), ECG200 (600 time-series of length 319), and Two_Patterns (512 time-series of length 1,024), which were suitable for evaluating clustering or classification algorithms on time-series data. We implemented our four distortion methods and measured three evaluation metrics: (1) *uncertainty preservation*, how many time-series preserved their uncertainty after the distortion; (2) *distance order preservation*, how many time-series preserved their distance orders;

(3) *clustering accuracy preservation*, how many clusters preserved their original clusters. In particular, for the clustering accuracy preservation, we first obtained the actual clustering result from the original data and then compared it with the result of the distorted data. The hardware platform was SUN Ultra 25 workstation equipped with UltraSPARC IIIi 1.34GHz CPU, 1.0GB RAM, and an 80GB HDD; its software platform was Solaris 10. We used C/C++ language for implementing our solutions and k-means clustering algorithm.

5.1   Uncertainty Preservation

To evaluate the uncertainty preservation, we compared (1) the distorted time-series, which was generated by adding the noise to the original time-series, and (2) the recovered time-series, which was obtained by applying the wavelet filter to the distorted time-series. Through these two steps, we investigated how much noise was remained after the filtering attack; more specifically, for a given time-series X, we compared the remaining uncertainty $u(X, X^r)$ with the given uncertainty $u(X, X^d)$. In the experiment, we measured $u(X, X^d)$ and $u(X, X^r)$ for every time-series X, and used their average as the result.

Fig. 5 shows the experimental result on uncertainty preservation. First, in Fig. 5 (a) of CBF, we note that WAVE shows the best result while RAND shows the worst result. This is because WAVE extremely focuses on preserving the uncertainty while RAND extremely focuses on preserving distance orders. In Fig. 5 (a) we also note that our advanced solutions, SNIL and DAPI, are in between WAVE and RAND. In particular, SNIL, which tries to spread the noise to many wavelet levels, is comparable with WAVE in preserving the uncertainty. This is because, even though the given noise is spread to many wavelet coefficients, a large portion of the noise is still concentrated on a small number of higher wavelet levels, and it is not easily removed by the wavelet filter. On the other hand, DAPI is relatively worse than SNIL because it has difficulty in generating the full amount of noise due to the small size of individual pieces.

Figs. 5 (b) and (c) of ECG200 and Two_Patterns show the similar trend with Fig. 5 (a) of CBF. In summary, our advanced solutions take advantage of WAVE and show the better uncertainty preservation compared with RAND. In particular, SNIL is comparable with WAVE in preserving the uncertainty.

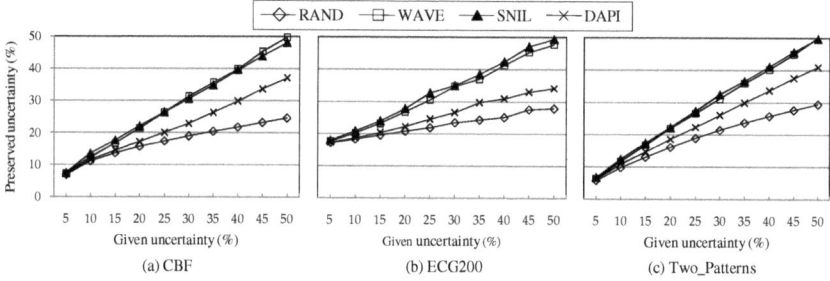

Fig. 5. Experimental result on uncertainty preservation

5.2   Distance Order Preservation

We constructed 10,000 triplets from each data sets. Each triplet consisted of three time-series as in Definition 2. We measured how much percent of triplets preserved their distance orders by investigating all 10,000 triplets for each method.

Fig. 6 shows the experimental result on distance order preservation. As shown in the figure, for all three data sets, RAND shows the best result, but WAVE shows the worst result. This trend is exactly opposite to that of uncertainty preservation. As we explained earlier, this is because RAND emphasizes distance orders while WAVE focuses on uncertainty. Similar to uncertainty preservation, our advanced solutions are in between RAND and WAVE by taking advatage of RAND in preserving distance orders. Unlike Fig. 5, however, DAPI is better than SNIL in Fig. 6 because DAPI well preserves the PAA distance by dividing a long time-series into smaller pieces.

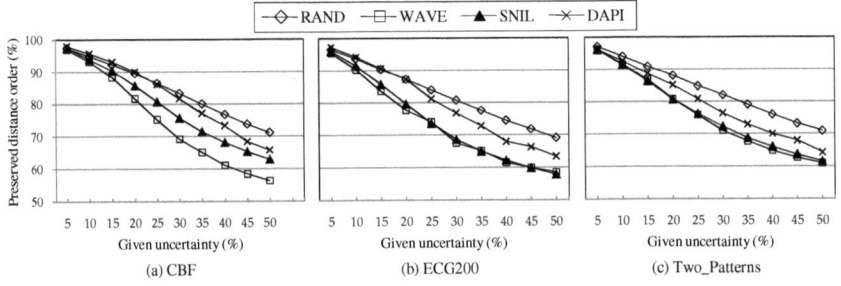

Fig. 6. Experimental result on distance order preservation

A notable point in comparing Figs. 5 and 6 is a tradeoff between uncertainty (i.e., privacy) and distance orders (i.e., mining accuracy). The uncertainty result of "RAND < DAPI < SNIL < WAVE" is exactly opposite to the distance order result of "WAVE < SNIL < DAPI < RAND." In case of SNIL and DAPI, we can emphasize one of uncertainty and distance orders by adjusting the input parameters ((l_s, l_e) in SNIL and p in DAPI). Thus, we can say SNIL and DAPI are more flexible than other methods in adjusting the tradeoff relationship.

5.3   Clustering Accuracy Preservation

We also experimented the actual clustering accuracy of the proposed distortion methods. As the measure of clustering accuracy, we used *F-measure* [11], which was widely used in information retrieval or data mining to evaluate the accuracy of retrieved or mined results. *F*-measure was computed by comparing the resulting clusters of original time-series and those of distorted time-series. In our experiment, the higher *F*-measure means the more accurate clustering result. For the detailed explanation about *F*-measures, readers are referred to [11].

After executing the k-means algorithm for the distortion methods, we obtain their F-measures, respectively. We omit the experimental result of F-measures since it is very similar to that of distance order preservation in Fig. 6. This means that preserving distance orders well reflects preserving clustering (or mining) accuracy. In other words, if a distortion method preserves distance orders well, it also preserves mining accuracy well. In summary, like the distance order preservation, RAND is the best while WAVE is the worst, and the advanced solutions are still in between RAND and WAVE.

6   Related Work

PPDM solutions can be classified into four categories [1]: random perturbation [6,13], k-anonymization [3], distributed privacy preservation [5], and privacy preserving of mining results [14]. We review these solutions w.r.t. our approach.

First, the random data perturbation adds white noise to the data in order to mask the sensitive values of data [1]. Agrawal and Srikant [2] first proposed random perturbation-based PPDM solutions. This random perturbation can be easily used for adding noise to time-series data, but it distorts distance orders as well and eventually incurs bad mining accuracy [13] (refer to RAND). Geometric transformation [12] and rotation perturbation [6] were proposed to get a *set* of distorted time-series from a *set* of original time-series. These solutions provide higher clustering/classification accuracy. However, they cannot deal with an individual time-series of a specific data source, and their related parameters should not be disclosed to preserve privacy. Papadimitriou et al. [13] proposed a novel perturbation solution which generated the wavelet-based noise to preserve privacy (i.e., uncertainty) against the filtering attack. This solution, on which our WAVE and advanced solutions are based, satisfies all three assumptions and the privacy preservation requirement, but it is still inadequate to the distance order preservation requirement as we explained in WAVE.

Second, k-anonymization increases anonymity of data by reducing the granularity of data representation with the use of generalization and suppression [1]. We can adopt the concept of k-anonymity in publishing time-series as follows: "Ensure at least k time-series should be similar." This anonymity problem is orthogonal to ours, and we may use our solutions to solve this problem.

Third, distributed privacy preservation provides secure mining protocols for the distributed environment. In general, those solutions use the cryptography-based secure multiparty computation (SMC) techniques to preserve data privacy of individual entities. Clifton et al. [5] proposed various types of SMC operations and used those operations for privacy preserving clustering and association rule mining. Those SMC-based solutions, however, are not suitable for our privacy model since in their solutions data sources need to co-work together, or some encryption parameters should be hidden from others.

Fourth, privacy preserving of mining results prevents the outputs (i.e., mining results) from disclosing data privacy [1]. For example, Verykios et al. [14] proposed solutions to hide the sensitive association rules that might disclose private data. We can also adopt this concept in publishing time-series as follows: "Ensure that the original time-series cannot be recovered even though the mining results are published." This output problem is also orthogonal to our distortion problem, and we may use our solutions to solve this problem.

Recently, Mukherjee and Chen [11] proposed a novel solution to privacy preserving clustering on time-series data. Their solution published a few Fourier coefficients instead of a whole time-series. Since the Fourier coefficients well preserve the Euclidean distance, their solution provides a higher clustering accuracy. However, it may cause privacy breach if positions of coefficients are revealed. To avoid this problem, they tried to hide the exact positions through the sophisticated permutation protocol [11]. However, the positions can be easily revealed if only one original time-series and its published coefficients are disclosed.

7 Conclusions

Time-series data are very sensitive since a time-series itself may disclose its corresponding private information (e.g., identifier). Thus, preserving both privacy and mining accuracy is an important issue in publishing time-series data. In this paper we presented naive and advanced solutions which considered mining accuracy preservation as well as privacy preservation. Our work can be summarized as follows. First, we proposed a privacy model of publishing sensitive time-series data and derived the related assumptions and requirements. Second, we analyzed the randomization-based solutions on the privacy model and presented their common problems. Third, we introduced a notion of the *noise averaging effect* of PAA and explained that the *PAA distance* might well preserve distance orders for the higher mining accuracy. Fourth, we proposed two naive randomization methods by exploiting the PAA distance. Fifth, to take advantages of both naive solutions, we proposed two more engineering solutions. Sixth, through extensive experiments, we showcased the superiority of our solutions.

Acknowledgments

This work was partially supported by Defense Acquisition Program Administration and Agency for Defense Development under the contract (UD060048AD).

References

1. Aggarwal, C.C., Yu, P.S.: Privacy-Preserving Data Mining: A Survey. In: Gertz, M., Jajodia, S. (eds.) Handbook of Database Security: Applications and Trends, pp. 431–460. Springer, Heidelberg (November 2007)
2. Agrawal, R., Srikant, R.: Privacy-Preserving Data Mining. In: Proc. of Int'l Conf. on Management of Data, ACM SIGMOD, Dallas, Texas, pp. 439–450 (May 2000)
3. Bayardo, R.J., Agrawal, R.: Data Privacy through Optimal *k*-anonymization. In: Proc. of the 21st Int'l. Conf. on Data Engineering, Tokyo, Japan, pp. 217–228 (April 2005)

4. Bertino, E., Lin, D., Jiang, W.: A Survey of Quantification of Privacy Preserving Data Mining Algorithms. In: Aggarwal, C.C., Yu, P.S. (eds.) Privacy-Preserving Data Mining: Models and Algorithms, June 2008, pp. 183–205. Kluwer Academic Publishers, Dordrecht (2008)
5. Clifton, C., Kantarcioglu, M., Vaidya, J., Lin, X., Zhu, M.Y.: Tools for Privacy Preserving Distributed Data Mining. SIGKDD Explorations 4(2), 28–34 (2002)
6. Huang, Z., Du, W., Chen, B.: Deriving Private Information from Randomized Data. In: Proc. of Int'l. Conf. on Management of Data, ACM SIGMOD, Baltimore, Maryland, pp. 37–48 (June 2005)
7. Inan, A., Kantarcioglu, M., Bertino, E.: Using Anonymized Data for Classification. In: Proc. of the 25th Int'l. Conf. on Data Engineering, Shanghai, China, pp. 429–440 (April 2009)
8. Keogh, E., Chakrabarti, K., Pazzani, M.J., Mehrotra, S.: Dimensionality Reduction for Fast Similarity Search in Large Time Series Databases. Knowledge and Information Systems 3(3), 263–286 (2001)
9. Keogh, E., Xi, X., Wei, L., Ratanamahatana, C.A.: The UCR Time Series for Classification/Clustering, http://www.cs.ucr.edu/~eamonn/time_series_data
10. Kim, B.-S., Moon, Y.-S., Kim, J.: Noise Control Boundary Image Matching Using Time-Series Moving Average Transform. In: Bhowmick, S.S., Küng, J., Wagner, R. (eds.) DEXA 2008. LNCS, vol. 5181, pp. 362–375. Springer, Heidelberg (2008)
11. Mukherjee, S., Chen, Z., Gangopadhyay, A.: A Privacy-Preserving Technique for Euclidean Distance-based Mining Algorithms using Fourier-related Transforms. The VLDB Journal 15(4), 293–315 (2006)
12. Oliveira, S.R.M., Zanane, O.R.: Privacy-Preserving Clustering by Data Transformation. In: Proc. of Brazilian Symp. on Databases, Amazonas, Brazil, pp. 304–318 (October 2003)
13. Papadimitriou, S., Li, F., Kollios, G., Yu, P.S.: Time Series Compressibility and Privacy. In: Proc. of the 33rd Int'l. Conf. on Very Large Data Bases, Vienna, Austria, pp. 459–470 (September 2007)
14. Verykios, V.S., Elmagarmid, A., Bertino, E., Saygin, Y., Dasseni, E.: Association Rule Hiding. IEEE Trans. on Knowledge & Data Engineering 16(4), 434–447 (2004)

Efficient Discovery of Generalized Sentinel Rules

Morten Middelfart[1], Torben Bach Pedersen[2], and Jan Krogsgaard[1]

[1] TARGIT A/S
[2] Aalborg University – Department of Computer Science
morton@targit.com, tbp@cs.aau.dk, jank@targit.com

Abstract. This paper proposes the concept of *generalized sentinel rules* (sentinels) and presents an algorithm for their discovery. Sentinels represent *schema level* relationships between *changes over time* in certain measures in a multi-dimensional data cube. Sentinels notify users based on previous observations, e.g., that revenue might drop within two months if an increase in customer problems combined with a decrease in website traffic is observed. If the vice versa also holds, we have a *bi-directional* sentinel, which has a higher chance of being causal rather than coincidental. We significantly extend prior work to combine multiple measures into better sentinels as well as auto-fitting the best warning period. We introduce two novel quality measures, *Balance* and *Score*, that are used for selecting the best sentinels. We introduce an efficient algorithm incorporating novel optimization techniques. The algorithm is efficient and scales to very large datasets, which is verified by extensive experiments on both real and synthetic data. Moreover, we are able to discover strong and useful sentinels that could not be found when using sequential pattern mining or correlation techniques.

1 Introduction

The Computer Aided Leadership and Management (CALM) concept copes with the challenges facing managers that operate in a world of chaos due to the globalization of commerce and connectivity [10]; in this chaotic world, the ability to continuously *react* is far more crucial for success than the ability to *long-term forecast*. The idea in CALM is to take the Observation-Orientation-Decision-Action (OODA) loop (originally pioneered by "Top Gun"[1] fighter pilot John Boyd in the 1950s), and integrate business intelligence (BI) technologies to drastically increase the speed with which a user in an organization cycles through the OODA loop. One way to improve the speed from observation to action is to expand the "time-horizon" by providing the user of a BI system with warnings based on "micro-predictions" of changes to an important measure, often called a Key Performance Indicator (KPI). A *generalized sentinel rule* (*sentinel* for short) is a causal relationship where changes in one or *multiple source measures*, are followed by changes to a *target measure* (typically a KPI), within a given time period, referred to as the *warning period*. We attribute higher quality to bi-directional sentinels that can predict changes in both directions, since such a relationship intuitively is less likely to be coincidental. An example of a sentinel for a company could be: "IF Number of Customer Problems go up and Website Traffic goes down THEN Revenue goes

---

[1] Colonel John Boyd was fighter instructor at Nellis Air Force Base in Nevada, the predecessor of U.S. Navy Fighter Weapons School.

P. García Bringas et al. (Eds.): DEXA 2010, Part II, LNCS 6262, pp. 32–48, 2010.

down within two months AND IF Number of Customer Problems go down and Website Traffic goes up THEN Revenue goes up within two months". Such a rule will allow a BI system to notify a user to take corrective action once there is an occurrence of, e.g., "Customer Problems go up and Website Traffic goes down", since he knows, based on the "micro-prediction" of the rule, that Revenue, with the probability stated by the rule's confidence, will go down in two months if no action is taken. In Section 4 we describe an example where a valuable, and not so obvious, sentinel was uncovered.

Compared to prior art, sentinels are mined on the measures and dimensions of multiple cubes in an OLAP database, as opposed to the "flat file" formats used by most traditional data mining methods. Sentinels find rules that would be impossible to detect using traditional techniques, since sentinels operate on *data changes* at the *schema level* as opposed to absolute *data values* at the *data level* such as association rules [1] and sequential patterns typically do [4]. This means that our solution works on *numerical data* such as measure values, whereas association rules and sequential patterns work on *categorical data*, i.e., dimension values. In [12] we specifically provide a concrete, realistic example where nothing useful is found using these techniques, while sentinel mining *do* find meaningful rules. In addition, *bi-directional* sentinels are *stronger* than both association rules and sequential patterns since such relationships have a greater chance of being causal rather than coincidental. The schema level nature of sentinels gives rise to the table of combinations (TC) and the reduced pattern growth (RPG) optimization (see Section 3), and such optimizations can therefore not be offered by sequential pattern mining or other known optimizations for simpler "market basket"-type data such as [6]. In addition to the TC and RPG optimizations, the auto-fitting of the warning period, and the ability to combine source measures into better sentinel rules, adds to the distance between our solution and optimizations offered in prior art such as [3,8,14,15,16,17,18].

Gradual rule mining [5] is a process much like association rules, where the categorical data are created by mapping numerical data to fuzzy partitions, and thus this technique works on numerical data similar to our solution. However, similar to association rules and sequential patterns, gradual rule mining does not have the schema level property of sentinels that allows our solution to create the strong bi-directional rules. Moreover, the primary objective of gradual rules is to describe the absolute values of a measure, whereas our solution operates on *changes* in the measure values. Therefore, for similar reasons as mentioned above, gradual rule mining does not have the ability to use the TC and RPG optimizations, and neither does it have the ability to auto-fit a warning period for a given rule.

Other approaches to interpreting the behavior of data sequences are various regression [2] and correlation [7,19] techniques which attempt to describe a functional relationship between one measure and another. Similar to gradual rules, these techniques are also concerned with the absolute values of a measure, as opposed to sentinels that are based on changes in the measure values. With regards to the output, sentinels are more *specific "micro-predictions"*, and are thus complementary to these techniques. Sentinels are useful for discovering strong relationships between a smaller subset within a dataset [11], and thus they are useful for detecting warnings whenever changes (that would otherwise go unnoticed) in a relevant source measure occur.

The novel contributions in this paper are as follows: First we generalize the concept of sentinel rules from previous work into generalized sentinel rules, that allow multiple source measures to be combined in sentinels, and that can facilitate auto-fitting of the best warning period. In this context, we define two new qualitative measures for sentinels, namely: *Balance* and *Score*, and we expand the previous notation to support our generalization. Secondly, we present an algorithm for sentinel discovery that can combine multiple source measures and auto-fit the optimal warning period within a given range. In the algorithm, we introduce and define the optimization technique called *Reduced Pattern Growth*, and we apply *Hill Climbing* for further optimization. In addition, our algorithm uses a so-called *Table of Combinations* that efficiently supports these optimization techniques. Third, we assess the computational complexity and conduct extensive experiments to validate our complexity assessment, and we verify that our optimized algorithm scales well on large volumes of real-world and synthetic data.

The remainder of the paper is structured as follows: Section 2 presents the formal definition, Section 3 presents the new SentHiRPG algorithm and its implementation. Section 4 presents a scalability study, and Section 5 presents our conclusions and proposals for future work.

2   Problem Definition

Running Example: We imagine having a company that sells products world-wide, and that we, in addition to the traditional financial measures such as revenue, *Rev*, have been monitoring the environment outside our organization and collected that information in three measures. The measure *NBlgs* represents the number of times an entry is written on a blog where a user is venting a negative opinion about our company or products. The measure *CstPrb* represents the number of times a customer contacts our company with a problem related to our products. The measure *WHts* represents the number of human hits on our website. We want to investigate whether we can use changes on any of the *external* measures (*NBlgs*, *CstPrb*, *WHts*) to predict a future change on the *internal measure* (*Rev*). To generalize our terminology, we call the external measures *NBlgs*, *CstPrb*, and *WHts source measures* and the internal measure, *Rev*, the *target measure*.

In Table 1 we see two subsets from our database, the source measures representing the external environment have been extracted for January to October 2008, and the target measure has been extracted for February to November 2008. For both source and target measures we have calculated the cases where a measure changes 10% or more, either up or down, from one month to another.

Formal Definition: Let C be a multi-dimensional *cube* containing a set of *facts*, $C = \{(d_1, d_2, ..., d_n, m_1, m_2, ..., m_p)\}$. The dimension values, $d_1, d_2, ..., d_n$, belong to the *dimensions* $D_1, D_2, ..., D_n$, and we refer to the "dimension part" of a fact, $(d_1, d_2, ..., d_n)$, as a *cell*. We say that a cell belongs to C, denoted by $(d_1, d_2, ..., d_n) \in C$, when a fact $(d_1, d_2, ..., d_n, m_1, m_2, ..., m_p) \in C$ exists. We say that a *measure value*, m_i, is the result of a partial function, $M_i : D_1 \times D_2 \times ... \times D_n \hookrightarrow \Re$, denoted by, $M_i(d_1, d_2, ..., d_n) = m_i$, if $(d_1, d_2, ..., d_n) \in C$ and $1 \leq i \leq p$. We assume, without loss of generality, that there is only one time dimension, T, in C, and that $T = D_1$, and subsequently $t = d_1$. In addition, we assume that measures $M_1, ..., M_{p-1}$ are source measures, and that measure M_p is the target measure. An *indication*, Ind_i,

Table 1. Source and target measure data with indications

Month	NBlgs	CstPrb	WHts	Indications
2008-Jan	80	310	1227	
2008-Feb	89	390	1101	$NBlgs\blacktriangle, CstPrb\blacktriangle, WHts\blacktriangledown$
2008-Mar	90	363	1150	
2008-Apr	99	399	987	$NBlgs\blacktriangle, WHts\blacktriangledown$
2008-May	113	440	888	$NBlgs\blacktriangle, CstPrb\blacktriangle, WHts\blacktriangledown$
2008-Jun	101	297	1147	$NBlgs\blacktriangledown, CstPrb\blacktriangledown, WHts\blacktriangle$
2008-Jul	115	323	1003	$NBlgs\blacktriangle, WHts\blacktriangledown$
2008-Aug	105	355	999	
2008-Sep	93	294	993	$NBlgs\blacktriangledown, CstPrb\blacktriangledown$
2008-Oct	100	264	1110	$CstPrb\blacktriangledown, WHts\blacktriangle$

(a) Source

Month	Rev	Indications
2008-Feb	1020	
2008-Mar	911	$Rev\blacktriangledown$
2008-Apr	1001	
2008-May	1015	
2008-Jun	900	$Rev\blacktriangledown$
2008-Jul	1025	$Rev\blacktriangle$
2008-Aug	1100	
2008-Sep	1090	
2008-Oct	970	$Rev\blacktriangledown$
2008-Nov	1150	$Rev\blacktriangle$

(b) Target

tells us whether a measure, M_i, changes by at least α over a period, o. We define $Ind_i(C, t, o, d_2, d_3, ..., d_n)$ as shown in Formula 1.

$$\text{if } (t, d_2, d_3, ..., d_n) \in C \wedge (t + o, d_2, d_3, ..., d_n) \in C \text{ then } Ind_i(C, t, o, d_2, d_3, ..., d_n) =$$

$$\begin{cases} M_i\blacktriangle & \text{if } \dfrac{M_i(t + o, d_2, d_3, ..., d_n) - M_i(t, d_2, d_3, ..., d_n)}{M_i(t, d_2, d_3, ..., d_n)} \geqq \alpha \\ M_i\blacktriangledown & \text{if } \dfrac{M_i(t + o, d_2, d_3, ..., d_n) - M_i(t, d_2, d_3, ..., d_n)}{M_i(t, d_2, d_3, ..., d_n)} \leqq -\alpha \\ M_i \blacktriangleright & \text{otherwise} \end{cases} \tag{1}$$

We refer to $M_i\blacktriangle$ as a *positive* indication, to $M_i\blacktriangledown$ as a *negative* indication, and to $M_i \blacktriangleright$ as a *neutral* indication. We define a wildcard, ?, meaning that M_i? can be either $M_i\blacktriangle$, $M_i\blacktriangledown$, or $M_i \blacktriangleright$. In addition, we define the *complement* of an indication as follows: $\overline{M_i\blacktriangle} = M_i\blacktriangledown$, $\overline{M_i\blacktriangledown} = M_i\blacktriangle$, and $\overline{M_i \blacktriangleright} = M_i \blacktriangleright$. We expand the complement to work for sets by taking the complement of each element, and we expand the wildcard to work for sets meaning that any member of the set can have any indication. An *indication set*, $IndSet(C, t, o, d_2, d_3, ..., d_n)$, as shown in Formula 2, is a set of all possible combinations of indications (of up to *RuleLen* source measures) that occur for one or more source measures in the same cell. We use *MaxSource* as a threshold for the maximum number of source measures we want to combine in an *IndSet*, and we denote the number of indications in a given *IndSet* by $RuleLen(IndSet)$.

$$IndSet(C, t, o, d_2, d_3, ..., d_n) = \{\{Ind_{i_1}(C, t, o, d_2, d_3, ..., d_n), ...,$$
$$Ind_{i_q}(C, t, o, d_2, d_3, ..., d_n), ..., Ind_{i_{RuleLen}}(C, t, o, d_2, d_3, ..., d_n)\}| \tag{2}$$
$$1 \leqq RuleLen \leqq MaxSource \wedge 1 \leqq i_q \leqq p - 1\}$$

A sentinel set, $SentSet$, is defined as all indications in a cube, C, given the offset, o, where the source measure indication sets, $IndSet_s$, are paired with the indications on the target measure, Ind_p, that occur a given *warning period*, w, later.

$$SentSet(C, o, w) = \{(IS_s, Ind_p(C, t + w, o, d_2, d_3, ..., d_n))|$$
$$IS_s \in IndSet(C, t, o, d_2, d_3, ..., d_n) \wedge$$
$$(t, d_2, d_3, ..., d_n) \in C \wedge (t + w, d_2, d_3, ..., d_n) \in C \tag{3}$$
$$\wedge \forall Ind_m \in IS_s : (Ind_m \neq M_m \blacktriangleright)\}$$

We say that $(Ind_{Source}, Ind_{Target}) \in SentSet(C, o, w)$ *supports* the *indication rule* denoted $Ind_{Source} \rightarrow Ind_{Target}$. The *support* of an indication rule, denoted by

$IndSupp_{Source \rightarrow Target}$, is the number of $(Ind_{Source}, Ind_{Target}) \in SentSet(C, o, w)$ which support the rule. Similarly, the support of $Source$, $IndSupp_{Source}$, is the number of $(Ind_{Source}, M_p?) \in SentSet(C, o, w)$. In Table 1, we have calculated the indications (Formula 1) with $\alpha = 10\%$ as well as arranged the indications for the source measures in the largest possible indication set (Formula 2) for each month. The combination of the source and target measures is equivalent to a sentinel set (Formula 3) where o and w are both set to 1 month. We see that for example the indication rule $NBlgs\blacktriangle \rightarrow Rev\blacktriangledown$ has a support, $IndSupp_{NBlgs\blacktriangle \rightarrow Rev\blacktriangledown}$, of 2, and the indication $NBlgs\blacktriangle$ has a support, $IndSupp_{NBlgs\blacktriangle}$, of 4.

A generalized sentinel rule is an *unambiguous* relationship between *Source* and *Target*, that consists of one or two indication rules. Therefore, we say that there are only two potential sentinels between a set of source measures, *Source*, and a target measure, *Target*, namely: $Source \rightsquigarrow Target$ or $Source \rightsquigarrow inv(Target)$, where *inv* represents an inverted relationship (intuitively, when source changes up, target changes down and vice versa). The relationships between the two potential generalized sentinel rules and their indication rules are defined in Formula 4.

$$Source \rightsquigarrow Target = \{Ind_{Source} \rightarrow Ind_{Target}, \overline{Ind_{Source}} \rightarrow \overline{Ind_{Target}}\}$$
$$Source \rightsquigarrow inv(Target) = \{Ind_{Source} \rightarrow \overline{Ind_{Target}}, \overline{Ind_{Source}} \rightarrow Ind_{Target}\} \tag{4}$$

If two contradicting indication rules are both supported in *SentSet*, e.g. $Ind_{Source} \rightarrow Ind_{Target}$ and $Ind_{Source} \rightarrow \overline{Ind_{Target}}$, we use the *contradiction elimination process* (Formula 5) to eliminate the indication rules with the least support that have the same premise, but a different consequent, and vice versa. However, in order to reflect the contradiction between the indication rules as a less desired feature, we reduce the support of the "cleansed rule" by deducting the support of the rules we eliminated from the support of the rules we preserve.

$$ElimSupp_{Source \rightsquigarrow Target} = IndSupp_{Source \rightarrow Target} - IndSupp_{Source \rightarrow \overline{Target}}$$
$$+ IndSupp_{\overline{Source} \rightarrow \overline{Target}} - IndSupp_{\overline{Source} \rightarrow Target} \tag{5}$$

Essentially, we force our generalized sentinel rule to be either $Source \rightsquigarrow Target$ or $Source \rightsquigarrow inv(Target)$, and thereby we effectively eliminate both *contradicting* (same premise but different consequent) and *orthogonal* (different premise but same consequent) indication rules. *ElimSupp* represents the sum of the support for the indication rule(s) in a sentinel after elimination of its contradictions, and if $ElimSupp_{Source \rightsquigarrow Target}$ is positive, it means that the sentinel $Source \rightsquigarrow Target$ contains the strongest indication rules (as opposed to $Source \rightsquigarrow inv(Target)$). Subsequently, $SentRules(C, o, w)$ (Formula 6) conducts the elimination process and extract the generalized sentinel rules from C with the offset o and the warning period w. We note that *SentRules* only contain rules where $ElimSupp > 0$, this way we eliminate sentinels composed by indication rules that completely contradict each other ($ElimSupp = 0$).

$$SentRules(C, o, w) =$$
$$\begin{cases} \{Source_s \rightsquigarrow Target_p \mid (Source_s?, Target_p?) \in SentSet(C, o, w)\} \\ \text{if } ElimSupp_{Source_s \rightsquigarrow Target_p} > 0 \\ \{Source_s \rightsquigarrow inv(Target_p) \mid (Source_s?, Target_p?) \in SentSet(C, o, w)\} \\ \text{if } ElimSupp_{Source_s \rightsquigarrow inv(Target_p)} > 0 \end{cases} \tag{6}$$

$$Balance_{Source \rightsquigarrow Target} = \frac{4 \times |A| \times |B|}{(|A| + |B|)^2}$$

$$\text{where } A = IndSupp_{Source \rightarrow Target} - IndSupp_{Source \rightarrow \overline{Target}} \tag{7}$$

$$B = IndSupp_{\overline{Source} \rightarrow \overline{Target}} - IndSupp_{\overline{Source} \rightarrow Target}$$

$$SentSupp_{Source \rightsquigarrow Target} =$$

$$\begin{cases} IndSupp_{Source}, & \text{if } Balance_{Source \rightsquigarrow Target} = 0 \wedge IndSupp_{Source} > 0 \\ IndSupp_{\overline{Source}}, & \text{if } Balance_{Source \rightsquigarrow Target} = 0 \wedge IndSupp_{\overline{Source}} > 0 \\ IndSupp_{Source} + IndSupp_{\overline{Source}}, & \text{otherwise} \end{cases} \tag{8}$$

$$Conf_{Source \rightsquigarrow Target} = \frac{ElimSupp_{Source \rightsquigarrow Target}}{SentSupp_{Source \rightsquigarrow Target}} \tag{9}$$

To determine the quality of a sentinel, $Source \rightsquigarrow Target \in SentRules(C, o, w)$, we define Formulae 7 to 9. *Balance* (Formula 7) is used to determine the degree to which a generalized sentinel rule is uni-directional (*Balance=0*) or completely bi-directional (*Balance=1*), meaning that there are exactly the same amounts of positive and negative indications on the target measure in the data used to discover the rule. *SentSupp* (Formula 8) tells us how often the premise of the sentinel occurs, and *Conf* (Formula 9) tells us how often, when the premise occurs, the consequent occurs within w time. We denote the minimum threshold for *SentSupp* by σ, the minimum threshold for *Conf* is denoted by γ, and the minimum threshold for *Balance* is denoted by β. With these definitions, we say that a sentinel, $A \rightsquigarrow B$, with an offset, o, and a warning period, w, exists in C when $SentSupp_{A \rightsquigarrow B} \geqq \sigma$, $Conf_{A \rightsquigarrow B} \geqq \gamma$, and $Balance_{A \rightsquigarrow B} \geqq \beta$. We use the following notation when describing a generalized sentinel rule: we use *inv* relative to the first source measure which is never inverted. The order of the source measures in a rule is unimportant for its logic, thus source measures can be ordered as it is seen most fit for presentation purposes. In the case where the rule is uni-directional, we add ▲ or ▼ to both the source and the target measure to express the distinct direction of the sentinel. We add $\wedge$ between the source measures, when there is more than one source measure in a rule, e.g., $A \wedge B \wedge C \rightsquigarrow D$, $A \wedge inv(B) \rightsquigarrow C$, and $A\blacktriangle \wedge B\blacktriangledown \rightsquigarrow C\blacktriangle$.

If we revert to our running example and apply Formulae 4, 5, and 6 to the data, we get Table 2 as output. Using Formulae 7, 8, and 9, we test each rule to see if it meets the thresholds $MaxSource = 3, \sigma = 3, \gamma = 60\%$, and $\beta = 70\%$. The column

Table 2. Sentinels ordered by their respective *Conformance* and *Score*

SentRules	RuleLen	SentSupp (ElimSupp)	Conf	Balance	Score	Conformance
$CstPrb \wedge inv(WHts) \rightsquigarrow inv(Rev)$	2	4 (4)	100%	100%	0.83	OK
$WHts \rightsquigarrow Rev$	1	6 (4)	67%	100%	0.67	OK
$NBlgs \wedge CstPrb \wedge inv(WHts) \rightsquigarrow inv(Rev)$	3	3 (3)	100%	89%	0.47	OK
$CstPrb \rightsquigarrow inv(Rev)$	1	5 (3)	60%	89%	0.43	OK
$NBlgs \wedge inv(WHts) \rightsquigarrow inv(Rev)$	2	5 (3)	60%	89%	0.35	OK
$NBlgs \wedge CstPrb \rightsquigarrow inv(Rev)$	2	4 (2)	50%	0%	0.10	Failed
$NBlgs \rightsquigarrow inv(Rev)$	1	6 (2)	33%	0%	0.08	Failed

Conformance lists the result of this test. We use same order as Table 1 to ease the readability. With regards to the failing rules we should note that the uni-directional rules $NBlgs\blacktriangle \rightarrow Rev\blacktriangledown$ and $NBlgs\blacktriangle \wedge CstPrb\blacktriangle \rightarrow Rev\blacktriangledown$ have higher confidence than the bi-directional rules based on the same indications. However since our threshold for balance is greater than zero, these uni-directional rules would also fail and are thus not listed in the table. In Table 2, the generalized sentinel rules found can be ordered by either *RuleLen*, *SentSupp*, *Conf*, *Balance*, or a combination in order to describe the quality of the rules. However, when using an optimization algorithm, e.g, hill climbing, it is desirable to be able to describe the quality of a rule with just a single number. For this purpose we denote the maximal value of *ElimSupp* for any sentinel in $SentRules(C, o, w)$ by $MaxElimSupp(C, o, w)$.

$$
\begin{aligned}
Score(Source \rightsquigarrow Target) = (1 - wp + \frac{(1 + Maxw - w) \times wp}{Maxw}) \\
\times (\frac{1}{2} + \frac{1 + MaxSource - RuleLen(Source)}{MaxSource \times 2}) \times \frac{ElimSupp_{Source \rightsquigarrow Target}}{MaxElimSupp(C, o, w)} \\
\times Conf_{Source \rightsquigarrow Target} \times (\frac{1}{2} + \frac{Balance_{Source \rightsquigarrow Target}}{2})
\end{aligned} \tag{10}
$$

We define *Score* for a sentinel, $Source \rightsquigarrow Target \in SentRules(C, o, w)$, as shown in Formula 10. With this definition of *Score*, we introduce the threshold, *Maxw*, which is the maximum length of the warning period, *w*, we are willing to accept. The constant, *wp*, represents the warning penalty, i.e., the degree to which we want to penalize rules with a higher *w* (0=no penalty, 1=full penalty). The idea of penalizing higher values of *w* is relevant if a pattern is cyclic, e.g., if the indication of a sentinel occurs every 12 months, and the relationship between the indications on the source measure(s) and the target measure is less than 12 months, then a given rule with a warning period *w* is more desirable than the same rule with a warning period *w*+12. We also take into consideration that it is desirable to have shorter, general rules with low *RuleLen*. This prevents our algorithm from "overfitting" rules and thus generating very specific and therefore irrelevant rules. In addition, *Score* takes into consideration the actual number of times the rule occurs in a cube, adjusted for contradictions, *ElimSupp*, as well as the confidence, *Conf*, of the rule. Finally, we consider the *Balance* of the rule, since we have a preference for rules that are bi-directional. In Table 2 the ordering by *Score* has proven useful, and we note that the two bottom rules with the lowest *Score* are also the rules that fail to meet the thresholds we set. Given these thresholds, and constants set to $wp = \frac{1}{2}, Maxw = 10$, we would expect a conforming rule to have: $Score \geq (1 - \frac{1}{2} + \frac{(1+10-1) \times \frac{1}{2}}{10}) \times (\frac{1}{2} + \frac{1+3-3}{3 \times 2}) \times \frac{3 \times 0.6}{4} \times 0.6 \times (\frac{1}{2} + \frac{0.7}{2}) = 0.15$. We should note that this is only a "rule of thumb" since the values in Formula 10 may vary, thus the thresholds needs to be inspected individually to determine if a rule is conforming or not. With *Score* as a uniform way to assess the quality of a generalized sentinel rule, we can now define *Optimalw*(C,o), as shown in Formula 11, which is the value of *w*, $1 \leq w \leq Maxw$, where *SentRules*(C,o,w) contains the rule with the highest *Score*. The reason for the construction details of *Score* is elaborated further in [13].

$$
\begin{aligned}
Optimalw(C, o) = w \text{ such that } 1 \leq w, w' \leq Maxw \wedge \exists S \in SentRules(C, o, w) : \\
(\forall w' \neq w : (\forall S' \in SentRules(C, o, w') : (Score(S) \geq Score(S'))))
\end{aligned} \tag{11}
$$

$$SentRulesPruned(C, o, w) = S \in SentRules(C, o, w) \mid$$
$$\nexists S' \in SentRules(C, o, w) : (Score(S') \geq Score(S) \land Ind_{Source_{S'}} \subset Ind_{Source_S})\} \quad (12)$$

Having found the optimal w, it is also desirable to prune the generalized sentinel rules such that we only output the best rules in terms of *Score*, and the shortest rules in terms of number of source measures. For this purpose, we use the *SentRulesPruned* function, as shown in Formula 12, that eliminates rules with poor quality (lower *Score*) if a shorter rule exists with at least as good a *Score*, and where the indication set is a proper subset of the longer rule.

We say that $SentRulesPruned(C, o, Optimalw(C, o))$ ordered by their respective *Score* are the best sentinels in a database, C, with the offset, o. Using the *SentRulesPruned* function, we note that $NBlgs \land CstPrb \land inv(WHts) \rightsquigarrow inv(Rev)$ in third line in Table 2 would be eliminated since the shorter rule $CstPrb \land inv(WHts) \rightsquigarrow inv(Rev)$ has a better *Score*. In other words, we do not improve the quality by adding $NBlgs$.

3   Discovering Generalized Sentinel Rules

Preliminaries: To discover all generalized sentinel rules in a cube, C, we intuitively need to test all possible rule combinations where the number of source measures combined varies from 1 to *MaxSource*, and the warning period, w, varies from 1 to *Maxw*. However, as an alternative to this brute force approach, we apply two good approximations to improve performance, namely, *Reduced Pattern Growth* (RPG) to optimize the combining of source measures, and *hill climbing* to optimize the auto-fit of warning periods. Intuitively, it is not hard to imagine that the number of source measures has a significant impact on the performance of the algorithm since they can each have indications in two directions, and all of these directions can be combined. This means that the total number of combinations for k source measures is $\sum_{x=1}^{l} \frac{2k!}{(2k-x)!}$ where $l = MaxSource$. If we preserve the order of the source measures we can reduce the number of potential rules (permutations) to $\sum_{x=1}^{l} \frac{2k!}{l!(2k-x)!}$, however, the number of combinations still explode when a significant amount of source measures needs to be examined. Therefore, there is a performance reward if we can *prune* the source measures that are unlikely to become part of any good rule, at an early stage in the algorithm.

Table 3. Logical Table of Combinations (TC) Example

Ind_{M_1}	Ind_{M_2}	Ind_{M_3}	... $Ind_{M_{p-1}}$	CombSupp	ElimSupp $w = 1$	ElimSupp $w = 2$	ElimSupp $w = 3$	... ElimSupp $w = Maxw$
Neutral	Neutral	Dec	Dec	1	1	-1	-1	1
Neutral	Inc	Inc	Inc	2	0	-2	1	1
Neutral	Dec	Dec	Dec	3	-1	3	1	-1
Neutral	Dec	Neutral	Dec	1	1	1	-1	0
Neutral	Dec	Dec	Inc	1	-1	0	0	-1
Neutral	Neutral	Inc	Dec	1	0	0	-1	1
Neutral	Inc	Neutral	Dec	1	0	-1	1	1
Neutral	Inc	Inc	Inc	1	-1	1	1	-1
Inc	Inc	Dec	Dec	4	3	4	-1	0
Dec	Inc	Inc	Inc	3	-3	-3	0	0

From experiments on real-world data, we know that the likelihood of individual source measures being part of a good rule can be described as a *power law*, meaning that a few source measures are very likely to be part of many good rules for a given target measure, whereas the majority of source measures are not likely to be part of a good rule at all. Given this property of source measures, the ability to prune has a significant potential for performance improvement. In addition to reducing the number of source measures we examine, we can save time and memory by storing only the best rules in memory. The RPG optimization, described below, has these two abilities.

The Table of Combinations (TC) (Table 3) is an intermediate table used for optimization. The TC is generated in one pass over the cube, C. We use a sliding window of $Maxw + o$ rows in memory for each combination of $(d_2, d_3, ..., d_n)$ to update the indications (Formula 1) on source and target measures for $w \in \{1, 2, ..., Maxw\}$. For each occurrence of combined source measure indications, $(Ind_{m_1}, ..., Ind_{m_{p-1}})$, the target measure indication, Ind_{m_p}, is calculated for all values of $w \in \{1, 2, ..., Maxw\}$, and the indications, $ElimSupp_w$, are mapped to integers as follows: Inc→1, Dec→-1, and Neutral→0. The discretized values are added to the fields $ElimSupp_w$ on the unique row in TC for the combination $(Ind_{m_1}, ..., Ind_{m_{p-1}})$. In addition, the field $CombSupp$ on this row is increased by 1. If a combination, $(Ind_{m_1}, ..., Ind_{m_{p-1}})$, does not exist in TC, an additional row is appended with the new combination. The TC holds all indication rules with $RuleLen(Source) = p - 1$ with their respective $ElimSupp$ (Formula 5), denoted by $ElimSupp_{w=x}$ for all $x \in \{1, 2, ..., Maxw\}$. We store the additional information about the direction of the target measure in the sign of $ElimSupp_w$ (positive=up, negative=down). The field $CombSupp$ in TC holds the support of the given source measure indication combination, and is used for calculating $SentSupp$ (Formula 8). Once generated, the TC is a highly compressed form representing the information needed to mine all potential sentinel rules in C. In the TC, a rule, $SentRule \in SentRules(C, o, w)$, has $SentSupp = \sum CombSupp$ (Formula 8) when selecting the rows that support either $Ind_{Source_{SentRule}}$ or $\overline{Ind_{Source_{SentRule}}}$. Similarly, the components for *Balance* (Formula 7) can be found as $A = \sum_{x \in X} ElimSupp_w$ where X are the rows that support $Ind_{Source_{SentRule}}$, and $B = \sum_{y \in Y} ElimSupp_w$ where Y are the rows that support $\overline{Ind_{Source_{SentRule}}}$. We recall from Formulae 4 and 6 that a bi-directional sentinel has *Balance* > 0 and thus require a pair of indication rules with opposite directions.

In Table 3, the sentinel $M_1 \rightsquigarrow M_p$ has $SentSupp_{M_1 \rightsquigarrow M_p} = 7$, $A = 3$, and $B = -3$, when $w = 1$ (as seen in column $ElimSupp_{w=1}$), meaning that the rule has $ElimSupp_{M_1 \rightsquigarrow M_p} = |A| + |B| = 6$. We note that A and B have been cleansed for contradictions prior to insertion in the TC, and that the sign is a property used by the TC and should thus be omitted. We have $Conf_{M_1 \rightsquigarrow M_p} = 0.857$ and $Balance_{M_1 \rightsquigarrow M_p} = 1$ when the warning period, w, is 1. Similarly, we can find the values for $M_3 \wedge M_{p-1} \rightsquigarrow inv(M_p)$ by inspecting the rows where $M_3 \neq$ Neutral $\wedge$ $M_{p-1} \neq$ Neutral. If we set $w = 2$, we find $SentSupp_{M_3 \wedge M_{p-1} \rightsquigarrow inv(M_p)} = 14$, $A = -4$, and $B = 6$, $ElimSupp_{M_3 \wedge M_{p-1} \rightsquigarrow inv(M_p)} = 10$, thus $Conf_{M_3 \wedge M_{p-1} \rightsquigarrow inv(M_p)} = 0.714$. In addition, we have $Balance_{M_3 \wedge M_{p-1} \rightsquigarrow inv(M_p)} = 0.960$. A sentinel is therefore typically combined from multiple rows in the TC, i.e., a rule with $RuleLen(Source) \leq MaxSource$ will need a full scan of TC to identify $ElimSupp$, *Balance*, and $SentSupp$ because

Table 4. Source Measure Influence & Pareto Example

w	Inf_{M_1}	Inf_{M_2}	Inf_{M_3}	$Inf_{M_{p-1}}$	Inf_{All}	$Inf_{M_2+M_3+M_{p-1}}$	$Pareto$
1	2	6	6	7	22	19	86%
2	2	7	6	8	25	21	84%
3	1	6	6	8	24	20	83%
...							
$Maxw$	0	5	6	7	18	18	100%

$MaxSource \ll p$. Since we do not know which source measure indications occur at the same time, there is no generic sorting method that can optimize the scans further.

Reduced Pattern Growth (RPG) is a method that delivers a good approximation of the top sentinel rules, and which is much more efficient than a full pattern growth of all combinations of source measures. The quality of the approximation is examined in detail in Section 4, specifically in Figure 4(a). In the RPG process, we identify the source measures, that are *most likely* to be part of the best generalized sentinel rules for a given value of w, as an alternative to testing *all* combinations of source measures. We do this by inspecting the *influence* of the source measure in the TC, defined as the number of rows in the TC in which a source measure has indications different from neutral while at the same time the indication on the target measure is different from neutral (zero in Table 3). With this definition we can assess the influence, *Inf*, of all source measures for each value of w as shown in Table 4. We note that the source measure M_{p-1} is the most influential from our TC (Table 3), specifically it has an influence of 8 for values $w = 2$ and $w = 3$, and it has an influence of 7 for values $w = 1$ and $w = Maxw$. With the notion of the source measures "behaving" in accordance with a power law, we apply a Pareto principle to select only the most influential source measures, meaning that we select the source measures with a total influence that account for more than $RPGpareto$ % of the sum of the influence of all source measures. In Table 4 we see that source measures M_2, M_3, and M_{p-1} account for more than 80% of the influence for all values of w, i.e. setting $RPGpareto = 80\%$ would mean that we only consider these three measures for the values of w shown in the table. Alternatively, setting $RPGpareto = 85\%$ would mean that source measure M_1 would also be included in the pattern growth for values $w = 2$ and $w = 3$. From this point, we *grow* sentinels from the measures identified. Starting with 1 source measure, we add the remaining influential source measures one at a time to create longer rules until the maximum number of source measures we desire is reached. In this process we only store a sentinel, and continue to add source measures, if the added source measures give a higher *Score*.

Hill Climbing identifies the warning period, w, where the sentinel with the highest *Score* exists as an alternative to calculating all max(*Score*) for all w. We optimize the hill climbing process by changing w +22 in the direction of the local maximum while *Score* increased. Once *Score* decreases, we have passed a local maximum, and we test *Score* for $w - 1$ as well to ensure that we have not stepped over the local maximum. During the hill climb, the set of sentinels with the highest *Score* resides in memory until a better set for another value of w is found. Upon termination, the best set of generalized sentinel rules $SentRulesPruned(C, o, Optimalw(C, o))$ resides in memory.

Algorithm: SentHiRPG

Input: A list of facts from a cube, C, ordered by $(d_2, d_3, ..., d_n, t)$, an offset, o, a maximum warning period length, $Maxw$, a maximum number of source measures per rule, $MaxSource$, a warning penalty, wp, a threshold for RPG, $RPGpareto$, a threshold for indications, α, a minimum $SentSupp$ threshold, σ, a minimum $Conf$ threshold, γ, and a minimum $Balance$ threshold, β.

Output: Sentinel rules with a given warning period, $Optimalw$, and their respective $SentSupp$, $Conf$, $Balance$, and $Score$.

Method: Sentinel rules are discovered as follows:

Procedure $UpdateTC$. For each cell pair, $\{(t, d_2, d_3, ..., d_n), (t + o, d_2, d_3, ..., d_n)\}$ in memory, calculate the indications (Formula 1) using α on source measures $m_1...m_{p-1}$, discretize $Ind_{m_i} \in (Ind_{m_1}, ..., Ind_{m_{p-1}})$ as Inc, Dec, or Neutral. If combination $(Ind_{m_1}, ..., Ind_{m_{p-1}})$ does not already exist in *Table of Combinations (TC)*, append row to TC. For each $w \in \{1, 2, ..., Maxw\}$, for the combination $(Ind_{m_1}, ..., Ind_{m_{p-1}})$, update the value of the indication, $ElimSupp_w$, by adding the indication (Formula 1) based on the pairs, $\{(t + w, d_2, d_3, ..., d_n), (t + w + o, d_2, d_3, ..., d_n)\}$ in memory, discretized as 1, -1, or 0. In addition, increase the value of the combination support counter, $CombSupp$, by 1.

Function $RPGmeasures(w)$; Returns a set of source measures. For each source measure, $M_i \in \{M_1...M_{p-1}\}$, calculate *influence* as $\sum ElimSupp_w$ for all rows in TC where $m_i \neq$ Neutral. Return source measures in ascending order of *influence* until $\frac{\sum influence \text{ of source measures returned}}{\sum influence \text{ of all source measures}} \geq RPGpareto$.

1. Scan C, and when the sliding window of $Maxw + o$ rows are in memory perform $UpdateTC$ whenever a new row is read. From this point keep only $Maxw + o$ rows in memory by disregarding the oldest row whenever a new row is read until a new combination of $(d_2, d_3, ..., d_n)$ occurs, at this point flush all rows and load new $Maxw + o$ rows for the next combination of $(d_2, d_3, ..., d_n)$. Repeat $UpdateTC$ whenever a new row is read until the scan of C is complete, and while the sliding window of $Maxw + o$ rows exist in memory.
2. Find the value of w that corresponds to $Optimalw(C,o)$ (Formula 11) by *hill climbing* on the value of $max(Score(SentRules(C,o,w)))$ (Formulae 6 and 10). The generalized sentinel rules for each tested w, $SentRules(C,o,w)$, are "grown" from 1 to $MaxSource$ in $RuleLen$ by combining the source measures returned by $RPGmeasures(w)$. Only rules with source measure combinations where $Score$ improves when adding an additional source measure are stored, meaning that the output will be equivalent to $SentRulesPruned(C, o, w)$ (Formula 12). While testing different values of w, the sentinels with the highest score at any given time is stored in memory and not flushed until a better set of rules exists for another value of w.
3. Output the "best" generalized sentinel rules from memory, i.e., $SentRulesPruned(C, o, Optimalw(C, o))$ (Formula 12), where $SentSupp >= \sigma$, $Conf => \gamma$, and $Balance >= \beta$.

The SentHiRPG Algorithm: We assume without loss of generality that of the p measures in the cube, C, $M_1...M_{p-1}$ are the source measures and M_p is the target measure.

Step 1 scans the cube, C, and builds the *Table of Combinations (TC)* (Table 3). Since the data is received in sorted order by the time-dimension, t, for each combination of $(d_2, d_3, ..., d_n)$, we only need a sliding window of $Maxw + o$ rows in memory to update the indications (Formula 1) on source and target measures for $w \in \{1, 2, ..., Maxw\}$. Using the procedure $UpdateTC$, each unique combination of source measure indications, $(Ind_{m_1}, ..., Ind_{m_{p-1}})$, that exists in the cube, C, is added or updated in the TC. For each source measure combination, the indications on the target measure, Ind_{m_p}, are calculated for all values of $w \in \{1, 2, ..., Maxw\}$, and the indications, $ElimSupp_w$, are mapped to integers as follows: Inc→1, Dec→-1, and Neutral→0. The discretized values are added to the fields $ElimSupp_w$ on the unique row in TC for the combination $(Ind_{m_1}, ..., Ind_{m_{p-1}})$. In addition, the field $CombSupp$ on the row is increased by 1 each time the source measure combination occurs. In other words, as we scan the cube, C, new unique combinations of $(Ind_{m_1}...Ind_{m_{p-1}})$ make the number of rows in the TC grow. Every time one of these combinations are found during the scan, 1, -1, or 0 is added to the corresponding value for $ElimSupp_w$, and the value of $CombSupp$ is increased by 1. Following Step 1, we have the contribution to $ElimSupp$ (Formula 5) for all values of $w \in \{1, 2, ..., Maxw\}$ as well as $SentSupp$ (Formula 8) for each source measure combination in the TC.

When selecting the rows that support either $Ind_{Source_{SentRule}}$ or $\overline{Ind_{Source_{SentRule}}}$ in the TC, a rule, $SentRule \in SentRules(C, o, w)$, has $SentSupp = \sum CombSupp$ (Formula 8). Similarly, the components for $Balance$ (Formula 7) can be found as $A = \sum_{x \in X} ElimSupp_w$ where X are the rows that support $Ind_{Source_{SentRule}}$, and $B = \sum_{y \in Y} ElimSupp_w$ where Y are the rows that support $\overline{Ind_{Source_{SentRule}}}$. We should note that since we express the direction (Inc or Dec) with the sign of the indications on the target measure, $ElimSupp_w$, we need to use the *absolute values* for each direction when calculating $ElimSupp$, thus we have $ElimSupp = |A| + |B|$ (Formulae 5 and 7). In the conceptual description of the TC, we calculated $SentSupp_{M_1 \rightsquigarrow M_p} = 7$, $Conf_{M_1 \rightsquigarrow M_p} = 0.857$ and $Balance_{M_1 \rightsquigarrow M_p} = 1$ for the sentinel $M_1 \rightsquigarrow M_p$ when $w = 1$. Calculating $Score$ with these values allows us to compare the quality of any sentinel combined from the source measures and the target measure in C using the TC.

In Step 2 we identify the best value of w, $Optimalw(C,o)$ (Formula 11), which is defined as the value of w where the sentinel(s) with the highest $Score$ exist(s) (Formula 10). We use two optimization techniques for this purpose: hill climbing and Reduced Pattern Growth as explained above. Hill climbing is a well-known optimization technique [9] and it is an alternative to testing all values of $w \in \{1, 2, ..., Maxw\}$ to identify $max(Score(SentRules(C,o,w)))$ (Formulae 6 and 10). During the hill climbing process, whenever a value of w needs to be inspected to identify $max(Score(SentRules(C,o,w)))$, we apply the Reduced Pattern Growth (RPG) function, $RPGmeasures$. Having identified the most influential source measures, we "grow" the sentinels with $RuleLen=1$ to "longer" rules until $RuleLen=MaxSource$ by combining the source measures returned by $RPGmeasures$ for a given value of w. During this process we only store sentinels with greater $RuleLen$ if the extended $RuleLen$ translates into a higher $Score$. This means that we are growing a set of sentinels equivalent to $SentRulesPruned(C, o, w)$ (Formula 12). Once all sentinels have been grown for a particular value of w, the $max(Score(SentRules(C,o,w)))$ value is returned to the hill climbing process to determine whether to examine more values of w or not. During the entire hill climb, the set of sentinels with the highest $Score$ so far is stored in memory until a better set is found for another value of w. Upon termination, the best set of generalized sentinel rules $SentRulesPruned(C, o, Optimalw(C, o))$ resides in memory.

In Step 3 the sentinels that conform with the thresholds for $SentSupp$, $Conf$ and $Balance$ from the set $SentRulesPruned(C, o, Optimalw(C, o))$ are output from memory.

Computational Complexity: The algorithm has a complexity of $\mathcal{O}(n + c \times p(q)^l \times k^l \times m)$ where n is the size of C, c is the size of TC, p the percentage of remaining source measures expressed as a function of q, where q is *RPGpareto*, l is *MaxSource*, and m is *Maxw*. In Section 4 we verify this assessment through extensive experiments.

Implementation: The SentHiRPG algorithm variants were implemented in Microsoft C# and compiled into a stand-alone 64-bit executable file. The initial version loaded the data directly from a Microsoft SQL Server during Step 1. However, this approach was not able to feed the data fast enough to stress test the algorithm. As a consequence, we loaded all data into main memory and from here into Step 1. This approach is realistic (see Section 4) and it provided sufficient bandwidth to stress the algorithm in order to see the effect of the optimizations applied. The TC built in Step 1 is stored in a hash table where each source measure indication for a row is encoded into 2 bits. In Step 2, when testing a given value of w, we found that testing all combinations of source measures from $RuleLen = 1$ to $RuleLen = MaxSource$, and storing only longer rules if *Score* improved, was far more efficient than a genetic algorithm without mutation. We use the following algorithm variants: **Brute**: brute force, both optimization options are off. **Hi**: **Hi**ll climbing optimization activated, RPG off. **RPG**: **R**educed **P**attern **G**rowth activated, hill climb off. **HiRPG**: both **Hi**ll climb & **R**educed **P**attern **G**rowth activated.

4   Experiments

We use a range of synthetic datasets and a range of real-world datasets for the experiments. The synthetic datasets closely resemble our running example and have the same rule relationships, since the three source measures are duplicated to create a dataset with any number of source measures. The synthetic datasets range from 1,000,000 to 10,000,000 rows in 1,000,000 row intervals, with 50 source measures and one target measure. We note that the sizes of these datasets are *huge* compared to the real-world dataset. In general, we would expect any real application of sentinels to work on *significantly fewer rows* since we typically aggregate the data, e.g., into months or weeks, before finding sentinels. In addition, we have generated datasets with 1,000,000 rows and with 1, 10, 20, 50, 100 and 150 source measures and one target measure. The real-world datasets are produced from the operational data warehouse of TARGIT A/S. Based on experience with more than 3,600 customers worldwide, we will characterize this dataset as typical for a medium-sized company with a mature data warehouse. The original dataset contains 241 months (20.1 years) of operational data scattered across 148 source measures. Descendants of this dataset are produced by selecting a given number of source measures randomly to produce datasets with 10, 20, 30, ..., 140 source measures. When nothing else is specified, the synthetic dataset has 1,000,000 rows, and the algorithm has the following settings: *wp*=0.5, *MaxSource*=3, *Pareto*=85%, and thresholds: *SentSupp*=3, *Conf*=0.6, and *Balance*=0.7.

Scaling rows: In Figure 1 we validate that "HiRPG" scales linearly to 10 million rows of data. In Figure 1(a) a "simple Brute" with TC optimization alone was far more efficient than the baseline from prior art [11]. In Figure 1(b) we compare "simple Brute" with "HiRPG "; the distance between the lines is the cost of auto-fitting w over 50

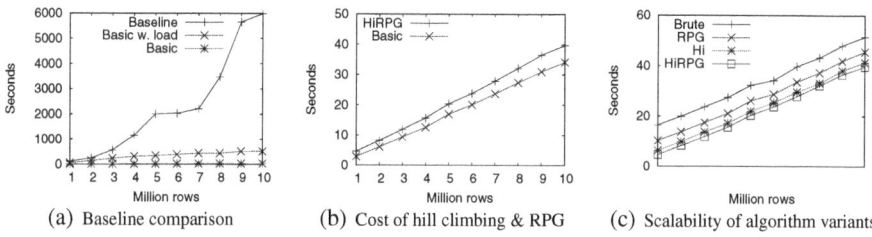

Fig. 1. Scaling the number of rows

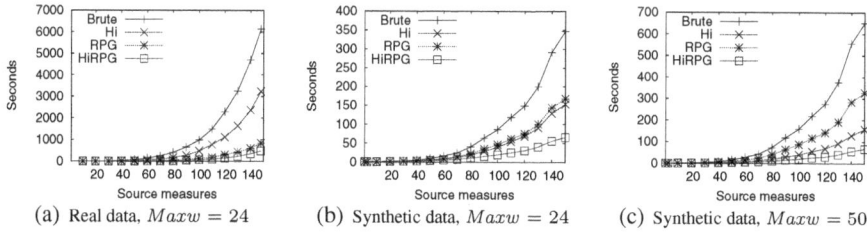

Fig. 2. Scaling the number of source measures

periods and combining 50 source measures. As expected, based on our assessment of computational complexity, we observe "simple Brute" and "HiRPG" to scale linearly. We observe the difference between "simple Brute" and "HiRPG" to be close to constant for an explanation of the slight increase in runtime. In Figure 1(c) the variants scale linearly as expected, and not surprisingly the fully optimized "HiRPG" is best.

Scaling source measures: Scaling the number of source measures on real data has an exponential impact on all variants, but "HiRPG" and "RPG" are very efficient in reducing this impact. On the comparable synthetic dataset in Figure 2(b) "RPG"and "Hi" are almost equal in efficiency, and we see "Hi" excelling when expanding the fitting period in Figure 2(c). We attribute this to the existence of a true power law in the real data, whereas in the synthetic data the relationships between source and target measures are simply repeated for every three measures which means that many measures have strong relationships. The fact that "Hi" improves further when increasing $Maxw$ is not surprising since hill climbing specifically reduces the cost of increasing $Maxw$. We note that for the synthetic data "HiRPG" is still by far the most efficient variant of the algorithm, and although the dominant computational complexity is cubic in the number of source measures ($MaxSource = 3$), the RPG optimization significantly reduces this impact.

Scaling the fitting period: In this experiment, we vary the $Maxw$ over which we fit the warning period, w. In Figure 3(a) and (b) we see that "HiRPG" is performing best when scaling $Maxw$, followed by "Hi", that lack the RPG optimization. Both variants scale linearly to the extreme fitting over 10,000 periods. In Figure 3(c) the same lack of RPG optimization is more evident on real data, given the power law explained earlier. The slight decrease in runtime on the higher values of $Maxw$ should be seen in the context

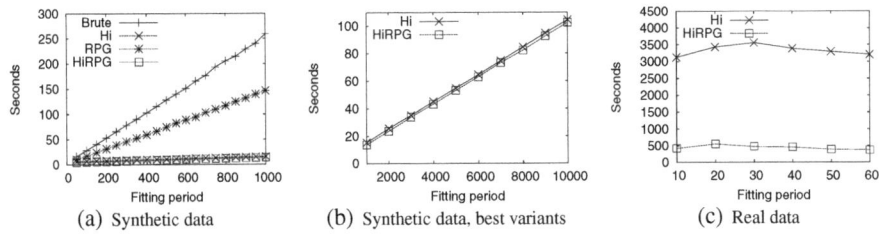

(a) Synthetic data   (b) Synthetic data, best variants   (c) Real data

Fig. 3. Scaling the fitting of warning period

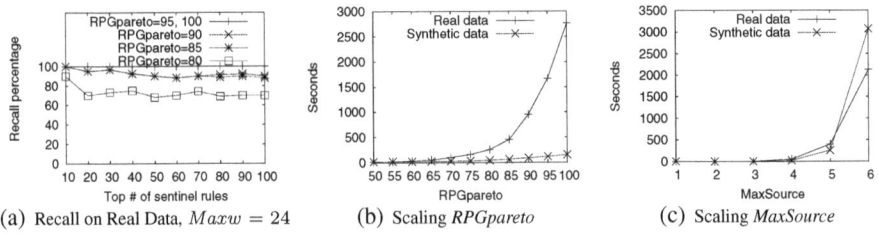

(a) Recall on Real Data, $Maxw = 24$ (b) Scaling *RPGpareto*   (c) Scaling *MaxSource*

Fig. 4. HiRPG Performance when scaling parameters *RPGpareto* and *MaxSource*

that the dataset has only 241 rows. Therefore, we should not interpret the findings on the real dataset as sub-linear scalability. In other words, we note that scaling the fitting of *w* scales linearly as expected from our assessment of computational complexity.

Scaling parameters *RPGpareto* and *MaxSource*: In Figure 4(a) we see the recall ($\frac{\text{Number of Sentinels - false negatives}}{\text{Number of Sentinels}}$) of the top 10 to 100 top-*Score* sentinels for "HiRPG". We note the significant drop in the recall at $RPGpareto = 80$. In Figure 4(b) we see performance for "HiRPG" when scaling *RPGpareto*. We notice the impact cost when $RPGpareto > 85$. Combining Figure 4(a) and (b) suggests a recall "sweet-spot" at $RPGpareto = 85$ (100% of top 10, and 88% of top 100) before the performance cost "explodes". In Figure 4(c) we scale *MaxSource* for "HiRPG". We note that performance on the real dataset passes the synthetic dataset as the complexity "explodes". We attribute this shift to the power law in the real dataset, i.e., as *MaxSource* increases, so does the effect of the RPG process.

Qualitative Experiment: Apart from assessing the performance of SentHiRPG, we also found interesting and business relevant sentinels on the real-world data, e.g, IF the number of people involved in a customer decision process decrease AND the revenue from training increase, both by 10% or more, THEN the total revenue for TARGIT A/S is expected to increase by 10% or more within three months; and vice versa. In this particular case for TARGIT A/S, it was surprising that the number of people involved in the decision process could be used as an indicator, whereas it has been known for some time that selling more training will typically make a customer expand his solution. Intuitively, it makes sense that if more people are involved in a decision process, then it takes more time, and therefore less revenue will be generated on the short-term. In [13] these sentinels and their business potential is described in greater detail.

5 Conclusion and Future Work

We have proposed a novel approach for discovering so-called generalized sentinel rules (sentinels) in a multi-dimensional data cube for business intelligence. We extended prior work to allow multiple measures to be combined into better sentinels using the novel qualitative measures *Balance* and *Score*. In addition, these measures were used to auto-fit the best warning period for the sentinel. We presented an algorithm that, given a target measure, could autonomously find the best warning period and output the best sentinels from this. In the algorithm we introduced a novel table of combinations (TC) and a reduced pattern growth (RPG) approach, and we demonstrated the optimization effect of these approaches in combination with a hill-climbing optimization to produce from ten to twenty times improvement in performance. We showed that our optimized algorithm scales linearly on large volumes of data and when fitting warning period over large period intervals, and that it scales close to linearly when combining large sets of source measures. We have previously demonstrated that sentinels can find strong and general rules that would not be found by sequential pattern mining or correlation techniques [11], this obviously holds even more for generalized sentinel rules.

For future work, a natural development would be to mine sentinels for multiple target measures simultaneously to improve performance. Secondly, we could exploit the multi-dimensional environment by having sentinel mining fit the aggregation level on dimensions as well as select the location and shape of the data area. Third, a parallelization of SentHiRPG could improve scaling to datasets with even more source measures.

Acknowledgments. This work was supported by TARGIT A/S, Cassiopeia Innovation and the European Regional Development Fund.

References

1. Agrawal, R., Imielinski, T., Swami, A.: Mining association rules between sets of items in large databases. In: Proc. of ACM SIGMOD, pp. 207–216 (1993)
2. Agrawal, R., Lin, K.I., Sawhney, H.S., Shim, K.: Fast similarity search in the presence of noise, scaling, and translation in timeseries databases. In: Proc. of VLDB, pp. 490–501 (1995)
3. Agrawal, R., Srikant, R.: Fast Algorithms for Mining Association Rules in Large Databases. In: Proc. of VLDB, pp. 487–499 (1994)
4. Agrawal, R., Srikant, R.: Mining Sequential Patterns. In: Proc. of ICDE, pp. 3–14 (1995)
5. Bosc, P., Pivert, O., Ughetto, L.: On Data Summaries Based on Gradual Rules. In: Proc. of Fuzzy Days, pp. 512–521 (1999)
6. Brin, S., Motwani, R., Ullman, J.D., Tsur, S.: Dynamic Itemset Counting and Implication Rules for Market Basket Data. In: Proc. of ACM SIGMOD, pp. 255–264 (1997)
7. Han, J., Kamber, M.: Data Mining Concepts and Techniques, 2nd edn. Morgan Kaufmann Publishers, San Francisco (2006)
8. Han, J., Pei, J., Mortazavi-Asl, B., Chen, Q., Dayal, U., Hsu, M.: FreeSpan: frequent pattern-projected sequential pattern mining. In: Proc. of KDD, pp. 355–359 (2000)
9. Lukatskii, A.M., Shapot, D.V.: Problems in multilinear programming. Computational Mathmatics and Mathmatical Physics 41(5), 638–648 (2001)
10. Middelfart, M.: CALM: Computer Aided Leadership & Management. iUniverse (2005)

11. Middelfart, M., Pedersen, T.B.: Discovering Sentinel Rules for Business Intelligence. In: Bhowmick, S.S., Küng, J., Wagner, R. (eds.) DEXA 2009. LNCS, vol. 5690, pp. 592–602. Springer, Heidelberg (2009)
12. Middelfart, M., Pedersen, T.B.: Discovering Sentinel Rules for Business Intelligence. DB Tech. Report no. 24, http://dbtr.cs.aau.dk
13. Middelfart, M., Pedersen, T.B.: Implementing Sentinel Technology in the TARGIT BI Suite (in submission)
14. Nakagaito, F., Ozaki, T., Ohkawa, T.: Discovery of Quantitative Sequential Patterns from Event Sequences. In: Proc. of ICDM, pp. 31–36 (2009)
15. Pei, J., Han, J., Mortazavi-Asl, B., Pinto, H., Chen, Q., Dayal, U., Hsu, M.C.: PrefixSpan: Mining Sequential Patterns by Prefix-Projected Growth. In: Proc. of ICDE, pp. 215–224 (2001)
16. Pei, J., Han, J., Mortazavi-Asl, B., Wang, J., Pinto, H., Chen, Q., Dayal, U., Hsu, M.: Mining Sequential Patterns by Pattern-Growth: The PrefixSpan Approach. IEEE TKDE 16(11), 1424–1440 (2004)
17. Srikant, R., Agrawal, R.: Mining Sequential Patterns: Generalizations and Performance Improvements. In: Apers, P.M.G., Bouzeghoub, M., Gardarin, G. (eds.) EDBT 1996. LNCS, vol. 1057, pp. 3–17. Springer, Heidelberg (1996)
18. Yang, J., Wang, W., Yu, P.S., Han, J.: Mining long sequential patterns in a noisy environment. In: Proc. of ACM SIGMOD, pp. 406–417 (2002)
19. Zhu, Y., Shasha, D.: StatStream: Statistical Monitoring of Thousands of Data Streams in Real Time. In: Proc. of VLDB, pp. 358–369 (2002)

Compound Treatment of Chained Declustered Replicas Using a Parallel Btree for High Scalability and Availability

Min Luo[1], Akitsugu Watanabe[1], and Haruo Yokota[1,2]

[1] Department of Computer Science, Tokyo Institute of Technology
2–12–1 Ookayama, Meguro-ku, Tokyo 152–8552, Japan
[2] Global Scientific Information and Computing Center, Tokyo Institute of Technology
2–12–1 Ookayama, Meguro-ku, Tokyo 152–8550, Japan

Abstract. Scalability and availability are key features of parallel database systems. To realize scalability, many dynamic load-balancing methods with data placement and parallel index structures on shared-nothing parallel infrastructure have been proposed. Data migration with range-partitioned placement using a parallel Btree is one solution. The combination of range partitioning and chained declustered replicas provides high availability while preserving scalability. However, independent treatment of the primary and backup data in each node results in long failover times. We propose a novel method for compound treatment of chained declustered replicas using a parallel Btree, called the Fat-Btree. In the proposed method, the single Fat-Btree provides access paths to both primary and backup data in all processor elements, which greatly reduces failover time. Moreover, it enables dynamic load balancing without physical data migration, and improves memory space utilization for processing the index. Experiments using PostgreSQL on a 160-node PC cluster demonstrate the effect.

1 Introduction

The explosive growth of digital information, together with the high performance and availability requirements, has driven a continuing interest in the research on database systems in shared-nothing parallel environments in which replication plays an important role in availability and scalability with load balancing among the different processing nodes. However, existing replicated-database systems have a weakness in scaling up under frequent update requests because of the high costs of synchronizing the replicas. Well-known approaches for cloud applications in large data centers, such as PNUTS, Dynamo and BigTable, sacrifice strong consistency for scalability. However, they lose opportunities of using the high system throughput for applications requiring stricter consistency, because the replicas are not consistent most of the time in these approaches. Moreover, the advantage of higher availability gained from the replication may also be lost in the long run because of the inconsistency [26]. Therefore, this trade-off between scalability, availability and consistency has long been an obstacle in efficient replication techniques.

P. García Bringas et al. (Eds.): DEXA 2010, Part II, LNCS 6262, pp. 49–63, 2010.

To attack this problem, efficient data access methods guaranteeing consistency between the replicas play an important role. The data access methods must also be capable of handling sophisticated failover and dynamic load-balancing for availability and scalability. Many parallel index structures have been introduced to access the data stored in a shared-nothing environment [16]. For instance, the Fat-Btree [24] and the aB$^+$-tree [13] are range-partition-based parallel indexes providing dynamic data allocation for high scalability. However, these parallel indexes are mainly focused on throughput or latency rather than availability.

Chained declustering data placement [9] adopts a low degree of replication to realize availability and scalability. From the availability point of view, only low degrees of replication are required. The range-partition-based index method is well suited to chained declustering. However, as far as we know, no valid indexing methods have been proposed to consider the use of replicas in chained declustering to enhance system availability and scalability.

We propose a database infrastructure for indexing range-partitioned data with a low degree of replication to achieve high scalability and availability without sacrificing data consistency. We first consider a parallel index structure on a range-partitioned chained replication database as a straightforward configuration for the infrastructure. We then propose a novel method of compound replica treatment utilizing the Fat-Btree index. It reduces the management cost of the index structure, balances load without data migration and enables shorter failover times. We also adopt the neighbor write-ahead log protocol (nWAL) [10] adapted to the proposed configuration, to reduce the synchronization cost between the replicas without data loss.

The key innovative points of this work are: a) it is the first proposal for managing both primary and backup within one directory structure; b) it has an efficient automatic load-balance algorithm for dynamic load skew without data migration; and c) it has an efficient failover algorithm for higher system availability without the cost of "promotion" backups. To the best of our knowledge, no previous work provides practical dynamic load-balancing or failover utilities in chained replication database systems, although they have long been claimed [8, 20]. We have evaluated our method on a PC cluster with 160 nodes. The experimental results demonstrate all the above-mentioned effects.

2 Background

We briefly review three technologies in shared-nothing parallel databases: data placement strategies, parallel indexing structures and synchronization methods.

2.1 Data Placement Strategies

Chained declustering [9] offers high availability, scalability and load balancing on shared-nothing parallel database systems [3]. Two physical copies of the relation are declustered by the same partitioning strategy, and the corresponding fragments of these two copies are stored on different Processing Elements (PEs), so data on a failed node will not be completely lost during the failure.

Basically, there are three types of partitioning strategies in chained declustering: hash, round-robin and value-range. Hash partitioning is ineffective for range queries and does not scale up [20], round-robin partitioning produces no skew, but is ineffective for queries, while value-range partitioning can treat range queries efficiently, but has a risk of load skew after repeated updates. Thus, each partition strategy has limitations, and it is important to choose an appropriate strategy and provide solutions to overcome the limitations.

2.2   Parallel Indexing Structures

To meet the demands of handling large amounts of data, parallel indexing structures have been proposed to provide dynamic data management, high throughput and efficient load skew handling via the index nodes for these systems [17].

The Fat-Btree is a parallel Btree structure that was proposed to provide dynamic data management, high throughput and efficient skew handling [24]. In a Fat-Btree, the leaf pages of the parallel Btree are distributed among PEs. Each PE has a subtree of the whole Btree containing the root node and intermediate index nodes between the root node and leaf nodes allocated to that PE. In the Fat-Btree structure, there are multiple copies of index nodes close to the root node, but they have a relatively low update frequency; on the other hand, leaf nodes have a relatively high update frequency but are not duplicated. Thus, the nodes with higher update frequencies have lower synchronization overhead. Therefore, the maintenance cost is much lower than other parallel Btree structures, such as Copy-Whole-Btree and Single-Index-Btree [24]. Moreover, the Fat-Btree has a higher cache hit rate [24] and more efficient concurrency control protocols LCFB [25] than conventional parallel Btrees [17, 14, 19, 25].

2.3   Synchronization with nWAL

Write-ahead log (WAL) [7] is widely used in database systems to ensure atomicity and durability. However, transactions must be suspended to wait for the forced log-write to stable storage before the new version replaces the previous one. In replication databases, this inefficiency is solved by the neighbor WAL (nWAL) protocol [10], which transfers the log of the transaction on PE_i into the memory of its replica nodes PE_{i+1} through the network before the transaction commits. Moreover, nWAL is naturally suitable for maintaining data consistency in range-partitioned chained replication systems, because the backup can always be synchronized with the latest version of its primary by using the nWAL in the memory. Thus, no extra synchronization messages are required.

3   A Straightforward Configuration

We present a straightforward configuration for our distributed replication database with existing techniques.

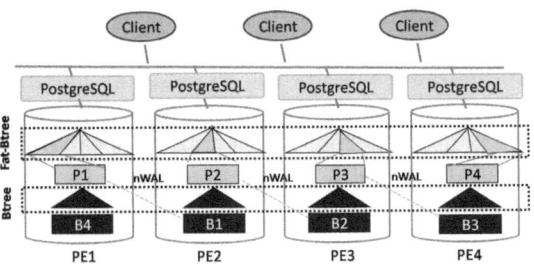

Fig. 1. Configuration of IndepIndexCDR

As mentioned in Section 1, the synchronization cost is the most serious obstacle to achieving high performance and scalability in replication databases. Considerable work has been done to reduce the costs [4, 6, 15, 11]. However, it is hard to achieve satisfactory results unless the number of replica copies is reduced. As we know, the chained replication system only maintains one replica for each piece of data, thus easily achieving higher throughput, availability and stronger consistency compared with other replication schemes [8, 20]. Therefore, we treat the chained replication scheme as the base configuration.

To provide a global access path for the primary data in a chained replication system, some parallel index structure is required. We adopt the Fat-Btree mainly because of its low cost in concurrency control [25] and high flexibility in the compound treatment discussed in the next section. Alternative indexes are discussed later in Section 6.

On the other hand, the backup parts may also be indexed to speed up synchronization in most distributed databases. Because they are not required to serve queries when no failure or skew happens, it is sufficient to maintain an independent Btree for the backup in each PE. Therefore, we adopt the independent Btree to index backups and name this configuration independent-index chained declustered replication *IndepIndexCDR*.

3.1 Implementation of IndepIndexCDR

We implemented IndepIndexCDR using PostgreSQL and the Fat-Btree. Each PE has an instance of PostgreSQL with a part of the Fat-Btree (a subFat-Btree) to access its primary. Following the chained declustering strategy, the backup is located in its primary's right-hand neighbor and it receives the nWAL from its primary synchronously, which may be applied to the backup asynchronously after the primary transaction commits. Figure 1 illustrates this configuration.

Because of space limitations, we omit the detailed data access process in IndepIndexCDR, which is almost the same as that in [24]. Note that for an update request, an nWAL message is sent to the right node before the result is returned to the user. The backup is then updated asynchronously by using the local Btree index on the backup.

3.2   Scalability in IndepIndexCDR

Because of the low degree of replication in IndepIndexCDR, the number of synchronization transactions is greatly reduced. Moreover, unlike the original chained replicated structure in [8, 20], in which only the head and tail PEs accepted query and update requests, IndepIndexCDR can access any data item in primary storage from any PE by the Fat-Btree; thus, the overhead for handling client requests is dispersed over all the PEs. Therefore, IndepIndexCDR should have good scalability. We will report a quantitative experimental evaluation of this later in Section 5.3.

3.3   Availability in IndepIndexCDR

We are not concerned with failure detection techniques here; we assume that a failure can be detected, and only consider the failover efficiency.

Almost all the replication databases claim failover capabilities, but the recovery time is seldom discussed. In addition, the metadata server that is required in many systems introduces a possible bottleneck and single point of failure, while no central node is required in IndepIndexCDR. Although a record of the neighbor in the backup circle is required in each PE, it is only updated when the adjacent PE fails, which is a low-probability event; thus, the cost may be negligible. In addition, the nWAL ensures no data loss whenever a failure occurs.

However, this structure has its weak points. As the backup are not indexed within the parallel index, they are not directly accessible from other PEs to share the workload if the primary is overloaded. To make the backup accessible, additional processes are required, such as merging the Btree with the subFat-Btree on the primary, or dumping the backup into the primary to build the Fat-Btree for these data. Obviously, the "promotion" process is very time consuming.

4   Compound Treatment

We now propose our solution, which overcomes the disadvantages while inheriting the advantages of IndepIndexCDR.

As mentioned, the above configuration can be improved if both primary and backup are managed by one parallel index. Fortunately, the chained declustered replication places continuous logical fragments in the range partitions as a primary for the current PE and a backup for its neighbor. As shown in Fig. 2, because the data are declustered and range partitioned, they can be coupled into the Fat-Btree structure without any intersection. Because the compound subFat-Btree on each PE also has overlapping intermediate paths to the subFat-Btrees located on its neighbors, as in the original Fat-Btree, it provides the access path from the root node to any data located in any PE either in primary or backup. Therefore, the independent Btree for the backup is no longer required. We name this configuration the compound-index chained declustered replication *CompIndexCDR*.

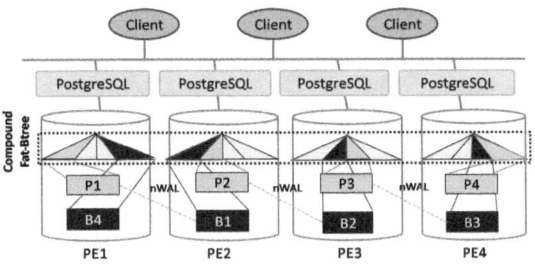

Fig. 2. Configuration of CompIndexCDR

Figure 3 shows an example of this configuration. The upper part shows a global view of the intermediate nodes in the Btree index for all the data over the range (1–60), which are evenly stored in four PEs. By using the original Fat-Btree, some of the nodes will be replicated in several PEs because they are overlapped. To help readers visualize this, we circle the nodes that have a copy in each PE. We use a dotted line for PE_1, a dashed line for PE_2, a solid line for PE_3, and a dashed-dotted line for PE_4. Note that the intermediate nodes that have more than one copy in the PEs have pointers to the leaf nodes located in other PEs. For example, the copy of node "1, 10" in PE_2 has the pointer to the leaf nodes "1, 7" and "10, 16" located in PE_1 and the leaf node "10, 16" located in PE_2. Thus, these overlapped intermediate nodes make an access path for tracing any data from the root among the PEs.

As shown in the lower part of Fig. 3, each PE has two subFat-Btrees for its backup (left) and primary (right), respectively. For the same volume of data, they may have similar index structure and intermediate nodes in their primary and backup PE, such as the index of PE_1's primary and PE_2's backup, while their leaf nodes have pointers to different data pages that are located in PE_1's primary and PE_2's backup, respectively.

To maintain the overlapped paths in this compound Fat-Btree, we also maintain pointers in the parent nodes to all the copies of their child nodes. An example of the overlapped paths for data in the range (31–40) is shown in Fig. 4. In this figure, the root node "1, 20, 46" has paths to all the copies of the second-level node "20, 31" and all the copies of this second-level node have the paths to the primary's and backup's leaf nodes. To distinguish these pointers, we store them with different flags, such as "P" and "B" in the intermediate node. Ordinarily, transactions are carried out by using the pointers that are marked with flag "P" in each intermediate node to access the primary data. nWAL will synchronize the backup after an update. Because the transactions started by the nWAL manager are local synchronization updates, they will use the index pointer marked by flag "B". Note that these pointers are kept available during structure modification operations (SMOs) [25], by referencing all the SMOs in any primary PE to its backup PE. Although the index structure may vary after the SMOs in each PE, the access paths still exist in the intermediate nodes.

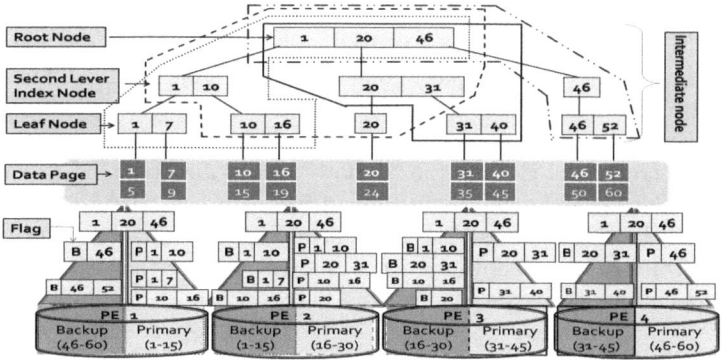

Fig. 3. A Compound Fat-Btree Model

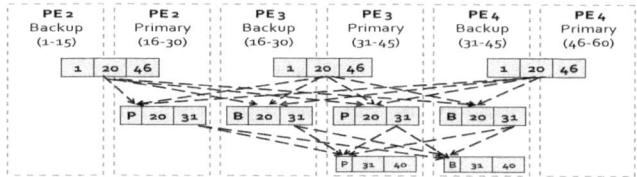

Fig. 4. Overlapped Paths in the Compound Fat-Btree

Our method does not introduce any additional overhead for keeping the backup's index up to date, even if indexing the backup for efficient synchronization is very common in other replication databases. We show that the double-index-sized CompIndexCDR does not reduce the throughput in Section 5.3.

4.1 Availability in CompIndexCDR

The time-consuming "promotion" process described in Section 3 is no longer necessary in CompIndexCDR. In practice, we only switch the flag values between "P" and "B", or modify the indexes of the adjacent PEs that have overlapped intermediate nodes with the failed one with value "P/B".

For example, if PE_3 fails, the new primary location information must be modified only in PE_2 and PE_4, which have intermediate nodes with PE_3. In this case, PE_2 and PE_4 will receive data records (31–45). Then, the flags of those intermediate nodes that covered this range of data records will switch in two steps: Step-A, switch the flags of the corresponding leaf nodes between "B" and "P"; Step-B, modify the flags of the other intermediate nodes having child nodes that are switched in Step-A, following the rules: if all the child nodes of an intermediate node are modified in Step-A, then switch the flag; else if the original flag in the intermediate node is the same as that of the modified leaf node in Step-A, then do nothing; else modify the flag to "P/B", which means some leaf nodes of this intermediate node are stored as backups while others can be accessed as primaries. After this process, node (P, 20, 31) in PE_2 will be (P/B, 20, 31) and node (B, 20, 31) in PE_4 will be (P/B, 20, 31), while the "31" in both nodes will have the primary point to the leaf node in PE_4. Thus, requests for the data located in this data range will be forwarded to PE_4's backup. If the original PE fails, there is no need to modify the new backup's location or the intermediate nodes on PE_3, and the "update" on the new primary does not write to backup, which is different from the load balancing described later. However, after the failed PE is replaced, the new backup's location will be registered.

We verified that this failover process is much quicker than that of IndepIndexCDR; a comparison experiment is presented in Section 5.4.

4.2 Scalability in CompIndexCDR

Load skew in a value-range partition may greatly degrade system scalability [3]. Ordinarily, data migration is required to handle the skew; this may take a long time and destroy the efficiency of load balancing [2].

The declustering replication scheme [9] provides the ability to balance skews without data migration. However, although there have been many studies on this topic, so far as we know, no one has provided an implementation of this capability in practice. In addition, the metadata servers introduced in these studies may decrease system availability. Our system, CompIndexCDR has no central node, and backups can share the workload with the primary PEs by some simple configuration rules.

For the above example of four nodes, the load can be balanced without data migration if, for example, the query frequency on PE_1, PE_3 and PE_4 is α, while the query frequency on PE_2 is 2α, as shown in Table 1. We also assume that P_i and B_i ($i \in [1, 4]$) in the table represent the same amounts of data. We assume that the query frequency is known merely for convenience; in practical terms, many methods have already been proposed to find the access probability on any PE, for example, [21] is one of them and is suitable for the Fat-Btree.

Table 1. Before Load Balancing

	PE_1	PE_2	PE_3	PE_4
Access rate	α	2α	α	α
Primary	P_1	P_2	P_3	P_4
Backup	B_4	B_1	B_2	B_3

Table 2. After Load Balancing

	PE_1	PE_2	PE_3	PE_4
Access rate	$\frac{5}{4}\alpha$	$\frac{5}{4}\alpha$	$\frac{5}{4}\alpha$	$\frac{5}{4}\alpha$
Primary	$P_1+\frac{1}{4}B_4$	$\frac{5}{8}P_2$	$\frac{1}{2}P_3+\frac{3}{8}B_2$	$\frac{3}{4}P_4+\frac{1}{2}B_3$
Backup	$\frac{3}{4}B_4$	$B_1+\frac{3}{8}P_2$	$\frac{5}{8}B_2+\frac{1}{2}P_3$	$\frac{1}{2}B_3+\frac{1}{4}P_4$

We first assume the workloads are "all-read". The balanced placement is shown in Table 2. Because of space limitations, we omit the detailed description of the balancing process. A general algorithm for load balancing is given in Fig. 5.

Next, we assume the workloads are "all-update". The method still balances the skew by using the same placement in Table 2. Note that each piece of data still keeps a copy of its backup by changing the "old primary" into "new backup" in the neighbor. For example, $\frac{3}{8}P_2$ on PE_2. Similarly to Section 4.1, all the

```
Dynamic Load Balancing
begin
    Let op, lp, and rp be the target, left, and right PE, respectively;
    Let pdm[i] be the primary data amount on node i;
    Receive a list of loads ll from lp;
    Update ll[op] to reflect the current workload of the PE;
    Calculate the average load al from ll;
    if ll[op] > al + Threashold then
        Calculate the new primary placement pp[n];  // if this skew can be balance without data migration
        for (i = 0; i < n; i++)     // n is the number of PEs
        {
            broadcast the modification msg to node[i] to change primary;
            if pdm[i] > pp[i] then
            {
                Demote_Primary(the leftmost data page in pdm[i], pdm[i] - pp[i]);
                set the 'Flag' for the new 'backup' at pdm[i] - pp [i];
                broadcast the modification msg to node[i + 1]{or node[0] if i = n}to change backup there;
                Promote_Backup(the leftmost data page in pdm[i + 1], pdm[i] - pp[i]);
                set the 'Flag' for the new 'primary' at pdm[i] - pp [i];
            }
            else
            {
                Promote_Backup(the rightmost data page in pdm[i], pp [i] - pdm[i]);
                set the 'Flag' for the new 'primary' at pp [i] - pdm[i];
                broadcast the modification msg to node[i - 1] {or node[n] if i = 0}to change primary there;
                Demote_Primary(the rightmost data page in pdm[i], pp[i] - pdm[i]);
                set the 'Flag' for the new 'backup' at pp [i] - pdm[i];
            }
        }
        Reset list ll ;
    if op is the rightmost PE then
        Send ll to the leftmost PE;
    else
        Send ll to rp;
end
```

Fig. 5. Algorithm for Handling Skews

"updates" on $\frac{3}{8}P_2$ will be forwarded through the intermediate path to PE_2 or PE_3, then the "updates" are carried out on $\frac{3}{8}B_2$ at PE_3, and the nWAL manager on PE_3 will send nWAL log messages to the backup PE_2. Because the data backups are always maintained in the system, future failures are still tolerable.

It is obvious that for update transactions, each node still has the same number of transactions to be executed even if the placement in Table 2 is adopted. However, as the backups are updated asynchronously, the backup overhead can be alleviated and the unexecuted updates can be preserved temporarily as nWAL records until the overhead is remitted. Therefore, small load skews (i.e., a 2α skew always arises after a node fails) can easily be balanced without data migration. Nevertheless, data migration is still necessary when the skew is more than $\frac{2(n-1)}{n-2}\alpha$ (n is the number of PEs). However, even so, the proposed method is able to halve the skew immediately, and reduce the data migration cost.

5   Experiments

It is hard to provide direct comparisons with former studies because, to the best of our knowledge, no previous work exists that manages both primary and backup within one directory or analyzes the failover efficiency issue in a

Table 3. Experimental environment

Blade servers:	Sun Fire B200x Blade Server
CPU:	AMD Athlon XP-M 1800+ (1.53 GHz)
Memory:	PC2100 DDR SDRAM 1 GB
Network:	1000BASE-T
Gigabit Ethernet Switch:	Catalyst 6505 (720 GB/s backbone)
Hard Drives:	TOSHIBA MK3019GAX
	(30 GB, 5400 rpm, 2.5 inch)
OS:	Linux 2.4.20
Java VM:	Sun J2SE SDK 1.5.0 03 Server VM

chained declustering database. Instead, we compare our work with the well-known Postgres-R (PG-R) and HBase replication DBMSs, as well as with the naive-structure IndepIndexCDR that treats primary and backup separately, to demonstrate the scalability and availability of our proposed approach.

5.1 Experimental Environment

The experimental system was implemented on a 160-node PC cluster system, and the experimental environment is summarized in Table 3. We used continuous integers to act as the primary key in each PE in these experiments.

5.2 Comparison with Postgres-R and HBase

At first, we compare the performance of CompIndexCDR with multi-replication DBMS PG-R and HBase to assess our system performance.

PG-R [23] is the first significant research prototype to provide fully functional database replication mechanisms based on an open-source database. HBase is an open-source, distributed database modeled after Google's BigTable; it provides BigTable-like capabilities and has recently attracted much interest in the distributed database community. We chose these two systems as the criteria for our comparisons. Although the configurations of these systems are different in replication strategies, what we focus on here is the comparison between full/partial replication with our approach.

We use the same experimental configuration as in [23], which gives the most recent performance results of PG-R, to make the first comparison. In this experiment, both PG-R and CompIndexCDR contain five or 10 nodes and each node stores 8000 tuples. The experiment is performed with 40 clients that are evenly distributed on PEs. Figure 6-a shows the experimental results.

For the "all-read" result, PG-R clearly outperforms our system because of its full replication scheme. However, as its authors declared [23], PG-R does not win by much and will lose in scalability even for a moderate number of updates when the system scales. In contrast, CompIndexCDR has a limitted update overhead and share the load along the cluster. We verify its scalability with up to 64 nodes in the next section.

We adopted the recently released HBase-0.20.2 to make the second comparison, which runs on our cluster system based on Hadoop-0.20.1 and jdk1.6.0.18.

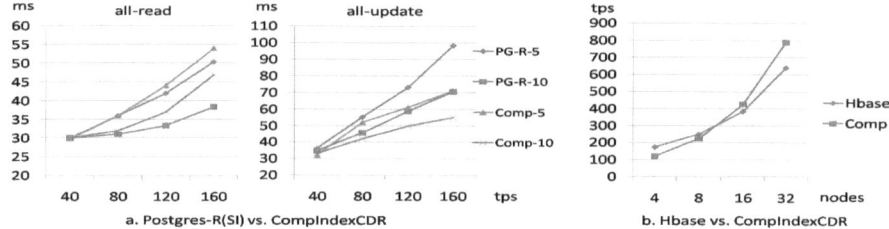

Fig. 6. Performance Comparison with Postgres-R and Hbase

We used the default settings in HBase and Hadoop throughout the experiments, except for changing the replica degree to two, which is the same as in our system. The "Datanode" number in our experiments is only up to 32 because the software in current computation clouds typically uses significantly fewer instances. For example, Amazon limits the number of nodes to 20 by default [27].

In this experiment, the dataset contains 10,000 tuples, and each tuple is 4 KBytes with four columns, and we have one client on each node to perform all-read transactions. The results are shown in Fig. 6-b. We do not provide all-update results here because they are almost the same as the all-read results. The figure shows that the column-oriented DBMS with BigTable has better performance for a small number of nodes. However, as our system has better scalability, it soon outperforms Hbase as the system grows.

We can conclude from these experiments that, for scalability and throughput, CompIndexCDR outperforms both the full-replication PG-R and partial-replication HBase. A further examination of the availability follows.

5.3   Throughput Comparison

We compare the throughput and scalability of our two structures to reflect their different index-traversing costs. The cost mainly consists of intranode and internode traverse costs. The first one is the cost for traversing a local index; larger datasets increase this cost. The second one is the cost for locating remote data as well as maintaining the distributed index; the low cost of this part is one of the strengths of our index. Because the throughput only reflects total traversing cost, and huge datasets increase the intranode cost in both systems, we show the difference in the internode costs more clearly by initializing each PE with a small dataset of 10,000 tuples, each of 134 bytes.

In the experiments, each PE receives the requests from their client nodes simultaneously; the key in each request is generated randomly from the whole data range. As for the 64 PE case, these queries will be: key = random$(1, 640000)$; select* from table where id = key; update table set value += 1 where id = key. Because we focus on the index cost, we do not use complex queries. In addition, each client sends 20 transactions to a PE serially and has a private thread on the target PE to process its requests.

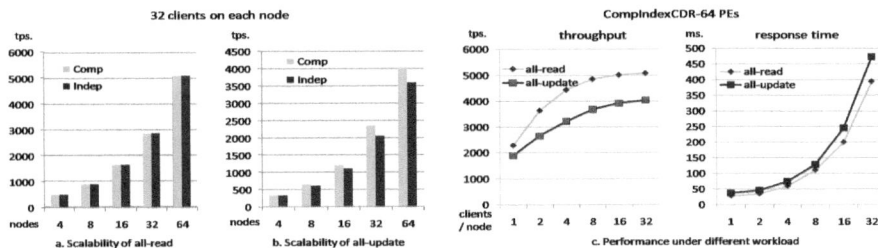

Fig. 7. Scalability Comparison

Figure 7-(a,b) shows the scalability of our systems with 32 clients on each node. Figure 7-a illustrates the all-read case. The performances of these two structures are almost the same and their scalability is quite good. The throughput increases by about 80% when the number of PEs is doubled. The missing 20% may result from the increasing amount of network communication caused by the growing number of remote transactions in the larger system. Figure 7-a also shows that although the index size of CompIndexCDR is twice that of IndepIndexCDR, their throughputs are almost the same. The reason may be that the backup is not accessed, so both systems have similar memory hit rates. Figure 7-b shows that CompIndexCDR has a better scalability for the all-update workload. This may be because the backup in IndepIndexCDR is updated by an independent Btree, while it is accessed within one compound index in CompIndexCDR, therefore the index is not switched in memory. We do not provide experiments for mixed read/write workloads, because they simply produce intermediate results between all-read and all-update workloads.

Figure 7-c shows the throughput and response time of CompIndexCDR at the scale of 64 PEs with various client numbers per node. We omit the results at other sizes, because those results are almost the same as this one. As the figure shows, the throughput of CompIndexCDR increases with the client numbers. When each node has 32 clients, the growth is moderate at a stable value that is defined by the limitations of the CPU and I/O performance in our cluster system. On the other hand, the response time also increases with the client numbers, which is normal in DBMSs and is usually solved by adding more PEs to scatter the workload. Figure 8-a shows the trend of response times as the number of PEs increases. For a fixed workload of 256 clients in total, the response time is effectively reduced because those requests are evenly scattered after new PEs are added into the system.

5.4   Failover Time Comparison

In this experiment, we used four PEs, each with 10,000 tuples and each serving two users. When a PE fails, we dump the backup into the primary and rebuild the Fat-Btree for those data for IndepIndexCDR, and we take the failover as described in Section 4 for CompIndexCDR. In Fig. 8-b, "Normal" stands for the

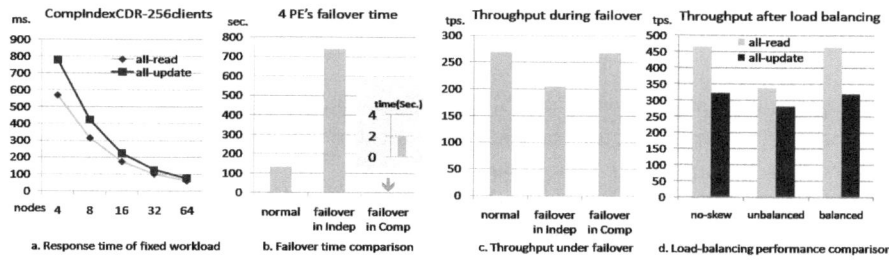

Fig. 8. Response Time and Failover, Load Balancing Performance

time required to initialize 10,000 tuples in one PE. "Failover in Indep" stands for the time taken by IndepIndexCDR to prepare a new primary. It is much higher than the value for "Normal" because the PE must process ordinary requests while making the new primary. As the graph shows, the failover in CompIndexCDR takes less than 2 seconds, which is much better than the time for IndepIndex-CDR. We also examined the throughput during failover. Figure 8-c shows that the throughput decreases only slightly during failover.

5.5   Load Balance Comparison

In this experiment, we used four PEs, each with 10,000 tuples and serving 32 users 40% of the requests access the data in one PE and the other three PEs share the remaining 60%, the same skew as in Sec. 4.2. Figure 8-d shows that this skew decreases the throughputs in "all-read" and "all-update" by 25% and 15%, respectively. After using the balancing method in Sec. 4.2, the "all-read" is almost the same as that of "no-skew" and the "all-update" is also much improved, which verifies the effectiveness as argued in Section 4.2.

6   Related Work

The primary/backup approach has been studied to provide system availability. The interleave declustering strategy provides an immediate load and space balance, but at the cost of reliability. Chained declustering strategy handles the range partition and reconcile high availability with load balancing without loss in reliability. Hot mirroring and RAID suggest the possibility of a hybrid approach, combining parity-based and primary-backup approaches.

On the other side, various parallel index structures have been proposed [16, 13, 24] to provide efficient access method for declustering database. In addition, [1] provides two alternatives for performing the necessary index modifications (OAT), however, they both lead to considerable SMOs cost [13]. Fat-Btree [24] vastly reduces the SMOs cost and improves the dynamic skew handling method. [5, 12] provide load balancing solutions with less data migration.

Unfortunately, all the above proposals mainly focused on only one or two targets among reducing synchronization cost, rapid recovery, efficient access and dynamic skew handling. They did not explore access methods or the use of replicas to promote scalability and availability. Note that, other parallel indexes that may be alternative in our method are not suitable for the compound treatment as ours. For example, the "FirstTierIndex" of the "aB⁺-tree" [13] must be updated immediately in all PEs after data reallocation. "GHT" in [16] requires an "AST" in all PEs to record the primary/backup switch, and [12] require much more modification of adjacent nodes' indexes in skew balancing.

7 Conclusions and Future Work

We proposed a compound index treatment of a chained, replicated declustering scheme CompIndexCDR for shared-nothing parallel database systems. So far as we know, this is the first research to support rapid recovery, dynamic skew handling and efficient data access, as well as reducing synchronization cost by the compound management of primary and backup in a chained replication system.

As the CPU and I/O performance of the PC cluster we used in our experiments are slower than those of current systems, we believe CompIndexCDR can achieve much better performance once the hardware is upgraded. Thus, we conclude that CompIndexCDR is a very good choice for highly scalable and available parallel databases.

In this work, we have not considered the time for restoring a backup after a failure. The adaptive overlapped declustering method [22] is a good way to reduce the restoration time. In future, we plan to combine adaptive overlapped declustering with CompIndexCDR. We also plan to consider the management of multi-replica schemes to improve availability and performance.

Acknowledgments

Part of this research was sponsored by CREST of Japan Science and Technology Agency (JST), and MEXT via a Grant-in-Aid for Scientific Research #19024028.

References

1. Achyutuni, K.J., Omiecinski, E., Navathe, S.B.: Two techniques for on-line index modification in shared nothing parallel databases. In: Proceedings of the ACM SIGMOD Int'l. Conf. on Management of data, June 04-06, pp. 125–136 (1996)
2. Arnan, R., Bachemat, E., Lam, T.K., Michel, R.: Dynamic data reallocation in disk arrays. ACM Transactions on Storage (TOS) 3(1), 2-es (2007)
3. Dewitt, D., Gray, J.: Parallel database systems: the future of high performance database systems. Communications of the ACM 35(6), 85–98 (1992)
4. Elnikety, S., Zwaenepoel, W., Pedone, F.: Database replication using generalized snapshot isolation. In: SRDS 2005, October 26-28, pp. 73–84 (2005)
5. Feelifl, H., Kitsuregawa, M., Ooi, B.C.: A fast convergence technique for online heat-balancing of btree indexed database over shared-nothing parallel systems. In: Ibrahim, M., Küng, J., Revell, N. (eds.) DEXA 2000. LNCS, vol. 1873, pp. 846–858. Springer, Heidelberg (2000)

6. Fekete, A.: Allocating isolation levels to transactions. In: Proceedings of 24th ACM SIG-(MOD/ACT/ART) symposium on principles of database systems (June 2005)
7. Haerder, T., Reuter, A.: Principles of transaction-oriented database recovery. ACM Computing Surveys (CSUR) 15(4), 287–317 (1983)
8. Hsiao, H.-I., DeWitt, D.J.: A performance study of three high availability data replication strategies. Distributed & Parallel Databases 1(1), 53–80 (1993)
9. Hsiao, H.-I., DeWitt, D.J.: Chained declustering: A new availability strategy for multiprocessor database machines. In: Proceedings of ICDE 1990, pp. 456–465 (1990)
10. Hvasshovd, S.-O.: Recovery in Parallel Database Systems, 2nd edn. Morgan Kaufmann Publishers, San Francisco (1999)
11. Kemme, B., Alonso, G.: Don't be lazy, be consistent: postgres-r, a new way to implement database replication. In: Proceedings of VLDB 2000, pp. 134–143 (September 2000)
12. Feelifl, H., Kitsuregawa, M.: RING: a strategy for minimizing the cost of online data placement reorganization for btree indexed database over shared-nothing machines. In: Proceedings of the 7th Int'l. Conf. on DASFAA 2001, pp. 190–199 (April 2001)
13. Lee, M.L.: Towards self-tuning data placement in parallel database. In: Proceedings of the ACM SIGMOD Int'l. Conf. on Management of data, pp. 225–236 (May 2000)
14. Miyazaki, J., Yokota, H.: Concurrency control and performance evaluation of parallel B-tree structures. IEICE Trans. INF. & SYST. E85-D(8), 1269–1283 (2002)
15. Ouyang, X., Yokota, H.: An efficient commit protocol exploiting primary–backup placement in a distributed storage system. In: Proceedings of the 12th Pacific Rim International Symposium on Dependable Computing, December 18-20, pp. 238–247 (2006)
16. Zezula, P., Amato, G., Dohnal, V., Batko, M.: Similarity Search The Metric Space Approach. In: Parallel and Distributed Indexes, ch. 5, Springer, US (2006)
17. Seeger, B., Larson, P.: Multi-disk B-trees. In: Proceedings of the 1991 ACM SIGMOD international conference on Management of data, May 29-31, pp. 436–445 (1991)
18. Taniar, D., Rahayu, J.W.: Global parallel index for multi-processor DB systems. Information Sciences 165(1-2), 103–127 (2004)
19. Yoshihara, T., Kobayashi, D., Yokota, H.: Mark-opt: A concurrency control protocol for parallel B-tree structures to reduce the cost of SMOs. IEICE Transactions on information and systems 90, 1213–1224 (2007)
20. Renesse, R.V., Schneider, F.B.: Chain replication for supporting high throughput & availability. In: Proceedings of the 6th USENIX Symposium, OSDI 2004, p. 7 (2004)
21. Watanabe, A., Yokota, H.: A directory traverse cost based skew handling for parallel data access. Transactions of IEICE (D-I) J85-D-I, 877–886 (2002)
22. Watanabe, A., Yokota, H.: Adaptive overlapped declustering: a highly available data-placement method balancing access load and space utilization. In: Proceedings of the 21st Int'l. Conf. on Data Engineering, pp. 828–839 (2005)
23. Wu, S., Kemme, B.: Postgres-r(si): Combining replica control with concurrency control based on snapshot isolation. In: Proceedings of ICDE 2005, pp. 422–433 (2005)
24. Yokota, H., et al.: Fat-Btree: An update conscious parallel directory structure. In: Proceedings of the 15st Int'l. Conf. on Data Engineering, p. 448 (1999)
25. Yoshihara, T., Kobayashi, D., Yokota, H.: A concurrency control protocol for parallel b-tree structures without latch-coupling for explosively growing digital content. In: Proceedings of the 11th Int'l. Conf. on EDBT 2008, vol. 261, pp. 133–144 (2008)
26. Yu, H., Vahdat, A.: The costs and limits of availability for replicated services. In: Proceedings of the 18th ACM SOSP 2001, October 21-24 (2001)
27. Amazon Web Service LLC (2009), http://aws.amazon.com/elasticmapreduce/

Query Reuse Based Query Planning for Searches over the Deep Web

Fan Wang and Gagan Agrawal

Department of Computer Science and Engineering
Ohio State University, Columbus OH 43210
{wangfa,agrawal}@cse.ohio-state.edu

Abstract. Nowadays, data dissemination often involves online databases that are hidden behind query forms, thus forming the *deep web*. Lately, there has been a lot of research interest on supporting query answering over the deep web. To answer a deep web query efficiently, the current approaches generate a query plan for each query independently.

However, in practice, deep web queries issued by a user over a short period of time can often share similarities. This, if properly utilized, can help us in generating more efficient query plan. In this paper, we have developed a solution for generating query plan for a deep web query based on the similarities between a given query and a set of earlier queries. Our algorithm systematically finds the reusable components of earlier query plans, and then develops a new query plan reusing these. While the resulting query plans may not be *optimal*, they are likely to enable more *data reuse*, and hence, speedup the execution.

1 Introduction

Nowadays, data dissemination often involves online databases that are hidden behind query forms, thus forming the *deep web*. To access data from the deep web, users need to submit queries on the input interfaces of deep web sources by specifying values for *input attributes*.

There has been a lot of work on supporting advanced query over deep web data sources [19,18,2]. Because answering deep web query involves data transmission over a network, current systems focus on finding an efficient query plan, either to use the least number of data sources [19,18] or to minimize the amount of data transmitted over the network[2,10]. Each of these methods, however, generates a plan for a query independently, disregarding any previous queries and their query plans.

A given user, or users from a particular organization, may issue several closely related queries over a short time. In fact, a recent study on the surface web [16] and our experience with a deep web query answering system [19,18], support this intuition. Thus, a query plan generated considering a single query in isolation cannot fully utilize the data cached from earlier queries. Instead, if one generates a query plan focusing specifically on maximizing the reuse, we are much more likely to be able to reuse data.

This paper presents a novel query planning algorithm for deep web queries. Our goal is to generate a query plan for a new query based on a set of similar (and reusable) previous queries and their query plans, achieve a high reuse of the cached data. While the

P. García Bringas et al. (Eds.): DEXA 2010, Part II, LNCS 6262, pp. 64–79, 2010.

resulting query plan may not be most efficient, its execution will replace many network operations with simple data reads. Thus, it is going to be very likely to significantly reduce the execution time. We illustrate our idea with the following example.

Example: A biologist, who is interested in the genes and Single Nucleotide Polymorphisms (SNPs) located in the two sex-determining chromosomes X and Y, issues two queries related with her research. The first query $Q1$ finds *"the IDs of all the SNPs, with a heterozygosity value between 0 and 0.4, located in chromosome X"*.

An efficient query plan for $Q1$ is shown in Figure 1(a). It shows that taking chromosome X as the input, SeattleSNP data source will return the SNPs located at chromosome X and the heterozygosity value of each SNP. By applying a filter on the heterozygosity data, we could obtain the answers for $Q1$. The input and output attributes of the data sources in this example is shown in Figure 1(d).

Now, another query, $Q2$, issued by the same researcher wants to find *"the IDs of all the SNPs, with a heterozygosity value between 0 and 0.3, located in gene OTC"*. If we consider query $Q2$ independently, the plan we could generate is shown in Figure 1(b). In this plan, given the gene OTC, we use dbSNP data source to obtain all the SNPs located in OTC, along with the heterozygosity values.

There is a similarity between the two queries, i.e., both of them request SNP IDs and heterozygosity values. Although SeattleSNP and dbSNP take different input attributes, they have *data redundancy*, as shown in Figure 1(d). In other words, we can obtain SNP IDs and heterozygosity data from both of the two data sources.

This motivates us to generate a new query plan for $Q2$ considering the already generated and executed plan for $Q1$. The *alternative* plan for $Q2$ is shown in the Figure 1(c). SeattleSNP is reused. Since $Q2$ is closely related with $Q1$, i.e., the gene OTC is located in chromosome X, the alternative plan could reduce execution time by reusing previous results from $Q1$. In this alternative plan, given the gene OTC, we first use the NCBI Gene data source to find the chromosome at which this gene is located, which is chromosome X. Then, we use chromosome X as input to SeattleSNP, as in the plan for $Q1$, to find the SNP IDs and their heterozygosity values. This alternative plan may be viewed as *sub-optimal* since it involves more data sources, compared with the original plan in sub-figure (b). However, it can effectively reuse the cached data from $Q1$ and speedup query execution. Particularly, many remote data accesses are replaced by local disk accesses, and the total execution time can be significantly reduced.

This paper presents algorithms to generate query plans for a deep web query, based on previous similar queries and their plans, with the goal of reusing results from previous queries aggressively. Our approach is motivated by the following observations. First, there is data redundancy across deep web sources [1]. The data redundancy discussed here focuses on *partial data overlapping*, i.e., two data sources may have overlapping data in terms of certain attributes. Data redundancy shows that a query can be answered by multiple query plans with different set of data sources, and our aim is to find the plan which could enable the most *data reuse* based on previous similar queries.

Second, deep web data sources return query answers in an *All-In-One* fashion, i.e., values of all the output attributes of a data source are returned, irrespective of the specific attributes requested in the query. For example, from Figure 1(d), we know that given a chromosome number, SeattleSNP could provide data about gene names,

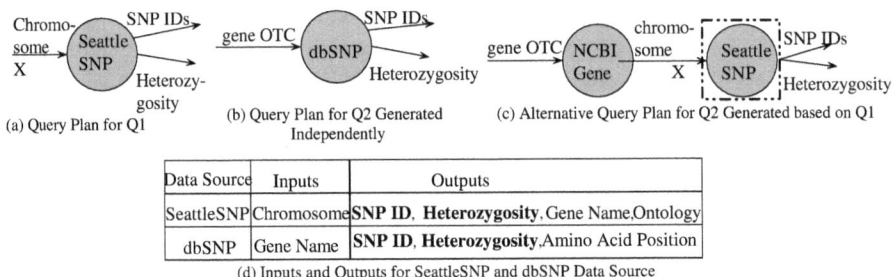

(a) Query Plan for Q1

(b) Query Plan for Q2 Generated Independently

(c) Alternative Query Plan for Q2 Generated based on Q1

Data Source	Inputs	Outputs
SeattleSNP	Chromosome	**SNP ID**, **Heterozygosity**, Gene Name,Ontology
dbSNP	Gene Name	**SNP ID**, **Heterozygosity**,Amino Acid Position

(d) Inputs and Outputs for SeattleSNP and dbSNP Data Source

Fig. 1. Motivating Example: (a) Query Plan for Q1;(b) Query Plan for Q2 Generated Independently; (c) Alternative Query Plan for Q2 Generated based on Q1; (d) Input and Output Attributes of `SeattleSNP` and `dbSNP` Data Source

ontology functions, SNP IDs and SNP heterozygosity. In the above example, although Q1 only asks for the SNP ID and the heterozygosity data, in fact, all the outputs of `SeattleSNP` will be returned, including the gene names and the ontology functions. This *All-In-One* feature of the deep web sources can facilitate data reuse. This is clearly in contrast with relational databases, where only the values of the attributes that are explicitly stated in the `select` clause of a query are returned.

Taking advantage of the above features of deep web sources and queries, our approach is as follows. We not only cache previous extracted data, but more importantly, cache the query plans for previous queries. Thus, we generate query plans for a new query based on the plans of the previous similar queries. If previous plans are reused, the likelihood of data reuse can be significantly increased. Because deep web queries are read-only queries and the content of such data sources is relatively static, we do not consider concurrency control issues or advanced data caching techniques [9].

The rest of the paper is organized as follows. In Section 2, we formulate our query reused-based query planning problem. We introduce the detail of selecting reusable previous sub-query plans in Section 3, and the query planning algorithm *QPReuse* in Section 4. In Section 5, we evaluate our system. We compare our work with related efforts in Section 6 and conclude in Section 7.

2   Problem Formulation

This section gives a formal description of the problem we are focusing on. Initially, we define the query and query plans we consider in our work.

2.1   Queries and Query Plans

Queries. We consider the queries in the select-project-join (SPJ) format. For example, the query Q1 in the motivating example can be represented as follows.

Example 2.1
SELECT s.SNPID, s.Heterozygosity
FROM SNP s, CHROMOSOME c
WHERE 0<s.Heterozygosity<0.4 AND s.chromsome=c.chromosome AND c.chromosome='X'

We extracts *search terms* from the *select* and *where* clause. *Search terms* consists of the *entity terms ET* and *attribute terms AT*. *Entity term* is used to *initiate* the answering of the query. When querying a deep web data source, a user must specify values for the input attributes of the data source to initiate the search, and these values are the *entity terms*. The search term(s) in the *where* clause with a value assignment are *entity term(s)*. In example 2.1, chromosome X is an *entity term*. Besides the *entity terms*, all other search terms are *attribute terms*. In answering the query, we need to obtain the *values* of these *attribute terms*. In example 2.1, SNPID and Heterozygosity are *attribute terms*. Overall, a deep web query considered in our work implies that we find the data records that are specified by the *entity term(s)*, and then obtain the values of the *attribute terms* from these records.

Query Plans. A query plan P of a query Q is defined as a graph $P = (V, E, V_0)$. V is a finite set of data sources in the query plan. E is a set of directed edges in the plan, each of which indicating an inter-dependence relation between a pair of data sources. An edge e pointing from node A to B implies that data source A provides the input attributes for B. V_0 is the set of starting data sources of the plan. Suppose P is the query plan of a query Q, we want the set of all output attributes of the nodes in P covers all *attribute terms* in Q, and the set of all input attributes of the nodes without incoming edges in P covers all *entity terms* in Q. The data source nodes covering *attribute terms* are called *target nodes*, and the data source nodes covering *entity terms* are called *starting nodes*.

The reason why such a graph P forms a valid query plan for Q is as follows. For an *attribute term* t_a in Q, we want to obtain its value from certain data sources, so we need at least one node in P that outputs (or covers) t_a. For an *entity term* t_e, t_e helps *initiate answering* of Q. Thus, we need the input of the *starting nodes* in P to cover t_e.

2.2 Query Reuse Problem

To formulate the query reuse problem, we use the following definitions.

Query Subplan: We define a query subplan $SubP$ of an original query plan P as a connected sub-graph of the original query plan graph. Formally, $SubP = (V', E', V_0')$ where $V' \subseteq V$, $E' \subseteq E$, $V_0' \subseteq V$ and $|V_0'| > 0$.

Query Subplan Set: The query subplan set $SubPSet$ of query plan P is the set containing all query subplans of P.

Ψ **Selection:** Given a new query and a previous query with plan P, we want to determine the subplan of P that, among all subplans of P, will enable *maximal reuse* for the new query, and of all subplans enabling such maximal reuse, will involve fewest data sources. This is captured through the Ψ selection operator. The selected query subplan is denoted as $\Psi(SubPSet(P))$.

Problem Definition: Using the above terms, our query reuse based query planning problem is formally stated as follows. We are given a list of n previous issued queries, each of which has a query plan P_i. Given a new query, we want to construct its query plan using a list of Ψ selected query subplans, $\Psi(SubPSet(P_1)), \Psi(SubPSet(P_2)), \ldots, \Psi(SubPSet(P_n))$ and other necessary data sources. We obtain the query plan for the new query, and while executing this query plan, we reuse previously cached data.

2.3   Solution Overview

To solve the problem formulated as above, we take the following steps. First, we define a *reusability* metric to identify the query plans that we would be beneficial to reuse. We consider these as *reusable* queries and query plans. Second, we select a list of such reusable previous queries and their plans, and then apply the Ψ selection function to obtain the sub-query plans we will like to reuse. Here, one challenge that we address is performing Ψ selection without explicit enumeration of all query subplans. Third, we modify the bidirectional query planning algorithm proposed in our earlier publications [17] so as to generate a query plan based on a list of selected reusable sub-query plans. Here, a particular issue is balancing the trade-off between reuse and plan quality. The first two issues are described in Section 3 and the last one is detailed in Section 4.

3   Reusable Sub-query Plan Selection

In this section, we describe how reusable query subplans are selected from cached queries and their plans.

3.1   Query Reusability Metric

For two queries $Q1$ and $Q2$, we want to know the potential of reusing $Q1$ in answering $Q2$. An intuitive method is to count how many common attribute terms they have. However, this method is not sufficient. For example, given a chromosome, the SeattleSNP data source returns four attribute terms as shown in Figure 1(d), which are t1=SNPID, t2=Heterozygosity, t3=Gene Name, and t4=Ontology Function. For a biologist only interested in the SNPs, his query $Q1$ would only contain two attribute terms, t_1, and t_2. For another biologist who is more interested in the genes, her query $Q2$ would contain the other two attribute terms, t_3 and t_4. However, due to the *All-In-One* feature of deep web data sources, when the first query is being answered, the values of t_3 and t_4 are also returned. Thus, $Q2$ can be answered completely by reusing the results of $Q1$. In the rest of this paper, we use the attribute term set to represent the query. For example, query $Q1$ can be represented as Q1={SNP ID, heterozygosity}.

To capture this observation in the query reusability metric, we formally define an *augmented query* as follows.

Augmented Query: Suppose we have a query plan P for a query Q. The *augmented query* of Q, Q^*, has the same entity term set as Q, but the attribute term set of Q^* is constructed as follows. For each $v_i \in V$, we extract all terms $t_{ij} \in Output(v_i)$ and the augmented query Q^* is defined to be $Q^* = \{t_{11}, t_{12}, \ldots, t_{1k_1}, \ldots, t_{|V|1}, \ldots, t_{|V|k_{|V|}}\}$ where $k_i = |Output(v_i)|$. Clearly Q^* is a superset of Q, and we say that Q^* is an *augmented query* of Q related to the plan P, and denote it as $A_P(Q)$. Considering the query $Q1$ in the motivating example, since the data source SeattleSNP is the query plan for answering Q1={SNP ID, heterozygosity}, the augmented query of $Q1$ is $A_P(Q1)$={ SNP ID, heterozygosity, Gene Name, ontology function}.

To compute the reusability of $Q1$ for answering $Q2$, we first obtain the augmented query of $Q1$, $A_P(Q1)$. The *query similarity* between $Q1$ and $Q2$ depends on the similarity between the attribute sets of $A_P(Q1)$ and $Q2$. We first define the similarity between two attribute terms. In general, an attribute term can be in the following format

$v_{low} < (=)$ `attribute term label` $< (=)v_{high}$, where v_{low} and v_{high} are the range constraint values. For two attribute terms a_1 and a_2, the similarity is defined as

$$Sim(a_1, a_2) = \begin{cases} 0 & \text{if } label(a_1) \neq label(a_2), \\ \frac{Range(a_1) \cap Range(a_2)}{Range(a_1) \cup Range(a_2)} & \text{if } label(a_1) = label(a_2). \end{cases}$$

For two queries $Q1$ and $Q2$, we use S_{reuse} to capture the similarities of the *common attribute terms* in $A_P(Q1)$ and $Q2$. We define $S_{reuse} = \sum_{a_i \in A_P(Q1)} \sum_{a_j \in Q2} Sim(a_i, a_j)$. We use $S_{penalty}$ to capture the attributes that $Q2$ contains but $A_P(Q1)$ does not. These attributes are requested by the new query $Q2$ but not retrieved by $A_P(Q1)$, as a result, they should be considered as a *penalty for reuse*. Specifically, we define $S_{penalty} = |\{a_i | a_i \in Q2 \text{ and } a_i \notin A_P(Q1)\}|$. Similarity, we use $S_{partialpenalty}$ to denote the number of attributes that $A_P(Q1)$ contains but $Q2$ does not. These attribute terms are retrieved by $A_P(Q1)$ but not requested by $Q2$ lower the similarity between $Q2$ and $A_P(Q1)$, even though they do not adversely impact the reuse of $Q1$ for $Q2$. We define $S_{partialpenalty} = |\{a_i | a_i \notin Q2 \text{ and } a_i \in A_P(Q1)\}|$.

Based on the above definitions, the similarity function between $Q1$ and $Q2$ is $Sim(Q1, Q2) = \mathcal{F}(S_{reuse}, S_{penalty}, S_{partialpenalty})$. The function $\mathcal{F}$ can be implemented in different ways, but it must meet several properties. First, $\mathcal{F}$ should be symmetric, second, it should be bounded within the interval [0,1]. Finally, $\mathcal{F}$ should be a non-decreasing function with respect to S_{reuse}, and a non-increasing function with respect to both $S_{penalty}$ and $S_{partialpenalty}$.

3.2 Ψ Selection of a Reusable Query Subplan

Suppose an earlier query $Q1$ has a high reusability score for a new query $Q2$. However, we cannot just reuse the entire plan of $Q1$, because the plan of $Q1$ may extract some terms which are not requested by $Q2$. We need to use a selection function to find a subplan that satisfies the following two conditions: 1) the subplan *maximally* covers $Q2$, and 2) the *size* of the subplan is minimal. We use the *maximal coverage* condition to select the previous subplan which could enable the maximal reuse. For the second condition, when we have a cache hit, the cost would be the number of local disk accesses, which is proportional to the number of data sources in the reused subplans. As a result, the *minimal size* condition is used to select the previous subplan which causes the lowest cost.

To explain the idea, we use the following example. The augmented query of $Q1$ is $A_P(Q1) = \{t_1, t_2, t_3, t_4, t_5, t_6, t_7, t_8\}$. The query plan of $Q1$ is shown in the subfigure (a) of Figure 2. The terms inside each nodes indicate the output set of the data source, and the terms located above the arrow of each link indicate the input terms of the pointed data source. The new query we consider is $Q2 = \{t_1, t_3, t_4\}$.

Maximal Coverage Query Subplan: We consider a query subplan $SubP$ to be *maximally covering* a new query $Q2$ iff the keywords covered by all the data sources in $SubP$ is a superset of the common terms in $Q2$ and $A_P(Q1)$.

In the example in Figure 2, sub-figures (b),(c) and (d) are all *Maximal Coverage Query Subplans* of the plan (a) w.r.t $Q2$. The subplan in sub-figure (d) shows *maximal coverage* subplans with *minimal size*.

Subplan Selection Problem Formulation:
The subplan selection problem can be converted to a graph problem as follows. The query plan of a previous query is considered as a graph $G = (V, E)$, where V is a set of data sources. Each $v_i \in V$ covers a set of terms S_i. We define the set of terms covered by any data source in the graph G as $GS = \bigcup_{\forall v_i \in V} S_i$. Given a new set T, which is a subset of GS, we want to find a *connected subgraph* of G, $SubG$, which covers *all* elements in T, while having the *minimal size*. We call this problem the *connected subgraph set cover* problem. We have established the following Lemma 1.

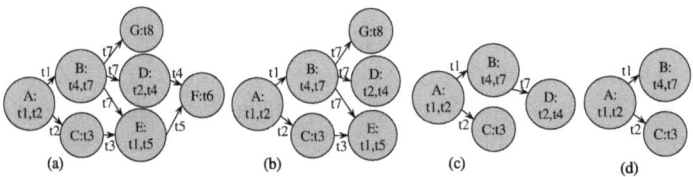

Fig. 2. Example for Ψ Selection: (a) Query Plan of Q1, (b) Ψ Selection Step 1, (c) Ψ Selection Step 2, (d) Ψ Selection Step 3 (minimal size obtained) (e) A query plan with only partial coverage.

Lemma 1. *The connected subgraph set cover problem is NP-hard.*

The proof of Lemma 1 is omitted due to lack of space. The basic idea is that we could reduce the *set cover problem* to the *connected subgraph set cover* problem.

In view of the above lemma, we have developed a polynomial-time but heuristic Ψ selection algorithm. The algorithm is based on two heuristics, The first heuristic is to *give preference to nodes covering search terms*, which implies that we prefer to include the node which can cover some search terms of the new query. The second heuristic is to give *preference to nodes close to each other*. The idea of this heuristic is shown in Figure 2(a). Suppose the set of elements we want to cover is $T = \{t_1, t_2, t_6\}$. We can see that the nodes A and F cover T with minimal size, which is 2. But, if we want to connect A and F using a subgraph, the smallest subgraph has a size of 6, (it will contain A, B, C, D, E and F). We can also observe that the nodes D, E, and F also cover T. Although here we use three nodes to cover T, but the subgraph connecting these three nodes only has a size of 3. The reason is that the nodes we selected were already connected to each other.

The Ψ selection algorithm scans the data sources in the query plan, in the reverse topological order. During the scan, a data source D in the plan P of previous query $Q1$ is removed if it meets any of the following two conditions: 1) D cannot provide any terms requested by $Q2$ (heuristic 1) 2) D can provide some terms requested in $Q2$, but is *subsumed* by one of its predecessors (heuristic 2).

To explain what we mean by *subsumed*, we use the following definitions. The data source D_i **contains** D_j according to a new query $Q2$ iff $Output(D_i) \supseteq Output(D_j) \cap Q2$. The data source D_i **topologically contains** D_j iff D_i **topologically precedes** D_j and D_i **contains** D_j. Finally, the data source D_i **subsumes** D_j if the following two conditions hold: 1) D_i **topologically contains** D_j, and 2) D_j has no descendants in the current subplan or although D_j has descendants, the output from D_i could provide the input of D_j's descendants. When D_i **subsumes** D_j, we can remove D_j, and if D_j has descendants, we can link D_i directly to D_j's descendants. From the above definition, we can know that if D_i *subsumes* D_j, D_i covers the required search terms that D_j can and meanwhile D_i topologically proceeds D_j. Since we scan data sources in the reverse topological order, if we find other data sources which covers other requested search terms later, these data sources would be *nearer* to D_i than D_j. The replacement of D_j for D_i shows the application of the second heuristic.

We use the example in Figure 2 to illustrate the Ψ selection method. The algorithm starts from node F in sub-figure(a). Since F cannot provide any terms in $Q2$, F is removed(sub-figure (b)). Then, similar as F, data sources E and G are removed. Node D provides term t_4, which is requested by $Q2$, so D is kept(sub-figure (c)). Data sources B and C are examined next. Both of them provide terms in $Q2$, and B subsumes D because: 1) B covers t_4(B *contains* D), 2) B topologically precedes D(B *topologically contains* D), and 3) D doesn't have dependents. As a result, D is *subsumed* by B and D is removed from the plan. Sub-figure(d) shows the final *subplan*.

3.3 Algorithm for Selecting Reusable Query Subplans

Our overall algorithm combines the ideas from the discussion above. One issue, however, is that for certain queries, we may want to reuse subplans from multiple prior queries. This is done through a greedy algorithm as follows.

Step 1: Given a new query $Q2$, from all previous cached queries, we find the query $Q1$, which has the highest reusability value. If the reusability value is greater than a threshold, the algorithm continues. Otherwise, the algorithm terminates.
Step 2: We obtain the augmented query of $Q1$ and invoke the Ψ selection function to obtain the corresponding *reusable query subplan* and add it into a reusable query subplans list.
Step 3: We update the query $Q2$ by deleting all search terms which are covered by the reusable query subplan generated in the previous step.
Step 4: We repeat steps 1 to 3 using the updated query $Q2$, until the algorithm terminates. The last updated $Q2$ contains the search terms which cannot be covered by any previous *reusable query subplan* and we define it as the *remainder query*.

4 Query Plan Generation Algorithm QPReuse

In the previous section, we described how we convert a new query $Q2$ into a list of reusable query subplans and a *remainder* query. Now, our goal is to generate a query plan for $Q2$. The planning algorithm QPReuse we present for this purpose is modified from the bidirectional deep web query planning algorithm proposed in our earlier publications [17]. Initially, we give an overview of this algorithm.

4.1 Background: Bidirectional Query Planning for Deep Web Query

There is a dependency graph capturing the dependency between data sources, i.e., if the output of data source A can be used as the input of data source B, there is an edge pointing from A to B. Given a query with n search terms and a dependency graph, we first identify the *target nodes* and *starting nodes* in the dependency graph as described in Section 2.1. Then, we want to find a subgraph which connects the starting nodes with the target nodes and at the same time covers all search terms. There maybe multiple such subgraphs can be valid query plans. Among them, we want to select the one with the *least execution time* and likely to give the *highest quality* of results. In our approach, we combine these two considerations into one *benefit model*.

Bidirectional Algorithm Overview: We explore the query plan in a bidirectional manner. We perform backward exploration from the target nodes to connect them with starting nodes. To accelerate this process, we also do forward exploration from the starting nodes. In this way, the bidirectional exploration can meet *mid-way*.

Bidirectional Exploration: Initially we add all *starting nodes* to a forward exploration queue, and all *target nodes* to a backward exploration queue. Then, the algorithm tries to find a optimal sub-graph to connect the target node set with the starting node set. At each iteration of the sub-graph exploration, the algorithm always selects the node with the *highest benefit*, CN, from the two queues. If CN belongs to the forward queue, all out-going neighbors of CN are explored. If CN belongs to the backward queue, all in-coming parents of CN are explored.

Edge Exploration: To build the sub-graph as the final query plan, paths (sequence of graph edges) must be explored to connect target nodes with starting nodes. Here, we always connect pair of nodes through the shortest path. This is realized by modifying the Dijkstra's shortest path algorithm.

Algorithm Termination: When every search term can be reached from at least one starting node with finite distance, a query plan is found.

4.2   Modified Algorithm for Enabling Reuse: QPReuse

We first give the intuition behind query planning using previous plans, and then we explain the detail of two modifications we made to our bidirectional planning algorithm.

The query planning algorithm QPReuse takes two input parameters. The first is the remainder query, denoted as $RQ = \{t_1, \ldots, t_k\}$, $k \geq 0$, where each t_i is a search term, The second parameter is the query template $QT = \{qt_1, \ldots, qt_n\}$, $n \geq 0$, where each qt_i is a reusable query subplan selected by the Ψ selection algorithm. If $n = 0$, there is no reusable query template, so we simply invoke our bidirectional planning algorithm. For other cases, we treat the remainder query as an ordinary query and do the query planning using the bidirectional algorithm. But we want to use the query templates whenever possible. To achieve this, we modify our bidirectional algorithm as follows.

In our QPReuse algorithm, we perform the forward and backward exploration in the normal fashion, till we reach a point where the current selected data source to be explored, denoted as CN, is in one of the templates in QT, qt_i. Now, we suspend the normal bidirectional planning, and instead, we do a depth first exploration from the node CN, along the edges in qt_i (the selected reusable subplan), to include all data sources in qt_i to be in the query plan of RQ. After the exploration of data sources in qt_i, we return to the normal bidirectional planning. The depth-first exploration of the query template is referred to as a *detour* exploration.

Detour Exploration: Suppose data source CN is currently being explored, and CN is involved in a query template qt. In detour exploration, we first do backward explorations from the node CN till we reach the starting node(s) of the query template qt. After the backward explorations, we start from node CN again to do forward explorations.

In the backward detour function, all the immediate parents of the current node are explored in a recursive fashion. The forward detour function works similarly and the immediate descendants of the current node are explored in a recursive fashion. An important difference between backward and forward detour exploration is that in the backward version, if we reach the starting node(s) of the query templates, we do not just stop. Instead, we do an extra backward exploration from the starting node(s) of the query template to its parents outside the query template. The reason is that the query template is only a detour, and we ultimately need to connect the starting node(s) of the detour with node(s) in the main query path that is generated by the original bidirectional algorithm.

Modified Edge Exploration: In our bidirectional planning algorithm, like the Dijkstra's shortest path algorithm, the shortest distance from a node to search term is updated whenever a shorter path is found through a newly explored node. However, in QPReuse, because we want to give *higher priority* to the *detour path* which normally have longer distance, we *lock* the shortest distance if it is achieved by going through a detour path. This ensures that it is not updated by a path generated by the normal bidirectional algorithm. We call this *shortest path locking for detour*. The other issue is that although we give priority to the detour path, we do not want to penalize the normal path severely. In other words, we do not want to reuse previous plans if it results in extremely large query plans. As a result, the lock is *released* if the distance coming from a normal path is *much* shorter than the locked shortest distance obtaining by a detour path. The release of the lock is controlled by a specified threshold.

4.3   Managing Cached Queries and Query Plans

The admission of a new query plan depends on the *newness* of the plan. The *Newness* of the plan is the benefit score of all nodes and edges not in any reused query templates. The score coming from the remaining nodes and edges is the *Re-usage* of the new plan. For a new plan np, if $Newness(np) \geq \alpha \times Re - usage(np)$, np is admitted. When a new plan is admitted, some previous cached plans may be *covered* by the new plan, and should be replaced. If the similarity score between the augmented query of the new query and the augmented query of a previous cached query is greater than a threshold, the previous cached query is removed.

5   Experimental Results

In this section, we describe the experiments we conducted to evaluate our techniques.

5.1 Experiment Setup

We used 12 biological deep web sources, which includes dbSNP[1], Entrez Gene[1], Protein[1], BLAST[1], SNP500[2], Seattle[3], SIFT[4], BIND[5], Human Protein[6], HGNC[7], Mouse SNP[8], and ALFRED[9]. A collaborating biologist provided us 24 real queries, which were divided into 4 groups, with each group intended to simulate a list of similar queries. The first query in the first, third, and the fourth group, and the first two queries in the second group, are considered *seeds*, and the rest of the queries in each group are considered as new coming queries. The reason we put 2 seeds in the second group is that we want to test our algorithm when there is a potential for reusing multiple previous queries.

Table 1 shows the *Cumulative Reusability Score* (CRS) for the 24 queries, which is computed as follows. We consider all previous queries in the same group as one single *existing* query, and then compute the reusability score between the new query and the existing query. In our evaluation, we compared three scenarios. They are Baseline with No Data Reuse (BaseNDR), Baseline with Data Reuse (BaseDR), and our Query Reuse Method with Data Reuse (ReuseDR). The BaseNDR refers to the method that both query plan generation and plan execution of a new query are independent of any previous query. The BaseDR refers to the case that we generate a query plan for the new query independent of any previous queries. But, during plan execution, we reuse any reusable data. The ReuseDR method refers to the strategy proposed in this paper.

Table 1. Experiment Queries

Group1	CRS	Group2	CRS	Group3	CRS	Group4	CRS
SeedA	N/A	SeedB	N/A	SeedD	N/A	SeedE	N/A
Q 1.1	43%	SeedC	N/A	3.1	60.3%	4.1	60%
1.2	37%	2.1	92%	3.2	60.5%	4.2	51.4%
1.3	43%	2.2	80.4%	3.3	55.6%	4.3	54.3%
1.4	37%	2.3	76%	3.4	51.4%	4.4	52.5%
1.5	41.5%	2.4	70%	3.5	46%	4.5	48%

5.2 Evaluation Metrics

Query Execution Time: Query Execution Time is the time for query planning and the time for actually issuing the query on each of the data sources in the plan.

---

[1] http://www.ncbi.nlm.nih.gov/
[2] http://snp500cancer.nci.nih.gov/home_1.cfm
[3] http://pga.gs.washington.edu/
[4] http://blocks.fhcrc.org/sift/SIFT.html
[5] http://www.bind.ca
[6] www.hprd.org
[7] www.genenames.org
[8] http://mousesnp.roche.com/
[9] http://alfred.med.yale.edu/alfred/

Query Plan Score: For each query, we compute the score of the query plan generated by the BaseNDR and our ReuseDR using the benefit model in the bidirectional planning algorithm [17]. We want to examine whether the desired trade-off between plan quality and plan execution time is achieved.

Actual Query Result: For the query plans generated using our method, we record the query results. We want to know whether the results from our method are *correct* and *complete* comparing with the true answers provided by our collaborating biologist.

5.3 Experimental Results

Comparing Query Execution Times: Figure 3 shows the comparison of query execution time for the three scenarios, BaseNDR, BaseDR and ReuseDR. The x axis is the query number in each group (the seed is excluded), and the y axis is the speedup that each scenario achieves compared to the BaseNDR method.

In the Figure 3, five queries are highlighted by ovals, and will be discussed below. For all the remaining 14 queries, both BaseDR and ReuseDR significantly outperformed BaseNDR. Our method ReuseDR outperformed BaseDR in 13 of the 14 queries, and in 8 of the 14 queries, our method achieved at least twice the speedup of the BaseDR method.

For the five queries highlighted by ovals, the two methods with data reuse, BaseDR and ReuseDR, do not obtain any speedups over the BaseNDR version. The reason is that for these five queries, our QPReuse algorithm decides not to reuse any previous cached query plans, because it is likely to give a query plan with low score (benefit).

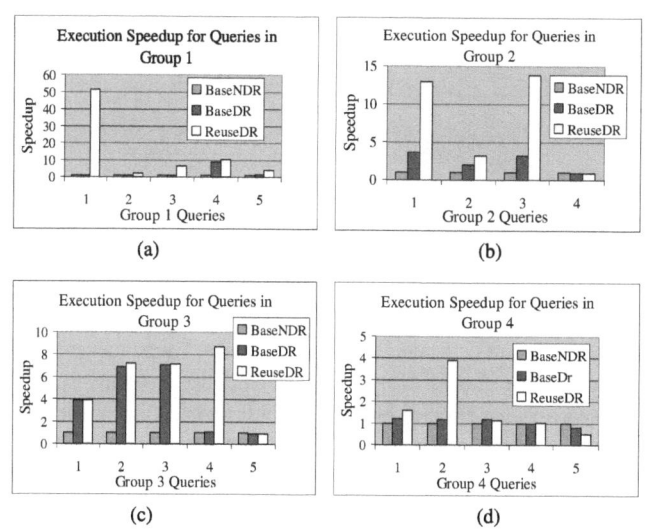

Fig. 3. Comparison of Query Execution Time: (a) Group 1 Queries;(b) Group 2 Queries; (c) Group 3 Queries; (d) Group 4 Queries

The overhead of checking local databases and QPReuse algorithm causes our method having lower performance than BaseNDR.

Comparing Query Plan Score: We now examine if our QPReuse algorithm achieves the desired trade-off between the execution speedup and the plan quality.

For each query, we first record the score of the plan, $Score_{ReuseDR}$, if it was generated with ReuseDR. Then, we record the score of the query plan, $Score_{BaseNDR}$, if generated with BaseNDR. The ratio of them are shown in Figure 4. The x axis represents the 19 queries (seeds excluded), and the y axis is the ratio $\frac{Score_{BaseNDR}}{Score_{ReuseDR}}$. The reference line is $Ratio = 2$, which means that the lowest score we accept for the query plan generated with reuse is half of the score of the plan generated by the original bidirectional algorithm.

From the figure, we observe that except for six queries, the ratios are below the reference line, though above 1. This shows that the query plans from our method indeed have lower score, but the score is not very low in most cases, and they still result in overall speedup. For the six queries for which the values are above the reference line, except for query number 13 (rectangle), all other five queries (circle) are the ones we highlighted by ovals in Figure 3. This shows that when the score of the plan generated by reusing tends to be very low, our algorithm correctly decides to generate the plan from the original method.

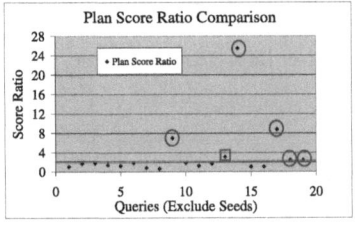

Fig. 4. Values of Plan Ratio $\frac{Score_{BaseNDR}}{Score_{ReuseDR}}$

Comparing Actual Query Results: Now we examine if the results from our ReuseDR method are *correct*(Crt) and *complete*(Cpt) w.r.t. our collaborating biologist provided answers.

Table 2 shows the results. First, for all the queries, the query results obtained by our strategy ReuseDR are always 100% *correct*. Second, except for 5 queries (in bold font), the results from ReuseDR are also 100% *complete*. Among the five queries with incomplete answers, 3 of them have nearly 100% complete answers (above 95%). This is because the query plans generated by ReuseDR reuse some data sources that have a low data coverage on some of the query terms.

Overall, the above statistics show that for most cases, the query plans generated by our ReuseDR method obtain *correct* and *complete* answers and at the same time have much shorter execution time.

Table 2. Query Results for Query Plans from ReuseDR

Grp1	Crt	Cpt	Grp2	Crt	Cpt	Grp3	Crt	Cpt	Grp4	Crt	Cpt
Seed A	N/A	N/A	Seed B	N/A	N/A	Seed C	N/A	N/A	Seed D	N/A	N/A
Query 1.1	100%	**95%**	Seed C	N/A	N/A	Query 3.1	100%	100%	4.1	100%	100%
1.2	100%	**69%**	2.1	100%	100%	3.2	100%	100%	4.2	100%	100%
1.3	100%	**95%**	2.2	100%	100%	3.3	100%	100%	4.3	100%	100%
1.4	100%	100%	2.3	100%	**97.7%**	3.4	100%	100%	4.4	100%	100%
1.5	100%	**80%**	2.4	100%	100%	3.5	100%	100%	4.5	100%	100%

6   Related Work

We now compare our work with existing work on a number of topics related to query answering, optimization and caching.

Answering Queries Using Views: The work on answering queries using views [7,4,6] is different from our work in the following aspects. First, in answering queries using views, a new query is answered using views *only*. However, in our work, we *reuse* previous plans to *accelerate* the answering of a new query, and the new query could still use data which is not cached (the remainder query). Second, the reusability of a materialized view is determined based on the exact matching between the table names in the view and the table names in the new query [8]. However, in our case, the new query could reuse data from *different* data sources as long as there is *data redundancy*. Finally, answering queries using views is based on *data reuse*, but our method is based on *query plan reuse*.

Multi Query Optimization: Sellis [15] proposes an algorithm which finds common tasks across sub-queries. Sub-queries sharing common tasks and data can be executed together. Roy *et al.* [14] represent multiple queries in a single DAG, sharing subexpressions. A greedy algorithm picks a set of nodes from the DAG to be materialized and then finds the optimal plan for the given set of materialized nodes. The common subexpressions in existing work are identified by direct comparison between queries. However, in our work, we utilize the *All-In-One* feature of deep web sources, and *augmented queries* of previous queries to identify the common data. Furthermore, existing algorithms work on relational database and queries are issued in a batch. It is clearly distinct from our context where deep web is the focus and queries are issued sequentially.

Query Caching for Dynamic Web Content: There are a number of efforts on optimizing query processing on data-intensive web sites on the server side [11,3,20,5] or the client side [13,9,12]. The server side caching mechanisms cache the dynamically generated web pages, shrinking the query processing time on the server side [3,11]. But, the network delivery time, a major component of the delay, is not reduced. Robinson and Lowden [13] developed a system that supports query results reuse for syntactically unrelated queries using range constraints. Common data in their system is identified by direct comparison of range constraints in queries. In our work, we proposed a more efficient method to identify reusable cached data based on augmented queries. Luo *et al.* [12] propose a proxy caching mechanism for database-backend web sites on keyword queries, which is *data-driven* caching. In their system, only one data source is considered. Our work is clearly distinct in considering multiple inter-dependent deep web data sources which can have data redundancy.

7 Conclusions

In this paper, we have presented a query reuse based query planning algorithm to generate query plans for deep web queries. In our algorithm, the query plan of a query is generated based on a list of similar previous queries and their query plans. Although the query plan thus generated may not be the optimal plan when considering the query independently, it could effectively take advantage of the cached data from previous queries so that the query execution time could be significantly reduced.

Our experimental results show that for 93% of the queries, the query plans generated by our algorithm runs faster than the plans generated based on a single query independently, and we achieve a least twice the speedup for more than 50% of the queries. For all experimental queries, our query plans obtain the correct results, and for about 90% of the queries, the answers obtained by our query plans is complete.

References

1. Bergman, M.K.: The deep web: Surfacing hidden value. Journal of Electronic Publishing 7 (2001)
2. Braga, D., Ceri, S., Daniel, F., Martinenghi, D.: Optimization of Multi-domain Queries on the Web. VLDB Endowment 1, 562–673 (2008)
3. Candan, K.S., Li, W.-S., Luo, Q., Hsiung, W.-P., Agrawal, D.: Enabling dynamic content caching for database-driven web sites. ACM SIGMOD Record 30, 532–543 (2001)
4. Das, G., Gunopulos, D., Koudas, N.: Answering top-k queries using views. In: Proceedings of the 32nd International Conference on Very Large Data Bases, pp. 451–462 (2006)
5. Datta, A., Dutta, K., Thomas, H., VanderMeer, D., Ramamritham, K., Fishman, D.: A comparative study of alternative middle tier caching solutions to support dynamic web content acceleration. In: Proceedings of the 27th International Conference on Very Large Data Bases, pp. 667–670 (2001)
6. Deutsch, A., Ludascher, B., Nash, A.: Rewriting queries using views wiht access patterns under integrity constraints. Theoretical Computer Science 371, 200–226 (2007)
7. Goldstein, J., Larson, P.: Optimizing queries using materialized views: A practical, scalable solution. In: Proceedings of the 2001 ACM SIGMOD International conference on Management of Data, pp. 331–342 (2001)
8. Halevy, A.: Answering queries using views: A survey. The International Journal on Very Large Data Bases 10, 270–294 (2001)
9. Keller, A.M., Basu, J.: A predicate-based caching scheme for client-server database architectures. The International Journal on Very Large Data Bases 5, 35–47 (1995)
10. Kementsietsidis, A., Neven, F., de Craen, D.V., Vansummeren, S.: Scalable multi-query optimization for exploratory queries over federated scientific databases. VLDB Endowment 1, 16–27 (2008)

11. Luo, Q., Krishnamurthy, S., Mohan, C., Pirahesh, H., Woo, H., Lindsay, B.G., Naughton, J.F.: Middle-tier database caching for e-business. In: Proceedings of the 2002 ACM SIGMOD International Conference on Management of Data, pp. 600–611 (2002)
12. Luo, Q., Naughton, J.F., Xue, W.: Form-based proxy caching for database-backed web sites. The International Journal on Very Large Data Bases 17, 489–531 (2001)
13. Robinson, J., Lowden, B.G.: Extending the re-use of query results at remote client sites. In: Ibrahim, M., Küng, J., Revell, N. (eds.) DEXA 2000. LNCS, vol. 1873, pp. 536–547. Springer, Heidelberg (2000)
14. Roy, P., Seshadri, S., Sudarshan, S., Bhobe, S.: Efficient and extensible algorithms for multi query optimization. ACM SIGMOD Record 29, 249–260 (2000)
15. Sellis, T.K.: Multiple-query optimization. ACM Transactions on Database Systems 13, 23–52 (1988)
16. Teevan, J., Adar, E., Jones, R., Potts, M.A.: Information re-retrieval: Repeat queries in yahoo's logs. In: Proceedings of the 30th Annual International ACM SIGIR Conference on Research and Development in Information Retrieval, pp. 151–158 (2007)
17. Wang, F., Agrawal, G.: Querying Deep Web Data Sources: A Structured Keyword Query Approach. Technical Report OSU-CISRC-6/09-TR33, The Ohio State University (June 2009)
18. Wang, F., Agrawal, G., Jin, R.: Query planning for searching inter-dependent deep-web databases. In: Ludäscher, B., Mamoulis, N. (eds.) SSDBM 2008. LNCS, vol. 5069, pp. 24–41. Springer, Heidelberg (2008)
19. Wang, F., Agrawal, G., Jin, R., Piontkivska, H.: Snpminer: A domain-specific deep web mining tool. In: Proceedings of the 7th IEEE International Conference on Bioinformatics and Bioengineering, pp. 192–199 (2007)
20. Yagoub, K., Florescu, D., Lssarny, V., Valduriez, P.: Caching strategies for data-intensive web sites. In: Proceedings of the 26th International Conference on Very Large Data Bases, pp. 188–199 (2000)

Efficient Parallel Data Retrieval Protocols with MIMO Antennae for Data Broadcast in 4G Wireless Communications*

Yan Shi[1], Xiaofeng Gao[2], Jiaofei Zhong[1], and Weili Wu[1]

[1] Department of Computer Science, The University of Texas at Dallas
{yanshi,fayzhong,weiliwu}@utdallas.edu
[2] School of Science and Technology, Georgia Gwinnett College
xgao@ggc.edu

Abstract. *Wireless Data Broadcast* is an efficient data dissemination method for public information to a large number of mobile/wireless clients. With the advance of the fourth-generation wireless communication system (4G), mobile devices may embed *multiple-input multiple-output* (MIMO) antennae to setup multi-connections to a base station. In this paper, we deal with data retrieval problem for mobile clients with MIMO antennae to retrieve a set of indexed data from parallel communication channels. Our purpose is to construct fast and energy efficient data retrieval protocols to reduce the response time and energy consumption. We name this problem as *parallel data retrieval scheduling with MIMO Antennae* (PADRS-MIMO), and propose two greedy heuristics named *Least Switch Data Retrieval Protocol* (*Least-Switch*) and *Best First Data Retrieval Protocol* (*Best-First*). We are the first work to deal with data retrieval with MIMO antennae for wireless data broadcast. We analyze the performances of *Least-Switch* and *Best-First* both theoretically and practically. Simulation results prove the efficiency of the two protocols.

1 Introduction

With the explosive increase of wireless/mobile clients and development of wireless network technologies, *wireless data broadcast* has become an important data dissemination method for public information, such as stock activities, traffic conditions, weather reports, and flight schedules. In a typical wireless data broadcast system, a set of data items are broadcasted over several channels at a base station repeatedly. Clients can access onto channels, search for their requested data (usually through index), and then download the corresponding data.

The most important issues in data broadcast are the energy efficiency and query response time for mobile clients, since the majority of mobile devices have limited battery power and constraint lifetime. As a result, *tuning time* and *access latency* are two significant criteria to evaluate the performance of a data broadcast system. According to the architectural enhancements, each

---

* This work is supported by NSF grant CCF-0829993 and CCF-0514796.

P. García Bringas et al. (Eds.): DEXA 2010, Part II, LNCS 6262, pp. 80–95, 2010.

mobile device has two mode: active mode and doze mode. They can only process operations in active mode, while "sleep" in doze mode to save energy. Consider a process from the time a client requires some data, to the time when this client finishes downloading, *tuning time* denotes the total time a client keeps active, while *access latency* denotes the time interval for the whole process. Intuitively, a critical problem to improve the performance of a wireless data broadcast system is to reduce the access latency and tuning time for clients.

Index technology is an efficient technique to reduce tuning time (e.g., B$^+$-Tree [6], Huffman Tree [4], Hash Table [14], Exponential Index [13], Signature Index [17], etc). With the help of index technique, clients can first get the estimated waiting time offset and channel information of their requested data, sleep during this offset, wake up and tune in corresponding channel right before the target data appear. Formally, in an indexed parallel wireless data broadcast system, it takes three steps to retrieve a data item:

1. *Initial Probing*: the client tunes in some broadcast channel and decide when the next index is arriving;
2. *Searching*: the client searches through the indices and locate the requested data item on the broadcast channels;
3. *Retrieving*: the client tunes in the channel where the requested data item is at and download the data item when it arrives.

Based on different index techniques, the searching process may vary in different broadcast system. However, no matter what index technique is used, by the end of the searching process, the client should have the knowledge of the time offset and resided channel of the requested data. Since in this paper, we focus on the data retrieving process, we omit the discussion of searching process and index constructions and assume we know the locations of requested data.

If a client requires more than one data items, it needs to get the location of every data item, order them as a permutation, and download them one by one sequentially. If the order of data items to retrieve is not appropriate, the client may spend unnecessarily extra time for downloading. Thus, a time efficient schedule for retrieving data from parallel channels which can reduce access latency is very important for the performance of the whole wireless data broadcast system. Moreover, during the data retrieving process, the tuning time is always the time needed for downloading the requested data. However, a client may switch among broadcasting channels several times (say, disconnect from one communication channel, and then construct connection to another channel). Switching among channels costs additional energy consumption [5], and thus the number of switchings during the data retrieving process also has notable impact on the energy efficiency of a data retrieving protocol. Therefore, a good data retrieving protocol should be able to reduce both access latency and number of switchings among channels during the retrieving process.

Some literatures discussed the data retrieval protocols to download a set of data from multiple broadcasting channels [3][5][9]. Their common assumption is that each client only has one antenna (or one retrieving process). Each time the antenna can access onto one communication channel to set up one connection. However, only one connection results in narrow bandwidth and small throughput, which is a crucial physical constraint for wireless networks to satisfy increasing requirements with Quality of Service (QoS) guarantees.

To solve this problem, the Fourth-Generation Wireless Communication System (4G) applies *multiple-input multiple-output* (MIMO) technology, which allows different data streams to be transmitted simultaneously from different transmitter antennae. 4G is a complete evolution which will become a total replacement of current Third-Generation Network System (3G) in the next few years. It is predicted that at that time, the majority of mobile clients will be equipped with MIMO antenna for high-speed communications. They can access onto multiple channels in parallel and shorten the query processing time significantly.

As a result, the focus of our research is to discuss how to schedule the retrieving process of a set of requested data, given their time offset and resided channels, using a client with multiple antennae. Our target is to minimize the access latency and number of channels switchings for the client. In other words, by the employing protocols proposed in this paper, a client should be able to download a set of requested data using multiple retrieving processes in parallel, with short response time and minimum energy consumption. We name this problem as *Parallel Data Retrieval Scheduling with MIMO Antennae* (PADRS-MIMO). In this paper, we present the communication model, formally define the PADRS-MIMO problem, and construct two greedy heuristics named *Least Switch Data Retrieval Protocol (Least-Switch)* and *Best First Data Retrieval Protocol (Best-First)*. We are the first work to discuss the data retrieval with MIMO antennae for wireless data broadcast problem and propose practical solutions. We analyze the performance of *Least-Switch* and *Best-First* both theoretically and practically, and prove their effectiveness and efficiency by simulation results.

The rest of our paper is organized as follows: in Sec. 2 we list the related works to PADRS-MIMO; In Sec. 3 we propose the communication model and formally define PADRS-MIMO. Section 4 analyzes the nature of PADRS-MIMO, constructs two data retrieval protocols for PADRS-MIMO which are illustrated with detailed examples and theoretical analysis. Section 5 evaluates the two algorithms' performances from several aspects by simulation results. Section 6 concludes the whole work and provides future directions for this topic.

2   Related Works

Multi-channel data broadcast has been a trend in the wireless data broadcast research area since it can significantly reduce the access latency by partitioning data onto multiple channels. Research works on multi-channel wireless data broadcast mainly focus on two aspects from the server side: how to schedule data on multiple channels to reduce access latency and how to design efficient indexing schemes to reduce tuning time.

Several literatures discussed the data scheduling problem on multiple channels in the wireless data broadcast environment. In [1], Ardizzoni et al. developed several algorithms based on dynamic programming techniques to optimally schedule skewed data on multiple channels while preserving the flat broadcast scheme of each channel. Prabhakara et al. [7] provided a wide range of design considerations for the server which broadcasts over the multi-level multi-channel air cache to improve server's performance. Saxena et al. [8] presented a balanced on-line broadcast scheduling scheme which adopts a hybrid push-pull broadcast schedule per channel. In [15], how to minimize the average access latency by optimally partitioning data among multiple channels was discussed by Yee et al., and an approximation algorithm that is less complex than optimal solution yet with near-optimal performance was developed.

Indexing techniques of multi-channel wireless data broadcast not only discuss how to index data, but also concern about how to allocate indices, given multiple channels. One popular way to allocate index is to assign certain channels as designated index channels and others as data channels. Jung et al. [4] presented a tree-structured index allocation method to allocate index on separate channels from data, which minimized average access latency by broadcasting hot data and their indices more frequently than less hot data and their indices. Waluyo et al. [11] presented a global index schemes where each index channel preserves a part of the index tree with replication among each other. Wang and Chen [12] adopted the distributed indexing technique proposed in [6] to multiple channel environment, by creating an virtual index tree for each data channel and multiplexing them onto one physical index channel.

Despite of various literatures on scheduling and indexing problems on server's side, there is very little research about how to schedule the data retrieval process more efficiently. In [9], Sun et al. presented two algorithms to retrieve a set of requested data from parallel broadcast channels. Hurson et al. [5] gave other two heuristic algorithms to schedule data retrieving from multiple broadcast channels to reduce the access latency and number of switchings among channels. However, both work assume there can be only one process to retrieve data, and data are evenly allocated on multiple channels.

3   Problem Formulation

In this section, we will discuss the communication model of MIMO wireless data broadcast system and formulate the PADRS-MIMO problem.

3.1   Communication Model

A *program* is a complete broadcast cycle which contains a set of data and possibly some index information. Suppose a set of data $D = \{d_1, \cdots, d_{|D|}\}$ are broadcasted in a program. Without the loss of generality, assume the keys of data items $d_i.key$ in D are monotonically increasing. The popularity of data items in D are represented by their access probability. Let $P = \{p_1, \cdots, p_{|D|}\}$ denote the access probability set of D, where each p_i is the access probability of d_i and we have $\sum_{i=1}^{|D|} p_i = 1$. Assume all data items in D have the same size, which fits in one *data bucket* on the broadcast channel.

D is allocated on N channels according to some data allocation method with respect to their access probability. Let $bcast_i$ represent one round of broadcasting all data on the ith channel, and $blength_i = |bcast_i|$. The $blength_i$ of channel i is decided by the data allocation method used. Different from [5] and [9], we assume that $blength_i$ is not necessarily of the same length, which is a common feature of most data allocation methods. Let $bcycle$ represent the total length of a program. Note that if any update is needed, it will take place between two consecutive programs. Therefore, within one program, the length of each broadcast channel needs to be the same. This can be achieved by making $bcycle$ equal to the least common multiple of $blength_i, i = 1, \cdots, N$. We define the time needed to broadcast one data bucket as one unit time. In a broadcast program, sequence numbers are assigned to each data bucket to represent its time offset from the beginning of the program, denoted as t, $1 \le t \le bcycle$.

Example 1. A data set D with 10 items are to be broadcasted. Their keys are $1, \cdots, 10$ respectively. The access probability is $P = \{0.1138, 0.0488, 0.0427, 0.3414, 0.0854, 0.0682, 0.0379, 0.0569, 0.0341, 0.1707\}$, which follows *zipf* distribution, a typical distribution used to model non-uniform access patterns of web data [4]. D is allocated on $N = 4$ broadcast channels using the *Dynamic Weight-Schedule* allocation method described in [3]. The data allocation result is illustrated in Fig. 1. Each data item is represented by its key value.

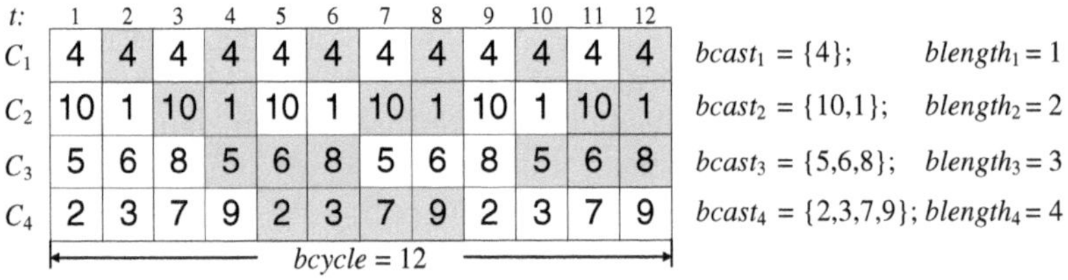

Fig. 1. Communication Model Example: Data Allocation of D

A mobile client has M antennae, which enables it to retrieve data using at most M processes in parallel. Due to technical constraints, the number of antennae is usually limited because of the size of mobile devices. The common number of antennae for mobile handsets is 2 or 3 [10]. On the other hand, the number of channels for a base station to broadcast data is relatively large. Therefore, we assume $M < N$.

As mentioned in Sec. 1, there are three steps to retrieve requested data from broadcasting channels. After *Initial Probing* and *Searching*, the client should know the location of each requested datum, which includes: key, resided channel, sequence number within the channel, and the channel blength, represented by a four-tuple $< key, ch, sq, blength >$ respectively. Based on the data allocation method, a data item d_i may be broadcasted multiple times in a program. The sq included in the tuple is defined as the sequence number of the first appearance of d_i in the broadcast program. For instance, if d_i is broadcasted on channel j, its sq should be $1 \le sq \le blength_j$. The sequence numbers of all data buckets broadcasting d_i in a program can be easily computed given its sq and $blength$.

3.2   Parallel Data Retrieval Scheduling Problem

The parallel data retrieval scheduling problem in a MIMO wireless data broadcast system can be addressed as follows. Given a base station broadcasting a set of data on multiple channels, a client with multiple antennae has a request of a subset of the data broadcasted. The client would like to start the data retrieving from a certain starting time. The problem is: 1) how to assign antennae to retrieve different data items in the request; and 2) how to order the retrieval of data items for each antenna, so that we can reduce

- the access latency of the data retrieval,
- number of switchings among broadcast channels.

Obviously, to reduce access latency is to reduce the time needed to retrieve a request. The reason why we want to reduce the number of switchings among channels is for the sake of energy efficiency. As discussed in [5], switching among channels also consumes energy. In general, one switching takes 10% of the active mode power consumption. However, the objectives of reducing access latency and reducing number of switchings can be contradictive to each other.

Example 2. In the broadcast program shown in Fig. 2, suppose the grey circled data items {1,2,3,4} are of request. The client has only one antenna and the starting point of retrieving process is at $t = 1$. If we want to minimize the access latency, the request should be retrieved in the order of "$3 \rightarrow 1 \rightarrow 4 \rightarrow 2$" which takes only 7 time units but needs 3 switchings. However, if we want to minimize the switchings, the best retrieving order should be "$3 \rightarrow 4 \rightarrow 2 \rightarrow 1$" which needs only 1 switching but takes 12 time units. This example shows that access latency and number of switchings can not be minimized at the same time.

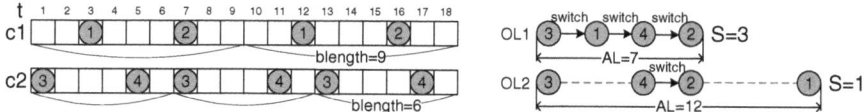

Fig. 2. Example of Possible Objective Contradiction

To balance the two objectives mentioned above, we introduce a cost function that takes both of them into consideration:

Definition 1. *The cost of a parallel data retrieving schedule is evaluated as* $c = \alpha \cdot AT + \beta \cdot S$, *where AT is the access latency of the data retrieving process from the starting time (initial probing and searching processes not included); S is the number of switches among channels during the retrieving process.*

α and β are adjustable parameters which can reflect the user's preference. For example, if the user's first priority is to get the requested data faster, α should be increased; if energy is of more concern, β should be increased.

Definition 2. *A request* $\mathbf{R} = \{R_1, \cdots, R_w\}$ *is a client request to retrieve* w *data items in a broadcast program. Each* $R_i \in \mathbf{R}$ *is a four-tuple:* $< key, ch, sq, blength >$. $\{R_i.key | R_i \in \mathbf{R}\} \subseteq \{d_j.key | d_j \in D\}$.

Given a request $\mathbf{R}$, there may be some channels that do not contain any data items in $\mathbf{R}$. During data retrieving, these channels will not affect the retrieving process at all, and thus can be ignored. In other words, when retrieving a request, we are only interested in broadcast channels which has requested data on them. Therefore, we have the following definition:

Definition 3. *Given* $\mathbf{R}$, *let* $CH = \{ch_1, \cdots, ch_K\}$, $K = |\bigcup_{R_i \in \mathbf{R}} \{R_i.ch\}|$. CH *is a set of requested channels where requested data reside.* $\forall ch_i \in CH$, *define:*

- $ch_i.begin = \min_{R_j.ch = ch_i} \{R_j.sq\}$: *starting sequence number of required data on* ch_i;
- $ch_i.end = \max_{R_j.ch = ch_i} \{R_j.sq\}$: *ending sequence number of required data on* ch_i;
- $ch_i.blength$: *blength of* ch_i.

Definition 4. *Given* $\mathbf{R}$, *a data retrieval schedule is a set of ordered lists of data items (represented by their key values)* $\mathbf{OL} = \{OL_1, \cdots, OL_M\}$ *for* M *processes, where each* OL_i *provides a schedule for one process. There should be* $\bigcup_{OL_j \in \mathbf{OL}} \{d_i.key | d_i \in OL_j\} = \bigcup_{R_j \in \mathbf{R}} \{R_j.key\}$. $\forall OL_i \in \mathbf{OL}$, *define:*

- $OL_i.t$ *as the first time slot after retrieving the last data item in* OL_i;
- $OL_i.ch$ *as the channel of the last data item in* OL_i;
- $OL_i.c$ *as the total downloading cost for* OL_i, *where* $c = \alpha AT + \beta S$.

In data retrieving, we need to pay attention to possible conflicts of retrieving data buckets on parallel channels. If a retrieving process is on ch_i at time t, there will be a conflict if it tries to download a datum on ch_j ($j \neq i$) which also appears at t, because the time needed to switch between channels is not neglectable, which is almost equivalent to broadcasting one data bucket [5]. More formally:

Definition 5. *For a retrieving process* OL_i, *a conflict will occur if it wants to retrieve* R_j *at* $OL_i.t$, *where* $R_j.sq = OL_i.t \% R_j.blength$, *and* $R_j.ch \neq OL_i.ch$.

Based on the above definitions, PADRS-MIMO can be more formally defined:

Definition 6. *Given* D *broadcasted on* N *channels, a client with* M *antennae has a request* $\mathbf{R}$. t_0 *is a predefined starting point in a program for the client to start retrieving* $\mathbf{R}$, *which is not necessarily the beginning of a program. The parallel data retrieval scheduling with MIMO Antennae problem (PADRS-MIMO) is to develop a function* $f : \{\mathbf{R}, t_0\} \rightarrow \mathbf{OL}$ *to produce a schedule* $\mathbf{OL}$ *without any conflicts, so that the cost function* c *defined in Def. 1 is as small as possible.*

Let us use an example to illustrate the above definitions.

Example 3. In the broadcast program in Example 1, assume a client with $M = 2$ antennae has a request $R = \{1, 2, 6, 8, 9\}$. The predefined starting time is $t_0 = 3$. The cost function parameters are $\alpha = 1$ and $\beta = 2$. The request channel set CH and a schedule $\mathbf{OL} = \{OL_1, OL_2\}$ are shown in Fig. 3. If OL_1 wants to retrieve d_4 at time 9, there will be a conflict because $OL_1.t = 9$ and $OL_1.ch = 2 \neq 1$.

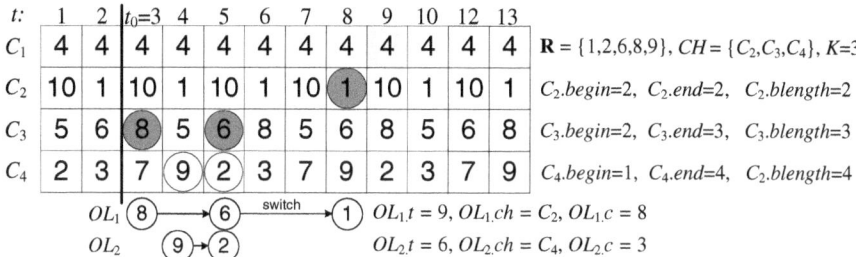

Fig. 3. Example of R, CH, $\mathbf{OL}$ and conflict

4   Efficient Parallel Data Retrieval Scheduling

As discussed in Sec. 3, how to schedule the request to multiple retrieving processes can significantly influence the access latency and energy consumption for clients. In this section, we introduce several algorithms to solve PARDRS-MIMO.

We first consider the relation between K: number of channels where requested data $\mathbf{R}$ locates; and M: number of available processes a client can use to retrieve data in parallel. When $K \leq M$, it is obvious that the optimal schedule is to assign a different process to retrieve requested data on each channel. This schedule requires only K processes. In the following discussion for algorithm construction, we assume that $K > M$. We will present two scheduling algorithms of parallel data retrieval for clients equipped with MIMO antennae. The first algorithm is *Least Switch Data Retrieval Scheduling (Least-Switch)*, which guarantees the least switching number S, and tries to minimize AL, while the second algorithm is *Best First Data Retrieval Scheduling (Best-First)*, which minimizes global cost function c.

4.1   Least Switch Data Retrieval Scheduling (*Least-Switch*)

For one retrieving process, switching among channels not only will consume energy, but also may introduce unnecessary conflicts. An intuitive thought is to minimize the total number of switchings among channels during the complete data retrieval. Therefore, we develop an algorithm which guarantees the least number of switchings among channels. Alg. 1 presents the *Least-Switch* protocol and Alg. 2 computes the cost function used in Alg. 1.

Least-Switch (Alg. 1) can be decomposed into two steps. The first step (Line 2 to 4) is to find the M channels to read. The criteria is the time needed for each channel to retrieve all requested data on the channel starting from t_0, which is evaluated by cost function c. M channels with the longest time will be assigned to the M processes. The second step (Line 5 to 8) is to read the rest $K - M$ channels. Choose the process with least cost, append to it the channel which requires the longest time to retrieve all requested data on it. If several processes have the same cost, randomly choose one to proceed. The M processes keep reading the remaining channels until all K channels are read and R is completely retrieved. Note that in Algorithm 1, once we append a channel ch'_j to an OL_i, it means appending every data item on this channel to OL_i with the order calculated from $OL_i.t$ and $ch'_j.blength$. $OL_i.t$, $OL_i.c$ and $OL_i.ch$ should also be changed correspondingly afterwards.

Algorithm 1. Least Switch Data Retrieval Scheduling (*Least-Switch*)

Input: R; CH; t_0.
Output: OL.
1: $\forall 1 \leq i \leq M$, $OL_i.t = t_0$, $OL_i.c = 0$, $OL_i.ch = \emptyset$.
2: Sort CH in descending order by $c(OL_1, ch_i, 1, 0)$ as CH'.
3: Append $ch'_1, \cdots, ch'_M$ to $OL_1, \cdots, OL_M$ correspondingly.
4: $CH' = CH' \setminus \{ch'_1, \cdots, ch'_M\}$.
5: **while** $CH' \neq \emptyset$ **do**
6:     $OL_i^* = \arg\min_{OL_i \in \mathbf{OL}}\{OL_i.c\}$; append $ch'^*_j = \arg\max_{ch'_j \in CH'}\{c(OL_i^*, ch'_j, 1, 0)\}$ to OL_i^*.
7:     $CH' = CH' \setminus \{ch'^*_j\}$.
8: **end while**

The reason why in *Least-Switch* we every time append the longest channel to the fastest process is that access latency of the complete data retrieval is determined by the process which takes the longest time. Therefore, we would like to balance the time needed by each process to avoid delay caused by some process which is much more slower than others. This can be achieved by keeping appending the "longest" channel remained to the fastest process.

Algorithm 2, the computation of the cost function plays an importance role in the *Least-Switch*. The input of the cost function is candidate process OL_i, channel ch_j, and parameters α and β. It will return the cost of downloading all requested data on ch_j starting from $OL_i.t$. Since *Least-Switch* always guarantees the minimum number of switchings, it is not necessary to consider switching during evaluating the cost function. Therefore, we set $\alpha = 1$ and $\beta = 0$.

For initial channel assignment to M processes, no switching is needed to tune in the target channel and thus no conflicts will occur. There are three possible cases (Line 3 to 7):

Case 1: If the starting point $local = OL_i.t\%ch_j.blength$ is before or at $ch_j.begin$ in a broadcast round on ch_j, the access latency is simply the distance between $ch_j.end$ and the starting point (Line 10).

Case 2: If the starting point is between $ch_j.begin$ and $ch_j.end$ or at $ch_j.end$, the access latency will be $ch_j.blength - local + g_2 + 1$, where g_2 is the requested data appeared right before or at $local$.

Case 3: If $local$ is after $ch_j.end$, the process needs to wait till the next $ch_j.end$ to finish downloading all requested data on ch_j, that is, $ch_j.blength - local + ch_j.end + 1$ time slots.

When $OL_i.ch \neq ch_j$ (Line 8 to 13), possible conflicts should be considered. The computation of cost is similar except for two differences: 1) If $local$ lies exactly at $ch_j.begin$, the access latency should be computed as Case 2. This is because the data item at $ch_j.begin$ will be the last requested data on ch_j available for OL_i due to the conflict; 2) Similarly, if $local$ lies at $ch_j.end$, the access latency should be computed as Case 3.

Algorithm 2. Cost Function $c(OL_i, ch_j, \alpha, \beta)$

Input: OL_i, ch_j, α, β.
Output: Cost for OL_i to download data on channel ch_j.
1: $local = OL_i.t \% ch_j.blength$ ▷ if $local = 0$, let $local = blength$
2: $g_1 = \max\limits_{R_k.ch=ch_j} \{R_k.sq | R_k.sq < local\}$; $g_2 = \max\limits_{R_k.ch=ch_j} \{R_k.sq | R_k.sq \leq local\}$.
3: **if** $OL_i.ch = \emptyset$ **then**   ▷ calculate cost at initial stage
4:     $S = 0$;
5:     **if** $local \leq ch_j.begin$ **then** $AL = ch_j.end - local + 1$;
6:     **else if** $ch_j.begin < local \leq ch_j.end$ **then** $AL = ch_j.blength - local + g_1 + 1$;
7:     **else** $AL = ch_j.blength - local + ch_j.end + 1$; **end if**
8: **else**   ▷ calculate cost with switch
9:     $S = 1$;
10:     **if** $local < ch_j.begin$ **then** $AL = ch_j.end - local + 1$;
11:     **else if** $ch_j.begin \leq local < ch_j.end$ **then** $AL = ch_j.blength - local + g_2 + 1$;
12:     **else** $AL = ch_j.blength - local + ch_j.end + 1$; **end if**
13: **end if**
14: Return $c = \alpha \times AL + \beta \times S$

Lemma 1. *The minimum number of switchings for a client with M processes to download requested data allocated on K channels is $K - M$ (when $K > M$).*

Proof. Consider one process first. If data are located on K channels, the process has to visit each channel at least once. Suppose it first accesses ch_i, it has to switch to the rest $K-1$ channels. Hence, $\min S = K-1$. If we have M processes, only the first M channels accessed do not need switchings, so $\min S = K - M$. □

Theorem 1. Least-Switch *guarantees minimum switchings to download* **R**.

Proof. From Line 3 in Alg. 1, M processes will access onto M channels without switching. According to the procedure between Line 5 and 8, and the fact that each data will only be downloaded once, each candidate channel $ch_i \in CH'$ will only be visited once. Therefore, in total *Least-Switch* has $S = K - M$. By Lemma 1, this algorithm guarantees the minimum number of switchings. □

Example 4. Fig. 4 is the result of *Least-Switch* with the setting in Example 3. The eclipses are the decision procedure and rectangles are the resulting schedules.

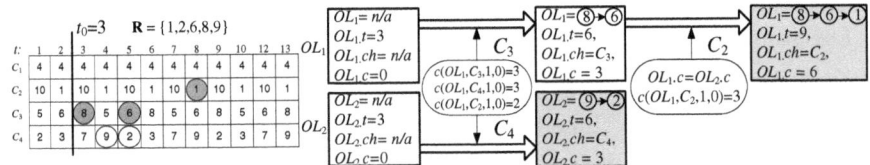

Fig. 4. Example of *Least-Switch* scheduling scheme

4.2 Best First Data Retrieval Scheduling (*Best-First*)

Least-Switch discussed in Section 4.1 can minimize the switching among channels during the entire data retrieval. However, if two requested data items on one channel are far from each other, the retrieving process reading that channel may miss the chance to retrieve available data items on other channels during waiting and thus increase the access latency. In order to take account of both the access latency and number of switching, we introduce the second algorithm which adopts the idea of best-first search, as illustrated in Alg. 3 and Alg. 4.

Algorithm 3. Best First Data Retrieval Scheduling(*Best-First*)

Input: R; CH; t_0.
Output: OL.

1: $\forall 1 \le i \le M$, $OL_i.t = t_0$, $OL_i.c = 0$, $OL_i.ch = \emptyset$.
2: $\forall 1 \le i \le K$, let $ch_i.R^*$ be the R_j that appears first on ch_i after t_0.
3: Sort CH ascendingly by $c(OL_1, ch_i.R^*, \alpha, \beta)$ as CH'. To break tie, select the one with longer $ch_i.blength$ if two channels have the same cost.
4: Append $ch'_1.R^*, \cdots, ch'_M.R^*$ to $OL_1, \cdots, OL_M$ correspondingly.
5: $\mathbf{R} = \mathbf{R} \setminus \{ch'_1.R^*, \cdots, ch'_M.R^*\}$; Update CH.
6: **while** $|CH| > M$ **do**
7:     $OL_i^* = \arg\min_{OL_i \in \mathbf{OL}} \{OL_i.c\}$; append $R_j^* = \arg\min_{R_j \in \mathbf{R}} \{c(OL_i^*, R_j, \alpha, \beta)\}$ to OL_i^*.
8:     $\mathbf{R} = \mathbf{R} \setminus \{R_j^*\}$; Update CH.
9: **end while**
10: Sort **OL** descendingly by $OL_i.c$.
11: **for** $i = 1$ to M **do**
12:     Append $ch_j'^* = \arg\min_{ch_j' \in CH'} \{c(OL_i, ch_j', \alpha, \beta)\}$ to OL_i; $CH = CH \setminus \{ch_j'^*\}$.
13: **end for**

Best-First (Alg. 3) can be interpreted as three phases:

Phase 1: *Initial Assignment* (Line 1 to 5). Starting from t_0, for each channel $ch_i \in CH$, $ch_i.R^*$ is its first appeared requested data item. All K channels will be sorted in ascending order by their $ch_i.R^*$ into CH'. $ch_i.R^*$ of the first M ch'_i will be assigned to M processes respectively and removed from request list R. $OL_i.t$, $OL_i.c$ and $OL_i.ch$ should also be updated correspondingly. If there is any tie occurred, the requested data whose channel has longer blength will be selected. This is out of the consideration that requested data on channels with longer blength appear less frequently in the broadcast program, and thus the client will need longer time to wait for it.

Phase 2: *Best First Assignment* (Line 6 to 9). The idea of best first search is adopted in this phase to choose the next data item to retrieve. We first choose the process OL_i^* with least cost, and append R_j^* which needs the least cost for OL_i^* to retrieve. Note that for OL_i^*, there might be multiple R_j^* with the same anticipated cost to retrieve. To break the tie, R_j^* with longer blength will be chosen. $OL_i.t$, $OL_i.c$ and $OL_i.ch$ should be updated after the assignment. If ch_i does not contain requested data any more, it should be removed from CH. This procedure will continue until there are only M channels with requested data left.

Phase 3: *Final Assignment* (Line 10 to 13). When there are only M channels with requested data on them, they will be assigned to the M processes with the same reason as if $K \leq M$. M processes will first be sorted by their total cost so far. Starting from the process with the most cost, it will append the channel that costs least for it to finish retrieving. Once again, the reason of doing this is to balance the cost of each process in order to avoid unnecessary delay caused by possible "super-costly" process.

Algorithm 4. Cost Function $c(OL_i, R_j, \alpha, \beta)$

    Input: OL_i, R_j, α, β.
    Output: Cost for OL_i to download data R_j.
1: $local = OL_i.t\%R_j.blength$ ▷ if $local = 0$, let $local = blength$
2: **if** $OL_i.ch = R_j.ch \vee OL_i.ch = \emptyset$ **then**   ▷ initial stage or no switch
3:     $S = 0$;
4:     **if** $local \leq R_j.sq$ **then** $AL = R_j.sq - local + 1$;
5:     **else** $AL = R_j.blength - local + R_j.sq + 1$; **end if**
6: **else**   ▷ switch from one channel to another
7:     $S = 1$;
8:     **if** $local \leq R_j.sq - 1$ **then** $AL = R_j.sq - local + 1$;
9:     **else** $AL = R_j.blength - local + R_j.sq + 1$; **end if**
10: **end if**
11: Return $c = \alpha \times AL + \beta \times S$

Different from Alg. 2, the cost computed in *Best-First* is related to single requested data item, instead of a channel. As described in Al. 4, For each OL_i and R_j pair, there are two possible cases: 1) OL_i is on the same channel as R_j or OL_i is still empty (Line 2 to 5) and 2) they are on different channels (Line 6 to 10). In the first case, no switching between channels is needed, and access latency can be easily computed. In the second case, one switching is required, and the possible conflict should be considered.

Example 5. With the setting in Example 3, *Best-First* is performed as shown in Fig. 5, with $\alpha = 1$ and $\beta = 2$. The eclipses are the decision procedure and rectangles are the resulting schedule.

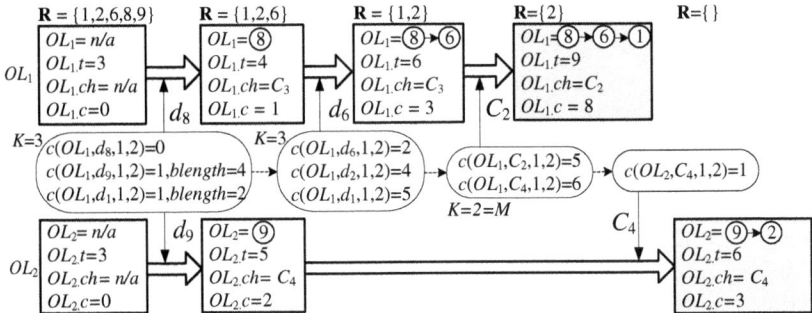

Fig. 5. Example of *Best-First* scheduling scheme

5 Performance Analysis

In this section, we will use simulation result to discuss the characteristics of PADRS-MIMO and evaluate the performances of *Least-Switch* and *Best-First* data retrieval scheduling schemes.

5.1 Simulation Setup

Simulation is implemented in Java 1.6.0 16 on an Intel(R) Xeon(R) E5520 computer with 6.00GB memory, with Windows 7 version 6.1 operating system. We simulate a base station with N broadcast channels, and multiple clients with various requests of data. The database to be broadcast has 10000 items [16], each of size 1KB. The access probability of the database follows zipf distribution [2], which is a typical model for non-uniform access patterns [4,12]. We adopt *Dynamic Weight-Schedule* [3] for data allocation. N varies from 5 to 30. (α, β) are set to $(1, 0)$ and $(1, 2)$ for *Least-Switch* and *Best-First*.

Clients have multiple antennae to retrieve the data from broadcast channels. The number of antennae varies from 1 to 10. The size of a request varies from 10 to 1000. For each experiment, we generate 100 requests to get their average access latency and number of switchings during data retrieval. Requests are generated according to data's access probability.

5.2 Simulation Results

When $N = 20$, $|\mathbf{R}| = 100$, we vary the number of antennae M. Access latency is measured in unit time (the time needed to broadcast one data bucket). Fig. 6 shows that when M is small ($M =1$, 2 or 3), *Best-First* takes much shorter access latency than *Least-Switch*, while when M is relatively large ($M \geq 4$), both protocol takes similar access latency. Due to current technique constraints of the number of antennae in mobile devices (usually 2 or 3 antennae [10]), we can claim that *Best-First* protocol has advantage in reducing response time of request retrieving. However, from Fig. 9, it is clear that *Best-First* needs much more

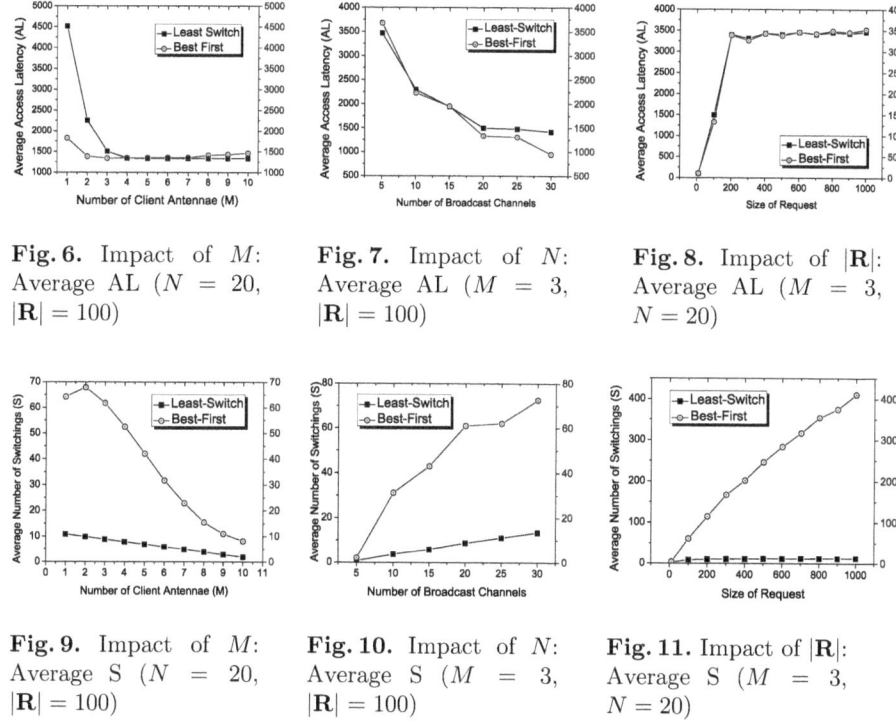

Fig. 6. Impact of M: Average AL ($N = 20$, $|\mathbf{R}| = 100$)

Fig. 7. Impact of N: Average AL ($M = 3$, $|\mathbf{R}| = 100$)

Fig. 8. Impact of $|\mathbf{R}|$: Average AL ($M = 3$, $N = 20$)

Fig. 9. Impact of M: Average S ($N = 20$, $|\mathbf{R}| = 100$)

Fig. 10. Impact of N: Average S ($M = 3$, $|\mathbf{R}| = 100$)

Fig. 11. Impact of $|\mathbf{R}|$: Average S ($M = 3$, $N = 20$)

switchings during retrieval than *Least-Switch*, which guarantees the minimum number of switchings. When M decreases, the number of switchings needed by *Best-First* is dropping dramatically, because more data will be retrieved during its third phase and thus it needs not switch among channels any more.

Next, we evaluate the impact of the number of broadcast channels on the two protocols' performances. The average access latency and number of switchings needed are shown in Fig. 7 and Fig. 10, given $M = 3$ and $|\mathbf{R}| = 100$. We observe that when N increases, access latency of both protocols decrease similarly. This shows the benefit of using more broadcast channels with respect to reducing the access latency. For *Least-Switch* protocol, the number of switchings increases only a bit when N increases. However, for *Best-First*, number of switchings increases significantly when N increases.

We are also interested in how the size of requests can influence the performances of two protocols. Fig. 8 presents the change of average access latency with the increasing of request size. Both protocols behave similarly. The access latency increases rapidly when request size first increase from 10 to 200. After that, it becomes relatively stable despite of the increasing request size. This is because when the number of requested data increases, the requested data will appear more frequently on broadcast channels and thus retrieval protocols can download them more continuously without having to wait extra time. As regards to number of switchings as shown in Fig. 11, *Best-First* have to switch much more times when the request size increases, while *Least-Switch* remains similar amount of switchings.

Based on the above observations, we can conclude that the two protocols proposed in this paper have their own advantages. With limited number of antennae in mobile devices, *Best-First* can significantly reduce the response time needed to download client requests, while *Least-Switch* can guarantee minimum energy consumption during data retrieving and also provide similar response time as *Best-First* when there are reasonable amount of antennae available.

6   Conclusions

In this paper, we are the first to propose data retrieval scheduling problem for mobile clients with MIMO antennae (PADRS-MIMO), which is a promising technique within the emerging 4G wireless network. We formally define PADRS-MIMO, analyze its nature, and then design two data retrieval scheduling protocols: *Least-Switch* and *Best-First* to minimize the access latency and energy consumption of clients. We proof that *Least-Switch* guarantees minimum number of switchings during the data retrieval process. The performance of two protocols are evaluated by simulation. Simulation results show the advantages of two protocols: *Least-Switch* is more energy effective while *Best-First* reduces response time significantly when the number of antennae in the mobile devices are limited. Our future work includes developing more advanced data retrieving scheduling scheme and provide more theoretical analysis on PADRS-MIMO.

References

1. Ardizzoni, E., Bertossi, A.A., Ramaprasad, S., Rizzi, R., Shashanka, M.V.S.: Optimal Skewed Data Allocation on Multiple Channels with Flat Broadcast per Channel. IEEE Transactions on Computers 54(5), 558–572 (2005)
2. Manning, C.D., Schütze, H.: Foundations of Statistical Natural Language Processing. MIT Press, Cambridge (1999)
3. Gao, X., Shi, Y., Zhong, J., Zhang, X., Wu, W.: SAMBox: A Smart Asynchronous Multi-Channel Black BOX for B^+-Tree based Data Broadcast System under Wireless Communication Environment. Distributed & Parallel Databases (2009) (in review)
4. Jung, S., Lee, B., Pramanik, S.: A Tree-Structured Index Allocation Method with Replication over Multiple Broadcast Channels in Wireless Environment. IEEE Transaction on Knowledge and Data Engineering 17(3), 311–325 (2005)
5. Hurson, A.R., Muñoz-Avila, A.M., Orchowski, N., Shirazi, B., Jiao, Y.: Power-Aware Data Retrieval Protocols for Indexed Broadcast Parallel Channels. Pervasive and Mobile Computing 2(1), 85–107 (2006)
6. Imielinski, T., Viswanathan, S., Badrinath, B.: Data on Air: Organization and Access. IEEE Transactions on Knowledge and Data Engineering 9, 353–372 (1996)
7. Prabhakara, K., Hua, K.A., Oh, J.: Multi-Level Multi-Channel Air Cache Designs for Broadcasting in a Mobile Environment. In: ICDE 2000, pp. 167–176 (2000)
8. Saxena, N., Pinottti, M.C.: On-line Balanced k-Channel Data Allocation with Hybrid Schedule per Channel. In: MDM 2005, New York, NY, USA, pp. 239–246 (2005)

9. Sun, B., Hurson, A.R., Hannan, J.: Energy-Efficient Scheduling Algorithms of Object Retrieval on Indexed Parallel Broadcast Channels. In: ICPP 2004, Montreal, Quebec, Canada, pp. 440–447 (2004)
10. Usman, M., Abd-Alhameed, R.A., Excell, P.S.: Design Considerations of MIMO Antennae for Mobile Phones. PIERS online 4(1), 121–125 (2008)
11. Waluyo, A.B., Srinivasan, B., Taniar, D.: Global Indexing Scheme for Location-Dependent Queries in Multi Channels Mobile Broadcast Environment. In: AINA 2005, pp. 1011–1016 (2005)
12. Wang, S., Chen, H.-L.: Tmbt: An Efficient Index Allocation Method for Multi-Channel Data Broadcast. In: AINAW 2007, vol. 2, pp. 236–242 (2007)
13. Xu, J., Lee, W.-C., Tang, X., Gao, Q., Li, S.: An Error-Resilient and Tunable Distributed Indexing Scheme for Wireless Data Broadcast. IEEE Transactions on Knowledge and Data Engineering 18(3), 392–404 (2006)
14. Yao, Y., Tang, X., Lim, E.-P., Sun, A.: An Energy-Efficient and Access Latency Optimized Indexing Scheme for Wireless Data Broadcast. IEEE Transactions on Knowledge and Data Engineering 18(8), 1111–1124 (2006)
15. Yee, W.G., Navathe, S.B., Omiecinski, E., Jermaine, C.: Efficient Data Allocation over Multiple Channels at Broadcast Servers. IEEE Transactions on Computers 51(10), 1231–1236 (2002)
16. Yee, W.G., Navathe, S.B.: Efficient data access to multi-channel broadcast programs. In: CIKM 2003, New York, NY, USA, pp. 153–160 (2003)
17. Zheng, B., Lee, W.-C., Liu, P., Lee, D.L., Ding, X.: Tuning On-Air Signatures for Balancing Performance and Confidentiality. IEEE Transactions on Knowledge and Data Engineering 21(12), 1783–1797 (2009)

Inferring Aggregation Hierarchies for Integration of Data Marts

Dariush Riazati, James A. Thom, and Xiuzhen Zhang

School of Computer Science and Information Technology
RMIT University,
Melbourne, Australia, 3001
dariush.riazati@student.rmit.edu.au,
{james.thom,xiuzhen.zhang}@rmit.edu.au

Abstract. The problem of integrating heterogeneous data marts is an important problem in building enterprise data warehouses. Specially identifying compatible dimensions is crucial to successful integration. Existing notions of dimension compatibility rely on given and exact dimension hierarchy information being available. In this paper, we propose to infer aggregation hierarchies for dimensions from a database instance and use these inferred aggregation hierarchies for integration of data marts. We formulate the problem of inferring aggregation hierarchies as computing a minimal directed graph from data, and develop algorithms to this end. We extend previous notions of dimension compatibility in terms of inferred aggregation hierarchies.

Keywords: Aggregation Hierarchy, Data Mart, Data Warehouse, OLAP, Summarizable, Compatible Dimensions.

1 Introduction

Data marts are based on the Star schema, a data model that allows implementation of relational data bases that can serve the purpose of multidimensional databases. In this model, quantitative data (i.e. *facts*) that can be aggregated are included in *fact tables*, and information by which facts are aggregated and filtered by (i.e. *dimension attributes*) are included in *dimension tables*. Dimension attributes are grouped into *levels* of *aggregation hierarchies* such that aggregated facts at a higher level of the hierarchy can be computed from facts at a lower level. This important property of aggregation hierarchies is known as *summarizability*.

Fig. 1(a) is an example of a simple data mart. Fig. 1(b) shows the hierarchy of levels on which the dimensions from Fig. 1(a) are organized. In this example, we can aggregate the sales for stores to get sales for localities and aggregate sales for localities to get the sales for cities. We say that the sales value is *rolled-up* from Store to Locality and from Locality to City.

Many autonomous data marts are developed over years in large enterprises. The integration of these data marts into an enterprise warehouse for enterprise-wide large scale analysis is a strategic business objective [6].

P. García Bringas et al. (Eds.): DEXA 2010, Part II, LNCS 6262, pp. 96–110, 2010.

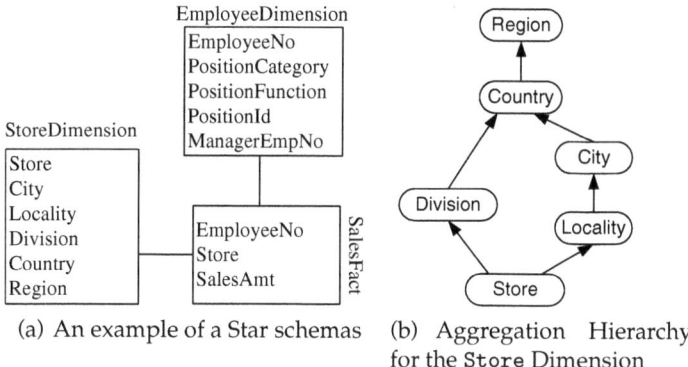

(a) An example of a Star schemas   (b) Aggregation   Hierarchy
for the Store Dimension

Fig. 1. Our running example of a data mart

The success of On-line Analytical Processing (OLAP) tools for exploration and summarization of information in the warehouse after integration relies on that the dimensions and aggregation hierarchies for the warehouse after integration accurately reflects those of the data marts before integration. An important problem in integration of data marts is the integration of dimensions. All existing work (e.g. [23] and [8]) assume that aggregation hierarchies defining the levels and the relationships between them, are given as part of the schema and the data is structured according to the given aggregation hierarchies. However, in many practical applications, especially in the situation of autonomous data marts, the aggregation hierarchy for dimensions are not always available.

In this paper, we propose to infer the aggregation hierarchies for dimensions from their data and use them for integration of data marts. In Section 2 we describe aggregation hierarchies. We formulate the problem of inferring aggregation hierarchies as computing from data a minimal directed graph for the roll-up relationship between levels, and develop algorithms to this end in Section 3. In Section 4 we describe the relationship between the schema-defined and inferred aggregation hierarchies. Aggregation hierarchies inferred from data have an important property, they guarantee the summarizability of data, which we prove in Section 5, alongside proving that inferred aggregation hierarchies can ensure the accuracy of integrating data marts, and therefore provide a sound basis for applications on top of the warehouse after integration. In Section 6 we discuss related works and finally in Section 7 we conclude and describe our future work.

2   Aggregation Hierarchies

In this section we introduce definitions for dimension, level and aggregation hierarchy in the context of relational multidimensional databases.

Definition 1. *A dimension D consists of an aggregation hierarchy H of levels. Given two levels l and l' of a dimension D, we say level l rolls-up to level l' (which we denote as $l \preceq l'$) if we can compute aggregated facts at level l' from facts at level l. The roll-up relationship $\preceq$ forms a partial order over the levels. The aggregation hierarchy H is an acyclic directed graph with no transitive edges, where the nodes of the graph are the levels and the edges are those roll-up relationships in the covering relation of the partial order between the levels.*

A dimension can be represented by a relational database schema.

Definition 2. *A dimension D can be represented by a relation $D(A_1, ..., A_n)$ of n attributes, where one or more attributes correspond to each level such that the levels partition all the attributes into disjoint subsets.*

Most levels correspond to only one attribute, in which case we will use the attribute name and the level name interchangeably, if a level corresponds to two (or more) attributes we will refer to the level by the combination of attribute names (using a '/' between the attribute names).

Definition 3. *A dimension table $T = \{t_1, ..., t_s\}$ is an instance of the relation $D(A_1, ..., A_n)$ representing the dimension, such that each tuple $t_a = \langle v_{a_1}, ..., v_{a_n} \rangle$ in T contains n values, where each value v_{a_i} is an element from the corresponding domain of attribute A_i.*

We can define a partial order over the attributes in a dimension table.

Definition 4. *Given an instance T of a dimension $D(A_1, ..., A_n)$, then we say $A_i \le A_j$ if for every pair of tuples $t_a = \langle v_{a_1}, ..., v_{a_n} \rangle$ and $t_b = \langle v_{b_1}, ..., v_{b_n} \rangle$ in T, $v_{a_i} = v_{b_i}$ implies $v_{a_j} = v_{b_j}$. The $\le$ relationship forms a partial order P over the set of dimension attributes.*

If there is a functional dependency $A_i \rightarrow A_j$ between two attributes of the relation $D(A_1, ..., A_n)$ representing the dimension, then it follows that $A_i \le A_j$ must hold in every instance T of dimension D. Furthermore, if two attributes A_i and A_j correspond to different levels l and l' such that $l \preceq l'$, then we require $A_i \le A_j$. Likewise, if two attributes A_i and A_j correspond to the same level, then we require $A_i \le A_j$ and $A_j \le A_i$.

Definition 5. *A Schema-defined Aggregation Hierarchy (SAH) is an aggregation hierarchy that is defined as part of the schema of D and constitutes a constraint on the tuples in any instance T of D.*

SAH describes roll-up relationships between levels as intended by the data modeler and/or the application. Ideally, an aggregation hierarchy must be defined as part of the data model and then implemented as constraints and enforced by the DBMS which ensures that the population of the dimension tables does not violate those constraints. For example, Oracle's syntax for creating a dimension allows specifying the aggregation hierarchy through explicit description of each level, attributes for each level and the relationship between the levels [18].

3   Inferring the Aggregation Hierarchy

Given that every instance of a dimension is constrained by its SAH, the data can be a reliable source from which we can infer the aggregation hierarchy. However the inferred aggregation hierarchies (IAH) may vary from the intended SAH. How similar the IAH for a dimension table is to the SAH for the same dimension table depends on how close the data represents the SAH. The difference is most often due to the fact that the dimension table is only partially populated, although this is rare in the real world. We will discuss this issue when we establish the viability of IAHs in testing for dimension compatibility in Section 5.

Definition 6. *Given an instance T of dimension D, the inferred partial order of attributes for D is the set of partial order relationships (P) inferred from the partial order relationships between attributes of D and inferred from T.*

In line with the definitions in Section 2, and in this section, we obtain the IAH in three steps. In the first step we obtain the partial order between dimension attributes. In the second step we remove the transitive partial order relationships and finally we obtain the levels and the inferred aggregation hierarchy.

3.1   Inferring the Partial Order of Attributes

We propose Algorithm 1 for inferring the partial order of attributes. We explain this algorithm using the following example: Let us suppose we wish to determine if Country $\leq$ City. The algorithm first sorts the tuples in T on Country. It then scans the values of Country and City. If for as long as the value in Country remains the same from one tuple to the next, then the value in City must also remain the same on the same tuples. If this holds true for the entire T then the roll-up relationship will hold. Given the sample data for Store dimension in Table 1 this relationship does not hold. By scanning T for Country against the remaining attributes we can see that only Country $\leq$ Region holds true. This process is applied for every attributes. The scan of T for each pair of attributes can however stop as soon as it is established that the partial order relationship does not hold. Fig. 2(a) shows the partial order of attributes inferred from the sample data for Store dimension in Table 1, where dashed lines represent transitive relationships.

The complexity of inferring partial order of attributes: The algorithm performs a sort for each attribute with the complexity in the order of $n\,p\log p$

Table 1. Sample data for Store and Shop dimensions

Region	Country	Division	City	Locality	Store
Asia Pacific	Australia	Div1	Sydney	Ryde	st1
Asia Pacific	Australia	Div1	Sydney	Ryde	st2
Asia Pacific	Australia	Div1	Melbourne	Epping	st3
Asia Pacific	Australia	Div1	Melbourne	Morang	st4
Asia Pacific	Australia	Div1	Melbourne	Brighton	st5
Asia Pacific	Australia	Div2	Geelong	Hill	st6

Country	City	Area	Suburb	Shop
Australia	Sydney	NE	Ryde	st1
Australia	Sydney	NE	Ryde	st2
Australia	Melbourne	NT	Epping	st3
Australia	Melbourne	NT	Morang	st4
Australia	Melbourne	SW	Brighton	st5
Australia	Geelong	NW	Hill	st6

Algorithm 1. Inferring the partial order relationships

Input Tuples $T = \{t_1, t_2, ..., t_p\}$ in the instance of a dimension $D(A_1, A_2, ..., A_n)$, and p is the number of tuples.
Output Partial order P of attributes.

```
 1:  P := {}
 2:  for each attribute A_i do
 3:      Sort T on A_i
 4:      for each attribute A_j do
 5:          for each tuple do
 6:              if A_i on current tuple equals A_i on the previous tuple then
 7:                  if A_j on current tuple does not equal A_j on the previous tuple then
 8:                      exit this loop
 9:                  end if
10:              end if
11:          end for
12:          if end of tuples was reached then
13:              P := P ∪ {(A_i, A_j)}
14:          end if
15:      end for
16:  end for
```

---

where n is the number of attributes and p is the number of tuples. We also scan T for every pair of attributes $(n^2 - 1)$ with the complexity in the order of $n^2 p$, though in some cases only a subset of T is scanned. Therefore, the complexity of the Algorithm 1 is $O(n^2 p + n p \log p)$.

Observe that Algorithm 1 obviously computes all partial order relationships between any pair of levels for any given dimension table.

3.2 Cover for Partial Order of Attributes

Definition of aggregation hierarchies does not includes transitive roll-up relationships as values at each level can be computed from the next immediate child level. In order to remove transitive partial order relationships, we can use existing algorithms ([1]) that remove transitive edges from a directed graph. Fig. 2(b) shows the partial order of attributes after removing transitive partial order relationships.

3.3 The Levels and the Inferred Aggregation Hierarchy

Based on Definition 4, the two attributes Country and Region in Fig. 2(c) correspond to the same level. We propose Algorithm 2 to obtain the levels that is disjoint subsets of attributes (L) and then associate each attribute with its corresponding level. The result is the inferred aggregation hierarchy over a set of levels with roll-up relationship between them. The resulting IAH is also an acyclic directed graph.

We assume that $q = |P|$ and $r = |L|$. Statements 2 to 8 of Algorithm 2 with complexity of q^2 add to L each attribute of any partial order (p_m) as a level unless there is another partial order (p_n) which makes their attributes to correspond to the same level in which case the added level will include both attributes. Statements 9 to 13 with the complexity of r^2 combine those subsets of L that have at least one common attribute. At this point, L contains disjoint subsets of attributes that correspond to the same level. Statements 14 to 16 with the complexity of $q\,r$ revisit the partial orders (copied into H_L as roll-ups) and assigns each attribute to a level name that is derived from attribute names in the corresponding level. Finally, duplicate roll-ups are removed. The complexity of statement 17 is q^2. The overall complexity for Algorithm 2 is $O(2\,q^2 + r^2 + q\,r)$.

Algorithm 2. Identifying levels and roll-ups

Input P is the partial order of attributes with no transitive relationship.
Output H_L is the inferred aggregation hierarchy.
        L is a set of levels corresponding to disjoint subsets of attributes.

1: $H_L := P, L = \{\}$
2: **for** each pair of partial order relationships ρ_m and ρ_n in P **do**
3:     **if** $\rho_m = A_i \leq A_j$ and $\rho_n = A_j \leq A_i$ **then**
4:         $L := L \cup \{A_i, A_j\}, H_L := H_L - \rho_m, \rho_n$
5:     **else**
6:         $L := L \cup \{A_i\}, \{A_j\}$
7:     **end if**
8: **end for**
9: **for all** $x \in L$ and $y \in L$ where $x \neq y$ **do**
10:     **if** $x \cap y \neq \varnothing$ **then**
11:         $L := L - \{x\}, L := L - \{y\}, L := L \cup \{(x \cup y)\}$
12:     **end if**
13: **end for**
14: **for** each partial-order $p_m = (A_i, A_j)$ in H_L and each subset of levels l_s in L **do**
15:     Replace any A_i and A_j that appear in l_s with a level name that is a combination of attribute names in l_s (and using a '/' between the attribute names).
16: **end for**
17: Remove any duplicate roll-up relationship from H_L.

---

4   Inferred Aggregation Hierarchies Subsuming Schema-Defined Aggregation Hierarchies

Based on the inferred partial order relationships in Fig. 2(a) we have Country $\leq$ Region which is a valid partial order relationship, and Region $\leq$ Country which may indeed be only a coincidence due to the incomplete instance of the Store dimension in Table 1. If the following tuple ⟨Asia Pacific,New Zealand,Div3,Wellington,Brooklyn,st7⟩ is added to the Store dimension table in Table. 1, the second partial order will become invalid. Other spurious partial orders are Locality $\leq$ Division and City $\leq$ Division. This is a general principle, as formulated in the theorem below.

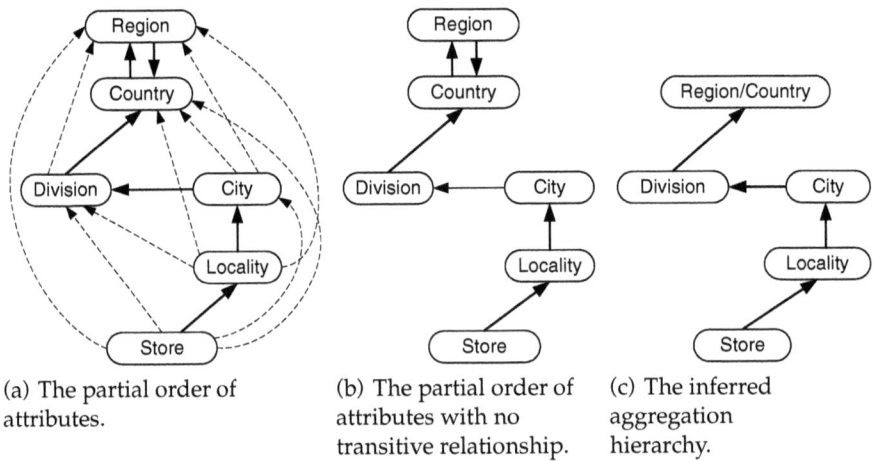

(a) The partial order of attributes.

(b) The partial order of attributes with no transitive relationship.

(c) The inferred aggregation hierarchy.

Fig. 2. Inferring the aggregation hierarchy

Theorem 1. *Given an instance T of a dimension table D and the inferred partial order of its attributes P as derived using Algorithm 1, the partial order of attributes P', and its covering relation, for the schema-defined aggregation hierarchy for D must be subgraph of P.*

Proof. It is possible to obtain a covering relation of the partial order over the attributes from the schema-defined hierarchy. We can also obtain the transitive closure of the partial order over the attributes from its covering relation.

- If the given dimension table is complete – in the sense that all base members are represented, given that all partial order relationships of attributes are captured in Algorithm 1, P' is equivalent to P.
- Otherwise suppose that some of the base members are removed. While some spurious roll-ups are added, none of the partial orders from P is removed. The latter point can be proved easily by the fact that the data from which the inferred partial order is derived is constrained by the schema-defined aggregation hierarchy.   □

From the above proof we have the following corollary.

Corollary 1. *If all base members are present in a given dimension table, the inferred aggregation hierarchy is the same as the schema-defined aggregation hierarchy.*

If not all base members are present in a given dimension table, all roll-up relationships in the schema-defined hierarchy are present or implied (by transitive roll-ups) in the inferred aggregation hierarchy.

  The inferred aggregation hierarchies derived for independent data marts are used to guide the process of integration. Intuitively two dimensions of autonomous data marts are compatible if their common information is consistent. Compatible dimensions are important for OLAP operations across multiple data marts. Another important issue to consider in data mart integration is fact summarizability [16,19,21]. Essentially the summarizability ensures the correctness of aggregations.

Definition 7. *Given two levels l_1 and l_2, and summary values for l_1, the roll-up $\rho = l_1 \preceq l_2$ is summarizable if using ρ yields correct summary values for l_2.*

Using either type of the aggregation hierarchies, summary values for each level can be computed by summing the values at the lower level. This is due to the fact that the roll-up relationships either constrain the data or are inferred from data. Consequently, using either of them guarantees the summarizability.

5   Integration of Matching Dimensions Using Inferred Aggregation Hierarchies

In this section, we discuss the properties of integration of matching dimensions using IAHs. Especially, we relate such properties with the integration using SAHs. We show that inferred aggregation hierarchies ensure the correctness of checking compatible dimensions and the summarizability of integrated facts.

5.1   Properties of Compatible Dimensions

To successfully integrate data marts, matching dimensions [1] must be compatible and the common information of the matching dimensions is used for OLAP operations after integration. A matching for two dimensions is a one-to-one injective partial mapping between levels of the two dimensions. According to Torlone [23], for a matching to be fully compatible, it must have three properties: i) *soundness*: the matching levels must have identical member values; ii) *coherence*: aggregation hierarchies must have matching roll-up relationships corresponding to their matching levels; iii) *consistency*: the data after the integration must be consistent with the original aggregation hierarchies. The tests for *coherence* and *consistency* achieve the same objective. The former uses schema information only and the latter uses the schemas and instances. A fully compatible matching ensures that the data after integration of dimensions with matching levels is summarizable.

Example 1. Fig. 3(a) shows a schema-defined matching between dimensions Store(Region, Country, Division, City, Locality, Store) and Shop(Country, Area, City, Suburb, Shop) and their matching levels. Suppose that the instance for the Store and Shop dimensions are as shown in Table 1. The matching is sound, coherent and consistent.

An important issue to consider is to ensure the summarizability of facts along the dimensions after integration. We have the following observation.

Theorem 2. *Integration based on coherent and consistent matchings of dimensions ensures summarizability.*

---

[1] The problem of identifying matching dimensions is beyond the scope of this paper.

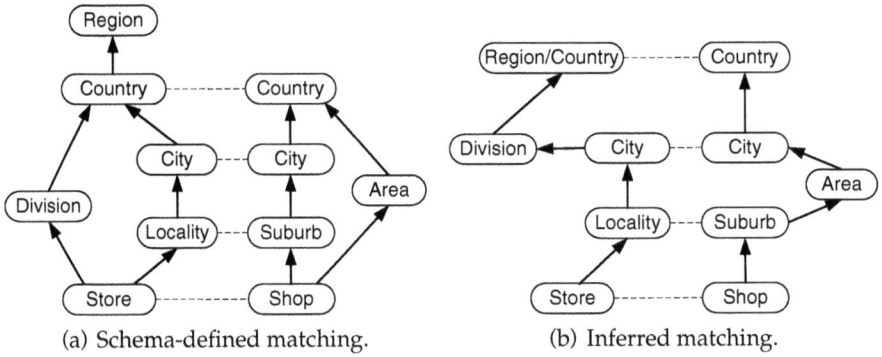

(a) Schema-defined matching.   (b) Inferred matching.

Fig. 3. Matchings between the Store and Shop dimensions

Proof. Following Definition 7, the roll-up relationships ensure the summariz-ability of facts along the dimensions before integration. The coherence and consistency for matching levels further ensure that the data after integration conforms to the original roll-up relationships. As a result, the facts after inte-gration are summarizable.   □

Let a matching defined using inferred hierarchies be called an inferred match-ing. Similar to a matching defined using schema-defined hierarchies, an in-ferred matching between dimensions comprises a set of one-to-one mappings between levels of the dimensions.

Soundness of inferred matchings: Testing the soundness of a matching be-tween (schema-defined or inferred) dimensions involves only checking that for a member of a level, its corresponding member in the mapped level is the same. For example, with the matching from Locality of the Store dimension to Suburb of the Shop dimension, the set of members of Locality are mapped to members of Suburb in one-to-one manner. Similarly the mapping of mem-bers for other levels is straightforward. Because checking the soundness of a matching does not depend on the rollup relationship between levels, obviously testing for soundness of a matching using inferred aggregation hierarchies is the same as that using given schema aggregation hierarchies.

    We will explain in the next two sections that testing for coherence and con-sistency for integration of dimensions using inferred aggregation hierarchies is also feasible.

5.2 The Coherence of Inferred Matchings

When inferred aggregation hierarchies are the same as the schema defined hi-erarchies, matchings defined for inferred hierarchies are the same as those de-fined for schema defined hierarchies. However, due to the incomplete data for dimensions, the inferred aggregation hierarchy for a dimension may contain spurious roll-up relationships that are not defined in the schema-defined hierarchy. As a result, a matching defined for inferred aggregation hierarchies may be different from the matching defined for schema-defined hierarchies.

True coherence: The coherence of an inferred matching defined on a set of levels is true coherence if the matching on these levels for the schema-defined matching is also coherent. This is present when the non-spurious as well as the spurious roll-ups (if any) over matching levels are the same in both inferred hierarchies.

Example 2. Based on the sample data for dimensions Store and Shop, the inferred aggregation hierarchies and their matching is shown in Fig. 3(b). The matching is sound and coherent. In comparison with the matching for schema-defined hierarchies shown in Fig. 3(a), this matching using inferred hierarchies is truly sound and coherent.

True incoherence: The incoherence of an inferred matching defined on a set of levels is true inconsistence if the matching on these levels for the schema-defined matching is also incoherent. This is present when the inferred hierarchies are the same as schema-defined hierarchies and/or the spurious roll-ups are different between the inferred hierarchies.

False coherence: The coherence of an inferred matching on a set of levels is false coherence if the matching defined on these levels for the schema-defined hierarchies is not coherent. False coherence is present when there are some spurious roll-up relationships in one of the inferred hierarchies that also exist in the other inferred hierarchy but as non-spurious roll-ups. In this case the matching is in indeed coherent for the dimension tables from which the hierarchies are inferred.

Example 3. The matching on the schema-defined hierarchies shown in Fig. 4(a) is incoherent. But the inferred matching shown in Fig. 4(b) is coherent. The inferred matching is a false coherent matching. This is made possible because of the spurious roll-up Suburb $\preceq$ Area.

False incoherence: The incoherence of an inferred matching on a set of levels is false incoherence if the matching defined on these levels for the schema-defined hierarchies is coherent.

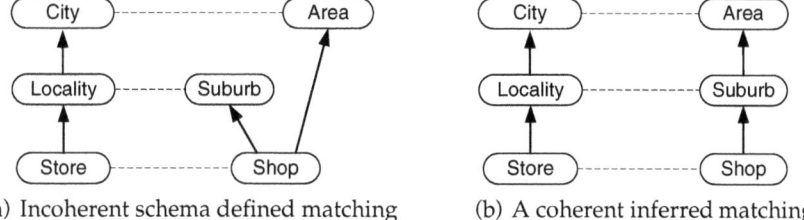

(a) Incoherent schema defined matching   (b) A coherent inferred matching

Fig. 4. A false coherent matching that uses inferred hierarchies

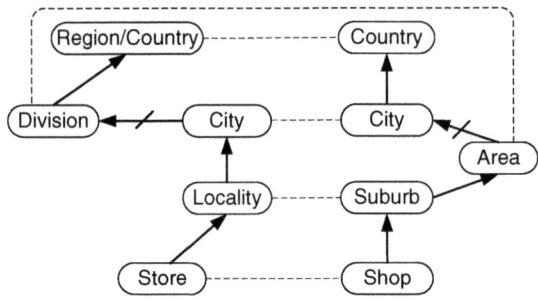

Fig. 5. False incoherent inferred matching

False incoherence is present when the spurious roll-up relationships relate the matching levels differently or some spurious roll-up relationships are missing in one of the inferred hierarchies.

Example 4. Suppose that `Division` and `Area` were matching levels in Fig. 3(a). In this case, the schema-defined matching between `Store` and `Shop` dimensions remains coherent. However, the inferred matching is incoherent. Fig. 5 shows spurious roll-ups (↛) relating matching levels differently.

5.3   The Consistency of Inferred Matchings

True consistency: The consistency of an inferred matching on a set of levels is true consistency if the matching defined on these levels for the schema-defined hierarchies is also consistent.

Example 5. Based on Table 1, the result of the integration of the two sample data for `Store` and `Shop` satisfy the constraints in the schema-defined and inferred hierarchies for these two dimensions.

True inconsistency: The inconsistency of an inferred matching on a set of levels is true inconsistency if the matching defined on these levels for the schema-defined hierarchies is also inconsistent.

Example 6. If the base members for `Store` included the following tuple:

`<Asia Pacific,Australia,Div3,Melbourne,Chelsea,st8>`

and the base members for `Shop` included the following tuple:

`<UK,London,WC,Chelsea,st9>`

then using the schema-defined hierarchies, they would be coherent but not consistent. The reason is that the data after integration does not reflect the roll-up Suburb $\preceq$ Country (for Shop) and Locality $\preceq$ Country (for Store) in the original hierarchies. The matching using inferred hierarchies would be also inconsistent if the two tuples were also present in the sample data in Table 1.

Table 2. True inconsistency: sample data for Store and Shop dimensions

Region	Country	Division	City	Locality	Store
Asia Pacific	Australia	Div1	Sydney	Ryde	st1
Asia Pacific	Australia	Div1	Sydney	Ryde	st2
Asia Pacific	Australia	Div1	Melbourne	Epping	st3
Asia Pacific	Australia	Div1	Melbourne	Morang	st4
Asia Pacific	Australia	Div1	Melbourne	Brighton	st5
Asia Pacific	Australia	Div2	Geelong	Hill	st6
Asia Pacific	Australia	Div3	Melbourne	Chelsea	st8

Country	City	Area	Suburb	Shop
Australia	Sydney	NE	Ryde	st1
Australia	Sydney	NE	Ryde	st2
Australia	Melbourne	NT	Epping	st3
Australia	Melbourne	NT	Morang	st4
Australia	Melbourne	SW	Brighton	st5
Australia	Geelong	NW	Hill	st6
UK	London	WC	Chelsea	st9

False consistency: The consistency of an inferred matching on a set of levels is false consistency if the matching defined on these levels for the schema-defined hierarchies is not consistent. In this case the matching is indeed consistent for the dimension tables from which the hierarchies are inferred.

Example 7. Using Example 6, if the base members included the additional tuples but were not present in the sample data in Table 1, then the data after integration would reflect the original hierarchies and the inferred matching would remain consistent.

False inconsistency: The inconsistency of an inferred matching on a set of levels is false inconsistency if the matching defined on these levels for the schema-defined hierarchies is consistent.

Example 8. If the sample data for Shop dimension in Table 1 included the tuple <Australia,Sydney,NT,Epping,st3> in place of <Australia,Melbourne,NT,Epping,st3> (resulting in Table 3) then the inferred hierarchy from the data after integration will not be consistent with either of the original hierarchies, since Epping rolls-up to contradicting values in City.

Table 3. False inconsistency: sample data for Store and Shop dimensions

Region	Country	Division	City	Locality	Store
Asia Pacific	Australia	Div1	Sydney	Ryde	st1
Asia Pacific	Australia	Div1	Sydney	Ryde	st2
Asia Pacific	Australia	Div1	Melbourne	Epping	st3
Asia Pacific	Australia	Div1	Melbourne	Morang	st4
Asia Pacific	Australia	Div1	Melbourne	Brighton	st5
Asia Pacific	Australia	Div2	Geelong	Hill	st6

Country	City	Area	Suburb	Shop
Australia	Sydney	NE	Ryde	st1
Australia	Sydney	NE	Ryde	st2
Australia	Sydney	NT	Epping	st3
Australia	Melbourne	NT	Morang	st4
Australia	Melbourne	SW	Brighton	st5
Australia	Geelong	NW	Hill	st6

Although false incoherence and false inconsistency may prevent the integration, what is critical is that where we do proceed with the integration, the result of the integration will be correct and summarizable. Based on the above, use of IAHs to test for compatibility guarantees the summarizability of data after integration and IAHs are viable for testing dimension compatibility for the instances from which the aggregation hierarchies are inferred. Compared to the situation where we are unable to ensure the accuracy of the integration due to the absence of SAHs, this is a significant outcome and a viable solution to the problem. This is summarized in the theorem below.

Theorem 3. *A coherent and consistent inferred matching for dimensions is sufficient but not necessary for the summarizability of integration.*

Proof. A summarizable integration on the matching levels of dimensions must have coherent and consistent inferred matching. But from the above discussions it can be seen that an inferred matching may present as incoherent or inconsistent based on the current dimension instances, but they are indeed coherent and consistent and thus can be integrated and the result is summarizable.   □

6   Related Work

Database integration is a well researched area [4,14,22], however integration of data warehouses and data marts has only attracted interest in recent years [10,23]. Kimball requires the integrating dimension tables to be *conformed*, meaning that they must be identical in terms of their semantics, structure and data [13]. This is clearly restrictive and difficult to achieve when integrating heterogeneous data marts [23].

In the absence of readily available SAHs, we could use a thesaurus to establish relativity of levels by identifying the super-ordination (Holonym) and sub-ordination (Meronym) for each pair of attributes based on their labels and/or data [3,23]. This is however a difficult task as labels and data are very often represented as acronyms. Another option would be to query the user to provide the aggregation hierarchy, but this may not be practical as the user may be accessing data from unfamiliar sources. It must be said that use of additional semantic information may help getting closer to the intended aggregation hierarchies, but the problem remains.

In automating the design of data warehouses based on multidimensional design, Romero et al. [20] have suggested using conceptual representations of the domain ontology in discovering of functional dependencies. In contrast, we aim to infer aggregation hierarchies from an existing database.

Smith et al. [2] have compared aggregation hierarchies with generalization and specialization in the context of class hierarchy. Akoka et al. [5] model aggregation hierarchy by mapping generalization hierarchies in dimensions to UML aggregations. Mózon et al. [17] capture additional semantic information on levels and the relationship between them as part of the definition for levels for aggregation hierarchies, these authors rely on given SAHs.

Several people have investigated the pre-integration requirements for dimensions [8,7,11,13,23] all of which rely on the prior knowledge of the SAH. The work by Torlone [23] offers a comprehensive analysis of these requirements, which we introduced in Section 5.

Coincidentally, functional dependency is similar to the partial order relationship. Discovering functional dependencies to design relational databases has been investigated by several people [9,12,15,20]. The complexity of inferring functional dependency between n levels over p tuples is $m(m\,n\,p\log p + n\,2^{2n})$ [15]. Unlike the roll-up relationship between levels, inferring functional dependency is concerned with the composition of attributes on the left as well as on the right hand side of the functional dependency and hence leading to higher order of complexity.

7   Conclusion and Future Work

In this paper we have discussed the problem with existing approaches for testing dimension compatibility. We have proposed a procedure for inferring the levels and the aggregation hierarchy using the current instance of a dimension. We have established the relationship between the schema-defined and inferred aggregation hierarchies as subsumed and subsuming hierarchies. We have established that using inferred hierarchies for testing for dimension compatibility is viable for the instances from which the aggregation hierarchies are inferred. In our future work we will test the theoretical aspect and algorithms introduced here in a semi-automated integration of data marts using real data.

References

1. Aho, A., Garey, M., Ullman, J.: The transitive reduction of a directed graph. SIAM Journal on Computing 1(2), 131–137 (1972)
2. Akoka, J., Comyn-Wattiau, I., Prat, N.: Dimension hierarchies design from uml generalizations and aggregations. In: Kunii, H.S., Jajodia, S., Sølvberg, A. (eds.) ER 2001. LNCS, vol. 2224, pp. 442–455. Springer, Heidelberg (2001)
3. Banek, M., Boris, V., Tjoa, A., Skocir, Z.: Automating the schema matching process for heterogeneous data warehouse. In: Song, I.-Y., Eder, J., Nguyen, T.M. (eds.) DaWaK 2007. LNCS, vol. 4654, pp. 45–54. Springer, Heidelberg (2007)
4. Batini, C., Lenzerini, M., Navathe, S.B.: A comparative analysis of methodologies for database schema integration. ACM Comput. Surv. 18(4), 323–364 (1986)
5. Bever, M., Ruland, D.: Aggregation and generalization hierarchies in office automation. In: Proceedings of the ACM SIGOIS and IEEECS TC-OA 1988 conference on office information systems, pp. 250–264. ACM, New York (1988)
6. Teradata BusinessObjects. Data mart consolidation and business intelligence standardization (2007), http://www.businessobjects.com/pdf/investors/data_mart_consolidation.pdf
7. Cabibbo, L., Torlone, R.: Dimension compatability for data mart integration. In: Proceedings of the 12th Italian Symposium on Advanced Database Systems, Universit'a degli studi Roma Tre, pp. 6–17. Dipartimento di Informatica e Automazione (2004)
8. Cabibbo, L., Torlone, R.: Integrating heterogeneous multidimensional databases. In: SSDBM 2005: Proceedings of the 17th international conference on Scientific and statistical database management, pp. 205–214. Lawrence Berkeley Laboratory, Berkeley (2005)
9. Carpineto, C., Romano, G., d'Adamo, P.: Inferring dependencies from relations: a conceptual clustering approach. International Journal of Intelligent Systems 15, 415–441 (2009)

10. Critchlow, T., Ganesh, M., Musick, R.: Automatic generation of warehouse media-tors using an ontology engine. In: Proceedings of the 5th International Workshop on Knowledge Represenation Meets Databases (KRDB 1998). CEUR Workshop Pro-ceedings, vol. 10, pp. 8-1–8-8 (1998)
11. Grossmann, W., Moschner, M.: Knowledge integration from multidimensional data sources. In: Moreno Díaz, R., Pichler, F., Quesada Arencibia, A. (eds.) EUROCAST 2007. LNCS, vol. 4739, pp. 345–351. Springer, Heidelberg (2007)
12. Kantola, M., Mannila, H., Räihä, K., Siirtola, H.: Discovering functional and inclu-sion dependencies in relational databases. International Journal of Intelligent Sys-tems 7, 591–607 (2007)
13. Kimball, R., Ross, M.: The Data Warehouse Toolkit. Wiley Computer Publishing, Chichester (2000)
14. Lenzerini, M.: Data integration: a theoretical perspective. In: PODS 2002: Proceed-ings of the twenty-first ACM SIGMOD-SIGACT-SIGART symposium on Principles of database systems, pp. 233–246. ACM, New York (2002)
15. Mannila, H., Räihä, K.: Algorithms for inferring functional dependencies from rela-tions. Data Knowl. Eng. 12(1), 83–99 (1994)
16. Mazón, J., Lechtenbörger, J., Trujillo, J.: A survey on summarizability issues in mul-tidimensional modeling. Data Knowl. Eng. (2009)
17. Mazón, J., Trujillo, J.: Enriching data warehouse dimension hierarchies by using se-mantic relations. In: Bell, D.A., Hong, J. (eds.) BNCOD 2006. LNCS, vol. 4042, pp. 278–281. Springer, Heidelberg (2006)
18. ORACLE,
    http://download.oracle.com/docs/cd/B19306_01/server.102/b14200/
    statements_5006.htm
19. Rafanelli, M., Shoshani, A.: Storm: a statistical object representation model. In: SS-DBM V: Proceedings of the fifth international conference on Statistical and scientific database management, pp. 14–29. Springer, New York (1990)
20. Romero, O., Calvanese, D., Abelló, A., Rodríguez-Muro, M.: Discovering functional dependencies for multidimensional design. In: DOLAP '09: Proceeding of the ACM twelfth international workshop on Data warehousing and OLAP, pp. 1–8. ACM, New York (2009)
21. Lenzand, H., Shoshani, A.: Summarizability in olap and statistical data bases. In: SSDBM 1997: Proceedings of the Ninth International Conference on Scientific and Statistical Database Management, pp. 132–143. IEEE Computer Society, Washington (1997)
22. Templeton, M., Henley, H., Maros, E., Van Buer, D.: Interviso: dealing with the com-plexity of federated database access. The VLDB Journal 4(2), 287–318 (1995)
23. Torlone, R.: Two approaches to the integration of heterogeneous data warehouses. Distrib. Parallel Databases 23(1), 69–97 (2008)

Schema Design Alternatives for Multi-granular Data Warehousing

Nadeem Iftikhar and Torben Bach Pedersen

Aalborg University, Department of Computer Science, Selma Lagerløfs Vej 300,
9220 Aalborg Ø, Denmark
{nadeem,tbp}@cs.aau.dk

Abstract. Data warehousing is widely used in industry for reporting and analysis of huge volumes of data at different levels of detail. In general, data warehouses use standard dimensional schema designs to organize their data. However, current data warehousing schema designs fall short in their ability to model the multi-granular data found in various real-world application domains. For example, modern farm equipment in a field produces massive amounts of data at different levels of granularity that has to be stored and queried. A study of the commonly used data warehousing schemas exposes the limitation that the schema designs are intended to simply store data at the same single level of granularity. This paper on the other hand, presents several extended dimensional data warehousing schema design alternatives to store both detail and aggregated data at different levels of granularity. The paper presents three solutions to design the time dimension tables and four solutions to design the fact tables. Moreover, each of these solutions is evaluated in different combinations of the time dimension and the fact tables based on a real world farming case study.

Keywords: Multi-granular data, data warehousing, data warehousing schema design.

1 Introduction

Data warehousing is commonly used by companies to store and analyze data. Data warehouses normally use dimensional schema designs to structure their data. A dimensional model consists of facts and dimensions. A fact is a single measurable record. A dimension, which is mostly textual and static in nature, is utilized to describe the fact. For example, in farm equipment related data; *parameters* are logged against *tasks* at certain *times* with certain *values*. A typical fact would be a *farming_equipment_data_log* with the value, for instance tractor speed as the measure, the task logging the parameter, the parameter being logged and the time of logging as the dimensions. The grain is thus *parameter per task per time per value*.

In data warehousing research, most of the work has concentrated on the performance and less work is done on schema design models [1]. A schema design model for data warehousing should has the ability to model the *multi-granular* data found in

P. García Bringas et al. (Eds.): DEXA 2010, Part II, LNCS 6262, pp. 111–125, 2010.

various real-world application domains. The leading schema designs for data warehousing, which are represented in Section 2, are evaluated against the multi-granular data storage requirement, and it is shown that none of these models possess the ability to store data at different levels of granularity. In this paper, we present and evaluate various schema design alternatives to structure both the detail and the aggregated data at different levels of granularity. The proposed alternatives consist of several combinations of the time dimension and the fact tables design approaches. In this work, the following two possible scenarios are considered.

- The detail data has a single level of granularity. To save storage and to enhance query performance, it is aggregated gradually to multiple levels of granularity. For example, in the farming business, the farm equipment related data is initially logged at *per parameter per task per second level,* when the data is older than three months it is aggregated to *per parameter per task per minute level,* next the data that is older than six months is aggregated to *per parameter per task per 2 minutes level* and so on. Thus in the example, the data will have different levels of granularity (second, minute, 2 minutes).

- Both the detail and the aggregated data consist of multiple levels of granularity. For instance, in extension to the above mentioned farming example, the data is initially logged at different levels of granularity for the reason that different parameters have different sampling frequencies. Some parameters are logged at *per parameter per task per second level* and some at *per parameter per task per minute,* followed by further higher levels of granularity due to gradual aggregation.

To the best of our knowledge, this paper is the first to present and evaluate schema design alternatives for multi-granular data warehousing. The paper is structured as follows. Section 2 presents the real-world farming case study, evaluates the standard schema design models against the requirements and explains the motivation behind the proposed schema design models. Section 3 defines the time dimension design solutions and Section 4 defines the fact table design solutions. Section 5 evaluates the proposed approaches in combinations with different time dimension and fact tables. Section 6 presents the related work. Finally, Section 7 summarizes and points to the future research.

2 Motivation

This section presents a real-world case study based on the farming business. The case study is a result of LandIT [2] that was an industrial collaboration project about developing technologies for integration, aggregation and exchange of data between farming devices and other farming-related IT systems, both for operational and business intelligence purposes. This section then discusses multi-granular data storage, both at the detail and the aggregated levels, according to the needs of the application domain and highlights the inappropriateness of the standard data warehousing design approaches to handle multi-granular data.

The main goal of this case study is to store and query data at different levels of detail or granularity, according to the needs of the application domain. We use the case study to illustrate the kind of challenges faced by storing multi-granular data, which are addressed by this paper. This case study concerns farm equipment related data in a field. The data is initially logged in order to comply with environmental regulations and is kept in detailed format at a single level of granularity or different levels of granularity, depending on the application domain. Further, as the data ages the detailed data may no longer be useful; however, the data could not be deleted due to the organizational and/or governmental data retention policies. This potentially excessive data volume would affect storage as well as query processing capacity of the data warehouse. Thus, the goal is to reduce the data by aggregating it. The fact table attributes used in this case study are: *Pid (Parameterid), Tkid (Taskid) Teid (Timeid)* and *Value*. The Task represents activities to distinguish all the work that is carried out by a contractor for a farmer in a particular field of a farm. The Parameter represents a variable code for which a data value is recorded. For example, parameter 247 represents the amount of chemical sprayed in liters; parameter 248 represents distance covered by a tractor in kilometers; parameter 1 represents tractor speed in km/h; parameter 41 represents area sprayed in hectares. Moreover, each parameter has a different data logging frequency, for example the logging frequency of parameter 1 and 247 are every 1 second, the logging frequency of parameter 41 and 248 are every 60 seconds and so on. The Time represents an instance of time when a data value is recorded. Lastly, the value is simply a numeric attribute. In order to list some example data, we used farm equipment related data. The snapshot of data consists of one task and four parameters with two different levels of time granularities. The data at the detailed level is represented as *(247, 10, 20, 13.44); (1, 10, 20, 11.20); (247, 10, 21, 13.57); (1, 10, 21, 11.20)...(247, 10, 80, 15.73); (1, 10, 80, 11.00); (248, 10, 80, 2.51); (41, 10, 80, 0.13)*. In this example data, parameter 247 and parameter 1 have granularity (logging frequency) equal to "second" and parameter 248 and parameter 41 have granularity equal to "minute". Each set of entries within parentheses represent a row. For instance, row number 1 reads as follows: Pid=247 (represents: amount of fertilizer used), Tkid=10 (represents: unique id for a task), Teid=20 (represents: time at a second granularity level 12.04.2007 16:25:01) and Value=13.44 (represents: current value of fertilizer used in liters). Further, the detailed data could be aggregated from the second level to the minute level if it is more than three months old. In that case, the data at the minute aggregated level is represented as *(247, 10, 80, 15.73); (1, 10, 80, 11.10); (248, 10, 80, 2.51); (41, 10, 80, 0.13)*. Note that parameters 248 and parameter 41 are not aggregated, since they are already at the minute granularity level, although, for parameter 247 and parameter 1 the aggregate functions MAX and AVG are used, respectively. Furthermore, the data could be aggregated from the minute level to the 2 minutes if it is more than six months old and so on. Two leading approaches to store data in a data warehouse – the star schema [3] approach and the snowflake schema [3] approach have been selected for analysis.

The star schema approach regarding multi-granular data is unable to represent data at different levels of detail due to the fact that it requires all the dimension attributes, to be given concrete values. For example, if a time dimension has the following attributes in the descending order: *Y (Year), M (Month), D (Day), H (Hour), M (Minute)* and *S (Second)*. In the case of data with uniform granularity, say at a second level, everything works well, where all the attributes ranging from the second to the year will be assigned values. However, in the case of multi-granular data, the levels of granularity for some data may vary; say to a minute level or even higher; however, the dimension table still requires the value for the second, since all the levels are mandatory, as shown below.

```
TIME(Teid, Y, M, D, H, M, S);
FACT(Pid, Tkid, Teid, Value);
```

Similarly, snowflake schema-based time dimensions are also unable to handle multi-granular data. Even though the association between a fact table and time dimensions is based on a separate dimension for each time granularity level (parent-child), however, the lowest granularity dimension (second) is pre-linked with the fact table. The following time dimension example represents a normalized dimensional structure based on a parent-child relationship. As structure is defined at the design level, it is not possible to make any changes to handle the data with granularity levels higher than second.

```
YEAR(Yid, Y); MONTH(Mid, M, Yid);
DAY(Did, D, Mid); HOUR(Hid, H, Did);
MINUTE(Mid, M, Hid); SECOND(Sid, S, Mid);
FACT(Pid, Tkid, Sid, Value);
```

In conclusion, based on the above mentioned case study, the existing data warehousing design solutions are not appropriate to handle multi-granular data both at the detail and the aggregated levels. For that reason, a number of schema design alternatives to handle multi-granular data are proposed and evaluated in this paper. We first propose time dimension design solutions followed by fact table design solutions, as we can consider the time dimension independently and not the vice versa. Furthermore, the reason to consider the time dimension only, rather than any other dimension(s), is that different levels of granularity always occur for time and the solution also works for other granularity phenomena.

3 Time Dimension Design Solutions

In this section, we describe several solutions for designing time dimension tables for multi-granular data warehousing. These time dimension tables may later be used in different combinations with fact tables presented in Section 4. In general, levels in dimension hierarchies can be arranged in various alternate orders. The one exception to this is the time dimension. In the time dimension, the hierarchal levels must follow a certain order, from the smallest value (a year) to the largest value (a second). The time dimension design solutions presented in this paper are based on single hierarchy (Fig. 1a) that is further composed of numerous one-to-many relationships. The proposed solutions can easily be applied to the fact tables having time as a dimension, in order to specify instances of fact table at different levels of granularity. In fact, three specific design solutions are presented for the time dimension, namely *the extended de-normalized time dimension solution, the extended normalized time dimension solution* and the *extended shrunken time dimension solution*. In the extended de-normalized time dimension solution, all the granularities are handled by a single dimension table. In the extended normalized time dimension solution, the granularities are handled by multiple dimension tables. Finally, in the extended shrunken time dimension solution, the granularities are handled by shrunken [4] dimension tables. These solutions are described in detail in the following subsections.

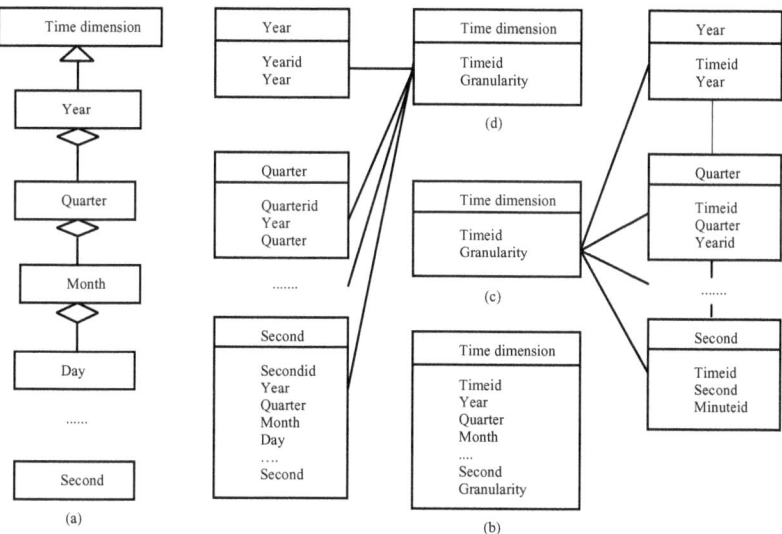

Fig. 1. Time dimension: (a) UML diagram, (b) Extended de-normalized time dimension, (c) Extended normalized time dimension and (d) Extended shrunken time dimension

3.1   The Extended De-normalized Time Dimension Solution

In the extended de-normalized time dimension, presented in Fig 1b, the standard star schema has been expanded with a new attribute *Granularity* to handle granularity at different levels rather than at a single level. The attribute granularity represents the level of detail of each time instance stored in the time dimension. Thus, with the inclusion of this new attribute in the time dimension table the associated fact tables are no longer restricted to store data only at the same single level of granularity. The snapshot of the time instances, presented in Table 1 is the result of the proposed extended de-normalized solution. In the snapshot, *Teid (Timeid)* is an auto generated primary key attribute. *Y (Year), Q (Quarter), M (Month)* and *D (Day)*, represent the standard calendar. For instance, row number 4 has a granularity at the day level and it is read as follows: year = 2009, 3rd quarter, month of August and day in month = 12. The level of granularity of the row is at day level for this reason rest of the attributes have NULL values. Further, *POD (Partofday)* represents an eight-hour time span. Hence each day, is divided into three eight-hour time spans, from 0000 to 0800 hours, 0800 to 1600 hours and 1600 to 0000 hours. The symbol shown in the time dimension for Partofday is Day[8-16[, which means that it is the second time span of the day, where 0800 hours are included but 1600 hours are excluded. 1600 hours will be included in the next time span. Furthermore, *4H (4Hour)* represents a four-hour time span, whereas, *H (Hour)* represents the standard hour in 24 hours time format. Moreover, *20M (20Minute), 10M (10Minute)* and *2M (2Minute),* represents twenty-minute, ten-minute and two-minute time spans. *M (Minute)* and *S (Second)* represent the standard minute and second. Finally, *Gran (Granularity)* represents the level of granularity of each row. One of the advantages of the proposed approach is: simple to use and to understand. However, one of the main disadvantages is redundancy, repeated attribute values especially null values.

Table 1. Snapshot of the extended de-normalized time dimension table proposed in our work

Teid	Y	Q	M	D	POD	4H	H	20M	10M	2M	M	S	Gran
1	2009	Null	Null	Null	Null	Null	Null	Null	Null	Null	Null	Null	Y
2	2009	3	Null	Null	Null	Null	Null	Null	Null	Null	Null	Null	Q
3	2009	3	8	Null	Null	Null	Null	Null	Null	Null	Null	Null	M
4	2009	3	8	12	Null	Null	Null	Null	Null	Null	Null	Null	D
5	2009	3	8	12	Day[8-16[	Null	Null	Null	Null	Null	Null	Null	POD
6	2009	3	8	12	Day[8-16[	[12-16[	Null	Null	Null	Null	Null	Null	4H
7	2009	3	8	12	Day[8-16[	[12-16[	15	Null	Null	Null	Null	Null	H
8	2009	3	8	12	Day[8-16[	[12-16[	15	[20-40[	Null	Null	Null	Null	20M
9	2009	3	8	12	Day[8-16[	[12-16[	15	[20-40[	[20-30[	Null	Null	Null	10M
10	2009	3	8	12	Day[8-16[	[12-16[	15	[20-40[	[20-30[	[24-26[	Null	Null	2M
11	2009	3	8	12	Day[8-16[	[12-16[	15	[20-40[	[20-30[	[24-26[	25	Null	M
12	2009	3	8	12	Day[8-16[	[12-16[	15	[20-40[	[20-30[	[24-26[	25	40	S

3.2   The Extended Normalized Time Dimension Solution

In the extended normalized time dimensions, presented in Fig. 1c, the standard snowflake schema has been modified to handle granularity at different levels. Like the standard snowflake schema, it consists of separate tables for Year, Quarter, Month, Day, Partofday, 4hour, Hour, 20minute, 10minute, 2minute, Minute and Second. This implies that each granularity level is represented in a separate table with its attributes. Furthermore, unlike the standard snow flake schema, the inclusion of a *primary time dimension* table and an extra attribute Granularity (Fig. 1c), allows the primary time dimension table to point to any level of granularity rather than to the lowest level only (which is the main problem with the snowflake schema in order to handle different levels of granularity). On the whole, similar to any normalized dimensional schema, each lower level dimension table provides supplementary information about the dimension table at a higher level of the hierarchy. The relational schema presented below is the result of the proposed extended normalized solution. The Time is the primary dimension table that is further linked to all of the dimension tables. In the primary Time dimension table, Teid (Timeid) is an auto generated primary key attribute and Gran (Granularity) represents the level of details related to this Timeid. Further, each lower level dimension table is also associated with higher level dimension table through their primary and foreign key combination. Some of the advantages of the normalized time dimensional approach are less storage space requirement due to the elimination of redundancy and easy maintenance due to the smaller nature of dimension tables. However, one of the main disadvantages is multiple join operations when querying, which may trigger performance and navigation concerns.

```
TIME(Teid, Gran);
YEAR(Teid, Y);
QUARTER(Teid, Q, Yid);
....
MIN(Teid, M, 2Mid);
SEC(Teid, S, Mid);
```

3.3 The Extended Shrunken Time Dimension Solution

In the extended shrunken time dimensions, presented in Fig 1d, the standard shrunken based dimension model has been enhanced by adding the primary time dimension table in order to handle granularity at different levels. The shrunken dimensions are subsets of the main dimensions. Each shrunken dimension is stored in a separate de-normalized dimension table with its own primary key. On the whole, this solution mainly consists of the primary Time dimension table and the multiple de-normalized shrunken dimension tables. The primary Time dimension table consists of an extra attribute Granularity to navigate the subsets of the time dimension tables when querying. Like the previous solution (sub-section 3.2), each subset of the Time dimension tables is linked with the primary Time dimension table in order to keep track of the level of granularity, however, unlike the previous solution each lower level dimension table is not connected with the dimension table at a higher level because of the de-normalized nature of the sub-dimension tables. The relational schema presented below is the result of the proposed extended shrunken solution. The Time dimension table is the primary dimension table that is further linked to all of the dimension tables. In the time dimension table, Teid (Timeid) is an auto generated primary key attribute and Gran (granularity) represents the level of details related to this Timeid. One of the advantages of the proposed approach is: fewer joins when querying. However, some of the main disadvantages are: complex navigation when querying and maintenance due to the existence of numerous sub-dimensions.

```
TIME(Teid, Gran);
YEAR(Yid, Y);
QUARTER(Mid, Y, Q);
....
MIN(Mid, Y, Q, M, D, POD, 4H, H, 20M, 10M, 2M, M);
SEC(Sid, Y, Q, M, D, POD, 4H, H, 20M, 10M, 2M, M, S);
```

4 Fact Table Design Solutions

In this section, we describe several solutions for designing fact tables for multi-granular data warehousing. These fact tables may later be used in different combinations with the time dimension tables presented in Section 3. In general, the fact table stores measures. The measures are factual data about the subject and normally contain numeric values. For instance, in the case study presented in Section 2, the measure is represented as value. The value represents quantitative data about farming equipment in the fields. For example, tractors speed in km/h. Further, each such measure is associated with the value at the lowest level of each dimension table involved. This association forces the fact

table to store data only at the lowest (same) level of granularity. To resolve this problem, four fact table design solutions are presented, namely *the single fact table solution,*
the multiple fact tables solution, the hybrid solution and *the single fact table without*
separate time dimension solution. In the single fact table solution, both detailed and
aggregated data (multiple granularities) are handled by a single table. In the multiple
fact tables solution differing data granularities are handled separately, one fact table for
each granularity level. Further, in the hybrid solution, the granularities are handled by
two fact tables; detail data is handled by one fact table, while the aggregated data is
handled by another fact table. Finally, in the single fact table without separate time
dimension solution all the granularities are handled by a single fact table with time as an
attribute rather than a separate dimension. In summary, the standard fact tables handle
only same single granularity, whereas, the proposed fact table design solutions along
with the associated proposed time dimension table solutions have the ability to specify
instances of a fact table at different levels of detail. These solutions are further described
in detail in the following subsections.

4.1 The Single Fact Table Solution

In this solution, multiple data granularities are handled by a single fact table. In the
case study presented in Section 2, each row of the fact table contains farming equipment related value for each parameter associated to each task each time. Thus, in this
case the grain [3] of the fact table is expressed as "value by parameter by task by
time" (note that time can be of any granularity ranging from the second to the year).
In the existing fact table design solutions, the grain refers to the single level of detail
of the information stored in each row of the fact
table, however in this solution, differing data
granularities are handled by adding "granularity"
as a column in the time dimensions, presented in
Section 3. The fact table (Table 2) has one measure Value. It has a primary key consisting of the
combined foreign keys (referencing the Parameter,
Task and Time dimensions respectively). The fact
table contains one row for every parameter logged
to each task each time. For example, the fact table
contains three different granularity levels (second,
minute and 2 minutes) and it is flexible enough to
hold both finer and coarser levels of granularities.
In Table 2, Teid (Timeid) 1200 and 1201 represent
granularity at a second level, 1260 represents

Table 2. Single fact table

Pid	Tkid	Teid	Value
1	29	1200	11.10
247	29	1200	13.44
1	29	1201	11.10
247	29	1201	13.57
..	..	..	..
248	29	1260	2.46
41	29	1260	0.11
..	..	..	
248	10	3020	5.42
1	10	3020	10.82
..	..	..	..

granularity at a minute level and Teid 3020 represents granularity at 2 minutes level.
Moreover, row number 1 (first row) in the fact table reads as follows: Pid=1 (tractor
speed), Tkid = 29 (unique id for a task), Teid=1200 (time at 1 second level
12.04.2007 16:25:01) and Value=11.10 (current value of the speed in km/h). Similarly, row number 7 (second last row) represents aggregated data and reads as follows: Pid=248 (distance covered by a tractor in km/h), Tkid=10 (a previous task),
Teid= 3020 (time at 2 minutes level 12.01.2006 13:10:00) and Value=5.42 (aggregated value of area sprayed). Some of the advantages of this solution are: simple

maintenance, transparent query processing, faster data loading and lower operational cost. However, some of the disadvantages are: a de-normalized structure and possibility of erroneous summarizability [5]. The erroneous summarizability may occur during higher levels of data aggregation process due to the presence of aggregated data at different levels of granularity. For example, if data warehouse contains aggregated data at the month and the day granularity levels using SUM function. In that case, to calculate the average extra care has to be taken, otherwise the aggregated data will be erroneous.

4.2 The Multiple Fact Tables Solution

In this solution, differing data granularities are handled by using multiple fact tables, one fact table for each granularity level. Each such fact table contains a single measure value, and the grain is expressed as "value by parameter by task by time" (note that time must be of same granularity ranging from the second to the year). In addition, the granularity in each of these multiple fact tables exactly refers to the same level of detail of the information stored in each row of the fact table. For example, fact table I (3a) contains time granularities equal to 1 second, fact table II (3b) contains time granularities equal to 1 minute and so on. The fact tables (Table 3a and 3b) have primary keys consisting of the combined foreign keys (referencing the Parameter, Task and Time dimensions respectively). Further, each of these fact tables contains one row for every parameter logged to each task each time. Some of the advantages of this solution are: smaller and more manageable pieces of data, simple to use and understand, easily manageable and scalable. However, some of the disadvantages are: uneven distribution of data (some fact tables might contain more data than others), high operational cost due to the increased number of physical tables and complex navigation due to large number of joins when querying.

Table 3a. Fact table I

Pid	Tkid	Teid	Value
1	29	1200	11.10
247	29	1200	13.44
1	29	1201	11.10
247	29	1201	13.57
..	..	..	..

Table 3b. Fact table II

Pid	Tkid	Teid	Value
248	29	1260	2.46
41	29	1260	0.11
248	29	1320	2.71
41	29	1320	0.18
..	..	..	..

4.3 The Hybrid Fact Table Solution

In this solution, the granularities are handled by two fact tables with the same set of attributes. The detail data are handled by the first fact table while the aggregated data with different levels of granularity are handled by the second fact table. Further, the detailed data may consist of single or multiple levels of granularity. Comparable to the previous fact table design solutions, each of the two fact tables has a measure Value, and has a primary key consisting of the combined foreign keys (referencing the Parameter, Task and Time dimensions respectively). Thus, the grain of each of the two fact tables is expressed as "value by parameter by task by time" (note that time can be of any granularity ranging from second to the year). Some of the advantages of this solution are: simple to manage, use, access and understand due to the separation of detail and aggregated data. However, some of the disadvantages are: possibility of erroneous summarizability and uneven data distribution between the tables.

4.4 The Single Fact Table without Separate Time Dimension Solution

In this solution, the different levels of granularity are handled by a single fact table without a separate time dimension. The fact table consists of five measures: *Pi (Parameterid), Tkid (Taskid), Ts (Timestamp), L (Label)* and *Value*. Timestamp represents the logging time of the value in UTC [6] format, whereas, Label represents the logging frequency of the value. Label is the level of granularity of each row in the fact table. It is first equal to logging frequency, then to aggregation granularity. For example in Table 4, L = 1 represents that the level of granularity of the concerned rows is equal to one second. Similarly, the L = 60 represents that the level of granularity of the concerned rows is equal to sixty seconds. The fact table has a primary key consisting of the combined foreign keys (referencing the Parameter, Task and Timestamp). It contains one row for every parameter logged to each task. As a result, the grain of the fact table is stated as "value by label by parameter by task by time" (note that time can be of any granularity ranging from second to the year). For example, row number 4 in the fact table reads as follows: Pid=41 (area sprayed in hectares) Tkid=29 (unique id for a task), Ts=1155208120 (time at 1 minute level 10.08.2005 11:08:40) and Value=0.14 (value of area sprayed in hectares). Some of the advantages of this solution are: less storage, fast query processing and efficient data loading and management due to the non-existence of a separate time dimension. However, one of the disadvantages is that this model is not common among data warehouse schema design solutions.

Table 4. Single fact table without separate time dimension

Pid	Tkid	Ts	L	Value
1	29	1155208060	1	11.10
247	29	1155208060	1	2.46
248	29	1155208120	60	12.42
41	29	1155208120	60	0.14
..	..	..	..	..

5 Evaluation

An evaluation of the schema design alternatives has been done with the following combinations of the proposed time dimension and fact tables.

Combination 1 (C1): single fact table without a separate time dimension table
Combination 2 (C2): extended de-normalized time dimension and single fact table
Combination 3 (C3): extended de-normalized time dimension and multiple fact tables
Combination 4 (C4): extended de-normalized time dimension and two fact tables
Combination 5 (C5): extended normalized time dimension and single fact table
Combination 6 (C6): extended normalized time dimension and multiple fact tables
Combination 7 (C7): extended shrunken time dimension and multiple fact tables

Performance tests have been carried out on single-level and multi-level aggregation queries based on the above mentioned combinations. The single-level queries aggregate data from a single level of granularity to a higher level of granularity. For example, to aggregate values from *per parameter per task per second* to *per parameter per task per minute*. The multi-level queries aggregate data from several different levels of granularity to a single higher level of granularity. For instance, to aggregate values from *per parameter per task per second* plus from *per parameter per task per minute* plus from *per parameter per task per 2 minutes* to *per parameter per task per 10 minutes*. The logical design of the fact and time dimension tables proposed in the paper and the queries used in the tests can be viewed at [7]. All the fact and dimension tables have primary and secondary indexes on all the primary and foreign keys, respectively.

The tests were designed to measure the query time in seconds, query complexity with respect to number of lines of code, overall aggregation time in seconds and storage used in MB. The tests were performed on a 2.0 GHz Intel® Core Duo with 512 MB RAM, running Ubuntu 8.04 (hardy) and MySQL 5.0.5. Every test was performed 5 times. The maximum and minimum values are discarded and an average is calculated using the middle three values. From the tests, it is observed that the processing rate of the single-granular queries for the above mentioned combinations is between 110,000 and 300,000 rows per second (Fig. 2). Combination 1, is the best, given that it is based on a single fact table without a separate time dimension table, whereas, rest of the combinations performed almost the same. In comparison, the processing rate of multi-granular queries (across three different levels of granularity) is between 20,000 and 300,000 rows per second (Fig. 3). Combination 1 again shows the best performance, whereas, combination 2 and 4 perform well for the reason that fewer JOIN and UNION clauses are required. Combinations (3, 5, 6 and 7) performed approximately identical since additional JOIN and UNION clauses are required. In conclusion, the multi-granular queries have an overhead of 275%, though; it is reasonable given the flexibility of the proposed schema design combinations to handle multi-granular data. Furthermore, the inclusion of the primary time dimension table, with Timeid and Granularity as attributes, in the extended normalized and shrunken time dimension schema design has matched the performance of the queries (Combination 5, 6 and 7) with Combination 3 that is based on a de-normalized time dimension and multiple fact tables. The primary time dimension table (sub-sections 3.2 and 3.3) acts as a front end table. Each entry in the table contains a search-key (Timeid) and a pointer (Granularity) to the sub-dimension table containing that value.

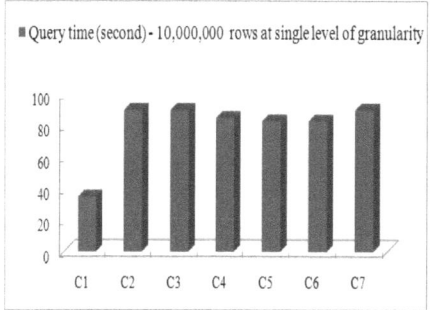

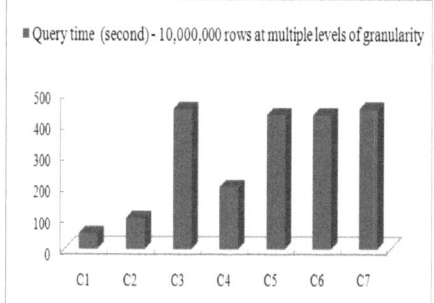

Fig. 2. Single-granular query time   **Fig. 3.** Multi-granular query time

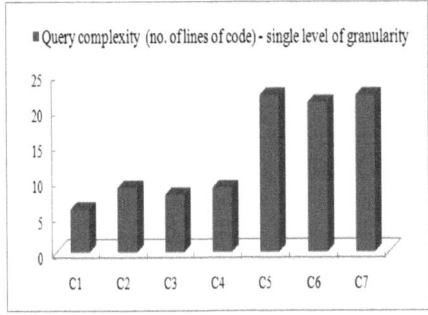

Fig. 4. Single-granular query complexity   **Fig. 5.** Multi-granular query complexity

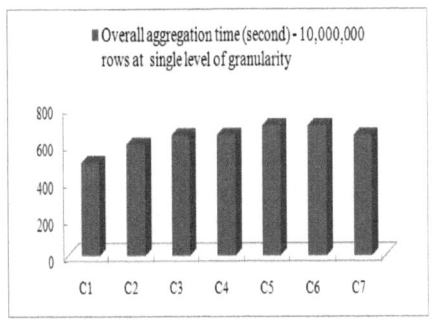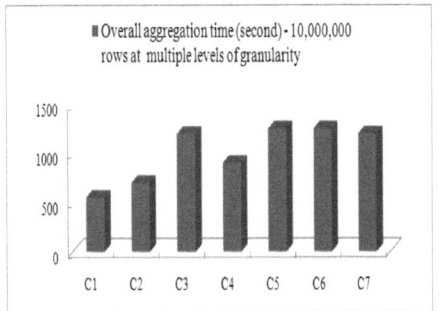

Fig. 6. Single-granular aggregation time   **Fig. 7.** Multi-granular aggregation time

Complexity of the single-granular queries is between 6 and 22 lines of code (Fig. 4) in comparison to approximately between 6 and 69 lines of code (Fig. 5) of the multi-granular queries. The number of lines of code for multi-granular queries are approximately three times higher than for the single-granular queries for the reason that to aggregate data across three different levels of granularity requires at least two UNION and numerous JOIN clauses. The complete aggregation process consists of 1) aggregating the existing rows based on single or different levels of granularity, 2) generating the new rows in the time dimension table(s) in order to point to the higher granularity rows in the fact table(s), 3) inserting the newly aggregated rows in the fact table(s) and 4) deleting the previous rows from the fact table(s). The total time of the aggregation process based on the single granular data performed really well. On average it takes about 650 seconds (Fig. 6) to aggregate 10,000,000 rows from the second granularity level to the minute granularity level, insert approximately 166,000 new aggregated rows and delete approximately 9,840,000 previous rows. Similarly, the total time of the aggregation process based on the multi-granular data (Fig. 7) is also performed reasonably well due to their flexibility in handling multi-granular data. Combination 2 and 4, performed better than the others due to fewer tables, whereas, combination 5 and 6 performed slightly worse than the average due to normalized nature of tables. Finally, the storage used by Combination 1 is approximately 457 MB (Fig. 8), which is the best, whereas, the rest of the combinations used between 1600 to 2000 MB to store 10,000,000 rows. Furthermore, Combination 5 and 6, which is based on normalized extended time dimension tables, demonstrates its worth in terms of storage used, whereas, Combination 7, which is based on shrunken time dimensions tables is slightly worse than average in terms of storage.

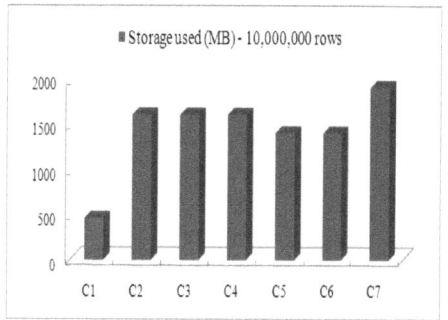

Fig. 8. Storage used

In conclusion, Combination 1 has proved to be the best with respect to query time, query complexity, overall aggregation time and storage used, given that it consists of only a single fact table with time as a "measure" rather than a separate dimension table. Combination 2 and 4 has also performed well in terms of query time, query complexity and over all aggregation speed, since that they are composed of a de-normalized time dimension table and at most two fact tables. Combination 5 and 6 has performed well as compared to rest of the combinations (excluding combination 1) in terms of storage used, for the reason that they consists of normalized time dimension tables. Combination 3 and 7 performed well in terms of query complexity as compared to combinations 4, 5 and 6, as they consists of de-normalized and shrunken time dimension tables, respectively. Based on the above mentioned facts, Combination 1 proved to be the best, however the level of granularity is not explained in the schema for that reason computations may require in queries, thus it may not be the ultimate choice for a data warehouse design solution. Combinations 2 and 4 proved to be the average, however in spite of this they can be the best option for a data warehouse design solution. The other reason is that they are also based on a schema that is similar to the star schema. Combinations 5 and 6 proved to be the worse than the average, given that they are based on a schema that is somewhat similar to the snowflake schema. Finally, Combinations 3 and 7 proved to be the worst, given that they are based on de-normalized and shrunken time dimension and multiple fact tables. Moreover, to the best of our knowledge this is the first paper to conduct a performance evaluation of the schema design alternatives for multi-granular data warehouses.

6 Related Work

Related work exists in several areas. In the context of gradual data aggregation, an efficient tree-based indexing scheme for dynamically and gradually maintaining aggregates is presented in [8]. Aggregates were maintained using multiple levels of temporal granularities. The focal point of this work is to introduce effective indexing schemes for storing aggregated data. A language for specifying a strategy to archive data and keep summaries of archived data in data warehouses is presented by [9]. The main motivation of this work is on archiving and generating aggregates depending on the age of archived data. Further, the semantic foundation for data reduction in dimensional data warehouses that permits the gradual aggregation of detailed data as the data gets older is provided by [10]. The work is purely theoretical and the main direction of this work is on querying multi-dimensional data that is being aggregated gradually. Furthermore, one specific gradual granular data aggregation based solution has been described in [11]. The work proposes an algorithm in order to achieve data aggregation effectively and to store multi-granular data efficiently. Lastly, a number of algorithms are proposed by [12] to enforce summarizability in dimensional hierarchies. The work is relevant in order to achieve summarizabiltiy in the case of dimensional hierarchies proposed in one of our design option. In contrast to most of the above mentioned approaches, the main focus of our work is to provide alternative relational multi-granular data warehousing schema design alternatives to store data at different levels of granularity.

In the context of data warehousing design quality and evaluation, a set of metrics are proposed by [13] to evaluate the quality of alternative design options of the data warehouse. Similarly, one survey paper [14] points out the need for the quality of design process and its consequences in data warehouses. However, both of these papers do not concentrate on data warehousing design alternatives with respect to multi-granular data. In comparison, our work concentrates on evaluating the schema design alternatives extensively both at the conceptual and the logical level with main focus on multi-granular data.

7   Conclusion

Motivated by the reality that data warehousing is widely used in industry for reporting and analysis of huge volume of data at different levels of detail, data warehouse schema design has become a major database research area. The two most popular schema designs are evaluated according to the requirement of storing data at different levels of granularity, and it is shown that both of them fail to satisfy the multi-granular data requirement fully or partially. Instead, we propose extended schema design alternatives, which addresses the multi-granular data storage and query requirements. The extended schema design alternatives have the ability to handle data at different levels of granularity. We present a real-world case study from the farming business, where we store multi-granular data both at the detail and aggregated levels. The schema design alternatives consist of three time dimension and four fact table design options to handle data at different levels of granularity. The proposed time dimension and fact tables are used in seven different combinations to evaluate their performance. The results show that the proposed schema design alternatives are capable of handling multi-granular data, effectively and efficiently. We also show how to represent the multi-granular data as relational tables, thus providing a basis for implementing the model using relational technology. In the future it should be investigated how the model can be efficiently implemented using standard data warehousing technologies. It is also interesting to investigate if the proposed schema design alternatives can be utilized with the help of a user interface. Finally, we believe that it is important to investigate how these schema design alternatives cope with multiple dimensions.

References

1. Samtani, S., Mohania, M., Kumar, V., Kambayashi, Y.: Recent Advances and Research Problems in Data Warehousing. In: Kambayashi, Y., Lee, D.-L., Lim, E.-p., Mohania, M., Masunaga, Y. (eds.) ER Workshops 1998. LNCS, vol. 1552, pp. 81–92. Springer, Heidelberg (1999)
2. LandIT, http://www.tekkva.dk/page326.aspx
3. Kimball, R.: The Data Warehousing Toolkit. Wiley Computer Publishing, NY (2002)
4. Kimball Design, http://www.kimballgroup.com/html/designtipsPDF/KimballDT61HandlingAll.pdf
5. Shoshani, A.: Summarizability. Encyclopedia of Database Sys. 19, 2880–2884 (2009)
6. UTC, http://en.wikipedia.org/wiki/Coordinated_Universal_Time
7. Logical Table Design and Queries, http://www.cs.aau.dk/~nadeem/queries.htm
8. Zhang, D., Gunopulos, D., Tsotras, V.J., Seeger, B.: Temporal and Spatio-Temporal Aggregations over Data Streams using Multiple Time Granularities. Info. Sys. 28(1-2), 61–84 (2003)
9. Boly, A., Hébrail, G., Goutier, S.: Forgetting Data Intelligently in Data Warehouses. In: IEEE Conference on Research, Innovat. and Vision for the Future, pp. 220–227. IEEE Press, NY (2007)
10. Skyt, J., Jensen, C.S., Pedersen, T.B.: Specification-based Data Reduction in Dimensional Data Warehouses. Info. Sys. 33(1), 36–63 (2008)
11. Iftikhar, N.: Integration, Aggregation and Exchange of Farming Device Data: A High Level Perspective. In: 2nd IEEE Conf. on the Applications of Digital Info., pp. 14–19. IEEE Press, NY (2009)
12. Rozeva, A.: Dimensional Hierarchies: Implementation in Data Warehouse Logical Schema Design. In: International Conference on Computer Systems and Tech., article 46. ACM Press, NY (2007)
13. Papastefanatos, G., Vassiliadis, P., Simitsis, A., Vassiliou, Y.: Design Metrics for DW Evolution. In: Li, Q., Spaccapietra, S., Yu, E., Olivé, A. (eds.) ER 2008. LNCS, vol. 5231, pp. 440–454. Springer, Heidelberg (2008)
14. Rizzi, S., Abello, A., Lechtenborger, J., Trujillo, J.: Research in Data Warehouse Modeling and Design: Dead or Alive? In: 9th ACM Int. Workshop on DW and OLAP, pp. 3–10. ACM Press, NY (2006)

An Agent Model of Business Relationships

John Debenham[1] and Carles Sierra[2]

[1] QCIS, University of Technology, Sydney, Australia
debenham@it.uts.edu.au
[2] Institut d'Investigació en Intel·ligència Artificial - IIIA,
Spanish Scientific Research Council, CSIC
08193 Bellaterra, Catalonia, Spain
sierra@iiia.csic.es

Abstract. Relationships are fundamental to all but the most impersonal forms of interaction in business. An agent aims to secure projected needs by attempting to build a set of (business) relationships with other agents. A relationship is built by exchanging private information, and is characterised by its intimacy — degree of closeness — and balance — degree of fairness. Each argumentative interaction between two agents then has two goals: to satisfy some immediate need, and to do so in a way that develops the relationship in a desired direction. An agent's desire to develop each relationship in a particular way then places constraints on the argumentative utterances. This paper describes argumentative interaction constrained by a desire to develop such relationships.

1 Introduction

Modelling long-term (business) relationships underpins the evolution of trust relationships. A basis for agent interaction is presented that manages the (business) relationships that an agent has with each agent that it interacts with. Our agent summaries its relationships using 'intimacy' and 'balance' measures. Its actions are then shaped by its desired values for these two measures that represent its foreseeable social aspirations, and are called the 'target intimacy' and 'target balance'. Given all of this, a particular interaction with another agent is approached both with the goal of negotiating towards a satisfactory conclusion, and as an opportunity to do so in a way that gradually develops the relationship towards its target. In this way the agent's target aspirations constrain and shape its argumentative behaviour in *relationship-based argumentation*.

Negotiation dialogues are traditionally organised around the basic illocutionary particles: *Offer*, *Accept* and *Reject*. Previous work has been centred on the design of negotiation strategies and on proposing agent architectures able to deal with the exchange of offers [1,2]. Game theory [3], possibilistic logic [4] and first-order logic [5] have been used for this purpose. Some initial steps in proposing rhetoric particles have been made, especially around the idea of *appeals*, *rewards* and *threats* [6]. Expanded negotiation dialogues, including these and other rhetoric moves, are known as *argumentation-based negotiations*. *Argumentation* in this sense is mainly to do with building (business) *relationships*. When we reward or threaten we refer to a future instant of time where the reward or threat will be effective, its scope goes beyond the current negotiation round.

P. García Bringas et al. (Eds.): DEXA 2010, Part II, LNCS 6262, pp. 126–140, 2010.

This paper is in the area labelled: *information-based agency* [7]. An information-based agent has an identity, values, needs, plans and strategies all of which are expressed using a fixed ontology in probabilistic logic for internal representation and in an illocutionary language for communication. We assume that such an agent resides in a electronic institution [8] and is aware of the prevailing norms and interaction protocols. An information-based agent makes no *a priori* assumptions about the states of the world or the other agents in it — these are represented in a world model, $\mathcal{M}^t$, that is inferred solely from the messages that it receives. The intuition behind information-based agency is that all illocutionary acts give away (valuable) information.

An agent's world model, $\mathcal{M}^t$, is a set of probability distributions for a set of random variables each of which represents the agent's expectations about some point of interest about the world or the other agents in it. Each incoming utterance is translated into a set of (linear) constraints on one or more of these distributions, and then the posterior state of the world model is estimated using entropy-based inference. These distributions are the foundation for the agent's reasoning.

A pair of agents interact by passing messages to each other. We assume that they share a common ontology and that their interactions are organised into dialogues, where a *dialogue* is a finite sequence of inter-related utterances. A *commitment* is a consequence of an utterance by an agent that contains a promise that the world will be in some state in the future. A *contract* is a pair of commitments exchanged between a pair[1] of agents. The set of all dialogues between two agents up to the present is their *relationship*. This discussion is from the point of view of an information-based agent α in a multiagent system where α interacts with negotiating agents, β_i, information providing agents, θ_j, and an *institutional agent*, ξ, that represents the institution where we assume that all interactions happen [8]: $\{\alpha, \beta_1, \ldots, \beta_o, \xi, \theta_1, \ldots, \theta_t\}$.

Our communication model is described in Section 2. Relationships are formalised in Section 3, and the agent architecture in Section 4. Section 6 describes an elaborate means of measuring the intimacy — degree of closeness — and balance – degree of fairness — that are based on measures of the information in any utterance. Section 7 describes the argumentation framework, and Section 8 concludes.

2   Communication Model

The communication language is detailed below; we assume that utterances in the communication language may be classified into unique illocutionary categories[2] $\mathcal{L} = \{l_i\}_{i=1}^L$. In order to structure agent dialogues we also need an ontology that includes a (minimum) repertoire of elements: a set of *concepts* (e.g. quantity, quality, material) organised in a is-a hierarchy (e.g. platypus is a mammal, Australian-dollar is a currency), and a set of relations over these concepts (e.g. price(beer,AUD)).[3] We model ontologies following an algebraic approach [9] and an ontology is a tuple $\mathcal{O} = (C, R, \leq, \sigma)$ where:

---

[1] Sets of commitments between more than two agents are not considered here.

[2] In a simple bargaining scenario these utterances could be: "propose", "accept" and "reject".

[3] Usually, a set of axioms defined over the concepts and relations is also required. We will omit this here.

1. C is a finite set of concept symbols (including basic data types);
2. R is a finite set of relation symbols;
3. $\leq$ is a reflexive, transitive and anti-symmetric relation on C (a partial order)
4. $\sigma : R \rightarrow C^+$ is the function assigning to each relation symbol its arity

where $\leq$ is the traditional *is-a* hierarchy. To simplify computations in the computing of probability distributions we assume that there is a number of disjoint *is-a* trees covering different ontological spaces (e.g. a tree for types of fabric, a tree for shapes of clothing, and so on). R contains relations between the concepts in the hierarchy, this is needed to define 'objects' (e.g. deals) that are defined as a tuple of issues. We then analyse dialogues in terms of the *dialogical framework* $\mathcal{L} \times \mathcal{O}$.

The semantic distance between concepts within an ontology depends on how far away they are in the structure defined by the $\leq$ relation. Semantic distance plays a fundamental role in strategies for information-based agency. How signed contracts, $Commit(\cdot)$, about objects in a particular semantic region, and their execution, $Done(\cdot)$, *affect* our decision making process about signing future contracts in nearby semantic regions is crucial to modelling the common sense that human beings apply in managing trading relationships. A measure [10] bases the *semantic similarity* between two concepts on the *path length* induced by $\leq$ (more distance in the $\leq$ graph means less semantic similarity), and the *depth* of the subsumer concept (common ancestor) in the shortest path between the two concepts (the deeper in the hierarchy, the closer the meaning of the concepts). Semantic similarity is then defined as:

$$\delta(c, c') = e^{-\kappa_1 l} \cdot \frac{e^{\kappa_2 h} - e^{-\kappa_2 h}}{e^{\kappa_2 h} + e^{-\kappa_2 h}}.$$

where $e = 2.71828$, l is the length (i.e. number of hops) of the shortest path between the concepts, h is the depth of the deepest concept subsuming both concepts, and κ_1 and κ_2 are parameters scaling the contributions of the shortest path length and the depth respectively.

The shape of the language that α uses to represent the information received and the content of its dialogues depends on two fundamental notions. First, when agents interact within an overarching institution they explicitly or implicitly accept the *norms* that will constrain their behaviour, and accept the established sanctions and penalties whenever norms are violated. Second, the dialogues in which α engages are built around two fundamental actions: (i) passing information, and (ii) exchanging proposals and contracts. A contract $\delta = (a, b)$ between agents α and β is a pair where a and b represent the actions that agents α and β are responsible for respectively. *Contracts* signed by agents and *information* passed by agents, are similar to norms in the sense that they oblige agents to behave in a particular way, so as to satisfy the conditions of the contract, or to make the world consistent with the information passed. Contracts and Information can thus be thought of as normative statements that restrict an agent's behaviour.

Norms, contracts, and information have an obvious temporal dimension. Thus, an agent has to abide by a norm while it is inside an institution, a contract has a validity period, and a piece of information is true only during an interval in time. The set of norms affecting the behaviour of an agent defines the *context* that the agent has to take into account.

α's communication language has two fundamental primitives: $\text{Commit}(\alpha, \beta, \varphi)$ to represent, in φ, the world that α aims at bringing about and that β has the right to verify, complain about or claim compensation for any deviations from, and $\text{Done}(\mu)$ to represent the event that a certain action μ^4 has taken place. In this way, norms, contracts, and information chunks will be represented as instances of $\text{Commit}(\cdot)$ where α and β can be individual agents or institutions. $\mathcal{C}$ is:

$$\mu ::= \; illoc(\alpha, \beta, \varphi, t) \mid \mu; \mu \mid$$
$$\textbf{Let } context \textbf{ In } \mu \textbf{ End}$$
$$\varphi ::= \; term \mid \text{Done}(\mu) \mid \text{Commit}(\alpha, \beta, \varphi) \mid \varphi \wedge \varphi \mid$$
$$\varphi \vee \varphi \mid \neg\varphi \mid \forall v.\varphi_v \mid \exists v.\varphi_v$$
$$context ::= \; \varphi \mid id = \varphi \mid prolog\_clause \mid context; context$$

where φ_v is a formula with free variable v, $illoc$ is any appropriate set of illocutionary particles, ';' means sequencing, and *context* represents either previous agreements, previous illocutions, the ontological working context, that is a projection of the ontological trees that represent the focus of the conversation, or code that aligns the ontological differences between the speakers needed to interpret an action a. Representing an ontology as a set predicates in Prolog is simple. The set *term* contains instances of the ontology concepts and relations.[5]

For example, we can represent the following offer: "If you spend a total of more than €100 in my shop during October then I will give you a 10% discount on all goods in November", as:

Offer(α, β,spent(β, α, October, X) $\wedge$ X $\geq$ €100 $\rightarrow$
    $\forall$ y. Done(Inform(ξ, α, pay(β, α, y), November)) $\rightarrow$
        Commit(α, β, discount(y,10%)))

ξ is an institution agent that reports the payment.

3   Relationships

There is evidence from psychological studies that humans seek a *balance* in their negotiation relationships. The classical view [11] is that people perceive resource allocations as being distributively fair (i.e. well balanced) if they are proportional to inputs or contributions (i.e. equitable). However, more recent studies [12,13] show that humans follow a richer set of norms of distributive justice depending on their *intimacy* level: equity, equality, and need. *Equity* being the allocation proportional to the effort (e.g. the profit of a company goes to the stock holders proportional to their investment), *equality* being the allocation in equal amounts (e.g. two friends eat the same amount of a cake cooked by one of them), and *need* being the allocation proportional to the need for the

---

[4] Without loss of generality we will assume that all actions are dialogical.
[5] We assume the convention that $C(c)$ means that c is an instance of concept C and $r(c_1, \ldots, c_n)$ implicitly determines that c_i is an instance of the concept in the i-th position of the relation r.

resource (e.g. in case of food scarcity, a mother gives all food to her baby). For instance, if we are in a purely economic setting (low intimacy) we might request equity for the Options dimension but could accept equality in the Goals dimension.

The perception of a relation being in balance (i.e. fair) depends strongly on the nature of the social relationships between individuals (i.e. the intimacy level). In purely economical relationships (e.g., business), equity is perceived as more fair; in relations where joint action or fostering of social relationships are the goal (e.g. friends), equality is perceived as more fair; and in situations where personal development or personal welfare are the goal (e.g. family), allocations are usually based on need.

We believe that the perception of balance in dialogues (in negotiation or otherwise) is grounded on social relationships, and that every dimension of an interaction between humans can be correlated to the social closeness, or *intimacy*, between the parties involved. According to the previous studies, the more intimacy across the illocutionary categories the more the need norm is used, and the less intimacy the more the equity norm is used. This might be part of our social evolution. There is ample evidence that when human societies evolved from a hunter-gatherer structure[6] to a shelter-based one[7] the probability of survival increased when food was scarce.

In this context, we can clearly see that, for instance, families exchange not only goods but also information and knowledge based on need, and that few families would consider their relationships as being unbalanced, and thus unfair, when there is a strong asymmetry in the exchanges (a mother explaining everything to her children, or buying toys, does not expect reciprocity). In the case of partners there is some evidence [14] that the allocations of goods and burdens (i.e. positive and negative utilities) are perceived as fair, or in balance, based on equity for burdens and equality for goods. See Table 1 for some examples of desired balances along five illocutionary categories.

Table 1. Some desired balances (sense of fairness) for five illocutionary categories

Illoc. Category	A new trading partner	my butcher	my boss	my partner	my children
Legitimacy	equity	equity	equity	equality	need
Options	equity	equity	equity	mixed[a]	need
Goals	equity	need	equity	need	need
Independence	equity	equity	equality	need	need
Commitment	equity	equity	equity	mixed	need

[a] equity on burden, equality on good

The perceived balance in a negotiation dialogue allows negotiators to infer information about their opponent, about its stance, and to compare their relationships with all negotiators. For instance, if we perceive that every time we request information it is provided, and that no significant questions are returned, or no complaints about not

---

[6] In its purest form, individuals in these societies collect food and consume it when and where it is found. This is a pure equity sharing of the resources, the gain is proportional to the effort.

[7] In these societies there are family units, around a shelter, that represent the basic food sharing structure. Usually, food is accumulated at the shelter for future use. Then the food intake depends more on the need of the members.

receiving information are given, then that probably means that our opponent perceives our social relationship to be very close. Alternatively, we can detect what issues are causing a burden to our opponent by observing an imbalance in the information or utilitarian senses on that issue.

4   Agent Architecture

A multiagent system $\{\alpha, \beta_1, \ldots, \beta_n, \xi, \theta_1, \ldots, \theta_t\}$, contains an agent α that interacts with other *argumentation agents*, β_i, *information providing agents*, θ_j, and an *institutional agent*, ξ, that represents the institution where we assume the interactions happen [8]. The institutional agent reports promptly and honestly on what actually occurs after an agent signs a contract, or makes some other form of commitment. In Section 5 this enables us to measure the difference between an utterance and a subsequent observation. Agents have a probabilistic first-order *internal language* $\mathcal{L}$ used to represent a *world model*, $\mathcal{M}^t$. A generic *information-based* architecture is described in detail in [7].

The agent architecture is shown in Figure 1. Agent α acts in response to a *need* that is expressed in terms of the ontology. Needs trigger α's goal/plan proactive reasoning, while other messages are dealt with by α's reactive reasoning.[8] Each plan prepares for the negotiation by assembling the contents of a 'briefcase' that the agent 'carries' into the negotiation[9]. The *relationship strategy* determines which agent to negotiate with for a given need; it uses risk management analysis to preserve a strategic set of trading relationships for each mission-critical need — this is not detailed here. For each trading relationship this strategy generates a *relationship target* that is expressed in the dialogical framework as a desired level of *intimacy* to be achieved in the long term.

Each negotiation consists of a dialogue, Ψ^t, between two agents with agent α contributing utterance μ and the partner β contributing μ'. Each dialogue, Ψ^t, is evaluated using the dialogical framework in terms of the *value* of Ψ^t to both α and β — see Section 6.2. The *negotiation strategy* then determines the current set of offers $\{\delta_i\}$, and then the *tactics*, guided by the *negotiation target*, decide which, if any, of these offers to put forward and wraps them in argumentation dialogue — see Section 7. We now describe two of the distributions in $\mathcal{M}^t$ that support offer exchange.

$\mathbb{P}^t(\mathrm{acc}(\alpha, \beta, \chi, \delta))$ estimates the probability that α should accept proposal δ in satisfaction of her need χ, where $\delta = (a, b)$ is a pair of commitments, a for α and b for β. α will accept δ if: $\mathbb{P}^t(\mathrm{acc}(\alpha, \beta, \chi, \delta)) > c$, for level of certainty c. This estimate is compounded from subjective and objective views of acceptability. The *subjective estimate* takes account of: the extent to which the enactment of δ will satisfy α's need χ, how much δ is 'worth' to α, and the extent to which α believes that she will be in a position to execute her commitment a [7]. $S_\alpha(\beta, a)$ is a random variable denoting α's estimate of β's subjective valuation of a over some finite, numerical evaluation space.

---

[8] Each of α's plans and reactions contain constructors for an initial *world model* $\mathcal{M}^t$. $\mathcal{M}^t$ is then maintained from percepts received using *update functions* that transform percepts into constraints on $\mathcal{M}^t$ — for details, see [7].

[9] Empirical evidence shows that in human negotiation, better outcomes are achieved by skewing the opening offer in favour of the proposer. We are unaware of any empirical investigation of this hypothesis for autonomous agents in real trading scenarios.

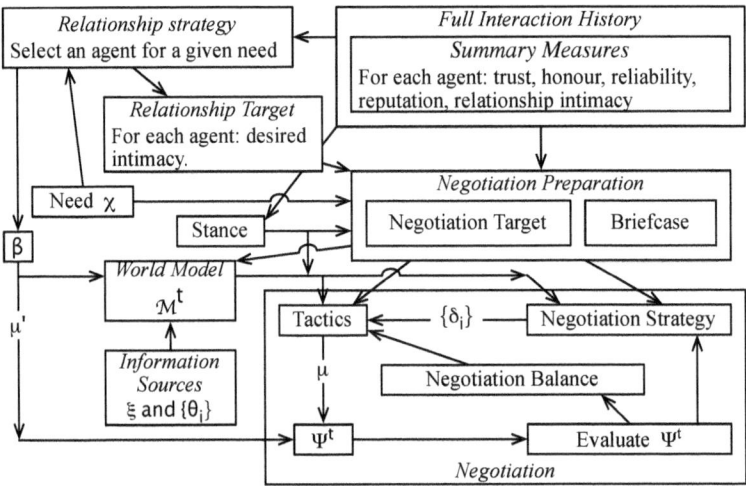

Fig. 1. The agent architecture

The *objective estimate* captures whether δ is acceptable on the open market, and variable $U_\alpha(b)$ denotes α's open-market valuation of the enactment of commitment b, again taken over some finite numerical valuation space. We also consider needs, the variable $T_\alpha(\beta, a)$ denotes α's estimate of the *strength* of β's motivating *need* for the enactment of commitment a over a valuation space. Then for $\delta = (a, b)$: $\mathbb{P}^t(\mathrm{acc}(\alpha, \beta, \chi, \delta)) =$

$$\mathbb{P}^t\left(\left(\frac{T_\alpha(\beta, a)}{T_\alpha(\alpha, b)} \right)^h \times \left(\frac{S_\alpha(\alpha, b)}{S_\alpha(\beta, a)} \right)^g \times \frac{U_\alpha(b)}{U_\alpha(a)} \geq s \right) \qquad (1)$$

where $g \in [0, 1]$ is α's *greed*, $h \in [0, 1]$ is α's degree of *altruism*, and $s \approx 1$ is derived from the *stance*[10] described in Section 7. The parameters g and h are independent. We can imagine a relationship that begins with $g = 1$ and $h = 0$. Then as the agents share increasing amounts of their information about their open market valuations g gradually reduces to 0, and then as they share increasing amounts of information about their needs h increases to 1. The basis for the acceptance criterion has thus developed from equity to equality, and then to need.

$\mathbb{P}^t(\mathrm{acc}(\beta, \alpha, \delta))$ estimates the probability that β would accept δ, by observing β's responses. For example, if β sends the message $\mathrm{Offer}(\delta_1)$ then α derives the constraint: $\{\mathbb{P}^t(\mathrm{acc}(\beta, \alpha, \delta_1)) = 1\}$ on the distribution $\mathbb{P}^t(\beta, \alpha, \delta)$, and if this is a counter offer to a former offer of α's, δ_0, then: $\{\mathbb{P}^t(\mathrm{acc}(\beta, \alpha, \delta_0)) = 0\}$. In the not-atypical special case of multi-issue bargaining where the agents' preferences over the individual issues *only* are known and are complementary to each other's, maximum entropy reasoning can be applied to estimate the probability that *any* multi-issue δ will be acceptable to β by enumerating the possible worlds that represent β's "limit of acceptability" [16].

---

[10] If α chooses to inflate her opening offer then this is achieved in Section 7 by increasing the value of s. If $s \gg 1$ then a deal may not be possible. This illustrates the well-known inefficiency of bilateral bargaining established analytically by [15].

5   Updating the World Model $\mathcal{M}^t$

α's world model consists of probability distributions that represent its uncertainty in the world state. α is interested in the degree to which an utterance accurately describes what will subsequently be observed. All observations about the world are received as utterances from an all-truthful institution agent ξ. For example, if β communicates the goal "I am hungry" and the subsequent negotiation terminates with β purchasing a book from α (by ξ advising α that a certain amount of money has been credited to α's account) then α may conclude that the goal that β chose to satisfy was something other than hunger. So, α's world model contains probability distributions that represent its uncertain expectations of what will be observed on the basis of utterances received.

We represent the relationship between *utterance*, φ, and subsequent *observation*, φ', by $\mathbb{P}^t(\varphi'|\varphi) \in \mathcal{M}^t$, where φ' and φ may be ontological categories in the interest of computational feasibility. For example, if φ is "I will deliver a bucket of fish to you tomorrow" then the distribution $\mathbb{P}(\varphi'|\varphi)$ need not be over *all* possible things that β might do, but could be over ontological categories that summarise β's possible actions.

In the absence of in-coming utterances, the conditional probabilities, $\mathbb{P}^t(\varphi'|\varphi)$, should tend to ignorance as represented by a *decay limit distribution* $\mathbb{D}(\varphi'|\varphi)$. α may have background knowledge concerning $\mathbb{D}(\varphi'|\varphi)$ as $t \to \infty$, otherwise α may assume that it has maximum entropy whilst being consistent with the data. In general, given a distribution, $\mathbb{P}^t(X_i)$, and a decay limit distribution $\mathbb{D}(X_i)$, $\mathbb{P}^t(X_i)$ decays by:

$$\mathbb{P}^{t+1}(X_i) = \Delta_i(\mathbb{D}(X_i), \mathbb{P}^t(X_i)) \tag{2}$$

where Δ_i is the *decay function* for the X_i satisfying the property that $\lim_{t\to\infty} \mathbb{P}^t(X_i) = \mathbb{D}(X_i)$. For example, Δ_i could be linear: $\mathbb{P}^{t+1}(X_i) = (1 - \nu_i) \times \mathbb{D}(X_i) + \nu_i \times \mathbb{P}^t(X_i)$, where $\nu_i < 1$ is the decay rate for the i'th distribution. Either the decay function or the decay limit distribution could also be a function of time: Δ_i^t and $\mathbb{D}^t(X_i)$.

Suppose that α receives an utterance $\mu = illoc(\alpha, \beta, \varphi, t)$ from agent β at time t. Suppose that α attaches an epistemic belief $\mathbb{R}^t(\alpha, \beta, \mu)$ to μ — this probability takes account of α's level of personal *caution*. We model the update of $\mathbb{P}^t(\varphi'|\varphi)$ in two cases, one for observations given φ, second for observations given ϕ in the semantic neighbourhood of φ.

First, if $\{\varphi_1, \varphi_2, \ldots, \varphi_m\}$ is the set of all possible observations and φ_k is observed then α may set $\mathbb{P}^{t+1}(\varphi_k|\varphi)$ to some value d. We estimate the complete posterior distribution $\mathbb{P}^{t+1}(\varphi'|\varphi)$ by applying the principle of minimum relative entropy as follows. Let $\boldsymbol{p}_{(\mu)}$ be the distribution: $\arg\min_{\boldsymbol{x}} \sum_j x_j \log \frac{x_j}{\mathbb{P}^t(\varphi'|\varphi)_j}$ that satisfies the constraint $\boldsymbol{p}_{(\mu)k} = d$. Then let $\boldsymbol{q}_{(\mu)}$ be the distribution:

$$\boldsymbol{q}_{(\mu)} = \mathbb{R}^t(\alpha, \beta, \mu) \times \boldsymbol{p}_{(\mu)} + (1 - \mathbb{R}^t(\alpha, \beta, \mu)) \times \mathbb{P}^t(\varphi'|\varphi)$$

and then let:

$$\boldsymbol{r}_{(\mu)} = \begin{cases} \boldsymbol{q}_{(\mu)} & \text{if } \boldsymbol{q}_{(\mu)} \text{ is more interesting than } \mathbb{P}^t(\varphi'|\varphi) \\ \mathbb{P}^t(\varphi'|\varphi) & \text{otherwise} \end{cases}$$

A measure of whether $\boldsymbol{q}_{(\mu)}$ is more interesting than $\mathbb{P}^t(\varphi'|\varphi)$ is: $\mathbb{K}(\boldsymbol{q}_{(\mu)}\|\mathbb{D}(\varphi'|\varphi)) > \mathbb{K}(\mathbb{P}^t(\varphi'|\varphi)\|\mathbb{D}(\varphi'|\varphi))$, where $\mathbb{K}(\boldsymbol{x}\|\boldsymbol{y}) = \sum_j x_j \ln \frac{x_j}{y_j}$ is the Kullback-Leibler distance between two probability distributions $\boldsymbol{x}$ and $\boldsymbol{y}$.

Finally incorporating Equation 2 we obtain the method for updating a distribution $\mathbb{P}^t(\varphi'|\varphi)$ on receipt of a message μ:

$$\mathbb{P}^{t+1}(\varphi'|\varphi) = \Delta_i(\mathbb{D}(\varphi'|\varphi), \boldsymbol{r}_{(\mu)}) \tag{3}$$

This procedure deals with integrity decay, and with two probabilities: first, the probability z in the utterance μ, and second the belief $\mathbb{R}^t(\alpha, \beta, \mu)$ that α attached to μ.

Second we consider the update of $\mathbb{P}^t(\phi'|\phi)$ given φ. Given $\mu = illoc(\alpha, \beta, \varphi, t)$ and the observation φ_k we define the vector $\boldsymbol{t}$ by

$$t_i = \mathbb{P}^t(\phi_i|\phi) + (1 - |\operatorname{Sim}(\varphi_k, \varphi) - \operatorname{Sim}(\phi_i, \phi)|) \cdot \operatorname{Sim}(\varphi_k, \phi)$$

with $\{\phi_1, \phi_2, \ldots, \phi_p\}$ the set of all possible observations in the context of ϕ and $i = 1, \ldots, p$. $\boldsymbol{t}$ is not a probability distribution. The multiplying factor $\operatorname{Sim}(\varphi', \phi)$ limits the variation of probability to those formulae whose ontological context is not too far away from the observation. The posterior $\mathbb{P}^{t+1}(\phi'|\phi)$ is obtained with Equation 3 with $\boldsymbol{r}_{(\mu)}$ defined to be the normalisation of $\boldsymbol{t}$.

6   Measuring the Confidence in a Relationship

A *dialogue*, Ψ^t, between agents α and β is a sequence of inter-related utterances in context. A *relationship*, Ψ^{*t}, is a sequence of dialogues. We first measure the *confidence* that an agent has for another by observing, for each utterance, the difference between what is said (the utterance) and what subsequently occurs (the observation). Second we *evaluate* each dialogue as it progresses in terms of the dialogical framework — this evaluation employs the confidence measures. Finally we define the *intimacy* of a relationship as an aggregation of the value of its component dialogues.

6.1   Confidence

Confidence measures generalise what are commonly called *trust*, *reliability* and *reputation* measures [17] into a single computational framework that spans the illocutionary categories $\mathcal{C}$. In Section 6.2 confidence measures are applied to valuing fulfilment of promises, to the execution of commitments, and to valuing dialogues.

Ideal observations. Consider a distribution of observations that represent α's "ideal" in the sense that it is the best that α could reasonably expect to observe. This distribution will be a function of α's *context* with β denoted by e, and is $\mathbb{P}_I^t(\varphi'|\varphi, e)$. Here we measure the relative entropy between this ideal distribution, $\mathbb{P}_I^t(\varphi'|\varphi, e)$, and the distribution of expected observations, $\mathbb{P}^t(\varphi'|\varphi)$. That is:

$$\mathbb{C}(\alpha, \beta, \varphi) = 1 - \sum_{\varphi'} \mathbb{P}_I^t(\varphi'|\varphi, e) \log \frac{\mathbb{P}_I^t(\varphi'|\varphi, e)}{\mathbb{P}^t(\varphi'|\varphi)} \tag{4}$$

where the "1" is an arbitrarily chosen constant being the maximum value that this measure may have. This equation measures confidence for a single statement φ. It makes sense to aggregate these values over a class of statements, say over those φ that are in the ontological context o, that is $\varphi \leq o$:

$$\mathbb{C}(\alpha, \beta, o) = 1 - \frac{\sum_{\varphi : \varphi \leq o} \mathbb{P}_\beta^t(\varphi) \left[1 - \mathbb{C}(\alpha, \beta, \varphi)\right]}{\sum_{\varphi : \varphi \leq o} \mathbb{P}_\beta^t(\varphi)}$$

where $\mathbb{P}_\beta^t(\varphi)$ is a probability distribution over the space of statements that the next statement β will make to α is φ. Similarly, for an overall estimate of β's *confidence* in α:

$$\mathbb{C}(\alpha, \beta) = 1 - \sum_\varphi \mathbb{P}_\beta^t(\varphi) \left[1 - \mathbb{C}(\alpha, \beta, \varphi)\right]$$

Preferred observations. The previous measure requires that: $\mathbb{P}_I^t(\varphi' | \varphi, e)$, has to be specified for each φ. Here we measure the extent to which the observation φ' is preferable to the original statement φ. Given a predicate $\text{Prefer}(c_1, c_2, e)$ meaning that α prefers c_1 to c_2 in environment e. Then if $\varphi \leq o$:

$$\mathbb{C}(\alpha, \beta, \varphi) = \sum_{\varphi'} \mathbb{P}^t(\text{Prefer}(\varphi', \varphi, o)) \mathbb{P}^t(\varphi' | \varphi)$$

and:

$$\mathbb{C}(\alpha, \beta, o) = \frac{\sum_{\varphi : \varphi \leq o} \mathbb{P}_\beta^t(\varphi) \mathbb{C}(\alpha, \beta, \varphi)}{\sum_{\varphi : \varphi \leq o} \mathbb{P}_\beta^t(\varphi)}$$

Certainty in observation. Here we measure the consistency in expected acceptable observations, or "the lack of expected uncertainty in those possible observations that are better than the original statement".

If $\varphi \leq o$ let: $\Phi_+(\varphi, o, \kappa) = \{\varphi' \mid \mathbb{P}^t(\text{Prefer}(\varphi', \varphi, o)) > \kappa\}$ for some constant κ, and:

$$\mathbb{C}(\alpha, \beta, \varphi) = 1 + \frac{1}{B^*} \cdot \sum_{\varphi' \in \Phi_+(\varphi, o, \kappa)} \mathbb{P}_+^t(\varphi' | \varphi) \log \mathbb{P}_+^t(\varphi' | \varphi)$$

where $\mathbb{P}_+^t(\varphi' | \varphi)$ is the normalisation of $\mathbb{P}^t(\varphi' | \varphi)$ for $\varphi' \in \Phi_+(\varphi, o, \kappa)$,

$$B^* = \begin{cases} 1 & \text{if } |\Phi_+(\varphi, o, \kappa)| = 1 \\ \log |\Phi_+(\varphi, o, \kappa)| & \text{otherwise} \end{cases}$$

As above we aggregate this measure for observations in a particular context o, and measure confidence as before.

Computational Note. The various measures given above involve extensive calculations. For example, Equation 4 contains $\sum_{\varphi'}$ that sums over *all* possible observations φ'. We obtain a more computationally friendly measure by appealing to the structure of the ontology described, and the right-hand side of Equation 4 may be approximated to:

$$1 - \sum_{\varphi' : \text{Sim}(\varphi', \varphi) \geq \eta} \mathbb{P}_{\eta, I}^t(\varphi' | \varphi, e) \log \frac{\mathbb{P}_{\eta, I}^t(\varphi' | \varphi, e)}{\mathbb{P}_\eta^t(\varphi' | \varphi)}$$

where $\mathbb{P}^t_{\eta,I}(\varphi'|\varphi, e)$ is the normalisation of $\mathbb{P}^t_I(\varphi'|\varphi, e)$ for $\text{Sim}(\varphi', \varphi) \geq \eta$, and similarly for $\mathbb{P}^t_\eta(\varphi'|\varphi)$. The extent of this calculation is controlled by the parameter η. An even tighter restriction may be obtained with: $\text{Sim}(\varphi', \varphi) \geq \eta$ and $\varphi' \leq \psi$ for some ψ.

6.2   Valuing Negotiation Dialogues

Suppose that a negotiation commences at time s, and by time t a string of utterances, $\Phi^t = \langle \mu_1, \ldots, \mu_n \rangle$ has been exchanged between agent α and agent β. This negotiation dialogue is evaluated by α in the context of α's world model at time s, $\mathcal{M}^s$, and the environment e that includes utterances that may have been received from other agents in the system including the information sources $\{\theta_i\}$. Let $\Psi^t = (\Phi^t, \mathcal{M}^s, e)$, then α estimates the *value* of this dialogue to itself in the context of $\mathcal{M}^s$ and e as a $2 \times L$ array $V_\alpha(\Psi^t)$ where:

$$V_x(\Psi^t) = \begin{pmatrix} I^{l_1}_x(\Psi^t) & \cdots & I^{l_L}_x(\Psi^t) \\ U^{l_1}_x(\Psi^t) & \cdots & U^{l_L}_x(\Psi^t) \end{pmatrix}$$

where the $I(\cdot)$ and $U(\cdot)$ functions are information-based and utility-based measures respectively as we now describe. α estimates the *value* of this dialogue to β as $V_\beta(\Psi^t)$ by assuming that β's reasoning apparatus mirrors its own.

In general terms, the information-based valuations measure the reduction in uncertainty, or information gain, that the dialogue gives to each agent, they are expressed in terms of decrease in entropy that can always be calculated. The utility-based valuations measure utility gain are expressed in terms of "some suitable" utility evaluation function $\mathbb{U}(\cdot)$ that can be difficult to define. This is one reason why the utilitarian approach has no natural extension to the management of argumentation that is achieved here by our information-based approach.

The *balance* in a negotiation dialogue, Ψ^t, is defined as: $B_{\alpha\beta}(\Psi^t) = V_\alpha(\Psi^t) \ominus V_\beta(\Psi^t)$ for an element-by-element difference operator $\ominus$ that respects the structure of $V(\Psi^t)$. The *intimacy* between agents α and β, $I^{*t}_{\alpha\beta}$, is the pattern of the two $2 \times L$ arrays V^{*t}_α and V^{*t}_β that are computed by an update function as each negotiation round terminates, $I^{*t}_{\alpha\beta} = \left(V^{*t}_\alpha, V^{*t}_\beta \right)$. If Ψ^t terminates at time t:

$$V^{*t+1}_x = \nu \times V_x(\Psi^t) + (1 - \nu) \times V^{*t}_x \qquad (5)$$

where ν is the learning rate, and $x = \alpha, \beta$. Additionally, V^{*t}_x continually decays by: $V^{*t+1}_x = \tau \times V^{*t}_x + (1 - \tau) \times D_x$, where $x = \alpha, \beta$; τ is the decay rate, and D_x is a $2 \times L$ array being the decay limit distribution for the value to agent x of the intimacy of the relationship in the absence of any interaction. D_x is the *reputation* of agent x. The *relationship balance* between agents α and β is: $B^{*t}_{\alpha\beta} = V^{*t}_\alpha \ominus V^{*t}_\beta$. In particular, the intimacy determines values for the parameters g and h in Equation 1. As a simple example, if both $I^O_\alpha(\Psi^{*t})$ and $I^O_\beta(\Psi^{*t})$ increase then g decreases, and as the remaining information-based components increase, h increases.

The notion of balance may be applied to pairs of utterances by treating them as degenerate dialogues. In simple multi-issue bargaining the *equitable information revelation* strategy generalises the tit-for-tat strategy in single-issue bargaining, and extends to a tit-for-tat argumentation strategy by applying the same principle across the dialogical framework.

7 Strategies and Tactics for Building Relationships

Each negotiation has to achieve two goals. First it may be intended to achieve some contractual outcome. Second it will aim to contribute to the growth, or decline, of the relationship intimacy.

We now describe in greater detail the contents of the *"Negotiation"* box in Figure 1. The negotiation literature consistently advises that an agent's behaviour should not be predictable even in close, intimate relationships. The required variation of behaviour is normally described as varying the negotiation *stance* that informally varies from "friendly guy" to "tough guy". The stance is shown in Figure 1, it injects bounded random noise into the process, where the bound tightens as intimacy increases. The stance, $S^t_{\alpha\beta}$, is a $2 \times L$ matrix of randomly chosen multipliers, each ≈ 1, that perturbs α's actions. The value in the (x, y) position in the matrix, where $x = I, U$ and $y \in \mathcal{L}$, is chosen at random from $[\frac{1}{l(I^{*t}_{\alpha\beta}, x, y)}, l(I^{*t}_{\alpha\beta}, x, y)]$ where $l(I^{*t}_{\alpha\beta}, x, y)$ is the bound, and $I^{*t}_{\alpha\beta}$ is the intimacy.

The negotiation *strategy* is concerned with maintaining a working set of proposals. If the set of proposals is empty then α will quit the negotiation. α perturbs the acceptance machinery (see Section 4) by deriving s from the $S^t_{\alpha\beta}$ matrix. In line with the comment in Footnote 9, in the early stages of the negotiation α may decide to inflate her opening offer. This is achieved by increasing the value of s in Equation 1. The following strategy uses the machinery described in Section 4. Fix h, g, s and c, set the Proposals to the empty set, let $D^t_s = \{\delta \mid \mathbb{P}^t(\text{acc}(\alpha, \beta, \chi, \delta)) > c\}$, then:

- repeat the following as many times as desired: add $\delta = \arg \max_x \{\mathbb{P}^t(\text{acc}(\beta, \alpha, x)) \mid x \in D^t_s\}$ to Proposals, remove $\{y \in D^t_s \mid \text{Sim}(y, \delta) < k\}$ for some k from D^t_s

By using $\mathbb{P}^t(\text{acc}(\beta, \alpha, \delta))$ this strategy reacts to β's history of Propose and Reject utterances.

Negotiation *tactics* are concerned with selecting some offers and wrapping them in argumentation. Prior interactions with agent β will have produced an intimacy pattern expressed in the form of $\left(V^{*t}_\alpha, V^{*t}_\beta \right)$. Suppose that the relationship target is $(T^{*t}_\alpha, T^{*t}_\beta)$. Following from Equation 5, α will want to achieve a *negotiation target*, $N_\beta(\Psi^t)$ such that: $\nu \cdot N_\beta(\Psi^t) + (1 - \nu) \cdot V^{*t}_\beta$ is "a bit on the T^{*t}_β side of" V^{*t}_β:

$$N_\beta(\Psi^t) = \frac{\nu - \kappa}{\nu} V^{*t}_\beta \oplus \frac{\kappa}{\nu} T^{*t}_\beta \qquad (6)$$

for small $\kappa \in [0, \nu]$ that represents α's desired *rate of development* for her relationship with β. $N_\beta(\Psi^t)$ is a $2 \times L$ matrix containing variations in the dialogical framework's dimensions that α would like to reveal to β during Ψ^t (e.g. I'll pass a bit more information on options than usual, I'll be stronger in concessions on options, etc.). It is reasonable to expect β to progress towards her target at the same rate and $N_\alpha(\Psi^t)$ is calculated by replacing β by α in Equation 6. $N_\alpha(\Psi^t)$ is what α hopes to receive from β during Ψ^t. This gives a *negotiation balance target* of: $N_\alpha(\Psi^t) \ominus N_\beta(\Psi^t)$ that can be used as the foundation for reactive tactics by striving to maintain this balance across the dialogical framework. A cautious tactic could use the balance to bound the response μ to each utterance μ' from β by the constraint: $V_\alpha(\mu') \ominus V_\beta(\mu) \approx S_{\alpha\beta}^t \otimes (N_\alpha(\Psi^t) \ominus N_\beta(\Psi^t))$, where $\otimes$ is element-by-element matrix multiplication, and $S_{\alpha\beta}^t$ is the stance. A less neurotic tactic could attempt to achieve the target negotiation balance over the anticipated complete dialogue. If a balance bound requires negative information revelation in one dialogical framework category then α will contribute nothing to it, and will leave this to the natural decay to the reputation D as described above.

The following are a list of components that we have described that could be combined into an agent's negotiation strategy. These components all constrain the agent's actions. We assume that they are all soft constraints and that they operate together with a hard constraint $C^t(\alpha, \beta, x^t)$ on the message x^t that α may send to β at time t.

Information-based strategies. Every communication gives away information and so has the potential to contribute to the intimacy and balance of a relationship. Information-based strategies manage the information revelation process. Let $M_{\alpha\beta}^t$ be the set of time-stamped messages that α has sent to β, and $M_{\beta\alpha}^t$ likewise both at time t. $\mathcal{M}^t$ is α's world model at time t and consists of a set of probability distributions. x^t denotes a message received at time t. $\mathbb{I}^t(\alpha, \beta, x^t)$ is the information gain — measured as the reduction of the entropy of $\mathcal{M}^t$ — observed by α after receiving message x^t. $\mathbb{I}^t(\beta, \alpha, x^t)$ is α's estimate of β's information gain after receiving message x^t from α.

The complete *information history* of both the observed and the estimated information gain, $G^t(\alpha, \beta)$, is:

$$G^t(\alpha, \beta) = \{(x^s, \mathbb{I}^s(\alpha, \beta, x^s)) \mid x^s \in M_{\beta\alpha}^t\} \cup$$
$$\{(x^s, \mathbb{I}^s(\beta, \alpha, x^s)) \mid x^s \in M_{\alpha\beta}^t\}$$

respectively.

In [7] we described to the model that α constructs of β. In general α can not be expected to guess β's world model, $\mathcal{M}_\beta^t$, unless α knows what β's needs are — even then, α would only know $\mathcal{M}_\beta^t$ with certainty if it knew what plans β had chosen. However, α always knows the private information that it has sent to β — for example, in Propose($\cdot$) and Reject($\cdot$) messages. Such private information could be used by β to estimate α's probability of accepting a proposal: $\mathbb{P}_\beta^t(\mathrm{acc}(\alpha, \beta, \chi', z))$, where χ' is the need that β believes α to have.

α's information-based strategies constrain its actions, x^t, on the basis of $\mathbb{I}^t(\beta, \alpha, x^t)$ and its relation to $G^t(\alpha, \beta)$. For example, the strategy that gives β greatest expected information gain:

$$\arg\max_z \{ \mathbb{I}_\beta^s(\beta, \alpha, z) \mid C^t(\alpha, \beta, z)\}$$

More generally, for some function f:

$$\arg\max_z \{ f(\mathbb{I}_\beta^s(\beta, \alpha, z), G^t(\alpha, \beta)) \mid C^t(\alpha, \beta, z)\}$$

the idea being that the f 'optimises' in some sense the information gain taking account of the interaction history.

Ontology-based strategies. The structure of the ontology may be used to manage the information revelation process in particular strategic areas. For example, α may prefer to build a relationship with β in the context of the supply of particular goods only [9]. The structure of the ontology is provided by the $\mathrm{Sim}(\cdot)$ function. Given two contracts δ and δ' containing concepts $\{o_1, \ldots, o_i\}$ and $\{o'_1, \ldots, o'_j\}$ respectively, the (non-symmetric) distance of δ' from δ is the vector

$$\boldsymbol{\Gamma}(\delta, \delta') = (d_k : o''_k)_{k=1}^i$$

where $d_k = \min_x\{\mathrm{Sim}(o_k, o'_x) \mid x = 1, \ldots, j\}$, $o''_k = \sup(\arg\min_x\{\mathrm{Sim}(o_k, x) \mid x = o'_1, \ldots, o'_j\}, o_k)$ and the function $\sup(\cdot, \cdot)$ is the supremum of two concepts in the ontology. $\boldsymbol{\Gamma}(\delta, \delta')$ quantifies how different δ' is to δ and enables α to "work around" or "move away from" a contract under consideration. In general for some function g;

$$\arg\max_z\{\, g(\boldsymbol{\Gamma}(z, x^s)) \mid x^s \in M^t_{\alpha\beta} \cup M^t_{\beta\alpha} \wedge C^t(\alpha, \beta, z)\}$$

the idea being that the g 'optimises' in some sense the ontological relationship with the interaction history.

8 Discussion

The ability of agents to conduct business relies on their ability to build business relationships with each other. In this paper we have introduced a novel approach to negotiation that incorporates a rich model of relationships that is dimensioned by the structure of the ontology and a set of illocutionary categories. It is grounded on business and psychological studies and introduces the concepts of *intimacy* and *balance* as key elements in understanding what is a negotiation strategy and tactic. Relationships are strengthened by managing the agent's dialogical moves. Each dialogical move produces a change in an array structure. The current balance and intimacy levels and the desired, or target, levels are then used by the tactics to determine what to say next. The architecture is simple and the implementation of the agents straightforward using tools from information theory.

We are currently exploring the use of this model as an extension of a currently widespread eProcurement software commercialised by a spin-off company of the laboratory of one of the authors. This tool has only a utilitarian modelling of the negotiation interactions and has motivated some criticisms from its users about the lack of modelling of long-lasting relationships that our model could solve.

References

1. Jennings, N., Faratin, P., Lomuscio, A., Parsons, S., Sierra, C., Wooldridge, M.: Automated negotiation: Prospects, methods and challenges. International Journal of Group Decision and Negotiation 10, 199–215 (2001)
2. Faratin, P., Sierra, C., Jennings, N.: Using similarity criteria to make issue trade-offs in automated negotiation. Journal of Artificial Intelligence 142, 205–237 (2003)
3. Rosenschein, J.S., Zlotkin, G.: Rules of Encounter. The MIT Press, Cambridge (1994)

4. Giménez, E., Godo, L., Rodríguez-Aguilar, J.A., Garcia, P.: Designing bidding strategies for trading agents in electronic auctions. In: Proceedings of the Third International Conference on Multi-Agent Systems (ICMAS 1998), pp. 136–143 (1998)
5. Kraus, S.: Negotiation and cooperation in multi-agent environments. Artificial Intelligence 94, 79–97 (1997)
6. Sierra, C., Jennings, N., Noriega, P., Parsons, S.: A Framework for Argumentation-Based Negotiation. In: Rao, A., Singh, M.P., Wooldridge, M.J. (eds.) ATAL 1997. LNCS, vol. 1365, pp. 177–192. Springer, Heidelberg (1998)
7. Sierra, C., Debenham, J.: Information-based agency. In: Proceedings of Twentieth International Joint Conference on Artificial Intelligence, IJCAI 2007, Hyderabad, India, pp. 1513–1518 (2007)
8. Arcos, J.L., Esteva, M., Noriega, P., Rodríguez, J.A., Sierra, C.: Environment engineering for multiagent systems. Journal on Engineering Applications of Artificial Intelligence 18 (2005)
9. Kalfoglou, Y., Schorlemmer, M.: IF-Map: An ontology-mapping method based on information-flow theory. In: Spaccapietra, S., March, S., Aberer, K. (eds.) Journal on Data Semantics I. LNCS, vol. 2800, pp. 98–127. Springer, Heidelberg (2003)
10. Li, Y., Bandar, Z.A., McLean, D.: An approach for measuring semantic similarity between words using multiple information sources. IEEE Transactions on Knowledge and Data Engineering 15, 871–882 (2003)
11. Adams, J.S.: Inequity in social exchange. In: Berkowitz, L. (ed.) Advances in experimental social psychology, vol. 2. Academic Press, New York (1965)
12. Sondak, H., Neale, M.A., Pinkley, R.: The negotiated allocations of benefits and burdens: The impact of outcome valence, contribution, and relationship. Organizational Behaviour and Human Decision Processes, 249–260 (1995)
13. Valley, K.L., Neale, M.A., Mannix, E.A.: Friends, lovers, colleagues, strangers: The effects of relationships on the process and outcome of negotiations. In: Bies, R., Lewicki, R., Sheppard, B. (eds.) Research in Negotiation in Organizations, vol. 5, pp. 65–94. JAI Press (1995)
14. Bazerman, M.H., Loewenstein, G.F., White, S.B.: Reversal of preference in allocation decisions: judging an alternative versus choosing among alternatives. Administration Science Quarterly, 220–240 (1992)
15. Myerson, R., Satterthwaite, M.: Efficient mechanisms for bilateral trading. Journal of Economic Theory 29, 1–21 (1983)
16. Debenham, J.: Bargaining with information. In: Jennings, N., Sierra, C., Sonenberg, L., Tambe, M. (eds.) Proceedings of Third International Conference on Autonomous Agents and Multi Agent Systems, AAMAS-2004, pp. 664–671. ACM Press, New York (2004)
17. Sierra, C., Debenham, J.: Information-based reputation. In: Paolucci, M. (ed.) First International Conference on Reputation: Theory and Technology (ICORE 2009), Gargonza, Italy, pp. 5–19 (2009)

Pivot Selection Method for Optimizing both Pruning and Balancing in Metric Space Indexes

Hisashi Kurasawa[1], Daiji Fukagawa[2], Atsuhiro Takasu[3], and Jun Adachi[3]

[1] The University of Tokyo, 2-1-2 Hitotsubashi, Chiyoda-ku, Tokyo, Japan
[2] Doshisha University, 1-3 Tatara Miyakodani, Kyotanabe-shi, Kyoto, Japan
[3] National Institute of Informatics, 2-1-2 Hitotsubashi, Chiyoda-ku, Tokyo, Japan

Abstract. We researched to try to find a way to reduce the cost of nearest neighbor searches in metric spaces. Many similarity search indexes recursively divide a region into subregions by using pivots, and construct a tree structure index. A problem in the existing indexes is that they only focus on the pruning objects and do not take into consideration the tree balancing. The balance of the indexes depends on the data distribution and the indexes don't reduce the search cost for all data. We propose a similarity search index called the Partitioning Capacity Tree (PCTree). PCTree automatically optimizes the pivot selection based on both the balance of the regions partitioned by a pivot and the estimated effectiveness of the search pruning by the pivot. As a result, PCTree reduces the search cost for various data distributions. Our evaluations comparing it with four indexes on three real datasets showed that PCTree successfully reduces the search cost and is good at handling various data distributions.

Keywords: Similarity Search, Metric Space, Indexing.

1 Introduction

Finding similar objects in a large dataset is a fundamental process of many applications, such as image completion [7]. These applications can be speeded up by reducing the query execution cost of a similarity search.

Similarity search indexes are used for pruning objects dissimilar to a query, and reduce the search cost, such as the distance computations and the disk accesses. Most indexing schemes use *pivots*, which are reference objects. They recursively divide a region into subregions by using pivots, and construct a tree structure index. They prune some of the subregions by using the triangle inequality while searching. That is, the methods of selecting pivots and dividing up the space by using these pivots determine the index structure and pruning performance. The existing pivot selection studies have mainly focused on increasing the number of pruned objects at a branch in the tree [2]. However, most previous works do not take into account the balance of the tree. It is well known that a reduction in the tree height reduces the average search cost. Thus, even if the pivots of the methods are enough good for pruning objects at the

P. García Bringas et al. (Eds.): DEXA 2010, Part II, LNCS 6262, pp. 141–148, 2010.

branch, they may not reduce the average search cost for every data distribution. Therefore, we developed a new pivot selection for optimizing both the pruning and the balancing.

The main contributions of this paper are as follows.

– We propose a new information theoretic criterion called the *Partitioning Capacity (PC)* for pivot selection. The PC takes both the object pruning effect and index tree balance into account.
– We developed a metric space index called the *Partitioning Capacity Tree (PCTree)*. The PCTree provides the necessary functions for constructing an index tree based on PC and for efficiently retrieving similar objects using the index tree.
– We show the efficiency of PCTree empirically through the results from several experiments where we compared the PCTree with several metric space indexes using three real datasets.

2   Related Work

Let $M = (D, d)$ be a metric space defined for a domain of objects D and a distance function $d : D \times D \mapsto \mathbb{R}$. M satisfies the postulates, such as the triangle inequality. Our index deals with the space and the distance.

The early indexing schemes of similarity searches used balanced tree structures. Vantage Point Tree (VPT) divides a region into two subregions based on the distance from a pivot [15]. Its distance is set for equally partitioning the objects in the region, so that the two subregions contain the same number of objects. Thus, VPT is a completely balanced binary tree. However, the early indexes are weak at pruning objects while searching. Their pivots are selected by using simple statical features, such as the variance. As a result, they require a huge search cost because they have to traverse many nodes in the tree.

The recent indexing schemes focus on pruning dissimilar objects rather than balancing the tree. Some schemes use other partitioning techniques instead of Ball partitioning. Vantage Point Forest (VPF) [16] and D-Index [6] divide a region into three sub-regions according the distance from the pivot. They aim to exclude the middle area in the region because it is difficult to judge whether objects in this area are similar or dissimilar to a query. Generalized Hyper-plane Tree (GHT) [14] divides the space by using a hyper-plane equidistant from the two pivots.

Other schemes improve the pivot selection. Many studies have attempted to select pivots that are distant from the other objects and other pivots [15, 2]. These methods used the mean [6], the variance [15], and the sum [5]. iDistance [8] selects pivots based on the clustering result and reduces the number of regions accessed during a search. OMNI-Family [10] chooses a set of pivots based on the minimum bounding region. MMMP [12] has a pivot selection based on the maximal margin, and it classifies dense regions. Moreover, some methods combine different pivot selection techniques [17].

Table 1. Notation

Notation	Meaning
D	domain of objects
d	distance function
R	region
S	object set
S_R	object set in R
$S_{R,\mathrm{sub}}$	subset of S_R
m	sampling parameter for $S_{R,\mathrm{sub}}$
S_p	pivot candidate set
s	sampling parameter for S_p
p	pivot
r_p	partitioning distance of p
q	query
$r_{o,k,S}$	distance between o and o's k-nearest neighbor in S
PC	pivot capacity
X_{i,p,r_p}	*partitioning distribution*
Y_{i,p,r_p}	*pruning distribution*
S_r	result set
k	# results
r_q	query range

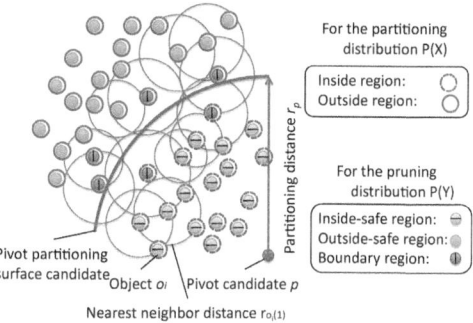

Fig. 1. Labels for measuring the PC

Still other schemes propose various index structures. Fixed Queries Array [3] is an array of distances between objects and pivots and it is used to save memory. List of Clusters (LC) [4] is a list of compact regions. It recursively divides the space into a compact region. iDistance consists of a list of pivots and a set of B^+-trees for storing objects. Thus, its index structure is partially balancing. Although the B^+-trees are balanced, the pivots are linearly accessed. Spatial Approximation Tree (SAT) [13] approximates a graph and uses pivots for approaching the query spatially. M-Tree [5] is a dynamic tree structure with object insertion and deletion. Slim-Tree [11] minimizes the overlap space managed by node in M-Tree.

As mentioned above, Almost all the recent methods do not take into consideration the tree balancing as VPT proposed. Our index thus aims to reduce the search cost by taking into account both the object pruning and tree balance.

3   Partitioning Capacity Tree

This section proposes PCTree, a similarity search index for a metric space. The index is designed for nearest neighbor searches and aims at reducing the search cost for various data distributions. Its indexing method selects pivots that optimize pruning objects and balance the index tree. Table 1 summarizes the symbols used in this article.

3.1   Partitioning Capacity

PC is a measure of the balance of the regions partitioned by a pivot and the estimated effectiveness of the pruning by the pivot. It represents the expected index

performance for a given query distribution. We assume that both the queries
and objects in the database are drawn from the same probability distribution.

For a metric space $M = (D, d)$ and a region $R \subseteq D$ in the space, suppose that
a pivot p and its partitioning distance r_p divide R into two subregions:

$$R_1(p, r_p) = \{o \in R \mid d(o, p) \leq r_p\}, R_2(p, r_p) = \{o \in R \mid d(o, p) > r_p\} .$$

We respectively refer to $R_1(p, r_p)$ and $R_2(p, r_p)$ as an *inside region* and an *outside
region* with respect to pivot p and distance r_p.

For a query q, let X be the random variable that represents the region in
which q is included, that is,

$$X = \begin{cases} X_{1, p, r_p} & \text{if } q \in R_1(p, r_p) \\ X_{2, p, r_p} & \text{if } q \in R_2(p, r_p) . \end{cases} \tag{1}$$

We call $P(X)$ a *partitioning distribution*. Since we assume a query is drawn from
the same distribution as the database, we estimate the partitioning distribution
as

$$P(X_{1, p, r_p}) = \frac{|S_{R_1(p, r_p)}|}{|S_R|}, \quad P(X_{1, p, r_p}) = \frac{|S_{R_2(p, r_p)}|}{|S_R|} , \tag{2}$$

where S_R denote the set of database objects that are in the region R.

Similarly, we define the *pruning distribution* of the query. For a set S of objects
in the database and an object o in the region R, let $r_{o,k,S}$ denote the distance
between o and o's kth nearest neighbor in S. Let us consider the following three
regions:

$$R_1'(p, r_p, k) = \{o \in R \mid d(o, p) + r_{o,k,S} \leq r_p\} , \tag{3}$$
$$R_2'(p, r_p, k) = \{o \in R \mid d(o, p) - r_{o,k,S} > r_p\} , \tag{4}$$
$$R_3'(p, r_p, k) = R - R_1'(p, r_p, k) - R_2'(p, r_p, k) . \tag{5}$$

Intuitively, $R_1'(p, r_p, k)$ is the set of the objects whose k-nearest neighbor objects are within $R_1(p, r_p)$. This means that, if a query q belongs to the region $R_1'(p, r_p, k)$, we can prune $R_2(p, r_p)$ when executing the k-nearest neighbor
search. Similarly, $R_2'(p, r_p, k)$ is the set of the objects whose k-nearest neighbor
objects are within $R_2(p, r_p)$. We respectively refer to $R_1'(p, r_p, k)$, $R_2'(p, r_p, k)$,
and $R_3'(p, r_p, k)$ as an *inside-safe region*, an *outside-safe region*, and a *boundary
region* w.r.t the pivot p, the distance r_p, and the number k of nearest neighbors.
Let Y be a random variable that represents the region in which the k-nearest
neighbor ranges of q is included, that is,

$$Y = \begin{cases} Y_{1, p, r_p} & \text{if } q \in R_1'(p, r_p, k) , \\ Y_{2, p, r_p} & \text{if } q \in R_2'(p, r_p, k) , \\ Y_{3, p, r_p} & \text{if } q \in R_3'(p, r_p, k) . \end{cases} \tag{6}$$

We call $P(Y)$ a *pruning distribution*. We estimate the distribution as

$$P(Y_{1,p,r_p,k}) = \frac{|S_{R'_1(p,r_p)}|}{|S_R|}, \tag{7}$$

$$P(Y_{2,p,r_p,k}) = \frac{|S_{R'_2(p,r_p)}|)}{|S_R|}, \tag{8}$$

$$P(Y_{3,p,r_p,k}) = \frac{|S_{R'_3(p,r_p)}|}{|S_R|}. \tag{9}$$

By using the two random variables X and Y, the PC for a pivot p is defined as

$$
\begin{aligned}
PC_k(p) \\
\equiv \max_{r_p} I(X;Y) \\
= \max_{r_p}\left(\sum_i \sum_j \left(P(X_{i,p,r_p}, Y_{j,p,r_p,k}) \cdot \log\left(\frac{P(X_{i,p,r_p}, Y_{j,p,r_p,k})}{P(X_{i,p,r_p})P(Y_{j,p,r_p,k})}\right)\right)\right)
\end{aligned}
\tag{10}
$$

where $I(\cdot;\cdot)$ is the mutual information. We choose the object p (resp. distance r_p) that maximizes Eq. (10) as a pivot (resp. partitioning distance).

The PC represents the mutual dependence of the random variables for the partitioning and the pruning distributions. We regard the value of the PC as the reduction in uncertainty when knowing either the partition or the pruning effect. That is, it is the amount of information shared by the partition and the pruning effect. The PC is less than the entropy of the partitioning distribution and that of the pruning distribution because of the mutual information theory. Partitions are effective for the search pruning result for a large PC value. Balanced partitions also tend to increase the PC value. The PC is almost the same as the channel capacity of a binary erasure channel in the information and coding theory [9].

3.2 PCTree Construction

PCTree has a tree consisting of *internal nodes* and *leaves*. An internal node is associated with one pivot and its partitioning distance whereas a leaf node is associated with one pivot and the objects.

We first select the most effective pivot candidates S_p because a database usually contains large amount of objects and calculating PC for all these objects is computationally infeasible. We select pivot candidates such that they are separated from each other by a certain distance d_p. We set d_p to be $d_{ave} \cdot s$, where d_{ave} is the approximate average distance between objects and s is a parameter. The approximate average distance is calculated by using randomly sampled objects. Let N (resp. n) denote the number of objects in the dataset (resp. pivot candidates), the cost is at most $O(N \cdot n)$. Note that the pivot candidate selection is done once before constructing an index.

Then, PCTree recursively divides a region into subregions by using pivots as well as VPT [15]. Each pivot is selected from the candidates S_p according to the criterion PC (Sec. 3.1 for details) for the predefined number m of samples from the region. When the PCs of the pivot candidates in a region are less than the minimum PC, the partitioning finishes and the region is set as a leaf node in the tree. we set the minimum PC as

$$\text{MinimumPC}(S) = -\frac{1}{|S|} \cdot \log \frac{1}{|S|} - \frac{|S|-1}{|S|} \cdot \log \frac{|S|-1}{|S|} , \qquad (11)$$

where S is the object set. The minimum PC represents the effect of the linear scan method. we regard the method as that in which all the objects in the dataset are pivots and each pivot prunes itself while searching. The indexing cost is $O(N^2)$ at most.

The PCTree requires three parameters during indexing. One parameter k is for calculating the PC in Eqs. (3), (4) and (5), and the other parameters s and m are for sampling objects. From preliminary experiments, we decided to set k, s, and m to 1, 0.8, and 500, respectively, in the index performance evaluations.

3.3   k-Nearest Neighbor Search

For the k-Nearest Neighbor searching, the PCTree receives a query q and the number of results k. The search procedure is almost the same as for VPT [15]. First, it creates an empty set as a result set S_r and sets the query range r_q to be ∞. It starts the search from the root node in the PCTree and recursively accesses the nodes. If the node is a leaf, it finds the objects whose distance to q are within r_q in the node while using the object-pivot distance constraint, and adds them to S_r. Then, it updates r_q to be the k-nearest neighbor radius of q in S_r. If the node is an internal node, it reads the pivot p and its partitioning distance r_p associated to the node. Then, if the inequality $d(p, q) \leq r_p$ is satisfied, it accesses the child node for the inside region and updates r_q. After this, if the inequality $d(p, q) + r_q > r_p$ is satisfied, it accesses the other child node and updates r_q. Similarly, if the inequality $d(p, q) \geq r_p$ is satisfied, it accesses the child node for the outside region and updates r_q. After this, if the inequality $d(p, q) - r_q \leq r_p$ is satisfied, it accesses the other child node and updates r_q. Finally, it answers the k closest objects. The search cost is $O(\log N)$ at best.

4   Performance Evaluation

We implemented PCTree on the Metric Space Library [1]. It provides several indexing algorithms, metric spaces, and datasets. We compared PCTree with GHT [14], MVP [15, 2], LC [4], and SAT [13], which are also in the library. As in the related work [13], the indexes were evaluated in terms of the distance computations. We measured the percentage of objects examined by the total number of distance calculations during the search. In the evaluation, 100 % of objects examined represents the cost of the linear scan algorithm. We conducted the experiment on a Linux PC equipped with an Intel(R) Quad Core Xeon(TM) X5492 3.40GHz CPU and that had 64GB of memory.

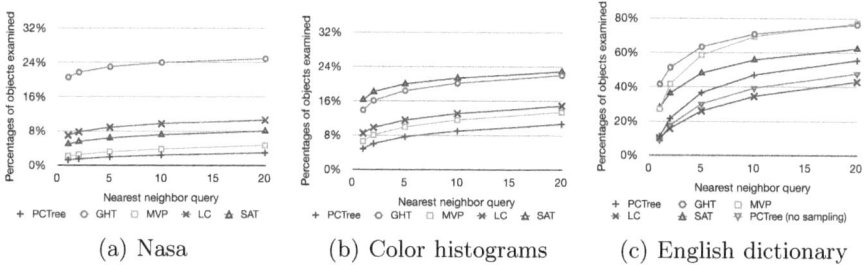

(a) Nasa   (b) Color histograms   (c) English dictionary

Fig. 2. Index Performance on Real Datasets

We used the following datasets.

Nasa is a set of feature vectors made by NASA [1]. It consists of $40,150$ vectors in a 20-dimensional feature space.

Color histograms are the color histograms of $112,544$ images represented by vectors in 112-dimensional space [1].

English dictionary consists of $69,069$ English words in the form of strings and that was generated by [1].

For each real dataset, we randomly selected $1,000$ objects for the queries and conducted the k-nearest neighbor queries on the remaining objects. All the real datasets are available from the Web.

We compared the PCTree with other methods for the real datasets. Figure 2 shows the index performances for the real dataset. The vertical axis represents the number of distance computations for the k-nearest neighbor queries where k ranges from 1 to 20. The horizontal axis is k.

The PCTree outperforms the other methods for the two vector datasets whereas it outperforms the others except for LC on the English dictionary dataset. Compared with the other methods, the PCTree is better than the other methods on a wide range of data distributions. For example, MVP is good for the NASA and Color histograms, but is weak for the English dictionary. On the other hand, LC is good for the English dictionary, but is weak for the NASA and Color histograms. We think that the PCTree automatically optimizes the index structure according the data distribution.

We guessed that the PCTree would need more samples for calculating the PC for English dictionary. Therefore, we plotted the search costs for the PCTrees and without sampling in Figure 2 (d). The figure shows that the search cost without sampling is close to that of the LC. It is difficult to determine the appropriate parameters without more knowledge about the data distribution. The parameter tuning of the PCTree remains a future topic of study.

5   Conclusion

We presented a similarity search index named PCTree. PCTree is based on maximizing both the pruning and balance. We defined the Partitioning Capacity (PC)

for selecting a pivot and its partitioning in PCTree. By using the PC, PCTree automatically optimizes the index structure according to the data distribution and reduces the search cost when using the PC.

We are currently attempting to improve the sampling scheme for pivot candidates. Having more pivot candidates can help to reduce the search cost. However, as the number of candidates increases, the indexing cost also increases. We have to reduce the indexing cost of PCTree before we can use it in practical situations.

References

[1] Metric spaces library, http://www.sisap.org/metric_space_library.html
[2] Bozkaya, T., Ozsoyoglu, Z.M.: Indexing large metric spaces for similarity search queries. ACM Trans. on Database Systems 24(3), 361–404 (1999)
[3] Chevez, E., Marroguin, J.L., Navarro, G.: Fixed queries array: A fast and economical data structure for proximity searching. Multimedia Tools Applications 14(2), 113–135 (2001)
[4] Chevez, E., Navarro, G.: A compact space decomposition for effective metric indexing. Pattern Recognition Letters 24(9), 1363–1376 (2005)
[5] Ciaccia, P., Patella, M., Zezula, P.: M-tree: An efficient access method for similarity search in metric spaces. In: VLDB (1997)
[6] Dohnal, V., Gennaro, C., Savino, P., Zezula, P.: D-index: Distance searching index for metric data sets. Multimedia Tools and Applications 21(1), 9–33 (2003)
[7] Hays, J., Efros, A.A.: Scene completion using millions of photographs. In: SIGGRAPH (2007)
[8] Jagadish, H.V., Ooi, B.C., Tran, K.L., Yu, C., Zhang, R.: idistance: An adaptive b+-tree based indexing method for nearest neighbor earch. ACM Trans. on Database Systems 30(2), 364–397 (2003)
[9] Jones, G.A., Jones, J.M.: Information and Coding Theory. Springer, Heidelberg (2000)
[10] Traina Jr., C., Santos Filho, R.F., Traina, A.J., Vieira, M.R., Faloutsos, C.: The omni-family of all-purpose access methods: a simple and effective way to make similarity search more efficient. The VLDB Journal 16(4), 483–505 (2007)
[11] Traina Jr., C., Traina, A.J.M., Seeger, B., Faloutsos, C.: Slim-trees: High performance metric trees minimizing overlap between nodes. In: Zaniolo, C., Grust, T., Scholl, M.H., Lockemann, P.C. (eds.) EDBT 2000. LNCS, vol. 1777, p. 51. Springer, Heidelberg (2000)
[12] Kurasawa, H., Fukagawa, D., Takasu, A., Adachi, J.: Maximal metric margin partitioning for similarity search indexes. In: CIKM (2009)
[13] Navarro, G.: Searching in metric spaces by spatial approximation. The VLDB Journal 11(1), 28–46 (2002)
[14] Uhlmann, J.K.: Satisfying general proximity/similarity queries with metric trees. Information Processing Letters 40(4), 175–179 (1991)
[15] Yianilos, P.N.: Data structures and algorithms for nearest neighbor search in general metric spaces. In: SODA (1993)
[16] Yianilos, P.N.: Excluded middle vantage point forests for nearest neighbor search. In: ALENEX (1999)
[17] Zhuang, Y., Zhuang, Y., Li, Q., Chen, L., Yu, Y.: Indexing high-dimensional data in dual distance spaces: a symmetrical encoding approach. In: EDBT (2008)

Minimum Spanning Tree on Spatio-Temporal Networks

Viswanath Gunturi[1], Shashi Shekhar[2], and Arnab Bhattacharya[1]

[1] Computer Science and Engineering, Indian Institute of Technology, Kanpur, India
[2] Computer Science and Engineering, University of Minnesota, USA
vgvm@cse.iitk.ac.in, shekhar@cs.umn.edu, arnabb@iitk.ac.in

Abstract. Given a spatio-temporal network whose edge properties vary with time, a *time-sub-interval minimum spanning tree* (TSMST) is a collection of minimum spanning trees where each tree is associated with one or more time intervals; during these time intervals, the total cost of this spanning tree is the least among all spanning trees. The TSMST problem aims to identify a collection of distinct minimum spanning trees and their respective time-sub-intervals. This is an important problem in spatio-temporal application domains such as wireless sensor networks (e.g., energy-efficient routing). As the ranking of candidate spanning trees is non-stationary over a given time interval, computing TSMST is challenging. Existing methods such as dynamic graph algorithms and kinetic data structures assume separable edge weight functions. In contrast, we propose novel algorithms to find TSMST for large networks by accounting for both separable and non-separable piecewise linear edge weight functions. The algorithms are based on the ordering of edges in edge-order-intervals and intersection points of edge weight functions.

1 Introduction

Given a spatio-temporal (ST) network, *a time-sub-interval minimum spanning tree* (TSMST) is a collection of distinct minimum spanning trees, where each tree is associated with one or more time intervals. During these time intervals, the total cost of this spanning tree is the least among all possible spanning trees. TSMST computation is important in ST network application domains such as wireless sensor networks. For example, energy-efficient transmission paths can be modeled as minimum spanning trees [1,2]. In many applications, the nodes are not stationary, i.e., they change their physical location with time, (e.g., sensor network among robots on a reconnaissance mission [3]). Usually in such scenarios, the sensor nodes physically move on a predetermined trajectory to collect data from an observation field. This network of sensors can be represented as a ST network where a sensor is represented as a node and the communication link between any two sensors is represented as an edge (see Figure 1(a)). Time dependent edge weights represent the cost of packet transmission between the sensors. Since the sensors are moving, the distance between two nodes changes with time. Since the energy required to transmit data from one node to another

P. García Bringas et al. (Eds.): DEXA 2010, Part II, LNCS 6262, pp. 149–158, 2010.

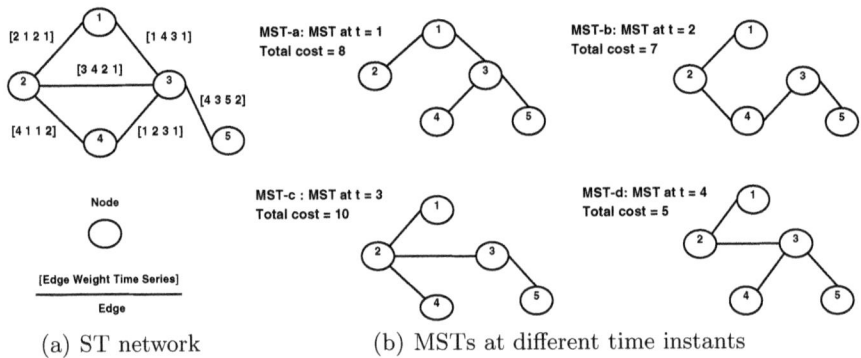

(a) ST network   (b) MSTs at different time instants

Fig. 1. Spatio-temporal network and its corresponding MSTs at various times

is directly proportional to the square of distance [1] between them, even a small change in the distance affects the cost of transmission significantly. The solution to the TSMST problem effectively determines the energy-efficient communication paths among these sensor nodes.

However, computing TSMST is expensive because of the non-stationary ranking of candidate spanning trees in a ST network. This is illustrated in Figure 1. Figure 1(b) shows the minimum spanning trees (MSTs) at different time instants for the ST network shown in Figure 1(a).

1.1   Related Work

Traditional methods for computing MSTs [4,5] developed for static networks assume stationary ranking of candidate trees, i.e., they assume that mutual ranking (on the basis of total cost) among the spanning trees does not change with time. Consequently, classical greedy algorithms (e.g., Kruskal's [5] and Prim's [5]) cannot be applied to TSMST problem on spatio-temporal networks.

Dynamic graph algorithms [6,7] and kinetic algorithms [8,9,10] incorporate non-stationary candidate ranking by making use of structures such as topology trees [6] and dynamic trees [11]. However, these dynamic data structures can model discrete changes such as single edge insertion, deletion or weight modification [6,7], but cannot handle piecewise linear edge weight functions.

Moreover, work done in field of non-stationary candidate ranking assume separable edge weights, i.e, they assume that there is no correlation between the different weights of an edge at different time instants. Due to this assumption the kinetic algorithms do not address the situation when change points and intersection points overlap. Change points are those time instants where an edge weight function changes its slope and intersection points are those time instants where two or more edge functions intersect, i.e., they have same edge weight. For example, in Figure 2 edge (3,5) has change points at $t = 3$ and $t = 2$ and edge weight functions of (1,3) and (2,4) intersect at $t = 1.5$, whereas the weight functions of edges (1,3) and (2,3) both change and intersect at the same point. At such points the MST cannot be changed even if it is required.

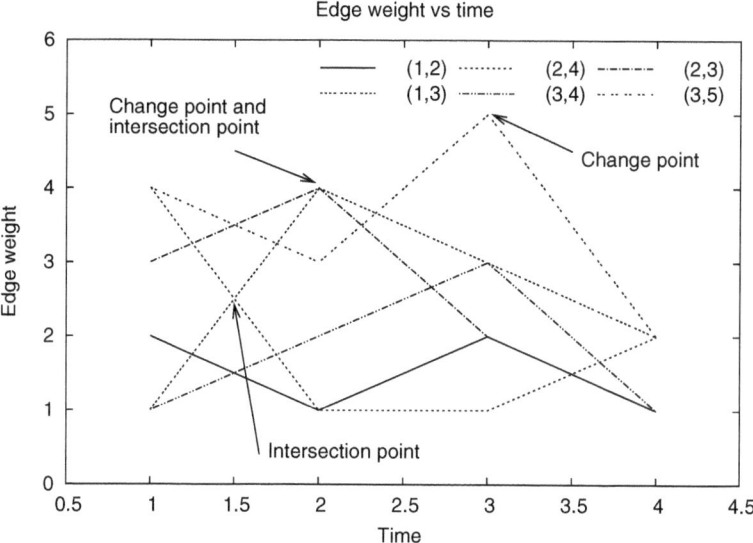

Fig. 2. Edge weight function plot

Our contributions: In this paper, we present the problem of time-sub-interval minimum spanning trees (TSMST) in a spatio-temporal network. We propose two algorithms to find the TSMST, and provide running time analyses. The algorithms allow the ST network to have both separable as well as non-separable edge weight functions.

The rest of the paper is organized as follows. Section 2 defines the basic concepts before presenting the formal problem definition. In Section 3, two algorithms for solving the problem of TSMST are described. We present the asymptotic complexity analysis of the algorithms in Section 4. Section 5 presents the experimental results before Section 6 concludes.

2   Basic Concepts and Problem Definition

This paper models a ST network as a *time-aggregated-graph* (TAG) [12,13]. A TAG is a graph in which each edge is associated with a edge weight function. These functions are defined over a time horizon and are represented as a time series. For instance, edge (3,5) of the graph shown in Figure 1(a) has been assigned a time series [4 3 5 2]; this implies that the weight of the edge at time instants t=1, 2, 3, and 4 are 4, 3, 5, and 2 respectively. The edge weight is assumed to vary *linearly* between two time instants. We also assume that no two edge weight functions have same values for two or more consecutive time instants of their time series. If such a case occurs, then the values of any one of the edges are increased (or decreased) by a small quantity ϵ to make them distinct. The time instant at which the weight function of an edge changes its slope called a *change point* and where two edge weight functions meet is called

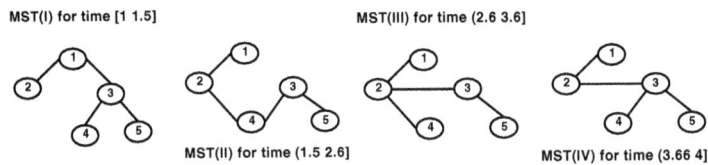

Fig. 3. TSMST for network shown in Figure 1(a)

an *intersection point*. The edge weight functions of the graph in Figure 1(a) are shown in Figure 2.

Definition 1 (Time-sub-interval). *A time-sub-interval,* $\tau = (\tau_s, \tau_e)$*, is a maximal sub interval of time horizon* $[1, K]$ *that has a unique MST, i.e., the MST does not change during this time interval.*

Definition 2 (Edge-order-interval). *An edge-order-interval,* $\omega = (\omega_s, \omega_e)$*, is a sub interval of time horizon* $[1, K]$ *during which there is a clear ordering of edge weight functions, i.e., none of the edges intersect with each other.*

Figure 3 shows the four time-sub-intervals and the corresponding MSTs. An edge-order-interval is guaranteed to have a unique MST (see [14] for detailed proof). Two or more consecutive edge-order-intervals may have the same MST. A time-sub-interval is usually composed of one or more edge-order-intervals. For example, in Figure 2, the interval (2.66 3.0) is an edge-order-interval whereas the interval (2.66 3.66] is a time-sub-interval which is a union of three consecutive edge-order-intervals (2.66 3.0), (3.0 3.5) and (3.5 3.66).

Problem Definition: Given an undirected ST network $G = (V, E)$ where V is the set of vertices of graph, E is the set of edges, and each edge $e \in E$ has a weight function associated with it. The weight function is defined over the time horizon $[1, K]$. The problem of $TSMST$ is to determine the set of distinct minimum spanning trees, $TMST_i$, and their respective time-sub-intervals.

The total cost of $TMST_i$ is least among all other spanning trees over its respective time-sub-intervals. We assume that for all edges $e \in E$, the edge weight function is defined for the entire time interval $[1, K]$. The weight of an edge is assumed to vary linearly between any two time instants of time series. The full version of the paper [14] describes how to relax the edge presence assumption.

In our example of an energy efficient communication network maintained by a group of sensors, the communication network is represented as a ST network shown in Figure 1(a). The collection of distinct minimum spanning trees and their corresponding time-sub-intervals is shown in Figure 3.

3   TSMST Computation Algorithms

Here we present two algorithms for computing the TSMST of a spatio-temporal network. Consider again the sample network shown in Figure 1(a) and its edge-weight function plot in Figure 2. The following observations can be inferred from edge weight function plot.

Observation 1. *Consider any two consecutive (with respect to time coordinate) intersection points of the edge weight functions. These time coordinates form an edge-order-interval. Within this time interval all the edge weight functions have a well defined order.*

Observation 2. *Using the ordering of edge weights within an edge-order-interval, an MST for this interval can be built using a standard greedy algorithm such as Kruskal's or Prim's.*

Observation 3. *There will be a single MST for the entire edge-order-interval.*

3.1   Time Sub-Interval Order (TSO) Algorithm

The time sub-interval order (TSO) algorithm is designed using the three previous observations. It starts by determining all edge-order-intervals by computing all the intersection points of the edge weight functions. The intersection points are then sorted with respect to time. Next, the algorithm computes a MST at each of these intersection points and outputs the set of distinct MSTs and their corresponding time-sub-intervals. The detailed pseudo code of the algorithm appears in the full version of the paper [14].

3.2   Edge Intersection Order (EIO) Algorithm

The TSO algorithm incurs an unnecessary overhead of computing a MST for each intersection point even if the MST does not change at a intersection point. This can happen when only tree or only non-tree edge weight functions intersect at a point. Moreover, if the edges involved in the intersection are from different bi-connected components, the MST will not change.[1] Furthermore, if only one tree and one non-tree edge (belonging to the same fundamental cycle) are involved in an intersection, we can exchange these edges in the tree provided they do not belong to any other fundamental cycle.[2] For example, in Figure 4, the MST for time $t > 1.5$ can be obtained by exchanging the edges involved in intersection in current MST. These ideas are presented formally in the following propositions. The proofs of these propositions appear in [14].

Proposition 1. *The intersection of the edge weight functions of two non-tree edges at any time instant will not affect the MST.*

Proposition 2. *The intersection of the edge weight function of two tree edges at any time instant will not affect the MST.*

Proposition 3. *The intersection of the edge weight functions of two edges belonging to different bi-connected components will not affect the MST.*

---

[1] A bi-connected component of a connected graph is a maximal set of edges such that the corresponding subgraph cannot be disconnected by deleting any vertex.

[2] Given a spanning tree and a non-tree edge, a fundamental cycle is the cycle created by adding the non-edge to the spanning tree.

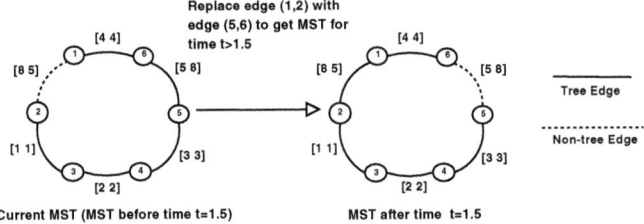

Fig. 4. Edge exchange inside a cycle

The above propositions are used in designing an incremental algorithm for computing the TSMST. The algorithm is called edge intersection order (EIO) algorithm. EIO starts by computing the MST of the network at time $t = 1$ and then continues to update the tree, only if necessary, at each intersection point. Through preprocessing, some additional information about the edges is stored while computing the MST at time $t = 1$. This information is used to save some computation while updating the MST for later intersection points. The modified reverse-delete algorithm (pseudo code in [14]) is used to compute the MST at time $t = 1$. The algorithm first computes the depth first search (DFS) tree [5] of the given ST network. A non-tree edge $ne = (f_s, f_e)$, where f_e is the ancestor of f_s, is chosen. Now, edge ne and edges seen while following the parent pointers from node f_s to f_e and ne form a cycle. The heaviest edge of this cycle is deleted. This cycle and its member edges are stored to save some computation while updating the MST at the intersection points. A DFS tree of the remaining edges is computed. A non-tree edge is again picked up and the heaviest edge in its cycle is deleted. This process continues until only $n - 1$ edges are left in the network. At this point there will no non-tree edges after constructing the DFS tree, i.e., all edges are tree edges. These edges form the MST of the network (proof in [14]). The bi-connected component information of all the edges is determined using the algorithm given in [15].

While considering an intersection point, two levels of filters are applied to prune the intersection point which cannot cause any change in MST. If all the edges involved in an intersection are either non-tree edges or tree edges, then it can be pruned using Proposition 1 and Proposition 2. Similarly, if all the edges involved in an intersection belong to different bi-connected components, then it can be pruned using Proposition 3. After applying these filters, the edges are grouped by their bi-connected component. Now, within each group (i.e., each bi-connected component), if the edges are only tree edges or non-tree edges, they can be again pruned using Proposition 1 and Proposition 2. After applying these filters, we check if the relative orders of edge weights before and after the intersection point are same. If so, again, the intersection point can be pruned.

If an intersection point is not pruned after applying all the filters, then a new MST is made by making changes to the previous MST. If only two edges (per bi-connected component) are involved in the intersection and they are part of only one common cycle, i.e., they are not part of any cycle except the one which is common, then we can directly exchange the edges in the tree, i.e., make the heavier edge as non-tree edge and the lighter edge as tree edge. The information regarding the cycles is collected while computing the MST at time $t = 1$. Note that in the two-edge intersection case discussed above, adding the non-tree edge to the tree and deleting the heaviest edge from its fundamental cycle would still give the correct MST. The information gathered while constructing the MST at $t = 1$ is used to save this unnecessary re-computation. In all other cases, we add each of the non-tree edges involved in the intersection to the MST and delete the heaviest edge from their respective fundamental cycles. The start time of the time-sub-interval of the new MST and end time of the time-sub-interval of previous MST are set to the time coordinate of intersection point at which the MST changed. The pseduo code of the EIO algorithm is presented in [14].

4   Analytical Evaluation

The correctness proofs of the TSO and the EIO algorithm appears in the full version of the paper [14].

Asymptotic Analysis of TSO: Since the edge weight is assumed to vary linearly between any two time instants in a time series, there can be at most $O(m^2)$ (where m is the number of edges) intersections among the edge weight functions. If this happens between all time instances in the entire time horizon $[1, \ldots K]$, the total number of intersections is $O(m^2 K)$. The time needed to sort all the intersection points is $O(m^2 K \log(m^2 K))$. For each intersection point, TSO recomputes the MST in $O(m \log m)$ time. Thus, the total time complexity of the TSO algorithm is $O(m^3 K \log m + m^2 K \log(m^2 K))$.

Asymptotic Analysis of EIO: The running time of the EIO algorithm is sensitive to the number of intersection points and number of edges involved per intersection point. Here, we consider two kinds of intersection points: (i) where all the edges are involved and (ii) where only two edges are involved.

First, consider the case of a two-edge intersection. The number of two-edge intersections between a pair of consecutive time instants is $O(m^2)$. All the filtering steps take $O(1)$ time. Similarly, sorting of edges involved in intersection (to check whether there is a change in relative order of edges) would take only constant time as there are only two edges. Finding the heaviest edge in the fundamental cycle of a non-tree edge can take $O(n)$ time for a graph of n nodes in the worst case (when the fundamental cycle involves all the nodes of the graph). Thus, the two-edge intersection case would take $O(m^2 n)$ in the worst case for one consecutive pair of time instances. This can happen for a maximum of $O(K)$ times, once between every two time instances of the time series.

Next, consider the case when $O(m)$ edges intersect at a single point. Sorting these edges takes $O(m \log m)$ time. This kind of intersection would involve a maximum of $O(m - n + 1)$ non-tree edges. Thus, finding the heaviest edge in the fundamental cycle would take $O(n)$ per non-tree edge, thereby incurring a total cost of $O(mn + m \log m)$ time in the worst case. Intersection of $O(m)$ edges can happen $O(K)$ times. This is because the edge weight functions vary linearly

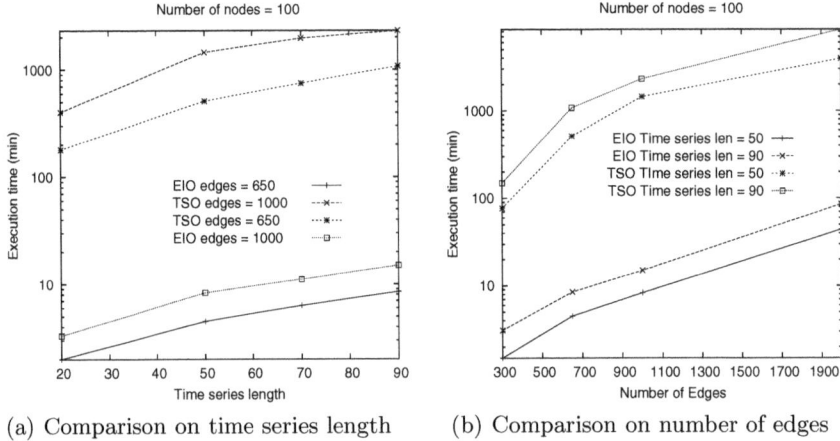

(a) Comparison on time series length   (b) Comparison on number of edges

Fig. 5. Comparison of EIO and TSO algorithms

between two time instances of the time series, and thus they can all meet at only one point between two time instants of the time series.

Suppose for a time interval of length K, two-edge intersections occur K_1 times and $O(m)$ edge intersections occur K_2 times. Here, $K_1 + K_2 = K$. The time required to sort the intersection points in two-edge intersection case is bounded by $O(m^2 K_1 \log(m^2 K_1))$, whereas it would take $O(K_2 \log K_2)$ time to sort when $O(m)$ edges are involved in the intersection. Therefore, the total time required for sorting all the intersection points is $O((m^2 K_1 + K_2) \log(m^2 K_1 + K_2))$. Thus, the total worst case time required by the EIO algorithm is the sum of times spent on two-edge intersections, $O(m)$-edge intersections, and the time required to sort all the intersection points. Hence, the overall time complexity of EIO is $O(m^2 n K_1 + mn K_2 + K_2 m \log m + (m^2 K_1 + K_2) \log(m^2 K_1 + K_2))$.

5   Experimental Analysis

We conducted experiments on both TSO and EIO algorithms in order to compare them as well as to see the effect of the different parameters on the running time. The experiments were run on synthetic graphs generated randomly (the details are in [14]). The experiments were conducted on an Intel Xeon workstation running Linux with 2.40GHz CPU, 8GB RAM.

Effect of length of time series: Figure 5(a) shows the performance of EIO and TSO algorithms as the length of the time series increases. The superior performance of the EIO algorithm over the TSO algorithm is due to the increase of intersection points that occurs with the increase in the length of time series.

Effect of number of edges: Figure 5(b) shows the performance of the algorithms as the number of edges increase. As expected, the execution time of TSO increases much more rapidly than EIO.

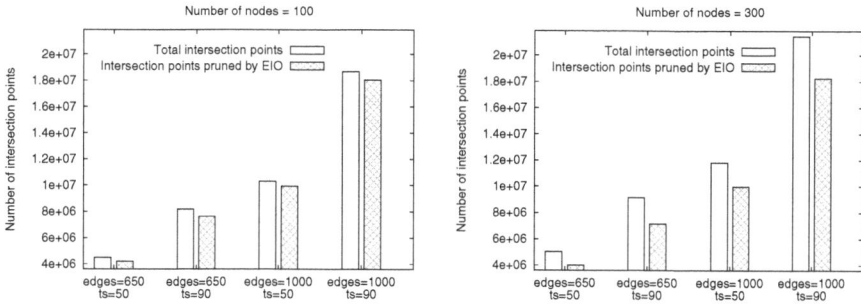

Fig. 6. Effect of filters on number of intersection points

Effect of filters: The next experiment measures the effect of the different filters. Figure 6 shows the total number of the intersection points and the number of intersection points pruned by the EIO algorithm. A large number of intersection points were pruned by the EIO algorithm, thereby clearly showing the superior performance of the filters used in the algorithm.

6   Conclusions

The time-sub-interval minimum spanning tree (TSMST) problem is a key component in various spatio-temporal applications. The paper proposes two novel algorithms for TSMST computation. The time sub-interval algorithm (TSO) computes the TSMST by recomputing the MST at all time points where there is a possible change in the ranking of candidate spanning trees (i.e., it recomputes MST at all the intersection points of edge weight functions) and then outputs the set of distinct MSTs along with their respective time-sub-intervals. The edge intersection order algorithm (EIO), on the other hand, updates the MST only if necessary at these time points. Complexity analysis shows that EIO is faster than TSO algorithm by a factor of almost $O(m)$. Experiments validate that EIO is faster than TSO algorithm by orders of magnitude. We plan to extend the algorithms to provide optimal solutions when the edge weight function is not constrained to be linear, but can be polynomial in nature.

Acknowledgments

We are particularly grateful to the members of the Spatial Database Research Group at the University of Minnesota for their helpful comments and valuable suggestions. We would like to extend our gratitude to Prof Shashank Mehta of Indian Institute of Technology Kanpur for his valuable comments. We would also like to extend our thanks to Kim Koffolt for improving the readability of this paper.

This work was supported by NSF grant (grant number NSF III-CXT IIS-0713214) and USDOD grant (grant number HM1582-08-1-0017 and HM1582-07-1-2035). The content does not necessarily reflect the position or the policy of the government and no official endorsement should be inferred.

References

1. Huang, G., Li, X., He, J.: Dynamic minimal spanning tree routing protocol for large wireless sensor networks. In: IEEE Conf. Industrial Electronics and Applications (IEA), pp. 1531–1535 (2006)
2. Muruganathan, S., Ma, D., Bhasin, R., Fapojuwo, A.: A centralized energy-efficient routing protocol for wireless sensor networks. IEEE Comm. Mag. 43, 8–13 (2005)
3. Yoon, S., Qiao, C.: A novel approach to reconnaissance using cooperative mobile sensor nodes. In: Military Communications Conf. (MILCOM), pp. 1–7 (2006)
4. Kleinberg, J., Tardos, E.: Algorithm Design. Pearson Education, London (2009)
5. Cormen, T.H., Leiserson, C.E., Rivest, R.L., Stein, C.: Introduction to Algorithms. MIT Press, Cambridge (2001)
6. Frederickson, G.: Data structures for on-line updating of minimum spanning trees. SIAM J. Computing 14, 781–798 (1985)
7. Henzinger, M.R., King, V.: Maintaining minimum spanning trees in dynamic graphs. In: Int. Coll. Automata, Languages, and Programming, pp. 594–604 (1997)
8. Agarwal, P., Eppstein, D., Guibas, L.J., Henzinger, M.R.: Parametric and kinetic minimum spanning trees. In: IEEE Symp. Foundations of Computer Science (FOCS), pp. 596–605 (1998)
9. Guibas, L.J.: Kinetic data structures: a state of the art report. In: Workshop on Algorithmic Foundations of Robotics (WAFR), pp. 191–209 (1998)
10. Basch, J., Guibas, L.J., Hershberger, J.: Data structures for mobile data. In: ACM-SIAM Symp. Discrete Algorithms (SODA), pp. 747–756 (1997)
11. Sleator, D.D., Tarjan, R.E.: A data structure for dynamic trees. In: ACM Symp. Theory of Computing (STOC), pp. 114–122 (1981)
12. George, B., Kim, S., Shekhar, S.: Spatio-temporal network databases and routing algorithms: A summary of results. In: Papadias, D., Zhang, D., Kollios, G. (eds.) SSTD 2007. LNCS, vol. 4605, pp. 460–477. Springer, Heidelberg (2007)
13. George, B., Shekhar, S.: Time-aggregated graphs for modelling spatio-temporal networks. J. Semantics of Data XI, 191 (2007)
14. Gunturi, V., Shekhar, S., Bhattacharya, A.: Minimum spanning tree on spatio-temporal networks (2010), arXiv:1003.1251v2 [cs.DS]
15. Alfred, V.A., Ullman, J.D., Hopcroft, J.E.: Data Structures and Algorithms. Addison-Wesley Longman, Amsterdam (1983)

Real-Time Temporal Data Warehouse Cubing

Usman Ahmed, Anne Tchounikine, Maryvonne Miquel, and Sylvie Servigne

Université de Lyon, CNRS
INSA-Lyon, LIRIS, UMR5205, F-69621, France
`firstname.lastname@insa-lyon.fr`

Abstract. Traditional data warehouses are built in an off-line periodic fashion which makes them less valuable in applications where the most up-to-date data is required. For these applications, data should be incorporated in the warehouse and made available as soon as possible in "Real Time Data Warehouse". In this paper we propose an indexing model named TiC-Tree, in order to simultaneously index and store multidimensional detailed and aggregated data. Our contribution exploits the temporal nature of data and focuses on range and/or group-by queries. We evaluate our proposal with the synthetic data set Star Schema Benchmark and advocate it in comparison with other existing solution.

Keywords: Data warehouse, OLAP, Real Time data, Graph based index.

1 Introduction and Motivation

The maintenance of data warehouse is usually carried out in an off-line fashion, after the facts are inserted via bulk incremental operations. This restriction makes them unsuitable for certain applications such as monitoring systems for natural risk management, traffic surveillance systems etc., that require the data to be always up to date. This type of applications raises the need of what can be called as Real Time Data Warehouses (RTDW), Active Data Warehouses [1] or Zero Latency Data Warehouses [2]. The main challenge is to integrate new data in the cube and make it available as soon as possible. Due to the type of application domain, query response time is of course a decisive key. Temporal OLAP analysis, i.e. drilling and slicing over time dimension, is of major interest. In this work: *First,* we propose a cubing model named TiC-Tree that provides fast update at arrival of each new fact and improves the response time of OLAP range and group-by queries over time dimension. The TiC-Tree indexes and stores detailed and aggregated data in the same structure and favors the grouping of data representing chronologically closed events even if the data delivery is delayed. *Then,* we implement our proposition and evaluate its performance with the synthetic data set Star Schema Benchmark [3] with slight modification, i.e. the addition of hierachy level *hour* in *Time* dimension. We examine the different cases when the input data set is chronologically ordered and when some of the data arrives with a delay.

P. García Bringas et al. (Eds.): DEXA 2010, Part II, LNCS 6262, pp. 159–167, 2010.
© Springer-Verlag Berlin Heidelberg 2010

2   Related Works

Cubing operation raises considerable challenges related to the complexity of calculation and storage of the data. An on-going and active research area includes numerous works that aim at defining strategies for selecting subsets of views to be materialized and efficient computation methods for the aggregates [4]. Graph based methods, such as Cubetrees [5] propose an alternative approach exploring hierarchical storage structures for the cube that may eventually be compressed (Dwarf [6], QC-Tree [7]). On the other hand, R-tree and variants [8,9] are widely used to index spatial and/or temporal data. These techniques are based on identification of Minimum Bounding Rectangle (MBR) for space partitioning. However, none of these methods support neither pre-defined dimension hierarchies nor pre-aggregation. Indices proposed in the special context of data warehouses include multidimensional array based methods, bit mapped indices, hierarchical and spatial indexing techniques based methods [10,5]. Among these, the DC-Tree[11] supports both pre-aggregation and pre-defined hierarchies and is based on X-Tree where MBR are replaced by MDS (Minimum Describing Sequence). Indeed, the use of MBR assumes that data is totally ordered in the referenced multidimensional space, while MDS uses the partial ordering of members induced by the dimension hierarchies.

3   Contribution

Time is a peculiar dimension of data warehouses. It is usually the only dimension whose instances grow continuously while all other dimensions are generally either static or slowly changing dimensions. Its members are naturally and totally ordered at each level of hierarchy, however, in context of real-time systems, this chronological order may be disregarded at the time of insertion because of network delays or system downtime. In this paper, we propose to take these features of time into consideration and propose a solution for cubing temporal data warehouses, thanks to a tree structure named TiC-Tree that uses a tailored definition of MDS with special handling for time dimension.

3.1   Multidimensional Model

Let $\mathcal{D}$ be a set of n dimensions $D_1, D_2, \ldots D_n$ of a multidimensional model. Each dimension D_i has a totally ordered set $\mathcal{L}_i$ of levels. A dimension hierarchy can be viewed as a directed acyclic graph of levels. We note $l_i^{k+1} \uparrow l_i^k$ the existing edge between 2 levels $l_i^{k+1}, l_i^k \in \mathcal{L}_i$ and $1 \le k \le |\mathcal{L}_i|$. An instance of a dimension D_i is defined by a set of values over $\bigcup domain(l_i^k)$, and a parent/child relation noted $\Uparrow$ between the members of levels such as $l_i^{k+1} \uparrow l_i^k$. A dimension instance can be viewed as a directed acyclic graph of members. If a path exists from n to m in the graph of members, then m is said to be an *ancestor* of n. Each dimension D_i contains a top most level ALL such that $domain(ALL) = \{all\}$.

Example 1. For time dimension D_t we define an ordered set of levels $\mathcal{L}_t$ as: $\mathcal{L}_t$ = {ALL, year, month, day, hour} *and* $|\mathcal{L}_t| = 5$ *and* All $\uparrow$ year $\uparrow$ month $\uparrow$ day $\uparrow$ hour *and* $2008 \Uparrow$ March $2008 \Uparrow$ 12 March 2008.

Definition 1. *(Fact Table)* *A fact table* $T_F(l_1^1, l_2^1, \ldots, l_n^1)$ *of a multidimensional model is a table with tuples* $< x_1, x_2, \ldots x_n, m_1, m_2, \ldots, m_p >$ *where* $x_i \in domain(l_i^1)$ *i.e. is a member of the lowest level of each dimension hierarchy and* m_j *is a numeric value called measure of the fact.*

Definition 2. *(Aggregate Table)* *An aggregate table* $T_A(l_1^{k_1}, l_2^{k_2}, \ldots l_n^{k_n})$ *of a multidimensional model is a table with tuples* $< x_1', x_2', \ldots x_n', a_1, a_2, \ldots, a_p >$ *where* $x_i' \in domain(l_i^{k_i})$ *and* $\exists i | k_i > 1$ *and* a_j *is an aggregate measure value obtained by some aggregation function.*

Table 1. Extracted tuples of the Fact Table with 4 dimensions and one measure value

ID	Customer	Supplier	Part	Time	Quantity
t1	Customer234	Supplier329	Part3432	04:00 05/03/1999	15
t2	Customer103	Supplier1023	Part862	12:00 23/04/2000	60
t3	Customer20	Supplier19	Part1322	17:00 11/07/1999	20
t4	Customer20	Supplier1360	Part1322	02:00 18/11/1999	15
t5	Customer293	Supplier1870	Part94	10:00 03/06/2000	10
t6	Customer293	Supplier329	Part94	13:00 13/02/1999	30
t7	Customer923	Supplier1870	Part647	03:00 13/12/1999	45

Definition 3. *(MDS)* *A Minimum Describing Sequence* $M(l_1^{k_1}, l_2^{k_2}, \ldots, l_n^{k_n})$ *is a sequence* $[S_1, S_2, \ldots, S_n]$ *of n sets where* $S_i \subset domain(l_i^{k_i})$ *and n is the number of dimensions.*

Two MDS $M(l_1^{k_1}, l_2^{k_2}, \ldots, l_n^{k_n})$ and $N(l_1^{k_1'}, l_2^{k_2'}, \ldots, l_n^{k_n'})$ are considered to be at same level if $\forall i = 1, 2, \ldots n$ $k_i = k_i'$, i.e. the corresponding sets have their members at the same levels of dimension hierarchies. An MDS can be regarded as a minimal representation of hyper-rectangle of members, in an n-dimensional space.

Example 2. M=[{Europe},{USA},{Part342},{1999}] and L=[{Europe},{USA, France}, {MFGR#5113}, {'Apr99','Jul99'}] are 2 MDS at different levels.

3.2 The TiC Tree

The TiC-Tree stores and indexes MDSs with associated aggregate or detailed values. Leaves of the tree are tuples of the fact table whereas internal nodes are aggregates computed on stored facts. Unlike the DC-Tree, the TiC-Tree processes the temporal data so as to help the grouping of closer values together. We define a set of metrics which are used to build, update and query the tree and allow optimization of MDS distribution in the nodes. These metrics take the totally ordered nature of time into account and facilitate this grouping. This grouping strategy will facilitate time range and group-by queries. For all non-temporal dimensions, the calculation of these metrics is based on the cardinality of dimension sets in the MDS, while for the temporal one these are calculated on the basis of time duration covered by the MDS.

Let M and N be two MDS with sequence $[S_1, S_2, \ldots, S_n]$ and $[T_1, T_2, \ldots, T_n]$ respectively. For time dimension D_t, we note:

$$int(S_t) = [\min_{m_i \in S_t} (m_i), \max_{m_i \in S_t} (m_i)] .$$

Definition 4. *(Contains) M contains N is true if $\forall i \neq t : T_i \subset S_i$ or $\forall n \in T_i, \exists m \in S_i \mid m$ is ancestor of n and for $i = t : int(T_t) \subset int(S_t)$ or $\forall n \in T_t, \exists m \in S_t \mid m$ is ancestor of n.*

MDS M contains another MDS N if: (1) for all non-temporal dimensions, all the sequence sets of N are included in those of M, or made of children of the members of the sequence sets of M, and (2) the time interval covered by N is contained in the interval covered by M.

Example 3. Let   M=[{France},{USA},{Part342},{2001,2002}];   N=[{USA}, {USA},{Part342},{2000,2002}];   O=[{Europe},{USA},{Part342},{2001,2005} then:

¬(M *contains* N) and (O *contains* M)

Definition 5. *(Overlap) The overlap of M and N denoted by overlap(M, N) is defined for 2 MDS at the same level of hierarchy, as:*

$$overlap(M, N) = \begin{cases} 0 & \text{if } int(S_t) \cap int(T_t) = \emptyset \\ \prod_{i=1, i\neq t}^{n} |S_i \cap T_i| & \text{else} \end{cases}$$

The overlap determines the volume of intersection between two MDSs. The overlap between two MDSs can be calculated, if and only if the MDSs are at same level.

Example 4. overlap(M,N) = 0*1*1=0 ; overlap(M,O) : Cannot be calculated.

Definition 6. *(Extension) The extension of M to accommodate N denoted by extension(M|N) is defined for 2 MDS at the same level of hierarchy, as:* $extension(M|N) = \sum_{i=1, i\neq t}^{n} |N_i - M_i| + df_l + df_u$ *where:*

$$df_l = \begin{cases} 0 \text{ if } \min_{m_i \in S_t} (m_i) - \min_{n_i \in T_t} (n_i) < 0 \\ \min_{m_i \in S_t} (m_i) - \min_{n_i \in T_t} (n_i) & \text{else} \end{cases} \text{ and } df_u = \begin{cases} 0 \text{ if } \max_{n_i \in T_t} (n_i) - \max_{m_i \in S_t} (m_i) < 0 \\ \max_{n_i \in T_t} (n_i) - \max_{m_i \in S_t} (m_i) & \text{else} \end{cases}$$

Extension of an MDS determines the enlargement needed in order to accommodate the newly coming MDS. Like *overlap*, *extension* also requires MDSs to be at same levels.

Example 5. extension(M|N)=1+0+0+1=2.

Tree Elements. TiC-Tree is composed of three different types of nodes, i.e. Data Nodes, Directory Nodes and Super Nodes. Let M be a MDS $M(l_1^{k_1}, l_2^{k_2}, \dots, l_n^{k_n})$ of sequence $[S_1, S_2, \dots, S_n]$.

Definition 7. *(Data Node)* *A data node N_{data} of the TiC-Tree is a tuple< $M, a_1, a_2, \dots a_p$ >where $|S_i| = 1$ and $k_i = 1, \forall i = 1, 2 \dots n$ and each a_j is a measure value of the node.*

Definition 8. *(Directory Node)* *A directory node N_{dir} of the TiC-Tree is a tuple< $M, \mathcal{E}, a_1, a_2, \dots a_p$ >,$\mathcal{E}$ is a set of pointers to other nodes whose MDS are contained in the MDS M. Size of $\mathcal{E}$ is limited and constant and determines the capacity of the node. Each a_j represents an aggregate measure value of the node.*

Definition 9. *(Super Node)* *A super node N_{super} of the TiC-Tree is a tuple< $M, \mathcal{F}, a_1, a_2, \dots a_p$ >where $\mathcal{F}$ is a set of pointers to other nodes whose MDSs are contained in the MDS M, and size of $\mathcal{F}$ is unlimited. Each a_j represents an aggregate measure value of the node.*

Figure 1 shows the TiC-Tree constructed with directory node capacity of 3. The leaves represent data nodes, while the rest are directory nodes. A data node encapsulates an MDS together with an associated measure value and represents a tuple of the fact table. A directory node, on the other hand, encapsulates an MDS and associated aggregate measure values of its children. This is also to be noted that the *root* of the TiC-Tree is always a directory node with MDS at level ALL for all the dimensions.

Tree Maintenance. In this section we illustrate the algorithms necessary for TiC tree maintenance. As the data warehouses do not generally have delete or update queries, we discuss only Insert and Search algorithms. Due to the lack of space, we donot give split and group-by query algorithms here. However we explain their working through the running example.

Insert. At the beginning, the tree has only one directory node called *root*. On arrival of a new fact, the data is packed into a data node and inserted into the *root*. The insert algorithm (see Algorithm 1) starts with updating the aggregate measure value of the node, and then searches for the node for subsequent insertion among its entries. This search is based on the metrics *Contain, Overlap* and *Extension* in order. This process continues recursively unless a directory or super node with entries of type "Data Node" is reached. Once the node for insertion is located, the data node is added to its entries (following the extension of the MDS of the directory/super node, if required). In case the directory node capacity is reached, an overflow occurs which induces the split of the directory node. This split may cause the parent node to overflow which will result in another split.

The split method of a directory node starts with selection of a split dimension and split level that are chosen on the basis of hierarchy level and cardinality of dimensions in the MDS. For example, in the MDS [{Europe}, {USA, France}, {MFGR#13}, {2000}], hierarchy level of Supplier and Part dimensions is 3 while

Algorithm 1. Insert Algorithm

TiCTDirectoryNode::Insert(TiCTDataNode *dataNode) {
/*Insert dtataNode into a Directory node "thisNode" */
update Aggregate Measure Value of thisNode;
followNode = ChooseSubTree(dataNode);
if (followNode != thisNode) **then** followNode.Insert(dataNode);
else {
   if (thisNode's Entries Type is "DataNode") **then**
     if (Number of thisNode's Entries < Maximum Entries Allowed) **then** add dataNode into the Entries
     else thisNode.Split(dataNode);
   else extend thisNode's MDS to accommodate dataNode and add it into the Entries}}

it's value is 4 for Customer and Time dimensions (see SSB [3]). First, the decision is taken on the basis of hierarchy level and one with the highest value is selected (in this example Customer and Time). In case of tie, the dimension with the higher cardinality is chosen as split dimension, which in our example is Supplier. After selecting the split dimension, hierarchy level of MDSs of all entries of the node and the node to split are adapted to splitLevel, two directory nodes are created and the entries are distributed in these newly created nodes on the basis of the metrics. If the overlap of the two MDSs exceeds the predefined limit, a new dimension is chosen for split. This process continues until a suitable split dimension is found; otherwise the overflowing node is adapted to a super node.

As a running example, the insertion of t5 (see Table 1) in figure 1(a), starts with the packing of t5 into a new data node and update of N1's aggregate measure. N3 is chosen for following insertion which requires zero extension as compared to 1 for N2. The data node is added to N3 and its aggregate measure

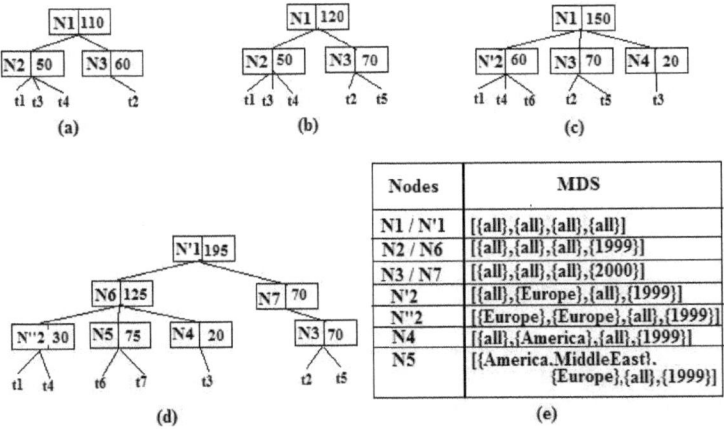

Fig. 1. TiC-Tree Updation: Insertion of tuples (a) t1, t2, t3 and t4 (b) t5 (c) t6 (d) t7 (e) MDSs corresponding the nodes

Algorithm 2. Range Query Algorithm

int TiCTNode::RangeQuery(range_MDS){
/*Query a node "thisNode" against a rangeMDS*/
result=0;
foreach dimension
   if(rangeMDS and thisNode's MDS are not at same level) **then** adapt MDS with lower level to the one with higher level;
if (Overlap(rangeMDS,thisNode.MDS)>0) **then** {
   if (thisNode's MDS is contained in rangeMDS) **then** result+=thisNode.AggregateMeasureValue;
   else if (thisNode.NodeType == "SuperNode" or "DirectoryNode")
      foreach "entry" in thisNode, result+=entry.RangeQuery(range_MDs); } return result; }

---

is updated. Subsequent insertion of t6 causes a split of N2 and produces the resulting tree structure shown in figure 1(c). Insertion of t7 results in the splitting of N'2 and consequently N1 that results in the final tree shown in figure 1(d).

Query. A range query is represented by an MDS that we call rangeMDS. For example, the query "Number of Parts sold tothe customers of Europe during the years 2000 to 2002" is represented by the rangeMDS [{Europe},{all},{all},{2000 ,2001,2002}]. The algorithm (see Algorithm 2) for range query takes a rangeMDS as input and queries the node, starting from the root. If the node's MDS is contained in the range_MDS, the node's aggregate value is added to the result. Else if node's MDS and rangeMDS overlap and the node is not a Data Node, then same algorithm is run recursively for all entries of the node.

For example, let us query the TiC-Tree in fig. 1(d) against above rangeMDS. The root's MDS is not contained in rangeMDS and overlap between them is greater than zero. Therefore, the algorithm continues searching the entries of N'1. MDSs of both the entries are not contained in rangeMDS and the overlap between the MDS of N6 and rangeMDS is also zero. However, N7's MDS and rangeMDS have overlap value greater than zero, and so as N3. The MDS of t2 is contained in the range MDS, so its measure value is added to the result while t5's MDS is not contained in the rangeMDS. As the leaves of the tree are reached, and no overlapping part of the tree is left, the method returns the result i.e. 60. The algorithm for group by query is almost the same, the only difference is in the input MDS. Group by query's MDS contains only one attribute per dimension at a time, e.g. "Number of Parts sold by region where region in (Europe, America)" is translated to group-by MDSs [{Europe},{all},{all},{all}] and [{America},{all},{all},{all}].

4   Experimentation and Results

In order to evaluate the performance, we developed the TiC-Tree as well as the DC-Tree. A 'csv' file containing the tuples of the fact table serves as input for

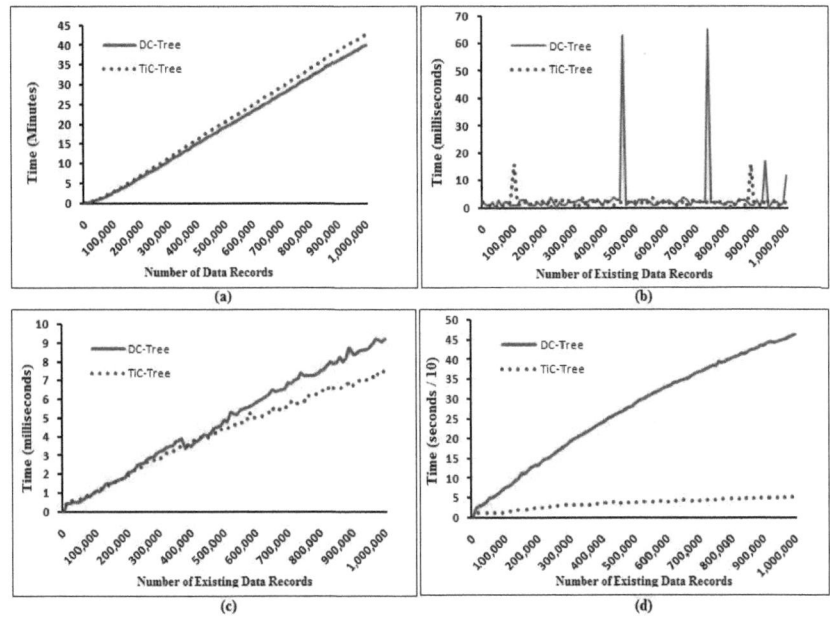

Fig. 2. Response time for: (a) Bulk Insertion (b) Single Record Insertion (c) Range Query (d) Group-by Query; on chronologically ordered data

these applications. All the tests were carried out on a system with 2.0 GHz Processor and 2GB RAM and evaluate the performance on the basis of insertion and query response time. The data warehouse schema for experimentation and performance evaluation is based on the Star Schema Benchmark. We use a fact table with 10,000 to 1,000,000 tuples, ordered chronologically and another data sets where some of the facts (5%, 10%,...) arrive out of time order, with the intention of studying the effect of these postponement. We execute a set of 100 range and group-by queries, for different levels (selected randomly) of dimension hierarchy on each state of the index resulting from the above insertion, average query execution time beeing recorded as result.

Figure 2(a and b) shows the comparitive results for bulk and single record insertion time, respectively, in DC-Tree and the TiC-Tree. In these cases, both indices show almost similar performance. Figure 2(c and d) summarizes the results of query response time for queries. In both cases TiC-Tree exhibits better performance than the DC-Tree. All the driven tests show that delay affects the insertion and query time only to a small extent and does not have any considerable effect on the performance of the index.

5   Conclusion

In this research work, we propose an index structure to store detailed and aggregated multidimensional data. In real time temporal data warehouses, members of time dimension grow rapidly over time, and we believe the TiC-Tree can be a solution for handling this special case. We propose to use and redefine Minimum Describing Sequences, in order to take advantage of special nature of time dimension and special nature of temporal OLAP queries. We propose to keep temporally close values together in tree nodes in order to facilitate range searches and group-by. Performance evaluation tests show a significant improvement for range and group-by queries, as compared to the reference index. TiC-Tree provides an idea of considering the nature of dimensions for data indexing, in our case, the total ordering of temporal data. The TiC-Tree shows performance improvement in query response time but it is still unable to deal with the costly split algorithm that will need futher investigation.

References

1. Polyzotis, N., Skiadopoulos, S., Alkis Simitsis, P.V., Frantzell, N.E.: Supporting Streaming Updates in an Active Data Warehouse. In: Proc. of the 23rd Int. Conf. on Data Engineering (2007)
2. Tho, M.N., Tjoa, A.M.: Zero-latency data warehousing for hetrogeneous data sources and continuous data streams. In: Proc. of the Fifth Int. Conf. on Information Integration and Web-based Applications Services (2003)
3. O'Neil, P.E., O'Neil, E.J., Chen, X., Revilak, S.: The Star Schema Benchmark and Augmented Fact Table Indexing. In: Nambiar, R., Poess, M. (eds.) TPCTC 2009. LNCS, vol. 5895, pp. 237–252. Springer, Heidelberg (2009)
4. Gupta, H.: Selection of Views to Materialize in a Data Warehouse. In: Afrati, F.N., Kolaitis, P.G. (eds.) ICDT 1997. LNCS, vol. 1186. Springer, Heidelberg (1996)
5. Roussopoulos, N., Kotidis, Y., Roussopoulos, M.: Cubetree: organization of and bulk incremental updates on the data cube. In: Proc. of ACM SIGMOD Int. Conf. on Management of Data, pp. 89–99. ACM, New York (1997)
6. Sismanis, Y., Deligiannakis, A., Kotidis, Y., Roussopoulos, N.: Hierarchical dwarfs for the rollup cube. In: Proc. of the 6th ACM Int. Workshop on Datawarehousing and OLAP, NY, USA (2003)
7. Lakshmanan, L.V.S., Pei, J., Han, J.: Quotient cube: how to summarize the semantics of a data cube. In: Proc. of the 28th Int. Conf. on Very Large Data Bases, VLDB Endowment, pp. 778–789 (2002)
8. Tao, Y., Papadias, D.: Efficient Historical R-Trees. In: Proc. of the 13th Int. Conf. on Scientific and Statistical Database Management, Washington, DC, USA (2001)
9. Berchtold, S., Keim, D.A., Kriegel, H.P.: The X-tree: An Index Structure for High-Dimensional Data. In: Proc. of 22th Int. Conf. on Very Large Data Bases, Mumbai (Bombay), India, pp. 28–39 (1996)
10. Papadias, D., Tao, Y., Kalnis, P., Zhang, J.: Indexing spatio-temporal data warehouses. In: Proc. of the 18th Int. Conf. on Data Engineering (2002)
11. Ester, M., Kohlhammer, J., Kriegel, H.P.: The DC-tree: A Fully Dynamic Index Structure for Data Warehouses. In: Proc. of the 16th Int. Conf. on Data Engineering, pp. 379–388 (2000)

PAGER: Parameterless, Accurate, Generic, Efficient kNN-Based Regression

Himanshu Singh, Aditya Desai, and Vikram Pudi

Center for Data Engineering, International Institute of Information Technology,
Hyderabad, India
{himanshusingh,aditya_desai}@students.iiit.ac.in, vikram@iiit.ac.in

Abstract. The problem of regression is to estimate the value of a dependent numeric variable based on the values of one or more independent variables. Regression algorithms are used for prediction (including forecasting of time-series data), inference, hypothesis testing, and modeling of causal relationships. Although this problem has been studied extensively, most of these approaches are not generic in that they require the user to make an intelligent guess about the form of the regression equation. In this paper we present a new regression algorithm PAGER – Parameterless, Accurate, Generic, Efficient kNN-based Regression. PAGER is also simple and outlier-resilient. These desirable features make PAGER a very attractive alternative to existing approaches. Our experimental study compares PAGER with 12 other algorithms on 4 standard real datasets, and shows that PAGER is more accurate than its competitors.

Keywords: Regression, prediction, K-nearest neighbours, Parameterless, Accurate.

1 Introduction

Regression analysis has been studied extensively in statistics [5, 4], there have been only a few studies from the data mining perspective. The algorithms studied from a data mining perspective mainly fall under the following broad categories - Decision Trees [7], Support Vector Machines [12], Neural Networks [13], Nearest Neighbour Algorithms [18] [19], Ensemble Algorithms [16] [17] among others. It may be noted that most of these studies were originally for classification, but have later been modified for regression [1].

In this paper we present a new regression algorithm **PAGER** – **P**arameterless, **A**ccurate, **G**eneric, **E**fficient kNN-based **R**egression, has the following features:

1. **Parameterless:** The parameterless nature of PAGER removes the burden from the user of having to set parameter values – a process that typically involves repeated trial-and-error for every application domain and dataset.
2. **Accurate:** Our experimental study in Section 4 shows that PAGER provides more accurate estimates than its competitors on several datasets. Among the algorithms we included for comparison are the best available algorithms from the Weka toolkit [1].

P. García Bringas et al. (Eds.): DEXA 2010, Part II, LNCS 6262, pp. 168–176, 2010.

3. **Generic:** Our approach assumes that dependent variable changes smoothly with change in independent variable, an assumption which is valid for most real-life datasets. Hence, it will work "out-of-the-box" and doesn't require to be tinkered with for every application domain and dataset.
4. **Efficient:** PAGER is based on the nearest neighbour (k-NN) approach and is thereby equally efficient, provided there are indexes available for easily finding the k nearest neighbours [2].
5. **Simple:** The design of PAGER is simple, as it is based on the k-NN approach. This makes it easy to implement, maintain, embed and modify as and when the situation demands.
6. **Outlier Resilient:** The output of PAGER for a particular input record R is dependent only on the nearest neighbours of R and is therefore insensitive to far-away outliers.

The remainder of the paper is organized as follows: In Section 2 we discuss related work and present the PAGER algorithm in Section 3. Then, in Section 4 we experimentally evaluate our algorithm and show the results. Finally, in Section 5, we summarize the conclusions of our study and identify future work.

2   Related Work

Traditional statistical approaches: Most existing approaches [4] [5] follow a "curve fitting" approach that requires the form of the curve in advance. Another problem with the curve fitting approaches is outlier (extreme cases) sensitivity which bias the results by pulling or pushing the regression curve in a particular direction. In addition to this they fit the regression equation to the entire plane and hence are unlikely to capture inherent relationships.

Global fitting methods: Generalized Projection Pursuit regression [6] is a statistical method which constructs a regression surface by estimating the form of the function in such a manner that it best fits the dataset without using any parameter. Lagrange by Dzeroski et. al. [9] generates the best fit equation over the observational data by constructing a large number of equation alternatives. Due to huge search space the approaches are computationally expensive. Lagramge [8], is a modification of Lagrange in which grammars of equations are generated from domain knowledge and the best-fit equations are determined by filtering the forms using the generated grammar. However, the algorithm requires prior domain knowledge. Support vector machine [12] (SVM) is involves complex abstract mathematics and thus resulting in techniques that are more difficult to implement, maintain, embed and modify as the situation demands. Neural networks are another class of approaches that have been used for regression [13] and dimensionality reduction as in Self Organizing Maps [15]. However, neural networks are complex "black box" models and hence an in depth analysis of the results obtained is not possible. Data mining applications typically demand an "white box" model where the prediction can be explained to the user, since it is to be used for decision support. Ensemble based learning [16] [17] is a new approach to regression where a number of machine learning algorithms are combined to build a learner having an accuracy better than the individual learners. A major problem associated with ensemble based learning is to determine the relative importance of each individual learner which is usually done by assigning weights to individual learners, where a high weight implies a higher relative However, all these algorithms suffer from the problem that they try to fit the entire data to a particular structure and and hence cannot capture inherent relationships in different localities of the dataset.

Decision Trees: *Regression trees* [7], are a variation of decision trees where the predicted output values are stored at the leaf node. These nodes are finite and hence the predicted output is limited to a finite set of values which is in contrast with the problem of predicting a continuous variable as required in regression.

k-**Nearest Neighbour:** Another class of data mining approaches that have been used for regression include *nearest neighbour techniques* [18] [19]. These algorithms are known to be simple and reasonably outlier resistant. Although these approaches have the desirable property of simplicity, they have relatively low accuracy because of the problem of determining the correct number of neighbours and the fact that they assume that all dimensions contribute equally. This is often not the case as some features may have a higher degree of correlation and some may not.

In this paper, we enhance the power of nearest neighbour predictors which intrinsically handle local variations. We eliminate the problems associated with nearest neighbour methods like choice of number of neighbours and difference in importance of dimensions. We use a novel weighting criterion which determines the relative importance of dimensions. This is used in combination with a unique stability criterion which determines the appropriate number of neighbours.

3   The PAGER Algorithm

In this section we present our algorithm.In section 3.1 we describe the regression problem and the variables used in   3.3, followed by the presentation of the pseudo-code for PAGER and a discussion of the same.

3.1   Problem Formulation and Definitions

Formally, the input to the problem consists of:

- A vector of d feature variables $X = (x_1, \ldots, x_d)$, forming an d-dimensional space
- A numeric target response variable y
- A training set of n data samples $D = \{(X_1, y_1), \ldots, (X_n, y_n)\}$, where X_i are points in X-space and y_i are the corresponding values of the response variable.

The output is an estimate of the value of y for new input points in X-space.

- **Training Data:** It is used for training the model. The training data has d-dimensions with feature variables $(A_1, \ldots, A_d)$ and the value of the feature variable A_j corresponding to the ith tuple (tuple-id i) can be accessed as $Data[i][j]$. The value of the dependant variable of the training tuple corresponding to id i, can be accessed as $y[i]$.
- **Test Tuple:** The test tuple consists of values of d-dependant variables where the value of feature variable A_i can be accessed as T_i.

3.2   Design of PAGER

The design of PAGER is based on the assumption that value of the dependant variable varies smoothly with the variation in values of dependant variable. Now, every smooth curve can be modelled to be a combination of piecewise linear curves. This intuition becomes the basis of our paper, where the task at hand is determining the nature of linearity in the locality of the test tuple. The approach to solving this problem is to construct a line using the two closest neighbours of the input tuple and thus approximating the linearity of the regression curve in this region to the linearity given by the line so constructed.

It may be noted that a standard kNN algorithm bounds values i.e. the values output by a kNN-algorithm are always between the minimum and maximum in the dataset. Our algorithm alleviates this problem by constructing a 1-dimensional predictor and hence approximates the linearity in the given region thus giving more accurate estimates. Now, corresponding to each dimension we have a separate line and hence there can be d different predictions for the output variable corresponding to each dimension. Assigning weights to predictors is thus the crux of the problem. Given a predictor, we determine mean error if the predictor was used in prediction of the k-neighbours (As their real dependant variable values are already known). If the mean error is large, the predictor for this dimension does not fit the neighbourhood well and hence should be labelled poor. We thus assign weights to be inversely proportional to the mean error. After this step the value that is predicted is the weighted sum of the value output by individual predictors. However, the task that now remains is determining the optimal k. A too small value of k may not capture the true nature of neighbourhood and hence may result in biased weights while an estimate using a large value of k may lose the local context. We thus get weighted predictions for all $k-2$ neighbours (the first two neighbours are used for drawing the line and hence error will be 0 and are skipped) and determine the mean error of prediction for these k neighbours. We call this error the Overall Mean Error. A low mean error is an indication that the local information has been properly encoded and hence we iterate on a range of k and then chose a k corresponding to minimum Overall Mean Error.

3.3   Pseudo-Code

The pseudo code for PAGER is as follows:

Algorithm 1. *Pseudo Code*

1: $MinimumError \leftarrow \infty$; $OutputValue \leftarrow 0$;
2: **for** $k = min$ to max **do**
3:     $ErrorUsingK \leftarrow 0\ Count \leftarrow 0$;
       $//Mean\ error\ in\ prediction\ using\ k-neighbours$
4:     $ClosestNeighbours \leftarrow GetNeighbours$
       $(Data, k, T)\ //Returns\ id\ of\ k-closest\ neighbours\ of\ T.$
5:     Inline 1-Dimensional Predictor;Weighting dimensions;Inline d-Dimensional
       Predictor
6:     $ErrorUsingK \leftarrow \frac{ErrorUsingK}{Count}$;
       $//\ Mean\ error\ in\ prediction\ of\ k\text{-}neighbours\ when\ k\text{-}neighbours\ used$
7:     **if** $MinimumError > ErrorUsingK$ **then**
8:         $OutputValue \leftarrow PredictVal$; $MinimumError \leftarrow ErrorUsingK$;
9:         $LIndex \leftarrow k\ //\ Value\ of\ k\ at\ which\ minimum\ is\ found$
10:    **end if**
11:    **if** $(k - LIndex) > log(n)^2$ **then**
12:        break
13:    **end if**
14: **end for**
15: $return\ OutputValue$;

Discussion of PAGER: For determining the correct value of number of neighbours (k) for a given test tuple T, we use an iterative procedure which takes three optional parameters, $Error\_Threshold$, lower (min) and upper bound (max) of k as input. However these parameters are optional and can be set to their optimal values automatically. The correct choice of k is a result of an exhaustive search in between min and max. If the parameters are not set, the algorithm automatically sets min to a sufficiently low value (about five), max to a sufficiently high value (tuples in training data/2) and $Error\_Threshold$ as ∞.

In Line 5 we build 1-dimensional predictors, which are d predictors, where each one is a line passing through the nearest two neighbours in one of the d dimensions. In line 6 we compute mean error for prediction of the k neighbours using these predictors and assign weights. In line 7, we predict the value of the dependant variable as a weighted sum of 1-dimensional predictors for the test tuple as well as the k neighbours. The sum of Overall Mean Errors for the k neighbours is assigned to $ErrorUsingK$. If we do not get a minimum for a relatively large run we stop the procedure in Line 14. The predicted value corresponding to the minimum $ErrorUsingK$ is then output as the predicted value.

4   Experimental Study

In this section, we evaluate the proposed PAGER algorithm. We describe the experimental setting and performance metrics in Section 4.1 and the experimental results in Section 4.2. The comparative results are shown in Table 3.

4.1   Experimental Settings

We compare our algorithm against 12 algorithms on 4 datasets. One of them is the weighted k-NN based approach [18,19]. The remaining eleven are available in the Weka toolkit [1] namely Additive Regression, Gaussian Regression, Isotonic Regression, Least Median Square Regression (LMS), Linear Regression, Multi Layer Perceptron based Regression (MLP), Pace Regression, RBF Regression, Simple Linear Regression, SMO Regression, and SVM Regression. Details of datasets are show in Table 1. Experiments have been done using the *leave one out* comparison technique which is a specific case of n-folds cross validation. Metrics used for comparison are RMSE (Root Mean Square Error) and ABME (Absolute Mean Error). The k-NN technique of [18, 19] was run on number-of-neighbours=10. The algorithms in Weka were run with the parameters as can be found in Table 3. The parameters of Weka are not described here due to lack of space and are available in [1].

Table 1. Dataset Description

Dataset	Number of tuples	Number of attributes	Source
CPU	299	6	[1]
Housing	506	13	[20]
Concrete	1030	8	[20]
BodyfatCPU	252	14	[21]

Table 2. Experimental Settings

Algorithm	Parameter settings
Additive Regression [22]	(AdditiveRegression -S 1.0 -I 10 -W trees.DecisionStump)
Gaussian Process	(GaussianProcesses -L 1.0 -N 0 -K supportVector.RBFKernel -C 250007 -G 0.5")
Isotonic Regression	(IsotonicRegression)
Least Median Square Regression [5]	(LeastMedSq -S 4 -G 0)
Linear Regression	(LinearRegression -S 0 -R 1.0E-8)
Multi Layer Perceptron [13] [14]	(MultilayerPerceptron -L 0.3 -M 0.2 -N 500 -V 0 -S 0 -E 20 -H a)
Pace Regression [3] [4]	(PaceRegression -E eb)
RBF Network	(RBFNetwork -B 2 -S 1 -R 1.0E-8 -M -1 -W 0.1)
Simple Linear Regression	(SimpleLinearRegression)
SMOreg Regression [10] [11]	(SMOreg -S 0.001 -C 1.0 -T 0.001 -P 1.0E-12 -N 0 -K supportVector.PolyKernel -C 250007 -E 1.0)
SVMReg Regression [10] [11]	(SVMreg -C 1.0 -N 0 -I supportVector.RegSMOImproved -L 0.001 -W 1 -P 1.0E-12 -T 0.001 -V -K supportVector.PolyKernel -C 250007 -E 1.0)

4.2   Results

In this section, we report the experimental results obtained for each dataset as can be seen in Table. 4. The bold figures indicate the best performers on the given dataset.

Table 3. Experimental Results on CPU, Housing, Concrete and Bodyfat Dataset

Regression	CPU Dataset		Housing Dataset		Concrete Dataset		Bodyfat Dataset	
Algorithm	ABME	RMSE	ABME	RMSE	ABME	RMSE	ABME	RMSE
PAGER	*9.28*	*31.04*	*2.10*	*3.47*	*5.87*	*7.71*	*0.38*	*0.49*
Additive	25.46	59.06	3.35	4.86	6.67	8.45	0.53	0.64
Gaussian	15.08	81.57	2.58	3.89	6.07	7.82	0.42	0.56
Isotonic	23.93	51.98	3.81	5.32	10.84	13.52	0.55	0.69
LMS	33.60	107.55	3.36	5.40	9.28	16.55	0.43	0.55
Linear	34.61	55.22	3.37	4.84	8.29	10.46	0.42	0.54
MLP	**6.28**	**16.70**	3.10	4.64	6.58	8.61	0.50	0.70
Pace	34.83	56.12	3.36	4.82	8.32	10.52	0.41	0.53
RBF	52.25	119.28	6.05	8.19	13.43	16.67	0.61	0.77
Simple Linear	43.13	70.46	4.52	6.23	11.87	14.50	0.50	0.62
SMO	20.70	64.22	3.25	5.09	8.23	10.97	0.43	0.56
SVM	20.71	64.24	3.24	5.08	8.23	10.97	0.43	0.56
*k*NN	18.92	74.83	2.97	4.63	6.55	8.57	0.45	0.58

4.3   Discussion of Results

From the experimental results it is evident that on the CPU dataset PAGER outperforms all other algorithms except the regression algorithm based on Multilayer Perceptron. The reason for this is analyzed in detail below. On the other three datasets *i.e.* the Housing Dataset, BodyFat Dataset and the Concrete Dataset, PAGER outperforms all other algorithms. It is particularly noteworthy that our algorithm performs very well on the Concrete Dataset which was claimed to be a challenging, highly non-linear function of its attributes. The success of our algorithm is due to our very valid assumption that the data variation may not be linear throughout but is usually linear in a very small neighbourhood of the given input tuple. This assumption is true for majority of the real life datasets as variations of the dependent variable based on variations in the values of independent variables typically show a smooth transition.

A positive point in this algorithm that is evident through the pseudo code is its simplicity and generic nature. The parameterless nature of this code makes it easy to apply it to any domain even if sufficient domain knowledge is not available. However, it should be noted that our algorithm works only for numeric data with no missing values.

5   Conclusions

In this paper we have presented and evaluated PAGER, a new algorithm for regression based on nearest neighbour methods. Evaluation was done against 12 competing algorithms on 4 standard real-life datasets. Although simple, it outperformed all competing algorithms on all datasets but one. Unlike most other algorithms, PAGER can be used "out-of-the-box" without having to extensively tune or tweak it for each application domain and dataset.

Future work includes determining high quality neighbours and the correct number of neighbours. Another future direction is to construct a curve or the closest fitting line from k neighbours instead of a line which is presently constructed from the two neighbours. It is also of interest to design algorithms that work when the independent variables are categorical, or come from a mixture of categorical and numeric domains.

References

1. Witten, I.H., Frank, E.: Data Mining: Practical machine learning tools and techniques, 2nd edn. Morgan Kaufmann, San Francisco (2005)
2. Jammalamadaka, N., Pudi, V., Jawahar, C.V.: Efficient Search with Changing Similarity Measures on Large Multimedia Datasets. In: Proc. of the International Multimedia Modelling Conference (2007)
3. Wang, Y.: A new approach to fitting linear models in high dimensional spaces, PhD thesis, Department of Computer Science, University of Waikato, New Zealand (2000)
4. Wang, Y., Witten, I.H.: Modeling for optimal probability prediction (2002)
5. Barreto, H.: An Introduction to Least Median of Squares. Chapter contribution to Barreto and Howland, Econometrics via Monte Carlo Simulation
6. Lingjaerde, O.C., Liestøl, K.: Generalized projection pursuit regression. SIAM Journal on Scientific Computing (1999)
7. Breiman, L., Friedman, J., Olshen, R., Stone, C.: Classification and regression trees. Wadsworth Inc. (1984)
8. Todorovski, L.: Declarative bias in equation discovery. M.Sc. Thesis. Faculty of Computer and Information Science, Ljubljana, Slovenia (1998)
9. Dzeroski, S., Todorovski, L.: Discovering dynamics: from inductive logic programming to machine discovery. Journal of Intelligent Information Systems 4, 89–108 (1995)
10. Smola, A.J., Scholkopf, B.: A tutorial on support vector regression. Technical Report NC2-TR-1998-030, NeuroCOLT2 Technical Report Series (1998)
11. Shevade, S., Keerthi, S., Bhattacharyya, C., Murthy, K.: Improvements to smo algorithm for svm regression. Technical Report CD-99-16, Control Division Dept of Mechanical and Production Engineering, National University of Singapore (1999)
12. Chu, W., Keerthi, S.S.: New approaches to support vector ordinal regression. In: Proc. of International Conference on Machine Learning (ICML 2005), pp. 142–152 (2005)
13. Ware, M.: Implementation of multilayer perceptron backpropagation (2005), http://weka.sourceforge.net/doc/weka/classifiers/functions/MultilayerPerceptron.html
14. Mielniczuk, J., Tyrcha, J.: Consistency of multilayer perceptron regression estimators. Neural Networks 53(2), 1019–1022 (1993)
15. Haykin, S.: Self-organizing maps. In: Neural networks - A comprehensive foundation, 2nd edn. Prentice-Hall, Englewood Cliffs
16. Breiman, L.: Bagging Predictors. Machine Learning 24(2), 123–140 (1996)
17. Schapire, R.E.: A Brief Introduction to Boosting. In: Proc. 16th International Joint Conf. Artificial Intelligence, pp. 1401–1406 (1999)
18. Fix, E., Hodges Jr., J.L.: Discriminatory analysis, non-parameteric discrimination: Consistency properties. Technical Report 21-49-004(4), USAF school of aviation medicine, Randolf field, Texas (1951)

19. Rousseeuw, P.J., Leroy, A.M.: Robust Regression and Outlier Detection. Wiley, Chichester (1987)
20. Asuncion, A., Newman, D.: UCI Machine learning repository (2007)
21. The body fat dataset (1985), `http://lib.stat.cmu.edu/datasets/bodyfat`
22. Friedman, J.H.: Stochastic Gradient Boosting. Technical Report Stanford University (1999), `http://www-stat.stanford.edu/~jhf/ftp/stobst.ps`

B2R: An Algorithm for Converting Bayesian Networks to Sets of Rules

Bartłomiej Śnieżyński[1], Tomasz Łukasik[2], and Marek Mierzwa

[1] AGH University of Science and Technology, Dept. of Computer Science
Krakow, Poland
bartlomiej.sniezynski@agh.edu.pl
[2] InventSoft sp. z o.o.
t.lukasik@inventsoft.pl

Abstract. In this paper B2R algorithm that converts Bayesian networks into sets of rules is proposed. It is tested on several data sets with various configurations and results show that accuracy is similar to original Bayesian networks even after pruning a high number of rules. It allows to exploit advantages of both knowledge representation techniques.

1   Introduction

Bayesian models provide very well founded uncertainty representation. If a knowledge base is created manually, tuning a set of rules takes a lot of time. However, if a knowledge base is generated automatically from data, it is not a problem. Another issue is the accuracy of the generated models. Probabilistic classifiers may be better than rule based ones, but the difference is usually small.

In many application domains, such as security systems, medical diagnosis, etc. very important issue is understanding of the knowledge, especially if it is generated automatically. In such domains, before the knowledge is used, it should be verified by a human expert. It is difficult to verify a knowledge that is in a form difficult to grasp. Next, generated classifiers are often used with a supervision of a human. To verify a decision of the system, the supervisor should have a possibility to check and understand the justification of the answer. What is also very important, using machine learning one can discover a new domain knowledge. This can not be done without feedback from a human expert. To make it possible, the knowledge generated has to have a form that is easy to interpret. Rules seem to correspond to human way of thinking very well [1].

To exploit advantages of both knowledge representation techniques, it would be good to have tools, which are able to transform knowledge between these formalisms. Such a conversion can be used for a knowledge visualization purposes and also to generate knowledge bases for diagnostic systems. It can be also considered as a pruning method, which may decrease complexity of the model.

This paper is a continuation of [2], in which a method for Naïve Bayes models was proposed. In this paper conversion algorithm is generalized, it takes as an input classifier that is a Bayesian network and a selected class node. As a result

P. García Bringas et al. (Eds.): DEXA 2010, Part II, LNCS 6262, pp. 177–184, 2010.
© Springer-Verlag Berlin Heidelberg 2010

we obtain a set of labeled rules that can be used to classify examples or visualize knowledge. This paper is based on [3], where idea of algorithm (without experimental results) was proposed.

In the following sections the conversion algorithm is described, its application example is shown. Next experimental results with use of several data sets and various configurations are presented. Next section discuses related works. Conclusions and plans of the future research conclude the paper.

2   B2R Conversion Algorithm

A Bayesian network is a pair (G, P), where G is a structure graph, which is directed and acyclic, and P is a set of local, conditional probability distributions between variables and its parents. Variables are denoted by $X_1, X_2, ..., X_n$. Set of parents of X_i is a denoted by $parents(X_i)$, and set of children by $children(X_i)$. Domains of the variables are denoted by $D_{X_1}, D_{X_2}, \ldots, D_{X_n}$. In this paper we assume that all domains are finite. We also assume, that the Bayesian network is used for classification. Therefore we distinguish one of the variables as a class variable. It is denoted by Y.

We begin the description of a conversion algorithm for Naïve Bayes (NB), which is a special case of a Bayesian network. It has a very simple structure: variable Y is connected to every variable $X_1, X_2, ..., X_n$. Such a network is used to predict value of root-variable (Y).

NB can be transformed into a set of rules by generating $|D_Y|$ rules for every value of every X_i variable. Let us call this procedure *A-transformation*. Rules would have the following form:

$$X_i = x_j^i \rightarrow Y = y_k, \tag{1}$$

where $x_j^i \in D_{X_i}, y_k \in D_Y$ are variable values. Similar technique can be used for the structure obtained from NB by reversing the direction of dependencies. In this case Y is a children of $X_1, X_2, \ldots, X_n$. For every $y_k \in D_Y$ and $(x_{j_1}^1, x_{j_2}^2, \ldots x_{j_n}^n) \in D_{X_1} \times D_{X_2} \times \ldots \times D_{X_n}$ – value combination of X_i variables we get a rule

$$X_1 = x_{j_1}^1 \wedge X_2 = x_{j_2}^2 \wedge \ldots \wedge X_n = x_{j_n}^n \rightarrow Y = y_k. \tag{2}$$

Let us call this procedure *B-transformation*.

Generated rules have various strength, therefore we need a way to represent it. Similarly, variable values can be inferred with various certainty. It can be represented by labeling rules and variable assignments. Choosing appropriate labels is very important, because it has a strong influence on the rule-based classifier performance. The choice can be formalized by defining a label algebra and a function transforming conditional probabilities into labels.

Label algebra can be defined as the following triple: $\mathcal{L} = (L, \star, \odot)$, where L is a set of labels with linear order (to choose the highest value during classification of examples), $\star : L^2 \rightarrow L$ is a rule aggregation operator, which is used during the classification, if two (or more) rules with the same consequence match the example. In such a case, rules are aggregated. $\odot : L^2 \to L$ is also used during classification to calculate a label of conjunction of conditions. $\star$ and $\odot$ should be associative and commutative to make the result of rule application independent from the rule and condition order, therefore $\star$ and $\odot$ can be extended to operate on a set of labels.

Labeled rule is a pair $r : l$, where r is a rule defined above, and $l \in L$. *Labeled variable assignment* is a triple $X_i = x_j : l$, where X_i is a variable, $x_j \in D_{X_1}$, and $l \in L$. Rule labels are calculated using *transformation function* $f : [0, 1] \to L$. Rule $X_i = x_j^i \to Y = y_k$ has a label $l = f(P(X_i = x_j^i | Y = y_k))$ and rule $X_1 = x_{j_1}^1 \wedge X_2 = x_{j_2}^2 \wedge \ldots \wedge X_n = x_{j_n}^n \to Y = y_k$ has label $l = f(P(Y = y_k | X_1 = x_{j_1}^1, X_2 = x_{j_2}^2, \ldots, X_n = x_{j_n}^n))$. If some variables are eliminated from the set of parents of Y during rule construction, conditional probability in label definition is derived from original probability distribution by eliminating the same variables. Variable assignment labels are calculated during rule application. See below for details.

Simple examples of algebras are: continuous ($[0, 1], \cdot, \min$) and discrete ($\{0, 1\}, \cdot, \min$), where $\cdot$ is a classical multiplication. Consequently, two transformation functions are defined: $f_1(p) = p$ (identity), and $f_2(p) = \mathrm{round}(p)$. The use of other algebras is also possible. One of the solutions is to scale probability values into the range $[-1, 1]$ and apply Certainty Factors style of aggregation.

The conversion described above has a serious drawback – it does not decrease model's complexity. To overcome this shortcoming, we introduce pruning to eliminate rules with low significance.

Let us start with rules obtained by A-transformation. A method of pruning in the case of $|D_{X_i}| = 2$ is very simple: probabilities $P(X_i = x_j^i | Y = y_k)$ close to 0.5 have lower influence on the hypothesis than ones with value close to 1 or 0. Therefore we can create rules for these probabilities, which have distance from 0.5 greater than a given threshold. If $|D_{X_i}| > 2$ we can use Entropy measure. It can be defined for a bunch of rules. The *bunch of rules* B_{ik} is a set of rules with the same variables in the premise and the same consequence: $B_{ik} = \{X_i = x_j^i \to Y = y_k : l\}_{j=1,2,\ldots,|D_{X_i}|}$. *Entropy* E is defined as follows:

$$E(B_{ik}) = \sum_{j=1}^{|D_{X_i}|} -P(X_i = x_j^i | Y = y_i) \log_2 P(X_i = x_j^i | Y = y_i). \qquad (3)$$

If $P(X_i = x_j^i | Y = y_i) = 0$ we assume that $P(X_i = x_j^i | Y = y_i) \log_2 P(X_i = x_j^i | Y = y_i) = 0$. To have normalized values, the *normalized Entropy* $En(B_{ik})$ is defined: $En(B_{ik}) = En(B_{ik})/E_{max}(|D_{X_i}|)$, where $E_{max}(n)$ is a maximal Entropy for a domain of size n.

High values of the normalized Entropy mean a high disorder and low information, therefore rules that belong to a bunch with such values of En are pruned. In order to have similar meaning of the threshold value t as in [4], i.e. to represent a low pruning by values close to 0 and a strong pruning by values close to 1, rules from B_{ik} are pruned if $En(B_{ik}) > (1 - t)$.

When rules are obtained using B-transformation, the bunch consists of rules
with the same premise and various values of the consequence variable:

$$B_{j_1,j_2,\ldots j_n} = \{X_1 = x_{j_1}^1 \wedge X_2 = x_{j_2}^2 \wedge \ldots \wedge X_n = x_{j_n}^n \rightarrow Y = y_k : l\}_{k=1,2,\ldots,|D_Y|}.$$
(4)

Entropy E is then defined as follows:

$$E(B_{j_1,j_2,\ldots j_n}) = \sum_{k=1}^{|D_Y|} -P_{k,j_1,j_2,\ldots j_n} \log_2 P_{k,j_1,j_2,\ldots j_n},$$
(5)

where $P_{k,j_1,j_2,\ldots j_n} = P(Y = y_k | X_1 = x_{j_1}^1, X_2 = x_{j_2}^2, \ldots, X_n = x_{j_n}^n)$.

The transformation defined above does not include probability distribution of
variables, which have no parent nodes and are not observed. It can be taken into
account by completing the set of rules with *default rules* of the form $\rightarrow X_i = x_j^i : l_j^i$, where $x_j^i \in D_{X_i}$, and $l_j^i = P(X_i = x_j^i)$. These rules have an empty
premise part and they always match examples during classification. $\star$ operator
domain should be extended to cover the $[0,1]$ range.

B2R Algorithm for conversion of Bayesian Networks into a set rules is presented in Fig. 1. Input data consists of a network to convert, and a class variable.
After conversion rules with significance lower than a given threshold t should be
eliminated, and default rules should be added. The algorithm always stops, because in every recursive execution one node is removed from the network.

Set of labeled rules produced (KB) can be used to calculate value of the class
variable for example $e = (X_{l_1} = e_{l_1}, X_{l_2} = e_{l_2}, \ldots, X_{l_n} = e_{l_n})$, where $\{X_{l_i}\}$ is a
subset of variables. A version of backward chaining can be used for this purpose.
Rules with the class variable in a consequence are selected. If values of their
premise variables are known, rules are fired, and value of the class variable is
calculated in the following way:

$$Y = \arg\max_{y_k} \star \{l \cdot \bigodot_{i=1}^{n} l_{k_i}$$

$$|X_{k_1} = e_{k_1} : l_{k_1} \wedge \ldots \wedge X_{k_n} = e_{k_n} : l_{k_n} \rightarrow Y = y_k : l \in \text{KB}\}.$$
(6)

It is assumed that if variable value is known, label for the appropriate assignment
is equal to maximal possible value (representing full certainty). If some variable

```
convert(Y − selected class node, Net − Bayesian network);
begin
    if Net is empty then return;
    p := parents(Y); c := children(Y);
    generate rules for structure Y ∪ c using A-transformation;
    generate rules for structure Y ∪ p using B-transformation;
    foreach X ∈ p ∪ c do  convert(X, Net − Y)
end
```

Fig. 1. B2R algorithm for converting a Bayesian network to a set of rules

Table 1. Performance of Bayesian Networks classifiers on the training data and number of generated rules (without pruning)

Data set	accuracy	#rules	Data set	accuracy	#rules
iris2	82.3	51	wine2	82.5	373
iris4	90.1	147	wine4	95.0	351
wbc2	94.5	4678	voting	98.6	199478
wbc4	95.7	746			

value is not known, a rule application procedure is executed recursively. Default rules are always fired if its conclusion variable is not known and is needed. They can be considered as initial labeling of variable assignments.

The following label algebras are used in the experiments: $\mathcal{L}_1 = ([0, 1], \cdot, \min)$, $\mathcal{L}_2 = ([0, 1], \oplus, \min)$, $\mathcal{L}_3 = ([0, 1], \max, \min)$. Symbol $\cdot$ is a standard multiplication, and $\oplus$ is a Certainty Factors style of aggregation: $l_1 \oplus l_2 = l_1 + l_2 - l_1 \cdot l_2$.

Let us call M_i a method of rule application using backward chaining and $\mathcal{L}_i$ label algebra. In method M_3, $\star = \max$ operation is done only on the top level of inference, for class variable. On the rest of levels, aggregation is not performed. Many different premise proofs are generated. Additionally, before calculating rule label, it is checked if premise proofs are contradictory (if for the same variable they assume different values). If they are, the label is not calculated and the next combination of premise proofs is taken into account. There is one more method used in the experiments. It also applies algebra $\mathcal{L}3$, but with a greedy approach. Proving the hypothesis, the rules with this hypothesis in a conclusion are selected from the knowledge base and ordered by decreasing value of the label. Then they are sequentially tried to be fired (including recursion) until the first successful attempt. Let us call this method M_3^g.

3   Experiments

The following data sets were used in experiments: Iris, Wisconsin Breast Cancer (WBC), Wine and Voting. All data sets were obtained from UCI Machine Learning Repository [5].

Continuous attributes of multivariate data sets were discretized (with equal width method) into domains with 2 and 4 values. These sets are respectively denoted by: iris2, iris4, wbc2, wbc4, wine2 and wine4. This transformation was performed using Weka data-mining software [6]. This package was also used to generate Bayesian Networks and test their performance. Accuracy measures of generated networks are presented in Tab. 1.

Generated networks were converted into sets of rules with continuous labels $(f(p) = p)$ with defaults, using the pruning threshold value t changing from 0 to 1. Then the resultant rule sets were tested on the training data with four rule application methods. Results of the experiments and complexity of generated classifiers, measured in number of rules before pruning, are shown in Fig. 2.

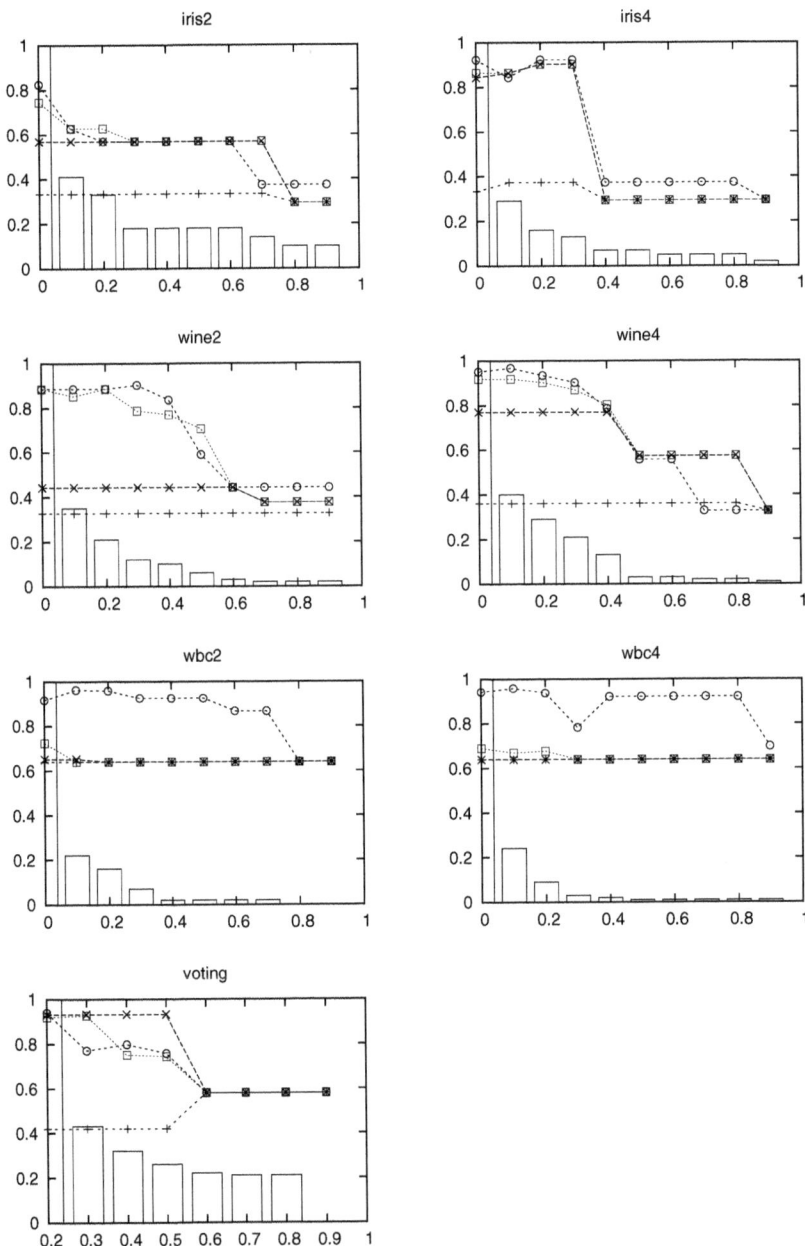

Fig. 2. Dependency of the performance of rule based classifiers and the number of rules generated form Bayesian Networks models from the threshold; bars (⊓) represent the relative number of rules (the number of rules generated with the given pruning threshold divided by the original number); the other symbols represent the accuracy of used reasoning methods: pluses (+) - M_3^g, exes (×) - M_3, circles (○) - M_2, squares (□) - M_1

Generally, results of experiments show the advantage of methods M_1 and M_2 over the others. These methods take into account the whole rule set in the reasoning process and · aggregation operator used corresponds well to the way how probability is calculated in Bayesian networks. The M_3 method is not much worse, but it is more sensitive to the number of ranges of discretization. The accuracy of classifiers using this rule application is therefore much higher for data sets known for their linear separability (like iris). As expected, the simplest method M_3^g had the lowest accuracy of all tested rule application methods.

In most cases the pruning with threshold between 0.1 and 0.3 narrowly increased the performance of classifiers. However the higher pruning largely decreases the accuracy. The results of classification with Bayesian network and M_1/M_2 classifiers are comparable – the difference for pruning threshold from 0.1 to 0.3 is about $\pm 5\%$. This is a very good result, especially if we take into account the fact that for threshold $t = 0.3$ pruning eliminates about 80% of rules. Accuracy of M_1 and M_2 methods for text data sets are also comparable with Bayesian network classifiers but the general performance of both was about 50%-60%. Unfortunately, because of the lack of space these results can not be presented here.

The relative number of rules drops significantly in range of pruning threshold 0.0 to 0.1 and asymptotically to 0 with its higher values. Exponential function seems to be a reasonable approximation of this relation.

4   Related Research

The most related paper to this research is [7], where possibility of using Certainty Factor model to represent Bayesian networks is analyzed. Methods for Noisy-OR, Noisy-AND, Noisy-MIN, Noisy-MAX and propagation of evidence are presented. It appears that many solutions used in practical applications of probabilistic models correspond to methods invented by Buchanan and Shortliffe.

The second closely related work is [8]. Experimental results of classification with belief networks (generated directly from data) and belief rules (generated from networks) are presented there.

Another interesting work is [9], where conversion of Bayesian networks into probabilistic horn abduction language is proposed. However, this formalism is more complicated than decision rules and resulting knowledge bases are not so easy to interpret.

Knowledge conversion methods in the opposite direction (from rule-based systems into probabilistic models) were investigated in a number of publications (e.g. [10,11]). There are also several works that aim at exploring problems that appear in probabilistic interpretation of Certainty Factor model (e.g. [12]).

5   Conclusions and Further Research

Transforming probabilistic models into decision rules can be very useful for visualization purposes. It allows to extract strong patterns appearing in a probabilistic models and present it in a user friendly way. These rules can be examined by the user or used for classification. One of the potential areas of application is a public security domain.

In the near future, we would like to make more experiments on a number of problem domains to compare accuracy of Bayesian networks and generated rule sets. Next task is to test several other label algebras, especially discrete ones, and with other ⊙ operation definitions. Also, other rule application algorithms should be examined.

Acknowledgments. The research leading to the results described in this paper has received funding from the Polish Ministry of Science and Higher Education Project number R00 0032 06.

References

1. Newell, A., Simon, H.: Human Problem Solving. Prentice-Hall, Englewood Cliffs (1972)
2. Śnieżyński, B.: Converting a naïve bayes models with multi-valued domains into sets of rules. In: Bressan, S., Küng, J., Wagner, R. (eds.) DEXA 2006. LNCS, vol. 4080, pp. 634–643. Springer, Heidelberg (2006)
3. Śnieżyński, B.: Conversion of a bayesian network into a set of rules: Initial results. Technical Report 1/2007, AGH University of Science and Technology, Krakow, Poland (2007)
4. Śnieżyński, B.: Converting a naïve bayes model into a set of rules. In: Klopotek, M., Wierzchon, S., Trojanowski, K. (eds.) Intelligent Information Processing and Web Mining. Advances in Soft Computing, vol. 5, pp. 221–229. Springer, Heidelberg (2006)
5. Asuncion, A., Newman, D.: UCI Machine Learning Repository (2007), http://www.ics.uci.edu/~mlearn/MLRepository.html
6. Witten, I.H., Frank, E.: Data Mining: Practical Machine Learning Tools and Techniques with Java Implementations. Morgan Kaufmann, San Francisco (1999)
7. Lucas, P.: Certainty-factor-like structures in bayesian belief networks. Knowl.-Based Syst. 14, 327–335 (2001)
8. Grzymala-Busse, J.W., Hippe, Z.S., Mroczek, T.: Belief rules vs. decision rules: A preliminary appraisal of the problem. In: Intelligent Information Systems, pp. 431–435 (2005)
9. Poole, D.: Probabilistic horn abduction and bayesian networks. Artificial Intelligence 64, 81–129 (1993)
10. Korver, M., Lucas, P.: Converting a rule-based expert system into a belief network. Medical Informatics 18, 219–241 (1993)
11. Middleton, B., Shwe, M., Heckerman, D.E., Henrion, M., Horvitz, E.J., Lehmann, H., Cooper, G.F.: Probabilistic diagnosis using a reformulation of the INTERNIST-1/QMR knowledge base ii: Evaluation of diagnostic performance. Methods of Information in Medicine 30, 256–267 (1991)
12. van der Gaag, L.: Probability-based models for plausible reasoning. PhD thesis, University of Amsterdam (1990)

Automatic Morphological Categorisation of Carbon Black Nano-aggregates

Juan López-de-Uralde[1], Iraide Ruiz[1], Igor Santos[1], Agustín Zubillaga[1],
Pablo G. Bringas[1], Ana Okariz[2], and Teresa Guraya[2]

[1] S³Lab, University of Deusto, Bilbao, Spain
{jlopezdeuralde,iraide.ruiz,isantos}@deusto.es,
{agustin.zubillaga,pablo.garcia.bringas}@deusto.es
[2] Universidad del País Vasco UPV/EHU, Bilbao, Spain
{ana.okariz,teresa.guraya}@ehu.es

Abstract. Nano-technology is the study of matter behaviour on atomic and molecular scale (i.e. nano-scale). In particular, carbon black is a nano-material generally used for the reinforcement of rubber compounds. Nevertheless, the exact reason behind its success in this concrete domain remains unknown. Characterisation of rubber nano-aggregates aims to answer this question. The morphology of the nano-aggregate takes an important part in the final result of the compound. Several approaches have been taken to classify them. In this paper we propose the first automatic machine-learning-based nano-aggregate morphology categorisation system. This method extracts several geometric features in order to train machine-learning classifiers, forming a constellation of expert knowledge that enables us to foresee the exact morphology of a nano-aggregate. Furthermore, we compare the obtained results and show that Decision Trees outperform the rest of the counterparts for morphology categorisation.

Keywords: aggregate morphology classifying, image processing, machine-learning, carbon black.

1 Introduction

Matter behaviour on nano-scale is subject to quantum mechanics where microscopic and macroscopic theories are no longer applicable [1]. On this scale, nano-technology is the science that studies the comportment of the matter. This science has experienced a great development in the last years. In fact, they are considered to be the *basis for the next industrial revolution* since they have been applied to different areas such as energy, health care, chemical industry and material production [2]. Therefore, these processes are leading material manufacturers to a new generation of nano-material based products [2].

In the particular case of rubber compounds, reinforced materials with nano-particles, such as carbon black, are of great interest to the material industry. Concretely, the latter modifies the mechanical and electrical properties of the former [3]. Although this process has been used in industrial production of rubber reinforced with carbon black [4] for the last years, the internal mechanisms that make that happen are not completely known.

P. García Bringas et al. (Eds.): DEXA 2010, Part II, LNCS 6262, pp. 185–193, 2010.
© Springer-Verlag Berlin Heidelberg 2010

In this way, there have been several studies about the morphology and micro-structure of carbonaceous particles, such as the ones produced by diesel combustion [5]. This engine-emitted particles were studied with the purpose of assessing their climate impact. Likewise, the waste-water treatment includes similar steps to the ones needed for carbon black characterization: microscopic image processing, object segmentation, morphological characterisation and fractal analysis [6]. Similarly, with CAT (Computerized Axial Tomography) scans the same procedure has been applied to evaluate the rank of a tumour [7]. Still, these methods are performed in a semi-automatic or manual way with the consequent time and resource consumption.

Against this background, we present the first automatic machine-learning-based nano-aggregate morphology categorisation method. This method, based upon geometrical and fractal features is able to train several machine-learning algorithms in order to correctly determine the morphology of these aggregates. Specifically, we contribute to the state of the art in two main ways. Firstly, it consists in automatically segmenting and characterising the carbon black aggregates within an image. This technique makes the geometrical characterisation of carbon black and other nano-particles easy and fast. Secondly, a machine learning based classifier sorts carbon black aggregates according to their morphology.

2   Carbon Black

As we mentioned before, one of the principal carbon black applications is the reinforcement of rubber. This process creates a material with notably increased tensile strength and better tear and abrasion resistance (i.e. the capacity of a material to withstand different forces). These changes are conditioned by molecular, chemical and rheological attributes of the elastomer, on the filler characteristics and on the mixing process and technology [8]. In addition, carbon black primary particles seem to be spherical, blended together forming aggregates[9]. Following the *Van der Waals* forces, aggregates connect forming agglomerates [9]. Fig. 1 shows a graphic representation of the size of particles, aggregates and agglomerates.

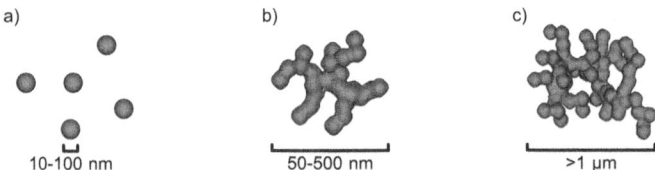

Fig. 1. Carbon black: a)particles; b)aggregate; c)agglomerate

Furthermore, the structure of carbon black particles ranges from crystalline to amorphous materials. Crystallite flat surfaces and amorphous carbon surfaces are less energetic areas, whereas crystallite edges are the most energetic ones [10].

Commonly, the aggregates can be divided into four different types of morphologies [11] (shown on Fig. 2). To this end, an estimation for discerning between the four categories is to calculate the *aggregate length/width* ratio and aggregate irregularity, however, this method is not an exact classification and includes a difficult value to measure: *irregularity* [11]:

- **Spheroidal:** Aggregates with a L/W ratio lower than 1.5 can be classified as spheroidal.
- **Ellipsoidal:** Aggregates with a L/W ratio between 2 and 3.5 can be classified as ellipsoidal.
- **Linear:** Linear ones have a L/W ratio greater than 3.5 and have low irregularity due to having elongated chains with few branches.
- **Branched:** Branched aggregates have also a L/W ratio greater than 3.5 but are highly irregular as a result of having more branches.

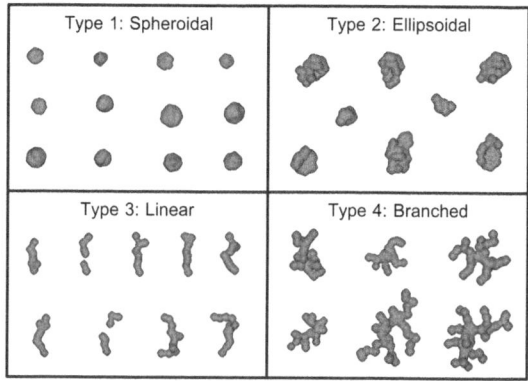

Fig. 2. Morphological categories for carbon black aggregates

Moreover, fractal dimension can also describe the aggregate structure [12]. Specifically, a fractal is a morphology that can be split into small copies of the whole [13]. To this end, Kaye [14] was the first one to apply fractal analysis to carbon black aggregates. He determined a perimeter fractal based upon the perimeter-area relationship of Mandelbrot [15] (shown on equation 1):

$$P \sim A^{Dp/2} \tag{1}$$

where P is defined as the projected aggregate perimeter, A as the projected area and D_p as the perimeter fractal. The greater the irregularity, the greater the D_p, however, highly acicular particles with a smooth perimeter may also give a high perimeter fractal [4].

Considering the scale of carbon black aggregates, *electron microscopes* are needed to analyse them. There are several types of microscope techniques based on the use of a particle beam of electrons such as *Transmission Electron Microscopy* (TEM) and *Scanning Electron Microscopy* (SEM).

3   Image Feature Extraction

Specifically, the aim of this treatment is to segment the aggregates and to extract several geometric features from them. Not only basic ones, such as area or perimeter, are considered but also more complex ones like perimeter fractal. Thereby, *machine-learning* classifiers will determine the morphology of unclassified aggregates using these features. Our algorithm follows the operations required by the *Standard Test Method for Carbon Black* [16] for analysing images captured by electron microscopes: background/noise elimination, thresholding, erosion and dilation.

In order to conduct the binarization, we start applying a Gaussian smoother [17], a 2-D convolution operator used to remove detail and noise. Second, we estimate a threshold for aggregate-background discrimination using Otsu's method [18]. We adjust this threshold to be more adequate for SEM images. We generate a binary image considering that pixels with value below the threshold correspond to background and pixels above it are part of the aggregate area.

Although we accomplish a smoothing process for noise reduction in the binarization phase, undesired elements may still be present in the image. These elements can be easily confused with the desired aggregates, thus, it is mandatory to eliminate them. To this end, we begin deleting minor areas and we continue filling holes inside aggregates. Moreover we improve the edge quality by dilating and eroding it with a disk shape morphological structuring element and we end deleting incomplete aggregates touching the edge of the image. Besides, we identify aggregates segmenting from the image the regions that surpass an specified area.

Based on the output image from the previous phase, we extract some geometric features, the ones marked with an '*' are the required ones according to the Standard Test Method for Carbon Black [16]. These parameters are measured in *nm* and when necessary estimated using stereological principles (i.e. the three-dimensional interpretation of two-dimensionally observed objects). To start with, the common ones are: perimeter*, area*, area-perimeter ratio, equivalent diameter, aggregate and particle volume, axis ratio, number of particles per aggregate, occlusion factor, absorption and circularity. In the second place are the parameters that require an explanation:

- **Feret diameters*:** A Feret diameter is defined as the distance between two tangents on opposite sides of the particle profile that are parallel to some fixed direction. So as to obtain valuable information related to the form of the particle, we extract 16 Ferets [16] separated by 11.5 degrees choosing the biggest (*Major Feret*), the smallest (*Minor Feret*) and the perpendicular one to the biggest.
- **Major and minor axis length:** Scalars specifying the length of the major and minor axis of the ellipse that have the same normalized second central moments as the region.
- **Centroid:** The center of mass of the region. It is formed by 2 values (x and y coordinates) normalized to the size of the *bounding box*, defined as the smallest rectangle containing the region.

- **Convex area:** The area of the *convex hull*, which is the smallest convex polygon that can contain the region.
- **Eccentricity:** The eccentricity is the ratio of the distance between the foci of the ellipse (i.e. the two points from which the distance to every point of the ellipse is constant) and its major axis length, taking values between 0 and 1.
- **Length-width ratio:** This ratio is computed with the maximum Feret and with the perpendicular Feret to the latter. Thereby, this commonly used parameter [11] is normalized.
- **Maximum Feret - minimum Feret ratio:** This ratio is similar to the previous one. However, considering that length and width are always orthogonal, it gives some extra information.
- **Area - convex area ratio:** Relation between the real area of the aggregate and the area of the *convex hull*. The smaller the ratio, the bigger the irregularity of the aggregate.
- **Extent:** Defined as the real area divided by the area of the *bounding box*.
- **Perimeter Fractal:** Determined by $P \sim A^{Dp/2}$ as explained in section 2.
- **Aggregation factor:** Defined by $13.092(\frac{P^2}{A})^{-0.92}$ where P is the perimeter and A is the area. If lower than 0.4 then it is equal to 0.4.

Finally, we generate a training vector $v = (v_1, v_2, ..v_{13})$ per aggregate containing all these characteristics. Concretely, each position v_n in the vector represents a geometric feature and has up to 6 decimals. The collection of vectors forms the *corpus I*, which provides the learning dataset for the classification system.

4 Experimental Evaluation

Initially, we obtained several images with three electron microscopes on different magnification scales. Thirteen images with 2 SEM microscopes, Hitachi S-3400N and Hitachi S4800, and eleven with a Transmission Electron Microscope, the Philips EM208S. After performing a preliminary evaluation of the aggregate segmenting process, we chose the second Scanning Electron Microscope (SEM).

In this way, we collected 102 images of carbon black aggregates with a Hitachi S-4800 Scanning Electron Microscope. Images were captured at 30000x magnification with an average of 3 aggregates per image resulting in 266 correctly segmented aggregates that have formed the case of study.

We segmented all the aggregates from the images, then we labelled them and finally, we generated a Comma-Separated Values (CSV) file with all the characteristics and finally we performed machine learning studies to classify the aggregates.

In these experiments, we extracted 26 variables from each aggregate. The dataset was not balanced for the four existing classes due to scarce data. Specifically, 9 aggregates were of type spheroidal, 86 ellipsoidal, 51 linear and 120 branched. To address both problems (scarce and unbalanced data) we applied Synthetic Minority Over-sampling TEchnique (SMOTE) [19], which is a combination of over-sampling the less populated classes and under-sampling the more populated ones. Nevertheless, the over-sampling is performed by creating synthetic minority class examples. In this way, instances were still unique and classes became more balanced.

More accurately, we conducted the next methodology in order to test the suitability of each machine-learning algorithm:

- **SMOTE:** We built a dataset that contains the result of applying SMOTE to the original dataset in order to compare the results of the machine-learning classifiers with and without this technique.
- **Cross validation:** This method is generally applied in machine-learning evaluation [20]. In our experiments, we performed a K-fold cross validation with $k = 10$. In this way, our dataset is 10 times split into 10 different sets of learning (90 % of the total dataset) and testing (10 % of the total data).
- **Learning the model:** For each fold, we accomplished the learning step of each algorithm using different parameters or learning algorithms depending on the specific model. In particular, we used the following models:
  - *Bayesian networks (BN):* With regards to Bayesian networks we utilize different structural learning algorithms: K2 [21], Hill Climber [22] and Tree Augmented Naïve (TAN) [23]. Moreover, we also performed experiments with a Naïve Bayes Classifier [20].
  - *Support Vector Machines (SVM):* We performed experiments with a polynomial kernel [24], a normalized polynomial Kernel [25] and Pearson VII function-based universal kernel [26].
  - *K-nearest neighbour (KNN):* We performed experiments with $k = 1$, $k = 5$, $k = 10$, $k = 15$, $k = 20$ and $k = 25$.
  - *Decision Trees (DT):* We performed experiments with J48(the *Weka* [27] implementation of the *C4.5* algorithm [28]) and Random Forest [29], an ensemble of randomly constructed decision trees.
- **Testing the model:** We evaluated the percent of correctly classified instances and the *area under the ROC curve* (AUC) that establishes the relation between false negatives and false positives [30].

Table 1 shows the obtained results in terms of accuracy percent. In this way, regarding the results without the use of SMOTE, most of the classifiers obtained only medium results, with the exception of Naïve Bayes method, which was the worst, with results lower than 50 %. Otherwise, when SMOTE technique was applied, every classifier improved its accuracy in a significant manner. Specially, Naïve Bayes increased its accuracy in more than 20 %. Furthermore, Random Forest, a type of Decision Tree, outperformed the rest of the classifiers with an accuracy of 83.61 %.

Nevertheless, focusing only on accuracy may be misleading and, therefore, we performed an analysis of the AUC. To this extent, Table 2 shows the results in terms of AUC. As occurred with accuracy, when SMOTE is omitted from the methodology the results are quite modest. Naïve Bayes was also the worst this time with an AUC of 0.81. Notwithstanding, we observed the same improvement using SMOTE, increasing the AUC of every classifier. Random Forest was also the best classifier in terms of AUC with a value of 0.94.

Table 1. Results of the machine-learning classifiers with regards to accuracy (%)

Machine-learning Model	Original Dataset	With SMOTE	
DT: J48	69.04	79.77	✓
DT: RandomForest with 1000 trees	73.40	83.61	✓
SVM: Polynomial Kernel	68.21	78.27	✓
SVM: Normalized Polynomial Kernel	67.30	75.68	✓
SVM: Pearson VII universal kernel	68.48	80.24	✓
KNN K=1	63.58	77.30	✓
KNN K=5	66.13	78.23	✓
KNN K=10	64.75	76.01	✓
KNN K=15	66.52	76.57	✓
KNN K=20	68.09	76.57	✓
KNN K=25	68.39	75.92	✓
Naïve Bayes	48.99	70.32	✓
BN: K2	56.37	77.33	✓
BN: Hill Climber	56.37	77.33	✓
BN: TAN	68.60	79.03	✓

✓, x, − statistically significant improvement, degradation or non significant change

Table 2. Results of the machine-learning classifiers with regards to AUC

Machine-learning Model	Original Dataset	With SMOTE	
DT: J48	0.76	0.81	−
DT: RandomForest with 1000 trees	0.89	0.94	−
SVM: Polynomial Kernel	0.82	0.91	✓
SVM: Normalized Polynomial Kernel	0.81	0.90	✓
SVM: Pearson VII universal kernel	0.81	0.90	✓
KNN K=1	0.70	0.71	−
KNN K=5	0.82	0.88	−
KNN K=10	0.83	0.90	−
KNN K=15	0.85	0.92	✓
KNN K=20	0.86	0.93	✓
KNN K=25	0.86	0.93	✓
Naïve Bayes	0.81	0.90	✓
BN: K2	0.85	0.92	✓
BN: Hill Climber	0.85	0.92	✓
BN: TAN	0.84	0.91	✓

✓, x, − statistically significant improvement, degradation or non significant change

Summarizing, by means of machine learning algorithms we were able to accomplish aggregate morphology classification. Besides, with the help of synthetic re-sampling more data was produced and the four classes became more balanced. Thereby, we overcame the imbalance problem without merging the dataset, an inappropriate option due to the size of our dataset.

5   Conclusions and Future Work

Nano-technologies have suffered a great development in the last years. Since nano-particles are able to modify the mechanical and electrical properties of materials [3], manufacturers have been led to a new generation of nano material-based production. Moreover, depending on the aggregate type [11,31] and the mixing process the obtained product varies [8].

In this paper, we have proposed the first automatic machine-learning-based nano-aggregate morphology categorisation method. This technique correctly determined the morphology of nano-aggregates, based on the use of geometrical and fractal characteristics as features for the training of several machine-learning classifiers. Furthermore, the empirical validation showed that this method is capable of classifying the morphology of aggregates with an accuracy of over 80%.

Future work will compare results based on original samples with the present results obtained with SMOTE re-sampling [19]. To this end, we will acquire more SEM images in order to generate a larger training dataset. In addition, we are planning to improve the image-processing algorithm so as to work with TEM images. On the other hand, we will focus on developing a 3-dimensional tool in order to accomplish *skeletonization* and 3D modelling of the aggregates.

Acknowledgements

We thank Mikel Salazar for the carbon black 3D simulation images created for this paper. In addition, we thank Ana Okariz and Teresa Guraya for their insight into carbon black and the provision of the TEM and SEM images. Last but not least we are grateful to Maria Carmen Huarte for her invaluable corrections.

References

1. Roco, M., Bainbridge, W.: Societal implications of nanoscience and nanotechnology. Kluwer Academic Pub., Dordrecht (2001)
2. Kiparissides, C., Clausen, B., Boehm, L., Wilkins, T., Kellermayer, M., Baraton, M., Hossain, K.: NMP expert advisory group (EAG) position paper on future RTD activities of NMP for the period 2010 - 2015. Technical report
3. Mather, P., Thomas, K.: Carbon black/high density polyethylene conducting composite materials: Part I Structural modification of a carbon black by gasification in carbon dioxide and the effect on the electrical and mechanical properties of the composite. Journal of materials science 32(2), 401–407 (1997)
4. Donnet, J., Bansal, R., Wang, M.: Carbon black: science and technology. CRC, Boca Raton (1993)
5. Soewono, A.: Morphology and microstructure of diesel particulates. Master's thesis
6. Amaral, A.: Image analysis in biotechnological processes: applications to wastewater treatment. PhD thesis, Universidadde Do Minho (2003)
7. Al-Kadi, O.: Tumour Grading and Discrimination based on Class Assignment and Quantitative Texture Analysis Techniques. PhD thesis, University of Sussex (2009)
8. Fröhlich, J., Niedermeier, W., Luginsland, H.: The effect of filler–filler and filler–elastomer interaction on rubber reinforcement. Composites Part A 36(4), 449–460 (2005)
9. Donnet, J.: Black and white fillers and tire compound. Rubber chemistry and technology 71(3), 323–341 (1998)
10. De, S., Naskar, K., White, J.: Rubber Technologist's Handbook, vol. 2. Smithers Rapra (2008)
11. Herd, C., McDonald, G., Hess, W.: Morphology of carbon-black aggregates: fractal versus euclidean geometry. Rubber chemistry and technology 65(1), 107–129 (1992)
12. Meakin, P.: Formation of fractal clusters and networks by irreversible diffusion-limited aggregation. Physical Review Letters 51(13), 1119–1122 (1983)

13. Mandelbrot, B.: The fractal geometry of nature. W.H. Freeman, San Francisco (1982)
14. Kaye, B.: Fractal description of fineparticle systems. Particle Characterization in Technology: Morphological analysis, 81 (1984)
15. Mandelbrot, B.: Form, chance, and Dimension. In: Chance and Dimension, pp. 1–234. Freeman, San Francisco (1977)
16. American Society for Testing and Materials: ASTM D3849-02 - Standard Test Method for Carbon Black - Morphological Characterization of Carbon Black Using Electron Microscopy (2002) (Testing method)
17. Pajares, G., de la Cruz, J.M.: Visión por Computador. Ra-Ma Publishers (2007)
18. Otsu, N.: A threshold selection method from gray-level histograms. Automatica 11, 285–296 (1975)
19. Chawla, N., Bowyer, K., Hall, L., Kegelmeyer, W.: SMOTE: synthetic minority over-sampling technique. Journal of Artificial Intelligence Research 16(3), 321–357 (2002)
20. Bishop, C.M.: Neural Networks for Pattern Recognition. Oxford University Press, Oxford (1995)
21. Cooper, G.F., Herskovits, E.: A bayesian method for constructing bayesian belief networks from databases. In: Proceedings of the 7th conference on Uncertainty in artificial intelligence (1991)
22. Russell, S.J., Norvig: Artificial Intelligence: A Modern Approach, 2nd edn. Prentice-Hall, Englewood Cliffs (2003)
23. Geiger, D., Goldszmidt, M., Provan, G., Langley, P., Smyth, P.: Bayesian network classifiers. In: Machine Learning, pp. 131–163 (1997)
24. Amari, S., Wu, S.: Improving support vector machine classifiers by modifying kernel functions. Neural Networks 12(6), 783–789 (1999)
25. Maji, S., Berg, A., Malik, J.: Classification using intersection kernel support vector machines is efficient. In: Proc. CVPR, vol. 1, p. 4 (2008)
26. Üstün, B., Melssen, W., Buydens, L.: Visualisation and interpretation of support vector regression models. Analytica chimica acta 595(1-2), 299–309 (2007)
27. Garner, S.: Weka: The Waikato environment for knowledge analysis. In: Proceedings of the New Zealand Computer Science Research Students Conference, pp. 57–64 (1995)
28. Quinlan, J.: C4. 5 programs for machine learning. Morgan Kaufmann Publishers, San Francisco (1993)
29. Breiman, L.: Random forests. Machine learning 45(1), 5–32 (2001)
30. Singh, Y., Kaur, A., Malhotra, R.: Comparative analysis of regression and machine learning methods for predicting fault proneness models. International Journal of Computer Applications in Technology 35(2), 183–193 (2009)
31. Ungár, T., Gubicza, J., Tichy, G., Pantea, C., Zerda, T.: Size and shape of crystallites and internal stresses in carbon blacks. Composites Part A 36(4), 431–436 (2005)

Towards Efficient Mining of Periodic-Frequent Patterns in Transactional Databases

R. Uday Kiran and P. Krishna Reddy

Center for Data Engineering
International Institute of Information Technology-Hyderabad
Hyderabad, India - 500032
uday_rage@research.iiit.ac.in, pkreddy@iiit.ac.in
http://research.iiit.ac.in/~uday_rage, http://iiit.ac.in/~pkreddy

Abstract. Periodic-Frequent patterns are an important class of regularities that exist in a transactional database. A pattern is *periodic-frequent* if it satisfies both minimum support (*minsup*) and maximum periodicity (*maxprd*) constraints. *Minsup* constraint controls the minimum number of transactions that a pattern must cover in a database. *Maxprd* constraint controls the maximum duration between the two transactions below which a pattern should reoccur in a database. In the literature an approach has been proposed to extract periodic-frequent patterns using single *minsup* and single *maxprd* constraints. However, real-world databases are mostly non-uniform in nature containing both frequent and relatively infrequent (or rarely) occurring items. Researchers are making efforts to propose improved approaches for extracting frequent patterns that contain rare items as they contain useful knowledge. For mining periodic patterns that contain frequent and rare items we have to specify low *minsup* and high *maxprd*. It is difficult to mine periodic-frequent patterns because the low *minsup* and high *maxprd* can cause combinatorial explosion. In this paper we propose an improved approach which facilitates the user to specify different *minsup* and *maxprd* values for each pattern depending upon the items within it. Also, we present an efficient pattern growth approach and a methodology to dynamically specify *maxprd* for each pattern. Experimental results show that the proposed approach is efficient.

Keywords: Data mining, frequent pattern, rare periodic-frequent pattern, multiple constraints.

1 Introduction

Periodic-frequent patterns [3] are an important class of regularities that exist in a database. In many real-world applications, these patterns provide useful information regarding the patterns which are not only occurring frequently, but also appearing periodically (or regularly) throughout a transactional database. The basic model of periodic-frequent patterns is as follows [3].

Let $I = \{i_1, i_2, \cdots, i_n\}$ be a set of items. A set $X = \{i_j, \cdots, i_k\} \subseteq I$, where $j \leq k$ and $j, k \in [1, n]$, is called a **pattern** (or an itemset). A transaction

P. García Bringas et al. (Eds.): DEXA 2010, Part II, LNCS 6262, pp. 194–208, 2010.
© Springer-Verlag Berlin Heidelberg 2010

$t = (tid, Y)$ is a tuple, where tid represents a transaction-id (or a timestamp) and Y is a pattern. A transactional database T over I is a set of transactions, $T = \{t_1, \cdots, t_m\}$, $m = |T|$, where $|T|$ is the size of T in total number of transactions. If $X \subseteq Y$, it is said that t contains X or X occurs in t and such transaction-id is denoted as t_j^X, $j \in [1, m]$. Let $T^X = \{t_k^X, \cdots, t_l^X\} \subseteq T$, where $k \leq l$ and $k, l \in [1, m]$ be the ordered set of transactions in which pattern X has occurred. Let t_j^X and t_{j+1}^X, where $j \in [k, (l-1)]$ be two consecutive transactions in T^X. The number of transactions or time difference between t_{j+1}^X and t_j^X can be defined as a **period** of X, say p^X. That is, $p^X = t_{j+1}^X - t_j^X$. Let $P^X = \{p_1^X, p_2^X, \cdots, p_r^X\}$, be the set of periods for pattern X. The **periodicity** of X, denoted as $Per(X) = max(p_1^X, p_2^X, \cdots, p_r^X)$. The **support** of X, denoted as $S(X) = |T^X|$. The pattern X is said to be periodic-frequent pattern, if $S(X) \geq minsup$ and $Per(X) \leq maxprd$, where $minsup$ and $maxprd$ are user-specified minimum support and maximum periodicity constraints. Both periodicity and support of a pattern can be described in percentage of $|T|$.

Table 1. Transaction database. Transactions are ordered based on timestamp.

TID	Items	TID	Items
1	bread, jam, pencil	7	bread, jam,
2	ball, bat, pen		ball, bat
3	bread, jam, ball	8	bed, pillow
4	bed, pillow	9	bread, jam
5	bread, jam	10	ball, bat
6	ball, bat		pencil

Table 2. Periodic-Frequent patterns having support ≥ 2 and $periodicity \leq 4$

Pattern	S	P	Pattern	S	P
bread	5	2	{bread,ball}	2	4
ball	5	3	{bread, jam}	5	2
jam	5	2	{ball,bat}	4	4
bat	4	4	{bed, pillow}	2	4
bed	2	4			
pillow	2	4			

Example 1. Consider the transactional database shown in Table 1. Each transaction in it is uniquely identifiable with a transactional-id (tid) which is also a timestamp of that transaction. Timestamp indicates time of occurrence of the transaction. The set of items, $I = \{bread, jam, ball, bat, bed, pillow, pencil, pen\}$. The set of $bread$ and jam i.e., $\{bread, jam\}$ is a pattern. This pattern occurs in $tids$ of $1, 3, 5, 7$ and 9. Therefore, $T^{\{bread, jam\}} = \{1, 3, 5, 7, 9\}$. Its support count (or support), $S(bread, jam) = |T^{\{bread, jam\}}| = 5$. The periods for this pattern are $1(= 1 - t_i)$, $2(= 3 - 1)$, $2(= 5 - 3)$, $2(= 7 - 5)$, $2(= 9 - 7)$ and $1(= t_l - 9)$, where $t_i = 0$ represents the initial transaction and $t_l = 10$ represents the last transaction in the transactional database. The periodicity of $\{bread, jam\}$, $Per(bread, jam) = maximum(1, 2, 2, 2, 2, 1) = 2$. If the user-specified $minsup = 4$ and $maxprd = 2$, the pattern $\{bread, jam\}$ is a periodic-frequent pattern because $S(bread, jam) \geq minsup$ and $Per(bread, jam) \leq maxprd$.

For this model, an efficient pattern growth approach based on a tree-structure, called Periodic-Frequent tree (PF-tree) was also discussed to discover complete set of periodic-frequent patterns [3]. The structure of PF-tree is different from the FP-tree [2]. It is because FP-tree was not proposed to consider the periodicity of a pattern.

Using only a single *minsup* and single *maxprd* constraints, it is easy to discover periodic-frequent patterns consisting of frequent items. However, real-world databases are mostly non-uniform in nature containing both frequent and relatively infrequent (or rarely) occurring items. More important, periodic-frequent patterns consisting of rare items i.e., rare periodic-frequent patterns can provide useful information.

Example 2. Generally in a supermarket, the set of items *bed* and *pillow* are rarely purchased than the set of items *bread* and *jam*. Also, the duration of two consecutive purchases of '*bed* and *pillow*' is relatively longer than the two consecutive purchases of '*bread* and *jam*'. However, the former set of items is more interesting as it generates more revenue per unit as in this case.

Rare periodic-frequent patterns have relatively low support (or frequency) and high periodicity (due to their sporadic nature). It is difficult to mine these patterns with a "single *minsup* and single *maxprd* model" because this model suffers from "rare item problem." That is, to mine rare periodic-frequent patterns, one has to specify low *minsup* and high *maxprd*. This may cause combinatorial explosion, producing too many periodic-frequent patterns, because, those frequent items will be associated with one another in all possible ways and many of them are uninteresting. Uninteresting periodic-frequent patterns are the patterns having low support and/or high periodicity and consist of only frequent items.

Example 3. Consider the transactional database shown in Table 1. To mine periodic-frequent patterns consisting of the rare items (*bed* and *pillow*), one has to set low *minsup* and high *maxprd*. Let *minsup* = 2 and *maxprd* = 4. Table 1 presents the discovered periodic-frequent patterns. It can be observed that along with the interesting patterns {*bread, jam*} and {*bed, pillow*}, the uninteresting patterns i.e., {*bread, ball*} and {*ball, bat*} (patterns represented in bold letters) have also been generated as periodic-frequent patterns. The patterns {*bread, ball*} and {*ball, bat*} are uninteresting because they contain only frequent items and have low support and/or high periodicity. These patterns can be considered as interesting if they have satisfied high *minsup* and low *maxprd*, say *minsup* = 4 and *maxprd* = 2. Like, the periodic-frequent pattern {*bread, jam*}.

In the literature, "rare item problem" was also confronted while mining frequent patterns using "single *minsup* model." Efforts are being made to propose improved approaches using "multiple *minsup* model" [4,5,6,7,8]. In this model, each item is specified with a support constraint, called minimum item support (MinIS), and *minsup* of a pattern is represented with the minimal *MinIS* value among all its items. Thus, each pattern can satisfy a different *minsup* depending upon the items within it. In [4,6], methodologies have been discussed to specify items' *MinIS* values depending upon their respective supports.

In this paper, we extend the existing "multiple *minsup* model" to "multiple *minsup* and multiple *maxprd* model" to efficiently mine periodic-frequent patterns consisting of both frequent and rare items. In the proposed model, each pattern can satisfy a different *minsup* and *maxprd* values depending upon the items within it. Specifically, the user specifies two types of constraints: (*i*) support constraint, called *minimum item support* (*MinIS*) and (*ii*) periodicity constraint, called *maximum item periodicity* (*MaxIP*). Thus, different patterns may need to satisfy different *minsup* and *maxprd* values depending upon the items within it.

The periodic-frequent patterns mined using the proposed model do not satisfy *downward closure property*. That is, not all non-empty subsets of a periodic-frequent pattern need be periodic-frequent. This increases the search space to discover complete set of periodic-frequent patterns. However, we propose an efficient pattern growth approach, which uses various techniques to minimize the search space for efficient mining of periodic-frequent patterns. Experimental results on both synthetic and real-world databases show that the proposed approach efficiently discover periodic-frequent patterns consisting of both frequent and rare items. However, it requires more runtime because the periodic-frequent patterns mined using the proposed model do not satisfy *downward closure property*.

The rest of the paper is organized as follows. In Section 2, we introduce the extended model of mining periodic-frequent patterns. For the proposed model, a pattern-growth approach based on a tree structure, called Multi-Constraint Periodic-Frequent tree (MCPF-tree) has been discussed in Section 3. We report our experimental results in Section 4. Finally, Section 5 concludes the paper.

2   The Extended Model

In the new model, each item in a transactional database has two types of constraints: a support constraint, called *minimum item support* (*MinIS*) and a periodicity constraint, called *maximum item periodicity* (*MaxIP*). A pattern is *periodic-frequent* if it satisfies lowest *MinIS* and maximum *MaxIP* values of all the items within it.

Continuing with the basic model of periodic-frequent patterns, let $MinIS(i_j)$ and $MaxIP(i_j)$ be the *minimum item support* and *maximum item periodicity* specified for an item $i_j \in I$. Then, a pattern $X = \{i_1, i_2, \cdots, i_k\} \subseteq I$ is periodic-frequent if:

$$S(X) \geq minimum(MinIS(i_1), MinIS(i_2), \cdots, MinIS(i_k)) \qquad (1)$$
$$and$$
$$Per(X) \leq maximum(MaxIP(i_1), MaxIP(i_2), \cdots, MaxIP(i_k))$$

Minimum item supports and maximum item periodicities enable us to achieve the goal of specifying higher *minsup* and lower *maxprd* for patterns that only involve frequent items, and specifying lower *minsup* and higher *maxprd* for patterns involving rare items.

2.1   Specifying *MaxIP* for an Item

If there exists numerous items within a transactional database, it will be very difficult for the user to manually specify *MinIS* and *MaxIP* values for every item. In the literature (multiple *minsup* based frequent pattern mining), there exists methodologies to specify items' *MinIS* values dynamically depending upon their respective supports [4,6]. In this paper, we propose a methodology to specify items' *MaxIP* values dynamically depending upon their support values. The methodology is as follows:

$$mip(i_j) = \beta \times S(i_j) + Per_{max}$$
$$MaxIP(i_j) = mip(i_j) \quad if \quad mip(i_j) \geq Per_{min} \quad\quad (2)$$
$$= Per_{min} \quad\quad otherwise$$

where, $S(i_j)$ is the support of the item i_j, Per_{max} and Per_{min} are the user-specified maximum and minimum periodicities such that $Per_{max} \geq Per_{min}$ and $\beta \in [-1, 0]$ is a user-specified constant. The above methodology has the following three properties.

Property 1. If $\beta = 0$ and $Per_{max} > Per_{min}$, each items' $MaxIP$ value will be equal to P_{max}. In such a scenario, the proposed model is same as mining periodic-frequent patterns with a single *maxprd* constraint, where $maxprd = Per_{max}$.

Property 2. If $Per_{max} = Per_{min}$, each items' $MaxIP$ value will be equal to Per_{max} or Per_{min}. The proposed model is same as mining periodic-frequent patterns with a single *maxprd* constraint, where $maxprd = Per_{max} = Per_{min}$.

Property 3. It is an order-reversing function. That is, in I, if $S(i_1) \leq S(i_2) \leq \cdots S(i_n)$, then $MaxIP(i_1) \geq MaxIP(i_2) \geq \cdots \geq MaxIP(i_n)$. Thus, as compared with frequent items, rare items will have high $MaxIP$ values.

2.2 Nature of the Periodic-Frequent Patterns

The periodic-frequent patterns mined using "single *minsup* and single *maxprd* model" satisfy *downward closure property*. That is, all non-empty subsets of a periodic-frequent pattern are periodic-frequent. However, the periodic-frequent patterns mined using the proposed model do not have to satisfy *downward closure property*.

Example 4. Let a, b and c be the three items in a transactional database. The user-specified *MinIS* values for these items be 10, 9 and 3 respectively. The user-specified *MaxIP* values for these items be 5, 6 and 5 respectively. After scanning the database, let the support of these respective items be 9, 8 and 4. Let the periodicity of these items be 4, 4 and 6 respectively. Clearly, the items a, b and c are non-periodic-frequent items (or 1-patterns) because $S(a) < MinIS(a)$, $S(b) < MinIS(b)$ and $P(c) > MaxIP(c)$. However, their superset i.e., $\{a, b, c\}$ with support=3 and periodicity=6 can be still be generated as a periodic-frequent pattern because supports of a, b and c are greater than or equal to $MinIS(c)$ and periodicities of a, b and c are less than or equal to $MaxIP(b)$. So, if we do not discard these non-periodic-frequent items, the *downward closure property* is lost.

2.3   Problem Definition

Given a transactional database T, items' $MinIS$ and $MaxIP$ values, discover complete set of periodic-frequent patterns that satisfy lowest $MinIS$ and maximum $MaxIP$ values of all the items within the respective pattern.

3   MCPF-Tree: Design, Construction and Mining

In this section, we describe the structure, construction and mining of periodic-frequent patterns using Multi-Constraint Periodic-Frequent Pattern-tree (MCPF-tree).

3.1   Structure of MCPF-Tree

The MCPF-tree consists of two components: MCPF-list and a prefix-tree. MCPF-list is a list with four fields: item, support (S), periodicity (P), $MinIS$ (mis) and $MaxIP$ (mip). The node structure of prefix-tree in MCPF-tree is same as the prefix-tree in PF-tree [3], which is as follows.

The prefix-tree in MCPF-tree explicitly maintains the occurrence information for each transaction in the tree structure by keeping an occurrence transaction-id list, called tid-list, only at the last node of every transaction. Two types of nodes are maintained in a MCPF-tree: ordinary node and $tail$-node. The ordinary node is similar to the nodes used in FP-tree, whereas the latter is the node that represents the last item of any sorted transaction. The structure of a $tail$-node is $N[t_1, t_2, ..., t_n]$, where N is the node's item name and $t_i, i \in [1, n]$, (n be the total number of transactions from the root up to the node) is a transaction-id where item N is the last item. Like the FP-tree [2], each node in a MCPF-tree maintains parent, children, and node traversal pointers. However, irrespective of the node type, no node in a MCPF-tree maintains support count value in it. We now explain construction and mining of MCPF-tree.

3.2   Constructing MCPF-Tree

Let id_l be a temporary array to record the $tids$ of the last occurring transactions of all items in the MCPF-list. Let t_{cur} and p_{cur} respectively denote the tid of current transaction and the most recent period for an item $i_j \in I$. The MCPF-tree is, therefore, maintained according to the process given in Algorithm 1.

Consider the transactional database shown in Table 1. Let the user-specified $MinIS$ values for the items $bread, ball, bat, jam, bed, pillow, pen$ and $pencil$ be 4, 4, 4, 4, 2, 2, 2 and 2, respectively. Let the user-specified $MaxIP$ values for these items be 2, 2, 2, 2, 4, 4, 4 and 4, respectively. Then, $L = \{bread, ball, bat, jam, bed, pillow, pen, pencil\}$.

In Fig. 1, we show how the MCPF-list is populated for the transactional database shown in Table 1. Fig. 1(a) shows the MCPF-list populated after inserting items in L order (Line 1 of Algorithm 1). Fig. 1(b), Fig. 1(c) and Fig. 1(d)

Algorithm 1. MCPF-tree (T: Transactional database, I: set of items, $MinIS$: items' minimum item support, $MaxIP$: items' maximum item periodicity)

1: Sort the items in descending order of their $MinIS$ values. Let this order of items be L. In L order, insert each item $i_j \in I$ into the MCPF-list with $S(i_j) = 0$, $P(i_j) = 0$, $mis(i_j) = MinIS(i_j)$ and $mip(i_j) = MaxIP(i_j)$.
2: **for** each transaction $t_{cur} \in T$ **do**
3:   **for** each $i_j \in t_{cur}$ **do**
4:     $S(i_j) + +$; $p_{cur} = t_{cur} - id_l(i_j)$;
5:     **if** $(p_{cur} > P(i_j))$ **then**
6:       $P(i_j) = p_{cur}$;
7:     **end if**
8:   **end for**
9: **end for**
10: At the end of T, calculate p_{cur} for each item by considering t_{cur} equal to the *tid* of the last transaction in T, and update their respective p value if $p_{cur} \geq P$. The purpose of this step is to reflect correct periodicity of each item in the MCPF-list.
11: **repeat**
12:   Let $i_k \in I$ be the item having lowest $MinIS$ (MIS_{min}) value among all frequent items.
13:   **for** each item i_j in MCPF-list **do**
14:     **if** $(S(i_j) < MIS_{min})$ **then**
15:       Remove i_j from the MCPF-list.
16:     **end if**
17:   **end for**
18:   Let $i_l \in I$ be the periodic item that has maximum $MaxIP$ (MIP_{max}) value among all the remaining items.
19:   **for** each item i_j in MCPF-list **do**
20:     **if** $((P(i_j) > MIP_{max})$ **then**
21:       Remove i_j from the MCPF-list.
22:     **end if**
23:   **end for**
24: **until** MCFP-list does not contain i_k
25: /* The above repeat step is necessary because i_k can be pruned if its periodicity is greater than the MIP_{max} value. */
26: Let L' be the sorted list of items in MCPF-list.
27: **for** each transaction $t \in T$ **do**
28:   Sort the items in L' order and create a branch in MCPF-tree as in PF-tree.
29: **end for**
30: For tree-traversal, maintain node-links in MCPF-tree as in PF-tree.

---

show the MCPF-list generated after scanning first (i.e., $t_{cur} = 1$), second (i.e., $t_{cur} = 2$) and every transaction (i.e., $t_{cur} = 10$), respectively (lines 2 to 9 in Algorithm 1). To reflect the correct periodicity for each item in the MCPF-tree, the whole MCPF-list is refreshed as mentioned in line 10 of Algorithm 1. The resultant MCPF-tree is shown in Fig. 1(e). In this figure, it can be observed that the periodicity of *pen* is changed from 2 to 8 as it did not appear in the database after $tid = 2$.

The periodic-frequent patterns mined using the proposed model do not satisfy *downward closure property*. This increases the search space for mining these patterns. To minimize the search space, we explore Lemma 1 and Lemma 2. The lowest MinIS value (MIS_{min}) among all frequent items in the MCPF-list is 2 (Line 12 in Algorithm 1). The item that has MIS_{min} value is *pencil*. Using $MIS_{min} = 2$, *pen* is pruned from the MCPF-list because its support is less than the MIS_{min} value (Line 13 to 17 in Algorithm 1). Among the remaining items, the periodic items are *bread, jam, ball, bat, bed* and *pillow*. The maximum $MaxIP$ value (MIP_{max}) among all these items is 4 (Line 18 in Algorithm 1). Using $MIP_{max} = 4$, *pencil* is pruned from the MCPF-list because its periodicity is greater than MIS_{min} (Line 19 to 23 in Algorithm 1). As the item *pencil* that represents the frequent item having lowest $MinIS$ is pruned, the above steps of finding new $MinIS$ is repeated (Line 14 in Algorithm 1). The new MIS_{min} is 2. The item *pillow* has MIS_{min} among all the remaining frequent items. Every item in MCPF-list has support than or equal to MIS_{min}. Therefore, no item is pruned from the MCPF-list. The MIP_{max} value among the remaining in MCPF-list is 4. Every item in the MCPF-list satisfies MIP_{max}. Therefore, no item is pruned from the MCPF-list. As the item *pillow* that represents the item having MIS_{min} is not pruned from the MCPF-list, the pruning process is completed. The resultant (compact) MCPF-list is shown in Fig. 1(f). Let L' be the new list of items in MCPF-list that are sorted in descending order of their MIS values.

Using L', perform second scan on the transactional database to construct prefix-tree in MCPF-tree. The construction of prefix-tree in MCPF-tree is same as the construction of prefix-tree in PF-tree (or FP-tree). Figure 2(a), Figure 2(b) and Figure 2(c) show the construction of MCPF-tree after scanning first, second and every transaction in the transactional database. In MCPF-tree, node-links are maintained as in FP-growth. For simplicity of figures, we are omitting them.

Lemma 1. *An item is* frequent *if its support is greater than or equal to its MinIS value. Any item which has support less than the lowest MinIS value among all* **frequent items** *cannot generate any periodic-frequent pattern.*

Proof. In the proposed model, an item which has the lowest $MinIS$ value within a periodic-frequent pattern is a frequent item (apriori property [1]). Therefore, every periodic-frequent pattern will have support greater than or equal to the lowest $MinIS$ value among all frequent items. Thus, any item which has support less than the lowest $MinIS$ value among all frequent items cannot generate any periodic-frequent pattern.

item	S	P	MIS	MIP	ld$_1$
bread	0	0	4	2	0
jam	0	0	4	2	0
ball	0	0	4	2	0
bat	0	0	4	2	0
bed	0	0	2	4	0
pillow	0	0	2	4	0
pencil	0	0	2	4	0
pen	0	0	2	4	0

a) Before scanning the transactional dataset

item	S	P	MIS	MIP	ld$_1$
bread	1	1	4	2	1
jam	1	1	4	2	1
ball	0	0	4	2	0
bat	0	0	4	2	0
bed	0	0	2	4	0
pillow	0	0	2	4	0
pencil	1	1	2	4	1
pen	0	0	2	4	0

b) After scanning first transaction

item	S	P	MIS	MIP	ld$_1$
bread	1	1	4	2	1
jam	1	1	4	2	1
ball	1	2	4	2	2
bat	1	2	4	2	2
bed	0	0	2	4	0
pillow	0	0	2	4	0
pencil	1	1	2	4	1
pen	1	2	2	4	2

c) After scanning second transaction

item	S	P	MIS	MIP	ld$_1$
bread	5	2	4	2	9
jam	5	2	4	2	9
ball	5	3	4	2	10
bat	4	4	4	2	10
bed	2	4	2	4	8
pillow	2	4	2	4	8
pencil	2	9	2	4	10
pen	1	2	2	4	2

d) After scanning last transaction

item	S	P	MIS	MIP
bread	5	2	4	2
jam	5	2	4	2
ball	5	3	4	2
bat	4	4	4	2
bed	2	4	2	4
pillow	2	4	2	4
pencil	2	9	2	4
pen	**1**	**8**	**2**	**4**

e) Updated the MCPF-list

item	S	P	MIS	MIP
bread	5	2	4	2
jam	5	2	4	2
ball	5	3	4	2
bat	4	4	4	2
bed	2	4	2	4
pillow	2	4	2	4

f) Final MCPF-list

Fig. 1. Construction of MCPF-list

Lemma 2. *An item is periodic if its periodicity is less than or equal to its MaxIP value. Let I' be the set of items which have support greater than or equal to lowest MinIS value among all frequent items. Any item which has periodicity greater than the maximum MaxIP value among all periodic items in I' cannot generate any periodic-frequent pattern.*

Proof. In the proposed model, an item which has maximum $MaxIP$ value in a periodic-frequent pattern is a periodic item (apriori property). Therefore, every periodic-frequent item will have periodicity less than or equal to maximum $MaxIP$ of all periodic items in I'. Hence, any item which has periodicity greater than the maximum $MaxIP$ value among all periodic items in I' cannot generate any periodic-frequent pattern.

The correctness of MCPF-tree is based on Property 4 and Lemma 3.

Property 4. A tail-node in MCPF-tree maintains the occurrence information for all the nodes in the path (from that tail-node to the root) at least in the transactions in its tid-list.

Lemma 3. *Let $P = \{i_1, i_2, \cdots, i_n\}$ be a path in a MCPF-tree where node i_n is the tail-node carrying the tid-list of the path. If the tid-list is pushed-up to node i_{n-1}, then i_{n-1} maintains the appearance information of the path $P' = \{i_1, i_2, \cdots, i_{n-1}\}$ for the same set of transactions in the tid-list without any loss.*

Proof. Based on Property 4, i_n maintains the occurrence information of the path P' at least in the transactions in its tid-list. Therefore, the same tid-list at node i_{n-1} exactly maintains the same transaction information for P' without any loss.

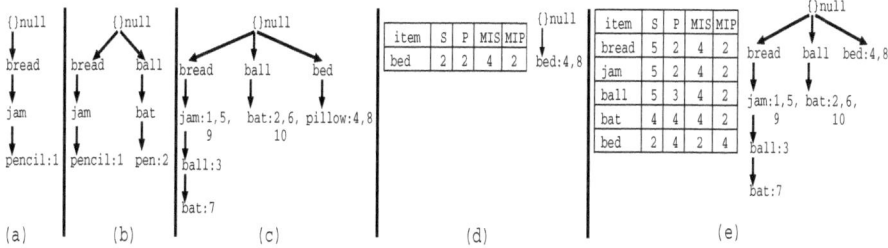

Fig. 2. Construction and mining MCPF-tree. (a) Prefix-tree after scanning first transaction (b) Prefix-tree after scanning second transaction (c) Prefix-tree after scanning every transaction (d) PT_{pillow} and CT_{pillow} and (e) MCPF-tree after pruning item *pillow*.

3.3   Mining of *MCPF-Tree*

Consider the bottom-most item, say i, of the *MCPF-list*. For i, construct a prefix-tree, say PT_i by accumulating only the prefix sub-paths of nodes i. Since i is the bottom-most item in the MCPF-list, each node labeled i in the MCPF-tree must be a tail-node. While constructing the PT_i, based on Property 4 we map the *tid*-list of every node of i to all items in the respective path explicitly in a temporary array (one for each item). It facilitates the periodicity and support calculation for each item in the MCPF-list of PT_i.

Lemma 4. *Let $S_i(j)$ be the support of an item j in PT_i. If $S_i(j) < MinIS(i)$, j cannot generate any periodic-frequent pattern in PT_i.*

Proof. The MCPF-tree is constructed in $MinIS$ descending order of items. So, all items in PT_i will have $MinIS$ values greater than or equal to the $MinIS$ of i. Thus, based on *apriori property*, an item j having $S_i(j) < MinIS(i)$ cannot generate any periodic-frequent pattern in PT_i.

Lemma 5. *In PT_i, let I'' be the set of items which have their support greater than or equal to $MinIS$ of i. In I'', any item which has periodicity greater than the maximum $MaxIP$ value among all $\{I'' \cup i\}$ items cannot generate any periodic-frequent pattern in PT_i.*

Proof. Any periodic-frequent pattern that can be generated from PT_i cannot have periodicity greater than the maximum $MaxIP$ value among the set of $\{I'' \cup i\}$. Therefore, any item in I'', which has periodicity greater than maximum $MaxIP$ value among all $\{I'' \cup i\}$ items cannot generate any periodic-frequent pattern in PT_i (*apriori property*).

Using Lemma 4 and Lemma 5, the conditional tree CT_i for PT_i is constructed by removing all those items which have support less than the $MinIS$ of i (say, $minsup_i$) or periodicity greater than the maximum $MaxIP$ of all items in $\{PT_i \cup i\}$ (say, $maxprd_i$). If a deleted node is a *tail*-node, its *tid*-list is pushed-up to its parent node. The contents of temporary array for the bottom item j in the MCPF-list of CT_i represents T^{ij} (i.e., set of all *tids* where items i and j occur together). From T^{ij}, it is a simple calculation of $S(ij)$ and $Per(ij)$ for the pattern $\{i, j\}$.

1. If $S(ij) \geq minsup_i$ and $Per(ij) \leq maximum(MaxIP(i), MaxIP(j))$, then $\{i, j\}$ is considered as a periodic-frequent pattern and the same process of creating prefix-tree and its corresponding conditional tree is repeated for further extensions of "ij".
2. Else, we will verify whether $S(ij) \geq MinIS(i)$ and $Per(ij) \leq maxprd_i$. If the above condition is satisfied, then the same process of creating prefix-tree and its corresponding conditional tree is repeated for further extensions of "ij" even though it is a non-periodic-frequent pattern. It is because higher orders of $\{i, j\}$ can still be periodic-frequent.

Next, i is pruned from the original *MCPF-tree*. (The procedure for pruning i is same as pruning non-periodic-frequent items in earlier steps.) The whole process of mining each item is repeated until MCPF-list $\neq \emptyset$. The correctness of this procedure is based Property 4 and Lemma 3.

Consider the item *pillow*, which is the last item in *MCPF-tree* of Fig. 1(f). The PT_{pillow} generated from *MCPF-tree* is shown in Fig. 2(d). Since, there is only one item *bed*, which is satisfying $MinIS_{pillow}$ and $MaxIP_{bed}(= minimum(- MaxIP_{pillow}, MaxIP_{bed}))$, $CT_{pillow} = PT_{pillow}$. From $T^{\{bed, pillow\}}$, the pattern $\{bed, pillow\}$ will be generated a periodic-frequent pattern. The MCPF-tree generated after pruning *pillow* is shown in Fig. 2(e). Similar process is repeated for other items in the MCPF-tree. Finally, the set of periodic-frequent patterns generated are $\{\{bread\}, \{jam\}, \{bed\}, \{pillow\}, \{bread, jam\}, \{bed, pillow\}\}$. The patterns $\{bread, ball\}$ and $\{ball, bat\}$, which are generated as periodic-frequent patterns at low *minsup* and high periodicity ($minsup = 2$ and $maxprd = 4$) in Example 3 have failed to be periodic-frequent patterns in the proposed model. It is because they failed to satisfy $minsup = 4$ and $maxprd = 2$.

In this way the proposed model is able to efficiently address *rare item problem* while mining rare periodic-frequent patterns.

One can assume that the structure of a MCPF-tree may not be memory efficient, since it explicitly maintains *tids* of each transaction. It was proven in [3] that such a tree achieves the memory efficiency by keeping transaction information only at the tail-nodes and avoiding the support field at each node.

3.4 Relationship among the Patterns Generated in Various Models

For a given transactional database T, let F be the set of frequent patterns generated at $minsup = x\%$. Let PF be the set of periodic-frequent patterns generated at $minsup = x\%$ and $maxprd = y\%$. Let $MCPF$ be the set of set of periodic-frequent patterns generated in the proposed model when items' $MinIS$ values were greater than or equal to $x\%$ and $MaxIP$ values were less than or equal to $y\%$. The relation between these patterns is $MCPF \subseteq PF \subseteq F$.

4   Experimental Results

In this section, we present the performance comparison of the proposed model against basic model discussed in [3]. All programs are written in $C++$ and run with Ubuntu 8.1 on a 2.66 GHz machine with 2GB memory. The runtime specifies the total execution time.

The experiments are pursued on two types of datasets: synthetic dataset ($T10I4D100k$) and real-world datasets (retail and mushroom). $T10I4D100k$ [1] is a large sparse dataset with 100,000 transactions and 870 distinct items. The mushroom dataset is a dense dataset containing 8,124 transactions and 119 distinct items. The retail dataset [9] is also a large sparse dataset with 88,162 transactions and 16,470 items. All of these datasets are available at Frequent Itemset MIning (FIMI) repository (http://fimi.cs.helsinki.fi/data/).

To specify items' $MaxIP$ values we used the methodology discussed in Equation 2. To specify items' $MinIS$ values we used the methodology proposed in [4]. It is as follows:

$$MinIS(i_j) = maximum(\gamma \times S(i_j),\ LS) \qquad (3)$$

where, LS is the user-specified lowest minimum item support allowed and $\gamma \in [0, 1]$ is a parameter that controls how the $MinIS$ values for items should be related to their supports.

4.1   Experiment 1

In synthetic, retail and mushroom datasets, both $minsup$ and LS values are fixed at 0.1%, 0.1% and 25% respectively. With $\gamma = \frac{1}{\alpha}$, and varying α, we present the number of periodic-frequent patterns generated in these databases at different $maxprd$ ($maxprd = P_{max} = P_{min} = x\%$) values in Fig. 3(a), Fig. 3(b) and Fig. 3(c) respectively.

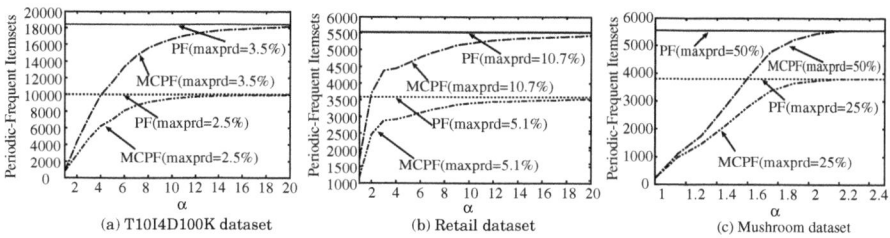

Fig. 3. Periodic-Frequent patterns generated at different $MinIS$ values

It can be observed that the number of periodic-frequent patterns significantly reduced by our model when α is not too large. When α becomes larger, the number of periodic-frequent patterns found by our model gets closer to that found by the traditional model (single $minsup$-$maxprd$ model). The reason is because when γ becomes larger more and more items' $MinIS$ values reach LS.

4.2   Experiment 2

In synthetic, retail and mushroom datasets, *minsup* and *LS* values are fixed at
0.1%, 0.1% and 25% respectively. Next, γ was set at 0.5%. With $\beta = -\frac{1}{\alpha}$ and
varying α from 0 to 20, we present the number of periodic-frequent patterns
generated in these databases at different *maxprd* values in Fig. 4(a), Fig. 4(b)
and Fig. 4(c) respectively.

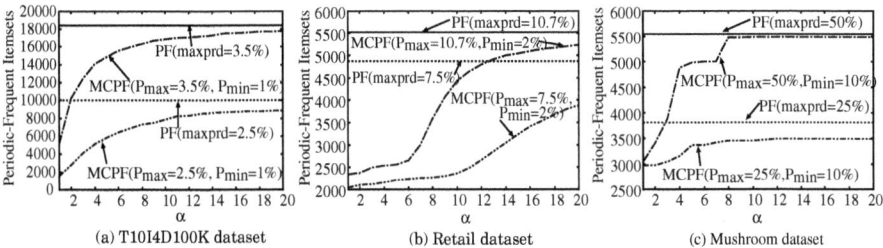

(a) T10I4D100K dataset   (b) Retail dataset   (c) Mushroom dataset

Fig. 4. Periodic-Frequent patterns generated at different *MaxIP* values

It can be observed that the number of periodic-frequent patterns significantly
reduced by our model when α is small. When α becomes larger, the number of
periodic-frequent patterns found by our model gets closer to that found by the
traditional model (single *minsup-maxprd* model). The reason is because when
α increases more and more items' *MaxIP* values reach P_{max}.

Fig. 5(a), Fig. 5(b) and Fig. 5(c) show the runtime taken by both of these
approaches at different *MaxIP* values. It can be observed that the runtime
taken by the proposed approach to generate periodic-frequent patterns increases
with increase in α value. It is because of the increase in number of periodic-
frequent patterns. (Similar observations can also be drawn for the previous experiments. Due to page limitations, we are not discussing them in this paper.)
Also, it can be observed that the proposed approach requires more runtime to
find periodic-frequent patterns. The reason is due to the increased search space
as periodic-frequent patterns mined using the proposed model do not follow
downward closure property.

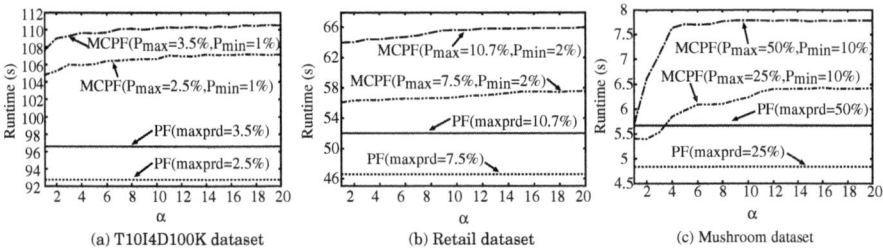

(a) T10I4D100K dataset   (b) Retail dataset   (c) Mushroom dataset

Fig. 5. Runtime taken for generating periodic-frequent patterns at different *maxprd*
values

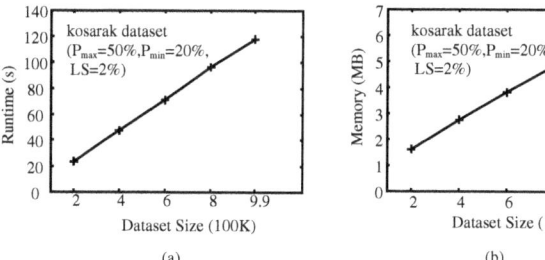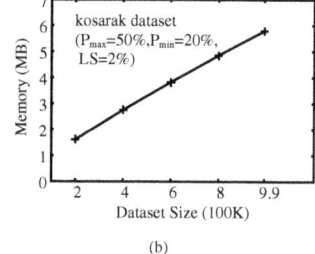

(a)   (b)

Fig. 6. Scalability of MCPF-tree. (a) Runtime requirements of MCPF-tree and (b) Memory requirements of MCPF-tree.

4.3   Experiment 3: Scalability Test

We studied the scalability of our MCPF-tree on execution time and required memory by varying the number of transactions in a database. We use real *kosarak* dataset for the scalability experiment, since it is a huge sparse dataset with 990,002 transactions and 41,270 distinct items. We divided the dataset into five portions of 0.2 million transactions in each part. Then we investigated the performance of MCPF-tree after accumulating each portion with previous parts and performing periodic frequent itemset mining each time. For calculation of items' $MinIS$ values in each experiment, the LS and γ values are fixed at 2% and 0.5%, respectively. Similarly, items' $MaxIP$ value, P_{min}, P_{max} and β values are fixed as 20%, 50% and -0.1 respectively. The experimental results are shown in Figure 6. The runtime and memory in y-axes of Figure 6(a) and Figure 6(b) respectively specify the total memory and execution time with the increase in database size. It is clear from the graphs that as the database size increases, overall tree construction and mining time, and memory requirement increase. However, MCPF-tree shows stable performance of about linear increase in runtime and memory consumption with respect to the database size. Therefore, it can be observed from the scalability test that MCPF-tree can mine periodic-frequent itemsets over large datasets and distinct items with considerable amount of runtime and memory.

5   Conclusion

In many real-world applications, rare periodic-frequent patterns can provide useful information. It is difficult to mine rare periodic-frequent patterns with a "single *minsup* and single *maxprd* model" due to rare item problem. In this paper, we have proposed an improved model to extract rare periodic-frequent patterns. In the proposed model each pattern can satisfy different *minsup* and *maxprd* values depending upon the items within it. For this model, we have also proposed an efficient pattern growth approach based on a tree structure, called MCPF-tree. As the periodic-frequent patterns mined using the proposed model do not satisfy *downward closure property*, the proposed pattern growth approach uses various other techniques to minimize the search space. We have also discussed a methodology to specify *maxprd* for each pattern dynamically. The experimental results demonstrate that the proposed approach efficiently finds periodic-frequent patterns consisting of frequent and rare items. However, it requires more runtime than single *minsup* and *maxprd* model, because the periodic-frequent patterns mined using the proposed model do not satisfy *downward closure property*.

References

1. Agrawal, R., Imielinski, T., and Swami, A.: Mining association rules between sets of items in large databases. In: SIGMOD, pp. 207-216 (1993)
2. Jiawei, H., Jian, P., Yiwen, Y., and Runying, M.: Mining Frequent Patterns without Candidate Generation: A Frequent-Pattern Tree Approach*. In: ACM SIGMOD Workshop on Research Issues in Data Mining and Knowledge Discovery, pp. 53-87 (2004)
3. Tanbeer, S. K., Ahmed, C. F., Jeong, B., and Lee, Y.: Discovering Periodic-Frequent Patterns in Transactional Databases. In: Pacific Asia Knowledge Discovery in Databases (2009)
4. Liu, B., Hsu, W., and Ma, Y.: Mining Association Rules with Multiple Minimum Supports In: Knowledge Discovery and Databases, pp. 337–241 (2009)
5. Hu, Y.-H., Chen, Y.-L.: Mining Association Rules with Multiple Minimum Supports: A New Algorithm and a Support Tuning Mechanism. Decision Support Systems 42(1), 1–24 (2006)
6. Uday Kiran, R., Krishna Reddy, P.: An Improved Multiple Minimum Support Based Approach to Mine Rare Association Rules. In: IEEE Symposium on Computational Intelligence and Data Mining (2009)
7. Uday Kiran, R., Krishna Reddy, P.: An Improved Frequent Pattern-growth Approach to Discover Rare Association rules. In: International Conference on Knowledge Discovery and Information Retrieval (2009)
8. Uday Kiran, R., Krishna Reddy, P.: Mining Rare Association Rules in the Datasets with Widely Varying Items' Frequencies. In: Kitagawa, H., Ishikawa, Y., Li, Q., Watanabe, C. (eds.) DASFAA 2010. LNCS, vol. 5981, pp. 49–62. Springer, Heidelberg (2010)
9. Brijs, T., Swinnen, G., Vanhoof, K., Wets, G.: The use of association rules for product assortment decisions - a case study. In: Knowledge Discovery and Data Mining (1999)

Lag Patterns in Time Series Databases

Dhaval Patel[1], Wynne Hsu[1], Mong Li Lee[1], and Srinivasan Parthasarathy[2]

[1] National University of Singapore
[2] Ohio State University
{dhaval,whsu,leeml}@comp.nus.edu.sg, srini@cse.ohiostate.edu

Abstract. Time series motif discovery is important as the discovered motifs generally form the primitives for many data mining tasks. In this work, we examine the problem of discovering groups of motifs from different time series that exhibit some lag relationships. We define a new class of pattern called *lagPatterns* that captures the invariant ordering among motifs. *lagPatterns* characterize localized associative pattern involving motifs derived from each entity and explicitly accounts for lag across multiple entities. We present an exact algorithm that makes use of the order line concept and the subsequence matching property of the normalized time series to find all motifs of various lengths. We also describe a method called *LPMiner* to discover *lagPatterns* efficiently. *LPMiner* utilizes inverted index and motif alignment technique to reduce the search space and improve the efficiency. A detailed empirical study on synthetic datasets shows the scalability of the proposed approach. We show the usefulness of *lagPatterns* discovered from a stock dataset by constructing stock portfolio that leads to a higher cumulative rate of return on investment.

1   Introduction

Time series motif discovery is an active research topic [1,8,11,12]. Time series motifs are the recurring patterns in single time series. Attempts have been made to generalize the notion of motifs from single time series to multi-dimensional time series data [16,10,13,14]. This generalization allows the handling of real world applications involving several data sources such as activity discovery using wearable sensor data, gene expression data showing the expression levels of multiple genes, stock market data giving the stock prices of diverse companies. However, none of these methods considers the ordering among the motifs in such an environment.

Fig. 1 shows the time series of QLogic, Intel and JP Morgan stocks. Motifs $m_1 = \{s_{11}, s_{12}, s_{13}\}$, $m_2 = \{s_{21}, s_{22}, s_{23}\}$ and $m_3 = \{s_{31}, s_{32}, s_{33}\}$ are highlighted in the time series of QLogic, Intel and JP Morgan stocks respectively. A closer examination of the motifs in Fig. 1 reveals that the subsequences from one motif occurs at a consistent lag relative to subsequences from other motifs. For example, s_{21} occurs with lag 6 relative to s_{11} while s_{31} occurs with lag 7 relative to s_{11}. This pattern is repeated for (s_{12}, s_{22}, s_{32}) and (s_{13}, s_{23}, s_{33}). In short, the lag relationship among the subsequences are repeated. The existence of such invariant ordering among the motifs suggests that there may exist some hidden relationships. Further investigation[1] reveals that QLogic

---

[1] Yahoo Finance - http://finance.yahoo.com

P. García Bringas et al. (Eds.): DEXA 2010, Part II, LNCS 6262, pp. 209–224, 2010.

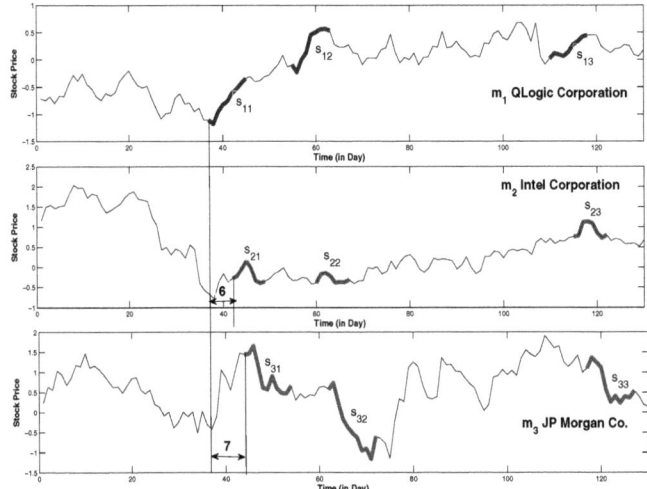

Fig. 1. Lag relationships among motifs m_1, m_2 and m_3 reflecting competitor/co-operative behavior

stock is competitor of Intel stock,while JP Morgan stock gives higher rating for investment in Intel Stock. Moreover, our experiments reveal that stock portfolio based on lag relationships leads to increase in the cumulative rate of return on investment.

In this paper, we define a new class of pattern called ***lagPatterns*** to capture the orderings among motifs from different time series. Unlike existing multi-dimensional motifs, *lagPattern* explicitly accounts for lags and the ordering among the multi-dimensional motifs. Finding *lagPattern* patterns involves two main steps:

1. Identify all motifs of various length in single time series.
2. Discover groups of multi-dimensional motifs with invariant orderings.

Both steps are computationally expensive. A time series of length L, without discretization, would have $O(L^2)$ subsequences of various length and hence $O(L^2)$ motifs. Thus, the naive enumeration based method for the first step is quadratic. With N time series, we would have $O(L^{2N})$ possible *lagPatterns*. As a result, an exhaustive search for *lagPatterns* is exponential. Here, we describe an efficient and scalable approach to prune the search space for both steps. The key contributions of this work are summarized as follows:

1. We define a new class of patterns to capture orderings among multi-dimensional motifs and prove that $lagPatterns$ satisfy the anti-monotonic property. This property allows us to prune the search space in the generation of $lagPatterns$. We design an efficient algorithm called $LPMiner$ that first aligns the motifs and utilizes an inverted index to quickly find multi-dimensional motifs with invariant orderings.
2. We extend the exact motifs discovery algorithm in [12] to discover motifs of all lengths. We take advantage of order line concept and subsequence matching property of normalized time series to reduce over 60% of the distance computations.

3. We evaluate the algorithms on both synthetic and real world datasets. Our experimental results show that the proposed approach is scalable. We show the usefulness of *lagPatterns* discovered from a stock dataset by constructing stock portfolio that leads to a two-fold increase in the cumulative rate of return on investment compared to the traditional mean variance analysis(MVA) portfolio selection strategy.

2 Preliminaries

Definition 1. A **time series** $T = \{v[1], v[2], ..., v[n]\}$ with length $|T| = n$ is a sequence of regularly sampled real value observations where $v[i]$ is observation value at time i.

Definition 2. A **subsequence of a time series**, denoted as $T[i, j]$, is a subset of **contiguous** observations starting at time i and ending at time j and has a length of $|T[i,j]| = j - i + 1$.

Definition 3. A subsequence $T[i, j]$ is **similar** to another subsequence $T[p, q]$ if they have the same length and $dist$(T[i,j], T[p,q]) $\leq \delta$, where $dist(.)$ is Euclidian distance and δ is a user-defined distance threshold.

Table 1. Running example

Time Series	Motifs m (correlation coefficient $coef = 0.95$)
T_1	$m_{11} = \{T_1[14,17], T_1[1,4], T_1[6,9], T_1[22,25]\}$
	$m_{12} = \{T_1[22,25], T_1[3,6], T_1[14,17]\}$
	$m_{13} = \{T_1[12,14], T_1[1,3], T_1[22,24]\}$
	$m_{14} = \{T_1[6,9], T_1[14,17], T_1[21,24]\}$
T_2	$m_{21} = \{T_2[15,17], T_2[2,4], T_2[7,9], T_2[23,25]\}$
	$m_{22} = \{T_2[17,20], T_2[6,9]\}$
T_3	$m_{31} = \{T_3[19,22], T_3[6,9], T_3[11,14]\}$
	$m_{32} = \{T_3[4,7], T_3[9,12], T_3[17,20]\}$
T_4	$m_{41} = \{T_4[20,23], T_4[7,10], T_4[12,15]\}$
T_5	$m_{51} = \{T_5[20,23], T_5[3,6], T_5[7,10], T_5[14,17]\}$

Definition 4. Given a time series T, a **time series motif** $m_{T[i,j]}$, having $T[i, j]$ as anchor subsequence, is the set of non-overlapping subsequences[2] from T that are **similar** to anchor subsequence. For simplicity, we will use m in place of $m_{T[i,j]}$ where $T[i, j]$ is obvious. The size of motif m, denoted as $|m|$, is the number of subsequences in m.

Definition 5. The **support of time series motif** m with anchor subsequence $T[i, j]$, denoted as $mSup(m)$, is defined as

$$mSup(m) = \frac{|T[i,j]| * |m|}{|T|} \tag{1}$$

For example, Table 1 shows a subset of motifs for five time series of length 25. The anchor subsequence in each motif is underlined. The support of m_{11} is given by $mSup(m_{11}) = (4 * 4)/25 = 0.64$.

---

[2] We can use the optimal greedy-activity-selector solution in [2] to discover the maximum set of non-overlapping subsequences.

Definition 6. Given N time series $T_1, T_2, \cdots, T_N$, let M_i be the set of motifs from time series T_i. A ***lagPattern*** of length k is a pattern template consisting of k motifs from different time series and their lags. Formally,

$$p = (\{m_{y_1}, m_{y_2}, \cdots, m_{y_k}\}, \{l_{y_1}, l_{y_2}, \cdots, l_{y_k}\}), m_{y_i} \in M_{y_i}$$

$y_i \neq y_j$ for $i \neq j$ and m_{y_i} lags m_{y_1} by l_{y_i}, $y_i, y_j \in [1, N]$ and $i, j \in [1, k]$.

For example, $p1 = (\{m_{11}, m_{21}, m_{41}\}, \{0,1,6\})$ is a *lagPattern* of length 3 but $p2 = (\{m_{11}, m_{12}\}, \{0,8\})$ is not a *lagPattern* as both motifs are from the same time series T_1. Note that, the lag between two motifs in *lagPattern* is a lag between their respective anchor subsequences.

Definition 7. A *lagPattern* $p1$ is a **sub-pattern** of another *lagPattern* $p2$ if all motifs in $p1$ also occurs in $p2$ with the same invariant ordering. For example, $p1 = (\{m_{11}, m_{41}\}, \{0,6\})$ is a sub-pattern of $p2 = (\{m_{11}, m_{21}, m_{41}\}, \{0,1,6\})$.

Definition 8. The **support of a *lagPattern*** $p = (\{m_1, m_2, \cdots, m_k\}, \{l_1, l_2, \cdots, l_k\})$, denoted as $pSup(p)$, is the size of the set $\{s_1 \in m_1, s_2 \in m2, \cdots, s_k \in m_k \mid s_y$ lags s_1 by $l_y, 1 \leq y \leq k\}$.

For example, consider $p = (\{m_{11}, m_{21}\}, \{0,1\})$. We observe that $T_2[7, 9] \in m_{21}$ lags $T_1[6, 9] \in m_{11}$ by 1. Similarly, $T_2[23, 25] \in m_{21}$ lags $T_1[22, 25] \in m_{11}$ by 1, $T_2[2, 4] \in m_{21}$ lags $T_1[1, 4] \in m_{11}$ by 1 and $T_2[15, 17] \in m_{21}$ lags $T_1[14, 17] \in m_{11}$ by 1. Hence, they support the *lagPattern* p. In this case, the support of p, $pSup(p)$, is 4.

Definition 9. Given a *lagPattern* p, the **participation ratio** of p is defined as

$$pRatio(p) = \frac{pSup(p)}{max_{m \in p}\{|m|\}} \tag{2}$$

For example, the $pRatio$ of $p = (\{m_{11}, m_{21}\}, \{0,1\}) = \frac{4}{max\{4,4\}} = 1$. The $pRatio$ is a variant of the well-known All_confidence measure [5] in association-based correlation analysis. The $pRatio$ measure is anti-monotonic. This property allows us to prune away a large part of the search space.

Theorem 1. *The participation ratio measure of a lagPattern is anti-monotonic, that is, if a lagPattern p satisfy $pRatio(p) \geq min\_ratio$, then any sub-pattern p' of p also satisfies $pRatio(p') \geq min\_ratio$.*

Proof. Let a length k *lagPattern* $p = (\{m_1, m_2, \cdots, m_k\}, \{l_1, l_2, \cdots, l_k\})$. We have

$$pRatio(p) = \frac{pSup(p)}{max_{m \in p}(|m|)}$$

Assume *lagPattern* p' is a sub-pattern of *lagPattern* p. It is obvious that $pSup(p') \geq pSup(p)$. Also, $max_{m' \in p'}(|m'|) \leq max_{m \in p}(|m|)$. Hence, $pRatio(p') \geq pRatio(p)$.

This implies we do not need to generate p if any sub-pattern p' of p does not satisfy the min_ratio constraint.

Definition 10. Given min_sup and min_ratio, a *lagPattern* p is **valid** if $pRatio(p) \geq min\_ratio$ and for all motifs m, $m \in p$, $mSup(m) \geq min\_sup$.

Problem Statement. Given min_sup and min_ratio, the problem of mining interesting *lagPatterns* across N time series is to discover all **valid *lagPatterns*** of length k, $2 \leq k \leq N$.

3   Discover Lag Patterns

The discovery of *lagPatterns* involves two main steps. We need to first identify all the motifs of various length in each time series, and then determine groups of motifs from different time series having invariant orderings. Algorithm 1 summarizes our overall approach to mine *lagPatterns*. We call Algorithm FindMotifs for each time series to find all its motifs(Line 4). Note that M_i denotes the set of motifs generated from time series T_i. Lines 6-8 remove motif m if it does not satisfy the minimum support. Otherwise, we align m to a reference time point and insert it into an inverted index(Lines 9-10). Next, we invoke Algorithm LPMiner to obtain the valid *lagPatterns* (Line 14). We will discuss the details of each algorithm in the following subsections.

Algorithm 1. Discover $lagPatterns$

Input: $N, L, min\_sup, min\_ratio, coef, minLen, maxLen$
Output: LP = set of $lagPatterns$

 1: $LP = \phi, invIndex = \phi$;
 2: $M = \phi$; // sets of motifs
 3: **for** $i = 1$ to N **do** {// N = Number of time series}
 4:     $M_i = $ **FindMotifs**$(T_i, coef, minLen, maxLen)$;
 5:     **for** each motif m in M_i **do**
 6:        **if** $mSup(m) < min\_sup$ **then**
 7:           $M_i = M_i - \{m\}$;
 8:        **else**
 9:           align m to a reference time point t_p;
10:           insert m into $invIndex$;
11:        **end if**
12:     **end for**
13: **end for**
14: $LP = $ **LPMiner**(N, L, min_sup, min_ratio, M); // L = Length of time series
15: return LP

3.1   Find All Motifs in a Time Series

To find all motifs from T, we consider each subsequence of length between $minLen$ and $maxLen$ from T as an anchor subsequence and discover it's similar subsequences from T. Here, we describe a method that uses order line[12,4] and subsequence matching property[9] to find all motifs. We use normalized time series subsequence[6].

Given a set D of normalized subsequences of length len from time series T and a pivot subsequence $s_p \in D$. We obtain an **order line** by sorting the subsequences in D according to their distance similarity from s_p. Recall, subsequence s_1

Table 2. (a) dataset of two-dimensional subsequences, (b) an ordering of subsequences with their distance value from subsequence 2 (c) distances of all subsequences from subsequence 7

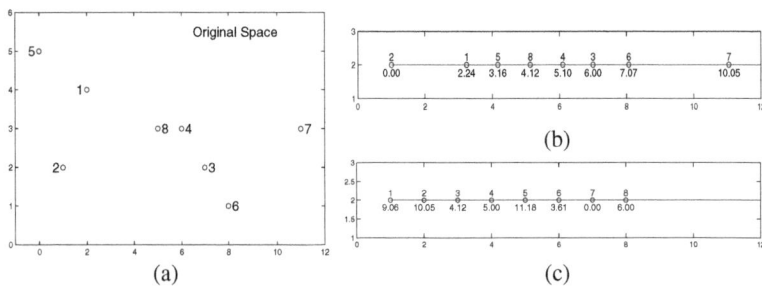

is similar to subsequence s_2 if $dist(s_1, s_2) \leq \delta$. Since we consider anchor subsequences of various lengths, this δ threshold should be length-invariant. Here we utilize the results in [17] which states that the Euclidian distance δ between two normalized time series of length len depends on their correlation coefficient $coef$, that is, $\delta = \sqrt{2 * (len - 1) * (1 - coef)}$. With this equation, we are able to employ the Euclidean measure in the similarity computation by setting the appropriate δ for varying length, given a fixed value of $coef$.

Table 2(a) shows the distribution of subsequences of length 2 in a two-dimensional space. Assuming that the subsequence 2 is pivot subsequence, Table 2(b) shows the order line. The number above the order line shows the subsequence id while the number below gives it's euclidian distance from pivot subsequence 2. Now, we discover similar subsequences for each anchor subsequence.

We traverse the order line (with pivot subsequence s_p) from left to right. Given a distance threshold δ, suppose s_i is the next subsequence on the order line. We determine the similar subsequences of s_i by checking all the subsequences that fall within δ distance from s_i on the order line. This is due to the reverse triangular inequality which states that $dist(s_i, s_j) \leq \delta$ if and only if $|dist(s_p, s_i) - dist(s_p, s_j)| \leq \delta$.

Consider Table 2(b). Let the subsequence we encounter be s_1 whose distance from the pivot subsequence s_2 is 2.24. If $\delta = 2$, then a subsequence s is similar to subsequence s_1 if $dist(s_2, s)$ falls within [2.24-δ, 2.24+δ], that is, [0.24, 4.24]. Hence, the set of candidate similar subsequences for s_1 is $c_{s_1} = \{s_5, s_8\}$. We compute the actual distances between s_1 and each subsequences in c_{s_1} to obtain the final set of subsequences that are similar to s_1(i.e., a motif having anchor subsequence s_1).

Similarly, the set of candidate similar subsequences for s_5, $c_{s_5} = \{s_1, s_8, s_4\}$. Note that, we do not need to compute the actual distance between s_5 and s_1 since $dist(s_5, s_1) = dist(s_1, s_5)$ and we have already obtained $dist(s_1, s_5)$ previously if s_1 and s_2 are similar. In other words, when traversing the order line from left to right, we need to perform the actual distance computations only for those candidates to its right.

Another observation is that multiple order lines can prune more candidates. Suppose we have a second order line with pivot subsequence s_7 (see Table 2(c)). Using the first order line(Table 2(b)), we have the set of candidate similar subsequences for s_5, $c_{s_5} = \{s_1, s_8, s_4\}$. From the second order line, we observe that $dist(s_7, s_8) = 6$

and $dist(s_7, s_5) = 11.18$. Hence, $dist(s_8, s_5) \geq 5.18$ which is more than δ. The same process is repeated for subsequence s_4. Thus, applying triangular inequality, we eliminate s_8 and s_4 from c_{s_5} without performing any distance computation. In summary, the first order line is used to obtain initial candidate set of similar subsequences for any subsequence while remaining order lines are used for further pruning.

The order line based algorithm efficiently finds all similar subsequences for a fixed length subsequences. In order to find similar subsequences for subsequence of length between $minLen$ to $maxLen$, we need to iterate the algorithm ($maxLen$ - $minLen$ + 1) times. We utilize the subsequence matching property[9] and reduce the number of iterations by 50%. The subsequence matching property states that,

$$dist(T[i, j+1], T[i_1, j_1 + 1]) \leq \epsilon \Rightarrow dist(T[i, j], T[i_1, j_1]) \leq \epsilon'$$

where, $\epsilon' = \sqrt{2\omega - 2\sqrt{\omega^2 - \omega.\epsilon^2.\dfrac{\sigma^2(T[\text{i,j+1}])}{\sigma^2(T[\text{i,j}])}}}$, $\omega = |T[i, j]|$.

This property is based on the observation that the occurrences of subsequences similar to $T[i, j+1]$ coincides with the occurrences of subsequences similar to $T[i, j]$ most of the time. Hence, we can discover the candidate set of subsequences similar to subsequence $T[i, j+1]$ while discovering set of subsequences similar to $T[i, j]$ by setting the appropriate distance threshold given by maximum$\{\delta, \epsilon'\}$. With this, we present an exact algorithm **FindMotifs**(See Algorithm 2).

FindMotifs finds similar subsequences of subsequence $T[i, j]$ of various length in a time series T. At each iteration, we set δ and prepares a database D(Lines 3-4). Line 5 prepares order lines. Next, it invokes *GenerateMotif* to obtain all matches of every anchor subsequences of length len as well as the candidate sets for anchor subsequences of length $len+1$. Line 10 prepares a database of subsequences of length $len+1$. Finally, we call *RefineMotif* to eliminate the false matches found in the candidate sets obtained by *GenerateMotif* for length $len + 1$(Line 11).

The *GenerateMotif* procedure discovers similar subsequences of length len subsequence, that is, $T[i, i + len - 1]$. At the same time, we also keeps track of the candidate sets for subsequences of length $len + 1$, that is, $T[i, i + len]$. We use s_j to denote the jth subsequence along the order line I. Next, we determine ϵ' and set the new distance threshold as $new\delta$(Lines 20-21). For each subsequence s_j on I, we obtain it's candidate set of subsequence similar to s_j using I(Line 22). Line 23 implements the triangular inequality based pruning and refine $canSet$. Finally, we compute the dist(s_j,s_k), $s_k \in canSet$. If $dist(s_k, s_j) \leq \delta$, we add s_k to the set of subsequence similar to s_j(i.e., m_{s_j}) and add s_j to m_{s_k} due to the **symmetry** property. In addition, if $dist(s_k, s_j) \leq \epsilon'$, then we add s_k to the candidate set c_{s_j}. Once all subsequences from I are processed, we return m_{s_j} and c_{s_j} discovered for all subsequences from D.

The *RefineMotif* procedure finds all similar subsequences for length $len + 1$. Again, we traverse the order line I from left to right(Line 34). To find subsequences similar to s_j, we use the candidate set c_{s_j} obtained by *GenerateMotif*. Line 37 calculates $dist(s_j,s)$, s in c_{s_j}. If distance $dist(s_j,s) \leq \delta$, we add s to m_{s_j} and s_j to m_s.

Algorithm 2. FindMotifs

Input: $T, coef, minLen, maxLen, numOrderLine$;
Output: M = set of motifs in T;

1: Set $M = \phi$ and $len = minLen$;
2: **while** $len \leq maxLen$ **do**
3:    $\delta = \sqrt{2 * (len - 1) * (1 - coef)}$;
4:    $D \leftarrow$ {normalized subsequences of length len from T};
5:    Prepare $numOrderLine$ order lines O;
6:    Let I denotes the first order line in O;
7:    $[M_{len}, C]$ = **GenerateMotif** (D, I, O, len, δ);
8:    Set $M = M \cup M_{len}$ **and** $len = len + 1$;
9:    $\delta = \sqrt{2 * (len - 1) * (1 - coef)}$;
10:   $D \leftarrow$ {normalized subsequences of length len from T};
11:   $[M_{len}]$ = **RefineMotif** (D, I, C, δ);
12:   Set $M = M \cup M_{len}$ **and** $len = len + 1$;
13: **end while**
14: return M;

  Procedure GenerateMotif$(D, I, pivotDist, len, \delta)$
15: Let M be the set of motifs m_s for all $s \in D$;
16: Let C be the set of candidate subsequences for all $s \in D$;
17: Set $m = \phi$ and $c = \phi$ for all $m \in M$ and $c \in C$;
18: **for** $j = 1$ to $|I|$ **do**
19:   select $s_j \in D$ as an anchor subsequence;
20:   Determine ϵ' using $len + 1$ and s_j;
21:   $new\delta = \max\{\epsilon', \delta\}$;
22:   $canSet$ = {candidate similar subsequences of s_j using I w.r.t. $new\delta$}
23:   $canSet$ = Refine $canSet$ using remaining orderlines
24:   **for** $s_k \in canSet$ **do**
25:     **if** $dist(s_k, s_j) \leq \delta$ **then**
26:       Add (s_k to m_{s_j}) and (s_j to m_{s_k})
27:     **end if**
28:     **if** $dist(s_k, s_j) \leq \epsilon'$ **then** Add (s_k to c_{s_j}) **end if**
29:   **end for**
30: **end for**
31: return M and C;

  Procedure RefineMotif(D, I, C, δ)
32: Let $M = \{m_s \forall s \in D\}$;
33: Set m to ϕ for $m \in M$;
34: **for** $j = 1$ to $|I|$ **do**
35:   **if** $s_j \in D$ **then**
36:     **for** each subsequence s in $c_{s_j} \in C$ **do**
37:       **if** $s \in D$ and $dist(s, s_j) \leq \delta$ **then**
38:         Add (s to m_{s_j}) and (s_j to m_s)
39:       **end if**
40:     **end for**
41:   **end if**
42: **end for**
43: return M;

3.2   Align Motifs

Having found the sets of motifs from each time series, the next step is to discover valid
$lagPatterns$. A naive approach is to enumerate all possible combinations of motifs
across multiple time series. Recall, this approach has an exponential time complexity.
The anti-monotonic property of $pRatio$ that we have proved in Section 2 allows us to
perform early elimination of $lagPatterns$ that cannot be valid.

In order to compute the $pRatio$ of a $lagPattern$ p, we need to obtain the $pSup(p)$.
We can speed up the computation of $pSup(p)$ for all patterns by aligning the motifs
to some reference time point t_p. Aligning motif m means aligning it's anchor subsequence to t_p and shifting all it's similar subsequences accordingly. We set t_p to be the
length of time series minus $minLen$(i.e., minimum length of motif). The alignment of
motifs provides us with information on which combination of motifs are likely to form
$lagPatterns$ that can satisfy the min_ratio.

In our example, we choose $t_p = 22$. Figures 2(a) and 2(b) show the anchor subsequences and its similar subsequences before and after alignment. The circled points
denote the anchor subsequences. After alignment, each time point will show a list of
motifs. We observe that the motifs, denoted by the symbols ◁, □ and ▽, occur together
at time points 9, 14, and 22. In other words, the $pSup(m_{21}, m_{31}, m_{41}, \{0, 4, 5\})$ is 3.
The $pRatio$ of this pattern is $\frac{3}{max\{4,3,3\}} = 0.75$.

To facilitate the support counting of $lagPattern$, we construct an inverted index for
the motifs occurring at each time point. Fig. 3 shows the inverted index obtained from
Fig. 2(b). Note that, at time point $t_p(=22)$, all the motifs are present. In other word,
all $lagPatterns$ are exists at time point t_p. We utilize this fact while calculating the
support of $lagPatterns$. Following the alignment, our method called $LPMiner$ utilizes
the inverted index and search for valid $lagPatterns$.

3.3   Algorithm LPMiner

Method LPMiner processes each motif and generates all length 2 $lagPatterns$ as follows. For each motif m, we obtain the start times of its similar matches after alignment.

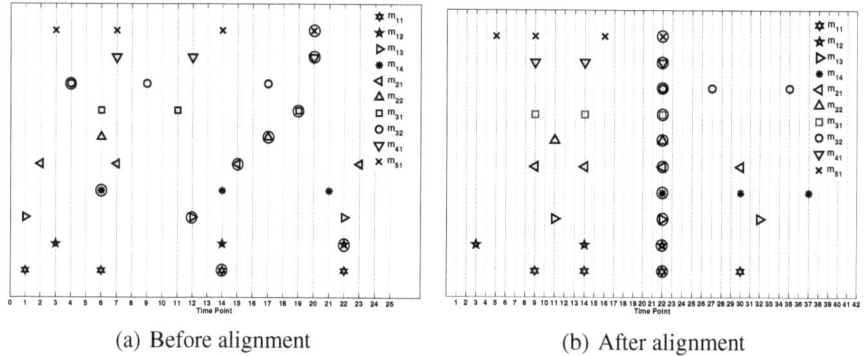

(a) Before alignment   (b) After alignment

Fig. 2. Motifs before and after alignment

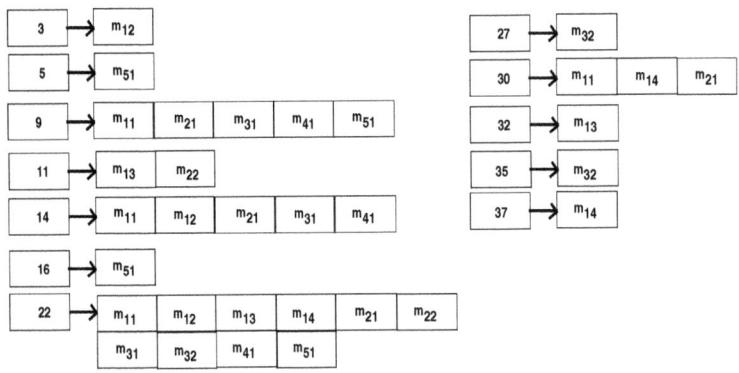

Fig. 3. Inverted index for motifs in Fig. 2(b)

These start times are used to probe the inverted index and to obtain all candidate motifs m'. Next, we form a $lagPattern$ between m and each candidate motif m', i.e., $p = (\{m, m'\}, \{l_1, l_2\})$. We also record the time points of the inverted index where the $lagPattern$ p is generated. Those $lagPatterns$ that satisfy the min_sup and min_ratio are valid and form the set of candidate patterns to generate longer $lagPatterns$ (since $lagPatterns$ are anti-monotonic).

Consider the motif m_{11}. After alignment, the start times of its matches are $\{9, 14, 22, 30\}$ (see Fig. 2(b)). We probe the inverted index at time points 9, 14 and 30 respectively and obtain candidate motifs. In this case, the set of candidate motifs are $canSet = \{m_{21}, m_{31}, m_{41}, m_{51}\}$[3]. Note that, there is no need to probe inverted index at the reference time point 22 since all motifs are aligned at this time point. In other word, any $lagPattern$ p is exists at this time point. The possible $lagPatterns$ are $(\{m_{11}, m_{21}\}, \{0,1\})$, $(\{m_{11}, m_{31}\}, \{0,5\})$, $(\{m_{11}, m_{41}\}, \{0,6\})$ and $(\{m_{11}, m_{51}\}, \{0,6\})$. For each $lagPattern$, we have recorded the time points of the inverted index from where it is generated. For example, the pattern $p = (\{m_{11}, m_{21}\}, \{0,1\})$ occurs at time points $\{9,14,22,30\}$. This implies $pSup(p)$ is 4. If $min\_ratio = 0.60$, then $pRatio(p) = \frac{4}{max\{4,4\}} = 1 \geq min\_ratio$. Hence, it can be used to generate the longer patterns. Note that, all $lagPatterns$ except $(\{m_{11}, m_{51}\}, \{0,6\})$ satisfy min_ratio constraints.

Let us consider the length 2 $lagPattern$ $p = (\{m_{11}, m_{21}\}, \{0,1\})$. For this pattern, we again probe the inverted indexes at time points $\{9, 14, 30\}$(again no need to probe inverted index at time point 22) and obtain the candidate motif m' from time series T' with $T' > T_2$ for extension. In this case, the set of candidate motifs $canSet = \{m_{31}, m_{41}\}$. Note that, motif m_{51} is not in $canSet$ as $lagPattern$ $(\{m_{11}, m_{51}\}, \{0,6\})$ does not satisfy the min_ratio. Hence, the possible length 3 $lagPatterns$ are $(\{m_{11}, m_{21}, m_{31}\}, \{0,1,5\})$ and $(\{m_{11}, m_{21}, m_{41}\}, \{0,1,6\})$ both of which are generated from time points $\{9, 14, 22\}$ and satisfy the min_ratio. The process is repeated until no new pattern is obtained.

---

[3] Without alignment method, all motifs from time series T_2, T_3, T_4 and T_5 are in $canSet$ for motif from time series T_1.

Algorithm 3. *LPMiner*

Input: N, L, min_sup, min_ratio, M
Output: LP = set of $lagPattern = \phi$

```
 1: for i = 1 to N − 1 do
 2:     motifSet = {motifs from M_i};
 3:     extSet = {time series from T_{i+1} to T_N};
 4:     for each motif m in motifSet do
 5:         Mine({m}, extSet);
 6:     end for
 7: end for
 8: return LP;
```

Procedure Mine(p, $extSet$)
```
 9: probeSet = {starting time points of p after alignment};
10: canSet = φ;
11: for each time point t in probeSet do
12:     for each m' in invIndex[t] do
13:         canSet = canSet ∪ {m', time point t};
14:     end for
15: end for
16: extPattern = φ, newExtSet = φ;
17: for each entry m' ∈ canSet do
18:     p' = form lagPattern between p and m' ;
19:     if pRatio(p') ≥ min_ratio then
20:         LP = LP ∪ p';
21:         newExtSet = newExtSet ∪ time series of m';
22:         extPattern = extPattern ∪ p';
23:     end if
24: end for
25: for each lagPattern lp ∈ extPattern do
26:     Mine(lp, newExtSet);
27: end for
```

Algorithm 3 shows the details of LPMiner. Line 2 obtains all the motifs from M_i. $extSet$ maintains the list of time series from which the candidate motifs are obtained for extension(Line 3). For each motif m, we call procedure **Mine** to discover $lagPatterns$. The **Mine** procedure recursively extends the given $lagPattern$ p. Line 9 obtains the time points of p to probe the inverted index. Lines 11-15 obtain all candidate motifs in $canSet$. Lines 17-24 generate the candidate $lagPattern$ between pattern p and each motif in $canSet$. The patterns satisfying min_ratio are stored in LP (Line 20) and $extPattern$ (Line 22). The $Mine$ procedure is called recursively for each generated pattern in $extPattern$ (Line 26).

Algorithm LPMiner utilizes the anti-monotone property and inverted index to speed up the generation of $lagPatterns$. We derive an upper bound estimate of the participation ratio to further improve efficiency of LPMiner by pruning infeasible candidate patterns early.

Optimization. This optimization uses $|m_{T[i,j]}|$ to estimate the maximum $pRatio$ of a $lagPattern$ $p = (\{m_1, m_2, ..., m_k\}, \{l_1, l_2, ..., l_k\})$. Since $pSup(p)$ must be less than or equal to $min_{m \in p}\{|m|\}$, the maximum $pRatio(p) \leq \frac{min_{m \in p}\{|m|\}}{max_{m \in p}\{|m|\}}$.

Consider $lagPattern$ $p = (\{m_{11}, m_{31}\}, \{0,5\})$. We have $|m_{11}| = 4$ and $|m_{31}| = 3$. Suppose the min_ratio is 0.80. Then the $pRatio(p)$ is $\frac{min\{3,4\}}{max\{3,4\}} = 0.75$ (< 0.80). Thus, this candidate is infeasible and can be removed from consideration for generating candidate $lagPatterns$.

For simplicity, LPMiner looks for exact lag among motifs. However, we can introduce a slack variable to relax this requirement. For example, LPMiner accesses inverted index at time points 11 and 32 to obtain candidates for m_{13}. However, with a slack value of 2, we now obtain possible candidates by accessing inverted index at time points $\{9,10,11,12,13\}$ and $\{30,31,32,33,34\}$. In this case, the pattern $(\{m_{13}, m_{21}\}, \{0,3\})$ will be in the output (See Fig. 2(b)).

4   Experimental Evaluation

We implement all our algorithms in C (compiled with GCC -O2). Our hardware configuration consists of a 3.2 MHz processor with 3GB RAM running Windows. We use synthetic datasets to verify the scalability of the proposed approach and real world datasets to demonstrate the usefulness of $lagPatterns$. A random walk generator [12,2] is used to generate synthetic datasets D with N=25 and L=100000.

4.1   Efficiency Experiments

FindMotifs Algorithm. We select one time series from dataset D and apply FindMotifs algorithm to find all the motifs. We compare the performance of FindMotifs with algorithm OrderLine. The OrderLine algorithm uses only order line concept. The number of order lines is 5[12]. Fig. 4(a) shows the results of varying L from 5000 to 100000. We set $minLen = 99$, $maxLen = 110$ and $coef = 0.95$. We observe that FindMotifs outperforms OrderLine, and the gap widens as the length of the time series increases.

Next, we set $L = 20000$ and vary the correlation coefficient $coef$ from 0.60 to 0.99. Fig. 4(b) shows the results in log scale. We observe that FindMotifs is much faster than

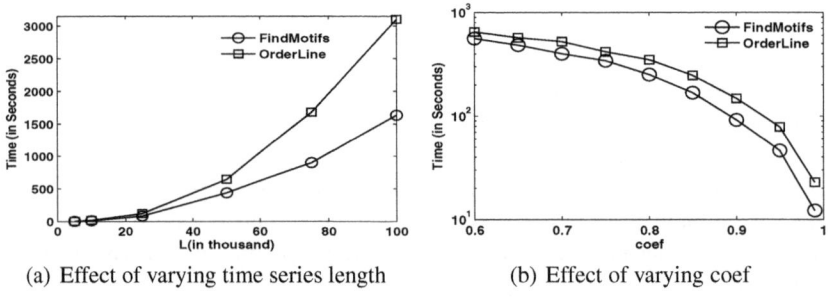

(a) Effect of varying time series length   (b) Effect of varying coef

Fig. 4. Runtime comparison between FindMotifs and OrderLine algorithms

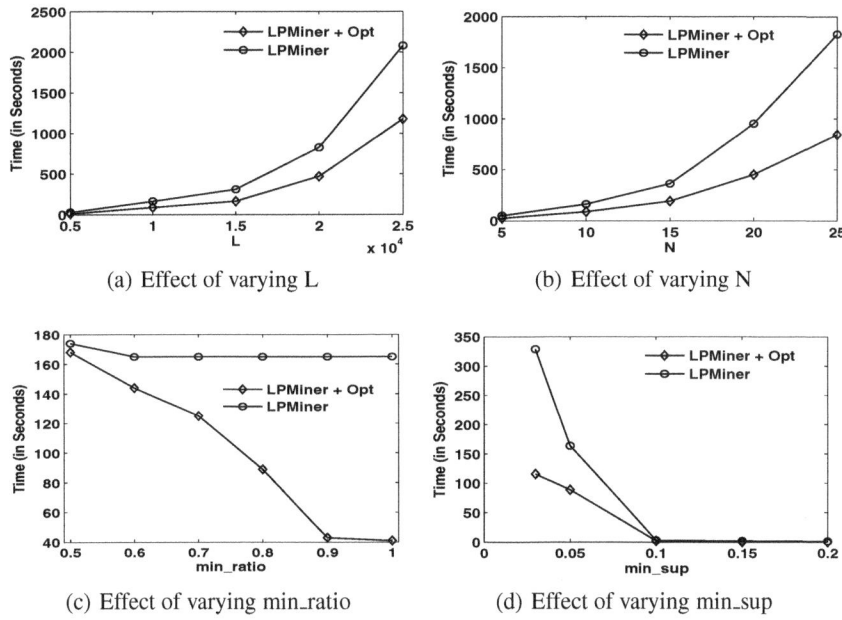

(a) Effect of varying L (b) Effect of varying N

(c) Effect of varying min_ratio (d) Effect of varying min_sup

Fig. 5. Evaluation of LPMiner on dataset D

OrderLine. In particular, when the correlation coefficient is greater than 0.9, FindMotifs is at least 50% faster than OrderLine. However, the gap narrows as $coef$ decreases. This is because FindMotifs estimates $new\delta(\geq \delta)$ in order to apply the subsequence matching property [9]. For low value of $coef$, $new\delta$ is much higher than δ resulting in a larger set of candidate subsequences for distance computation.

LPMiner Algorithm. Now, we report the results of our experiments on the datasets D. Unless otherwise stated, we set coef = 0.95, min_sup = 0.05, min_ratio = 0.80, N = 10, L = 10000, Min_Len = 99 and Max_Len = 110. Fig. 5 shows the results. Note that, running time does not include time required by FindMotifs algorithm. We observe that increasing L and N leads to an exponential increase in the runtime of LPMiner. This is expected since more *lagPatterns* will be generated with a large L and N. However, our optimization strategy is effective in cutting down the runtime. We also evaluate LPMiner by varying min_sup (see Fig. 5(d)) and min_ratio (see Fig. 5(c)). Increasing min_sup reduces the number of subsequences and results in smaller inverted lists. Hence, the runtime decreases. Increasing min_ratio reduces the total number of possible valid *lagPatterns*, hence the runtime also decreases. Also, LPMiner takes less than one second to build an inverted index in all experiments. We also observed similar trends of LPMiner algorithm on real stock dataset.

4.2 Effectiveness Experiments

In this section, we mine $lagPatterns$ from real dataset and discuss usability of the discovered patterns. We use S&P100 stock dataset(http://biz.swcp.com/stocks/, N=100,

222   D. Patel et al.

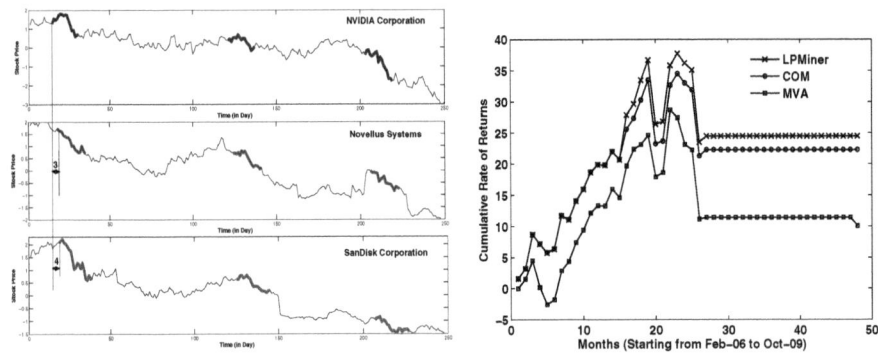

(a) Lag based motif association among Nvidia, (b) Cumulative monthly rate of returns on
Novellus and SanDisk stocks.   MSCI-G7 Index.

Fig. 6. Usability of $lagPatterns$ discovered from real world dataset

L=250) to find interesting localized associations among stock movements. Fig. 6(a)
and Fig. 1 show examples of the discovered patterns. We observe that there is cooperative behavior among Nvidia, Novellus and SanDisk stocks. All these stocks are from
semiconductor industry and none of them are competitor of each other. We use Yahoo
Finance to verify competitor/co-operative behavior. To obtain these results, we set $coef$
= 0.90, min_sup = 0.10, min_ratio = 0.75, Min_Len = 6 and Max_Len = 21.

   To further validate the effectiveness and utility of the discovered patterns, we construct a portfolio of equities selected from Morgan Stanley Capital International G7
(MSCI-G7) Index(www.mscibarra.com). We use the equity indices of seven countries
(Canada, France, Germany, Japan, Singapore, UK and USA) recorded daily over a 5
year period from March 2005 to October 2009(N=7, L=1260). The objective of a portfolio construction is to achieve a higher rate of return over a period of time (cumulative
rate of return). Existing methods such Mean Variance Analysis(MVA) determine the
investment weight for each equity indices from historical data.

   Recently, an alternative method that updates the investment weights based on analyzing the co-movements of equities (COM) has been reported[15]. In order to leverage
the $lagpatterns$, we first use the co-movement model to set the initial weights and
subsequently utilize our $lagPatterns$ to update the investment weights as described in
[15]. Our $lagPatterns$ are obtained using LPMiner with $coef$ = 0.95, min_sup = 0.10,
min_ratio = 0.80, $minLen$ = 3, $maxLen$ = 10, N = 7 and L = 240(one year window).

   We construct the portfolio for each month (March 2006 to October 2009) based on
the data from the previous 12 months. We consider four week as one month. Fig. 6(b)
presents the cumulative monthly rate of returns for MVA, COM and LPMiner. We observe that the cumulative rate of returns (over a period of 3 years) for LPMiner, COM
and MVA is **26.64%**, 22.26% and 11.41% respectively. It is also important to note that
this trend is observed across the board for most time points. The more than two-fold
increase of LPMiner over MVA highlights the utility of our approach.

Significance of *lagPatterns.* Now, we verify the significance of *lagPatterns* by shuffling the time series data using Fisher-Yates shuffle method [2]. The *lagPatterns* are mined from the original dataset and shuffled dataset for the same set of parameters (See Table 3). We observe that, introducing randomness in the data significantly reduce the number of motifs and or *lagPatterns*. This shows that the discovered motifs and *lagPattern* are not due to random chance, but that they are meaningful patterns from the original time series, as we have significantly fewer number of patterns in the shuffled data. Similar observation is also found for the other parameters and datasets.

Table 3. The number of Motifs and *lagPatterns*

Dataset	# Motifs		# *lagPatterns*	
	Original Data	Shuffled Data	Original Data	Shuffled Data
S&P100 stock	110862	9166	2145943	1321
MSCI-G7 index	3535	2100	22	0

5 Related Work

Existing motif discovery approaches in time series are either approximate[1,16,10,14] or exact[8,12,11]. In approximate motif discovery, time series is discretized into symbolic sequences and most recurring subsequences is discovered using variation of random projection based method[1]. Lin. et. al. in [8] introduces the notion of K-motifs, that is, a motif having K^{th} highest count of non-overlapping occurrences. The proposed algorithm hashes all subsequences into a table using their SAX word and then the promising buckets are processed to discover K-motifs. These works differ from ours in that they are approximate and dealing with fixed length motifs.

Recently, Mueen et. al. in [12,11] propose algorithm to find the exact motifs efficiently by limiting the motifs to just pairs of time series that are very similar to each other. Both algorithms use order line and triangular inequality to reduce the distance computations. Their methods discover motifs of the given length. These works differ from ours in that their motif is pair of most similar subsequence.

There are also works that extends [1] to discover approximate multi-dimensional motifs from multiple time series [16,10,13,14]. However, none of them consider time lag and invariant ordering among motifs. Further, we do not adapt time series subsequences clustering method[3] to discover *lagPattern*, since clustering time series subsequence is meaningless as suggested in [7]. Our work aims to discover groups of motifs that exhibit some invariant ordering among the motifs within each group and explicitly capture the lag among them. To the best of our knowledge, none of the existing methods are able to discover the *lagPattern* as motivated in our introduction.

6 Conclusion

In this paper, we have introduced new class of patterns called *lagPatterns* and presented an efficient solution to discover them. Our proposed approach extracts motifs from each time series, then aligns and index them. We have described an algorithm LPMiner to mine *lagPatterns*. Our experimental results demonstrate that the proposed approach is scalable and meaningful patterns can be discovered from real world dataset.

References

1. Chiu, B., Keogh, E., Lonardi, S.: Probabilistic discovery of time series motifs. In: SIGKDD, pp. 493–498 (2003)
2. Cormen, T., Leiserson, E., Rivest, L., Stein, C.: Introduction to Algorithms. The MIT Press, Cambridge (2001)
3. Das, G., Lin, K., Mannila, H., Renganathan, G., Smyth, P.: Rule discovery from time series. In: SIGKDD, pp. 16–22 (1998)
4. Gsli, H., Samet, H.: Properties of embedding methods for similarity search in metric spaces. In: PAMI, pp. 530–549 (2003)
5. Han, J., Kamber, M.: Data Mining: Concepts and Techniques. Morgan Kaufmann, San Francisco (2000)
6. Keogh, E., Kasetty, S.: On the need for time series data mining benchmarks: A survey and empirical demonstration. DMKD 7(4), 349–371 (2003)
7. Keogh, E., Lin, J.: Clustering of time-series subseq. is meaningless: implications for previous and future research. KIS 8(2), 154–177 (2005)
8. Lin, J., Keogh, E., Lonardi, S., Patel, P.: Finding motifs in time series. In: Temporal Data Mining (2002)
9. Loh, W., Kim, S., Whang, K.: A subsequence matching algorithm that supports normalization transform in time-series databases. DMKD, 5–28 (2004)
10. Minnen, D., Isbell, C.L., Essa, I., Starner, T.: Discovering multivariate motifs using subsequence density estimation and greedy mixture learning. In: AAAI (2007)
11. Mueen, A., Keogh, E., Bigdely-Shamlo, N.: A disk-aware algorithm for time series motif discovery. In: ICDM (2009)
12. Mueen, A., Keogh, E., Zhu, Q., Cash, S.: Exact discovery of time series motifs. In: SDM (2009)
13. Oates, T.: Peruse: An unsupervised algorithm for finding recurring patterns in time series. In: ICDM, pp. 330–337 (2002)
14. Vahdatpour, A., Amini, N., Sarrafzadeh, M.: Toward unsupervised activity discovery using multi-dimensional motif detection in time series. In: IJCAI (2009)
15. Wu, D., Fung, G.P.C., Yu, J.X., Liu, Z.: Mining multiple time series co-movements. In: Zhang, Y., Yu, G., Bertino, E., Xu, G. (eds.) APWeb 2008. LNCS, vol. 4976, pp. 572–583. Springer, Heidelberg (2008)
16. Yoshiki, T., Kazuhisa, I., Kuniaki, U.: Discovery of time-series motif from multi-dimensional data based on mdl principle. Machine Learning 58(2-3), 269–300 (2005)
17. Zhu, Y., Shasha, D.: Statstream: Statistical monitoring of thousands of data streams in real time. In: VLDB (2002)

An Efficient Computation of Frequent Queries in a Star Schema

Cheikh Tidiane Dieng[1,2], Tao-Yuan Jen[1], and Dominique Laurent[1]

[1] ETIS - CNRS - ENSEA, Université de Cergy Pontoise, F-95000, France
[2] Laboratoire d'Analyse Numérique et Informatique, Université Gaston-Berger, Saint-Louis, Senegal

Abstract. Although the problem of computing frequent queries in relational databases is known to be intractable, it has been argued in our previous work that using functional and inclusion dependencies, computing frequent conjunctive queries becomes feasible for databases operating over a star schema. However, the implementation considered in this previous work showed severe limitations for large fact tables. The main contribution of this paper is to overcome these limitations using appropriate auxiliary tables. We thus introduce a novel algorithm, called Frequent Query Finder (FQF), and we report on experiments showing that our algorithm allows for an effective and efficient computation of frequent queries.

Keywords: Frequent Queries, Functional Dependencies, Inclusion Dependencies, Query Comparison, Star Schemas.

1 Introduction

The problem of discovering frequent patterns in a (relational) database is one of a main topics in data mining. However, even when patterns are restricted to conjunctive queries, this problem is known to be intractable, because the size of the search space is exponential in the size of the database. Nonetheless, it is argued in [9,10] that mining all frequent conjunctive queries (*i.e.*, conjunctive queries whose answers have a cardinality greater than or equal to a predefined threshold) becomes tractable if the underlying database operates over a star schema, and if constraints such as functional and inclusion dependencies, are taken into account.

Indeed, it has been shown in [9,10] that such dependencies allow for comparing queries according to a pre-ordering with respect to which the support measure is anti-monotonic (the support of a query being the number of tuples in the answer of that query). As a consequence, a level-wise algorithm such as Apriori ([1]) can be used, with the basic additional feature that the considered pre-ordering induces an equivalence relation for which two equivalent queries have the same support. Consequently, one computation per equivalence class allows to determine the support of all queries of this class.

P. García Bringas et al. (Eds.): DEXA 2010, Part II, LNCS 6262, pp. 225–239, 2010.

In the present paper, similarly to [10], we consider a relational database operating over a star schema and we follow the approach of [10] for mining frequent projection-selection-join queries in which joins are performed along keys and foreign keys. In this setting, our contribution is to provide an *efficient* computation of such frequent queries.

Indeed, it is shown in [10] that the number of scans of the database in our algorithms is in $O(N \times |U|)$ (where N is the number of dimensions of the underlying star schema and $|U|$ is the number of attributes in this schema), and that, in order to have an efficient implementation, an appropriate indexing technique should be used to count the supports safely. The problem to solve is finding an *at most linear* technique for counting only once the duplicates occurring in the answers to projection queries; a problem which basically requires a *quadratic* scan of the table. Unfortunately, as such an indexing technique must work for *all* possible attribute sets, no solution could be found. In fact, in the implementation of [10], auxiliary data are stored in main memory, so as to keep track of the tuples computed so far for a given query. Therefore, this technique is still quadratic (since duplicates are still checked against the auxiliary data), and more importantly, experiments result in main memory overflow for large fact tables.

In order to cope with this important limitation, we propose a novel efficient and scalable algorithm, called *Frequent Query Finder* (FQF), for the computation of frequent queries. According to FQF, every table r to be mined is associated with an auxiliary table AUX(r), whose role is to associate every tuple t of r with all attribute sets S such that $t.S = t'.S$ for some tuple t' occurring in r *before* t, with respect to the scanning order of r. Assuming that these auxiliary tables are computed, it turns out that counting the supports becomes *linear* in the size of the table to be mined. We are thus provided with an *efficient* and *scalable* implementation, in the sense that runtime keeps very low and that no main memory overflow occurs, even for large datasets (up to 100,000 tuples in our experiments). We refer to Section 5 for experiments.

We also emphasize that, although the computation of an auxiliary table AUX(r) is still quadratic in the size of r, our experiments show that our new implementation, even when involving the computation of the auxiliary tables, outperforms that of [10] (when the comparison is possible). Moreover, it is also argued in the end of the paper that the computation of the auxiliary tables can be seen as a pre-processing phase, when mining frequent queries.

The paper is organized as follows: In Section 2, we briefly overview related work, and in Section 3, we recall from [10] the basic definitions and properties of our approach. Then, in Section 4, we present our algorithm FQF for mining conjunctive queries and in Section 5, we report on experiments showing that our algorithms are efficient. Section 6 concludes the paper and discusses future work.

2   Related Work

Early approaches dealing with frequent queries [2,3,8,12] consider a *fixed* set of "objects" to be counted when computing supports, meaning that in these approaches, all queries of interest are projections over a *fixed* attribute set. Moreover, apart from [4], none of these approaches consider constraints on the data, such as functional dependencies, for optimizing the computation. In [4], equivalent attribute sets with respect to functional dependencies are used for query optimization, based on materialized views, which is not the case in our approach.

To the best of our knowledge, [5] is the first approach for mining frequent queries in the general context where the set of objects to be counted is *not* fixed. However, in [5], equivalent queries are generated, which can not be tested efficiently (a problem that does not exist in our approach); and moreover, data dependencies are not taken into account.

The work of [6], dealing with mining tree queries in a graph, is also closely related to ours. Indeed, in [6], a graph is seen as a binary relation, and frequent tree queries are expressed as SQL projection-selection-join queries. This work is somehow generalized in [7] to the case of projection-selection-join queries, with the restriction that a given relation cannot occur more than once in the joins. Queries considered in [7] are *more general* than ours, since (*i*) all possible joins in which base relations occur at most once are considered in [7], whereas we only consider such joins along keys and foreign keys, and (*ii*) selection conditions of the form $(Y = Y')$ where Y and Y' are relation schemas are allowed in [7], which is not the case in our approach. However, in [7], dependencies are not taken into account, thus resulting in redundant computations.

In our previous work [9,10,11], we have considered conjunctive query mining in a star schema, focussing successively on projection queries ([11]), projection-selection queries ([9]), and projection-selection-join queries ([10]). The main contribution of this previous work is to show that taking dependencies into account in query comparison results in an efficient computation of frequent conjunctive queries. In particular, in [10], it is shown that if the database schema is a star schema, then the problem of mining frequent projection-selection-join queries where joins are performed along keys and foreign keys becomes tractable. However as previously mentioned, in [10], experiments show severe limitations, and the contribution of this paper is to propose an efficient and scalable implementation that overcomes these limitations.

3   Formal Model

3.1   Queries

We first recall that a database Δ over a *star schema* consists of a distinguished table φ with schema F, called the *fact table*, together with a set of other tables $\delta_1, \ldots, \delta_N$ with schemas $D_1, \ldots, D_N$, called the *dimension tables*, such that:

1. If $K_1, \ldots, K_N$ are the (primary) keys of $\delta_1, \ldots, \delta_N$, respectively, then, denoting by K the union of these keys (*i.e.*, $K = K_1 \ldots K_N$), K is the key of φ. In other words, for every $i = 1, \ldots, N$, δ_i satisfies $K_i \to D_i$ and φ satisfies $K \to F$. We denote by $\mathcal{F}$ the set of these functional dependencies.

2. For every $i = 1, \ldots, N$, $\pi_{K_i}(\varphi) \subseteq \pi_{K_i}(\delta_i)$ (thus each K_i is a foreign key in the fact table φ). The attribute set $M = F \setminus K$ is called the *measure* of the star schema.

As usual, we denote by $\mathcal{F}^+$ the set of all functional dependencies that can be inferred from $\mathcal{F}$, using the Armstrong's axioms, and we denote by X^+ the set of all attributes A such that $X \to A$ is in $\mathcal{F}^+$ ([13]).

In what follows, we consider a *fixed* database $\Delta = (\delta_1, \ldots, \delta_N, \varphi)$, along with projection-selection-join queries with the following specificities:

- the tuple in selection condition is either the *empty tuple*, denoted by $\top$ (in which case all tuples are selected), or a tuple y over Y (in which case all tuples t such that $t.Y = y$ are selected);
- the joins are performed along keys and foreign keys, that is, either the join is reduced to a single table, or it involves the fact table φ.

Definition 1. *Let $\Delta = (\delta_1, \ldots, \delta_N, \varphi)$ be a database over a star schema. The considered set of queries, denoted by $\mathcal{Q}$, is the set of all queries of the form $q = \pi_X(\sigma_y(r))$, or more simply $\pi_X \sigma_y(r)$, such that $XY \subseteq R$ (R denotes the schema of r), and where:*

- *r is either a table in Δ or a join of such tables containing φ;*
- *y is either the empty tuple $\top$ or a tuple over relation schema Y.*

For every query q in $\mathcal{Q}$, the support of q in Δ, denoted by $sup(q)$, is the cardinality of the answer to q. Given a support threshold min-sup, a query q is said to be frequent if $sup(q) \geq$ min-sup.

We illustrate our approach using the following example, borrowed from [10], and that we shall use as a running example throughout the paper.

Example 1. Consider the database Δ consisting of three tables and a set of functional and inclusion dependencies, as shown in Figure 1. The meaning of the attributes is as follows:

- *Cid, Cname* and *Caddr* stand for Customer Identifier, Customer Name and Customer Address,
- *Pid* and *Ptype* stand for Product Identifier and Product Type,
- *Qty* stands for Quantity (*i.e.*, number of products sold).

The schema of Δ is clearly a star schema, with *Sales* as its fact table, and *Cust* and *Prod* as its dimensional tables.

The queries $q_1 = \pi_{Cid}\sigma_{\mathtt{Paris}}(Cust)$ and $q_2 = \pi_{Cid}\sigma_{\mathtt{Paris\,beer}}(Cust \bowtie Prod \bowtie Sales)$ are in $\mathcal{Q}$, the answers of which being $\{c_1, c_2, c_3\}$ and $\{c_1, c_2\}$, respectively. Thus, if *min-sup* = 2, these queries are frequent.

On the other hand, $\pi_{Cid}(Cust)$ and $\pi_{Cid}(Cust \bowtie Sales)$ are also in $\mathcal{Q}$, and are written as $\pi_{Cid}\sigma_\top(Cust)$ and $\pi_{Cid}\sigma_\top(Cust \bowtie Sales)$, respectively. Their answers are respectively $\{c_1, c_2, c_3, c_4\}$ and $\{c_1, c_2\}$.

Cust	Cid	Cname	Caddr
	c_1	John	Paris
	c_2	Mary	Paris
	c_3	Jane	Paris
	c_4	Anne	Tours

Prod	Pid	Ptype
	p_1	milk
	p_2	beer

Sales	Cid	Pid	Qty
	c_1	p_1	10
	c_2	p_2	5
	c_2	p_1	1
	c_1	p_2	10

$$\mathcal{F}: Cid \rightarrow Cname\,Caddr$$
$$Pid \rightarrow Ptype$$
$$Cid\,Pid \rightarrow Qty$$

$$\mathcal{I}: \pi_{Cid}(Sales) \subseteq \pi_{Cid}(Cust)$$
$$\pi_{Pid}(Sales) \subseteq \pi_{Pid}(Prod)$$

Fig. 1. The database of the running example

3.2 Query Comparison

Definition 2. *Let $q = \pi_X \sigma_y(r)$ and $q_1 = \pi_{X_1} \sigma_{y_1}(r_1)$ be queries in $\mathcal{Q}$. q_1 is said to be more specific than q in Δ, denoted by $q \preceq q_1$, if one of the following holds:*

1. *$y_1 \notin \pi_{Y_1}(r_1)$*
2. *$y \in \pi_Y(r)$, $y_1 \in \pi_{Y_1}(r_1)$, and $Y_1 \rightarrow X_1 \in \mathcal{F}^+$*
3. *All of the following hold:*
   (a) either $r = r_1$ or r_1 involves the fact table φ,
   (b) $y \in \pi_Y(r)$, $y_1 \in \pi_Y(r_1)$, $Y_1 \rightarrow X_1 \notin \mathcal{F}^+$,
   (c) $XY_1 \rightarrow X_1 \in \mathcal{F}^+$ and $Y_1 \rightarrow Y \in \mathcal{F}^+$,
   (d) $yy_1 \in \pi_{YY_1}(r \bowtie r_1)$.

Example 2. In the context of Example 1, consider again the queries $q_1 = \pi_{Cid}\sigma_{\text{Paris}}(Cust)$ and $q_2 = \pi_{Cid}\sigma_{\text{Paris beer}}(Cust \bowtie Prod \bowtie Sales)$. Referring to Definition 2.3, we have: (a) $Cust \bowtie Prod \bowtie Sales$ involves the fact table $Sales$, (b) $\text{Paris} \in \pi_{Caddr}(Cust)$, $\text{Paris beer} \in \pi_{Caddr\,Ptype} (Cust \bowtie Prod \bowtie Sales)$ and $Caddr\,Ptype \rightarrow Cid \notin \mathcal{F}^+$, (c) $Cid\,Caddr\,Ptype \rightarrow Cid \in \mathcal{F}^+$, $Caddr\,Ptype \rightarrow Caddr \in \mathcal{F}^+$, and (d) $\text{Paris beer} \in \pi_{Caddr\,Ptype}(Cust \bowtie Prod \bowtie Sales)$. Therefore, $q_1 \preceq q_2$.

Consider now $q'_2 = \pi_{Cname}\sigma_{c_2\,\text{beer}}(Cust \bowtie Prod \bowtie Sales)$. Then, by Definition 2.2, $q_1 \preceq q'_2$, because $\text{Paris} \in \pi_{Caddr}(Cust)$, $c_2\,\text{beer} \in \pi_{Cid\,Ptype}(Cust \bowtie Prod \bowtie Sales)$ and $Cid\,Ptype \rightarrow Cname \in \mathcal{F}^+$.

For $q_3 = \pi_{Cid\,Cname}\sigma_\top(Cust \bowtie Sales)$, as above, $q_3 \preceq q_2$ holds. For $q'_3 = \pi_{Cid\,Cname\,Caddr}\sigma_\top(Cust \bowtie Prod \bowtie Sales)$, applying again Definition 2.3, it can be seen that $q_3 \preceq q'_3$ and $q'_3 \preceq q_3$ hold.

For $q_4 = \pi_{Qty}\sigma_{\text{beer 15}}(Prod \bowtie Sales)$, by Definition 2.1, we find $q_1 \preceq q_4$, $q_2 \preceq q_4$ and $q_3 \preceq q_4$, because $\text{beer 15} \notin \pi_{Ptype\,Qty}(Prod \bowtie Sales)$.

For $q_5 = \pi_{Qty}\sigma_{\text{beer 5}}(Prod \bowtie Sales)$, we have $q_1 \preceq q_5$, $q_2 \preceq q_5$ and $q_3 \preceq q_5$. Indeed, by Definition 2.2, $\text{Paris} \in \pi_{Caddr}(Cust)$, $\text{beer 5} \in \pi_{Ptype\,Qty}(Prod \bowtie Sales)$ and $Ptype\,Qty \rightarrow Qty \in \mathcal{F}^+$. By Definition 2.1, we also have $q_5 \preceq q_4$.

It has been shown in [10] that the relation $\preceq$ is indeed a pre-ordering (*i.e.*, reflexive and transitive), with respect to which the support is anti-monotonic, *i.e.*, $(\forall q, q_1 \in \mathcal{Q})(q \preceq q_1 \Rightarrow sup(q_1) \leq sup(q))$.

Clearly, this property is required when mining patterns according to a level-wise algorithm, such as Apriori ([1]). Moreover, the pre-ordering $\preceq$ induces an equivalence relation defined as follows: two queries q and q_1 in $\mathcal{Q}$ are said to be *equivalent*, denoted by $q \equiv q_1$, if $q \preceq q_1$ and $q_1 \preceq q$ hold. The equivalence class of a query q is denoted by $[q]$.

Referring back to Example 2, the queries q_3 and q_3' are equivalent, since it has been seen that $q_3 \preceq q_3'$ and $q_3' \preceq q_3$ both hold.

As a consequence of the anti-monotonicity property mentioned above, equivalent queries have the *same* support. Therefore, instead of computing the supports of individual queries, we consider *one* query per equivalence class.

The pre-ordering $\preceq$ is extended to the set of equivalence classes $\mathcal{C}$, and then becomes an *ordering* (*i.e.*, reflexive, anti-symmetric and transitive) over $\mathcal{C}$. Moreover, a class $[q]$ is said to be *frequent* if its support (*i.e.*, the support of all queries in $[q]$) is greater than or equal to *min-sup*.

It is easy to see that all queries $q = \pi_X \sigma_y(r)$ in $\mathcal{Q}$ such that $y \notin \pi_Y(r)$ are equivalent and have a support equal to 0, a value meant to be less than the support threshold *min-sup*. Similarly, all queries $q = \pi_X \sigma_y(r)$ in $\mathcal{Q}$ such that $y \in \pi_Y(r)$ and $Y \rightarrow X \in \mathcal{F}^+$ are equivalent, and have a support equal to 1, another value meant to be less than *min-sup*. Thus, these equivalence classes, respectively denoted by C_0 and C_1, are not considered in the computation of frequent queries.

Equivalence classes different than C_0 and C_1, whose set is denoted by $\mathcal{C}^*$, have been characterized in [10]. We simply recall that, given a query $q = \pi_X \sigma_y(r)$ such that $[q]$ is in $\mathcal{C}^*$, we consider the representative $q^+ = \pi_{X'} \sigma_{y'}(r')$ of $[q]$ such that:

1. $X' = (XY)^+$ and $Y' = Y^+$,
2. $r' = r$ if r is a dimension table, otherwise, $r' = J$ where J is the join of all tables in Δ,
3. y' is the tuple over Y^+ such that y is a subtuple of y' and $y' \in \pi_{Y^+}(r')$.

In the remainder of the paper, all considered queries are assumed to satisfy the properties above, and stand for their equivalence classes.

Example 3. In Example 1, we have $J = (Cust \bowtie Prod \bowtie Sales)$. As seen in Example 2, $q_3 = \pi_{Cid\,Cname} \sigma_\top (Cust \bowtie Sales)$ and $q_3' = \pi_{Cid\,Cname\,Caddr} \sigma_\top(J)$ are equivalent. As $(Cid\,Cname)^+ = (Cid\,Cname\,Caddr)$ and $\emptyset^+ = \emptyset$, $[q_3]$ is represented by q_3'. It can be seen that $[q_3]$ is the set of all queries $\pi_X \sigma_\top(r)$ such that $Cid \subseteq X \subseteq (Cid\,Cname\,Caddr)$, and either $r = J$ or $r = (Cust \bowtie Sales)$.

For $q = \pi_{Cname\,Ptype} \sigma_{p_2}(J)$, we have $(Cname\,Pid\,Ptype)^+ = (Cname\,Pid\,Ptype)$, $Pid^+ = (Pid\,Ptype)$, and $p_2\,\text{beer} \in \pi_{Pid\,Ptype}(J)$. Thus, $[q]$ is represented by $\pi_{Cname\,Pid\,Ptype} \sigma_{p_2\,\text{beer}}(J)$, and this class is the set of all queries $\pi_X \sigma_y(r)$ such that $Cname \subseteq X \subseteq (Cname\,Pid\,Ptype)$, and
  - either $y = p_2$ and $r = (Cust \bowtie Sales)$ or $r = J$
  - or $y = p_2\,\text{beer}$ and $r = J$.

Algorithm FQF

Input: The database Δ associated to an N-dimensional star schema and
a support threshold *min-sup*.
Output: The set *Freq* of all frequent classes.
Method:
$Freq = \emptyset$
for $i = 1, \dots, N$ do
   mine(δ_i, $Freq(\delta_i)$)
   $Freq = Freq \cup Freq(\delta_i)$
compute $J = \delta_1 \bowtie \dots \bowtie \delta_N \bowtie \varphi$
mine(J, $Freq(J)$)
$Freq = Freq \cup Freq(J)$
return $Freq$

Fig. 2. The main algorithm FQF

4 Algorithms

4.1 Main Algorithm: FQF

As in [10], frequent classes in $\mathcal{C}^*$ are computed by a level-wise algorithm, called
Frequent Query Finder (FQF), whose main steps are shown in Figure 2: all
dimension tables are first mined, and then the join J of all tables in Δ is mined.
Moreover, as in [10], we define the notion of *generic class* to avoid generating
classes that are processed in the same way.

Definition 3. *Given a class $q = \pi_X \sigma_y(r)$ in $\mathcal{C}^*$, the* generic class *associated to
q, denoted by $\langle X, Y, r \rangle$, is the set of all classes $\pi_X \sigma_{y'}(r)$ in $\mathcal{C}^*$ such that y' is a
tuple in $\pi_Y(r)$, i.e., $\langle X, Y, r \rangle = \{\pi_X \sigma_{y'}(r) \in \mathcal{C}^* \mid y' \in \pi_Y(r)\}$.*

Algorithm mine, shown in Figure 3, follows a level-wise strategy ([1]). Namely,
starting with the less specific generic class, that is r, the following steps are
iterated until no frequent classes are generated:

1. Generate and prune the set C of candidate generic classes, based on the
   current set L of frequent generic classes (see [10]);
2. Compute the supports of all classes associated with the remaining candidate
   generic classes in C;
3. Discard all classes whose support is less than the support threshold;
4. Assign L to the set of all generic classes that contain at least one frequent
   class.

However, the steps above require more attention than in Apriori, because (*i*)
we are dealing with equivalence classes, instead of individual itemsets, (*ii*) the
ordering over $\mathcal{C}^*$ is more difficult to handle than set inclusion, and (*iii*) computing
the supports requires to efficiently scan the database.

    Consequently, the main difficulties are first, generating and pruning generic
classes, and second, computing efficiently the supports of classes in $\mathcal{C}^*$. The first
point has been addressed in [10] (see Proposition 7 in [10]), but not the second
one, which is the main contribution of the present paper.

Algorithm mine

Input: A table r (either a dimension table δ_i or the join J) defined over R.
Output: The set $Freq(r)$ of all frequent classes in $\mathcal{C}^*$ of the form $\pi_X \sigma_y(r)$.
Method:
if $|r| < min\text{-}sup$ then
    //no computation since, for every q in $\mathcal{C}^*$ of the form $\pi_X \sigma_y(r)$, $|r| \geq sup(q)$
    $Freq(r) = \emptyset$
else //the computation starts with the generic class $\langle R, \emptyset, r \rangle$
    $L = \{\langle R, \emptyset, r \rangle\}$; $Freq(r) = \{\pi_R \sigma_\top(r)\}$
    while $L \neq \emptyset$ do
        //L is the set of frequent generic classes from the previous level
        $C = \texttt{generate}(L, r)$
        $C = \texttt{prune}(C, L, r)$
        $\texttt{scan}(C, \text{AUX}(r), L, L_{Freq}(r))$
        //L contains all frequent generic classes of the current level, and
        //$L_{Freq}(r)$ is the corresponding set of frequent classes
        $Freq(r) = Freq(r) \cup L_{Freq}(r)$
return $Freq(r)$

Fig. 3. Computing frequent queries on a table r

4.2   Algorithm scan

When scanning a given table r, the main difficulty is that every tuple in the answer to a query must be counted only *once*, whereas, due to projection, it might occur several times when scanning r. In order to cope with this difficulty, it is argued in [10] that indexing techniques are required. Unfortunately, considering such indexing techniques, which have to work for *all* possible attribute sets, is not realistic. In order to cope with this problem, in [9,10], each scan is associated with huge volumes of auxiliary data, resulting in main memory overflow for large fact tables.

Instead, in the present paper, before scanning r, we build an auxiliary table, denoted by $\text{AUX}(r)$, as follows. Assuming that r contains n tuples $t_1, \ldots, t_n$, the first row $\text{AUX}(r)[1]$ of $\text{AUX}(r)$ is set to the empty set, and for every $i = 2, \ldots, n$, the ith element of $\text{AUX}(r)$, denoted by $\text{AUX}(r)[i]$, contains all maximal (with respect to set inclusion) attribute sets S for which there exists $j < i$ such that $t_j.S = t_i.S$. Therefore, when considering t_i during a scan of r, knowing that S is in $\text{AUX}(r)[i]$ ensures that for every $X \subseteq S$, $t_i.X$ has already been processed.

The corresponding algorithm is shown in Figure 4, where $match(t_i, t_j)$ stands for the set of all attributes A such that $t_i.A = t_j.A$. We note that computing $match(t_i, t_j)$ amounts to compare t_i and t_j, which does not require any index.

Example 4. We illustrate the construction of the auxiliary table $\text{AUX}(J)$ in the context of Example 1, for $J = Cust \bowtie Prod \bowtie Sales$. The tables J and $\text{AUX}(J)$ are shown in Figure 5.

Since $match(t_2, t_1) = Caddr$, we obtain $\text{AUX}(J)[2] = Caddr$. Similarly, since $match(t_3, t_1) = (Pid\,Caddr\,Ptype)$ and $match(t_3, t_2) = (Cid\ Cname\ Caddr)$,

Algorithm aux

Input: A table r to be scanned containing tuples $t_1, \ldots, t_n$.
Output: The table AUX(r).
Method:
AUX$[1] = \emptyset$
for each $i = 2, \ldots, n$ do
   AUX$(r)[i] = \emptyset$
   for each $j = 1, \ldots, i - 1$ do
      compute $match(t_i, t_j)$
      if AUX$(r)[i]$ contains no super set of $match(t_i, t_j)$ then
         AUX$(r)[i] = $ AUX$(r)[i] \cup match(t_i, t_j)$
return AUX(r)

Fig. 4. Computing the auxiliary table AUX(r)

AUX$(J)[3]$ is the set of these two attribute sets. The computation for AUX$(J)[4]$ is similar, but although $match(t_4, t_3) = Caddr$, this schema does not appear in AUX$(J)[4]$. This is so because $Caddr$ is a subset of $(Cid\,Cname\,Caddr\,Qty)$ and $(Pid\,Caddr\,Ptype)$ that both belong to AUX$(J)[4]$.

J	Cid	Pid	$Cname$	$Caddr$	$Ptype$	Qty
t_1	c_1	p_1	John	Paris	milk	10
t_2	c_2	p_2	Mary	Paris	beer	5
t_3	c_2	p_1	Mary	Paris	milk	1
t_4	c_1	p_2	John	Paris	beer	10

AUX(J) i	AUX$(J)[i]$ $(1 \leq i \leq 4)$
1	$\emptyset$
2	$Caddr$
3	$(Pid\,Caddr\,Ptype), (Cid\,Cname\,Caddr)$
4	$(Cid\,Cname\,Caddr\,Qty), (Pid\,Caddr\,Ptype)$

Fig. 5. The table J and the associated table AUX(J) of Example 1

Now, given a table r and assuming that AUX(r) has been computed, the supports of equivalence classes over r are computed through parallel scans of r and AUX(r). The corresponding algorithm **scan** is shown in Figure 6. The input of Algorithm **scan** is a set C of candidate generic classes of the form $\langle X, Y, r \rangle$ for which r contains the tuples $t_1, \ldots, t_n$. All frequent classes associated with all generic candidate classes in C are computed as follows: For every $i = 1, \ldots, n$, the following actions are performed, for every $\langle X, Y, r \rangle$ in C:

1. If AUX$(r)[i]$ contains a super schema of X, then $t_i.X$ has been encountered for some $j < i$. Thus $t_i.X$ has already been processed for all classes with a projection over X. Otherwise, $t_i.X$ is encountered for the first time, and thus, has to be processed.

Algorithm scan

Input: The set C of candidate generic classes, the table AUX(r).
Output: The set L of frequent generic classes in C, and the associated frequent classes $L_{Freq}(r)$.
Method:
$L = \emptyset$; $L_{Freq}(r) = \emptyset$
for each $\langle X, Y, r \rangle \in C$ **do**
    $L(\langle X, Y, r \rangle) = \emptyset$
for each $i = 1, \ldots, n$ **do** //r contains tuples $t_1, \ldots, t_n$
    for each $\langle X, Y, r \rangle \in C$ **do**
        if $\exists\, X' \in$ AUX$(r)[i]$ such that $X \subseteq X'$ **then**
            //$t_i.X$ has been encountered before, and thus has been counted
            nothing to do
        else
            if $\exists\, Y' \in$ AUX$(r)[i]$ such that $Y \subseteq Y'$ **then**
                //$\pi_X \sigma_{t_i.Y}(r)$ has already been encountered, thus
                //$t_i.X$ must be counted for the support of $\pi_X \sigma_{t_i.Y}(r)$
                $sup(\pi_X \sigma_{t_i.Y}(r)) = sup(\pi_X \sigma_{t_i.Y}(r)) + 1$
            else
                //$\pi_X \sigma_{t_i.Y}(r)$ has not been encountered before,
                //thus either prune it or initialize its support
                if not(pruneQuery($\pi_X \sigma_{t_i.Y}(r)$)) **then**
                    $sup(\pi_X \sigma_{t_i.Y}(r)) = 1$
                    $L(\langle X, Y, r \rangle) = L(\langle X, Y, r \rangle) \cup \{\pi_X \sigma_{t_i.Y}(r)\}$
for each $\langle X, Y, r \rangle \in C$ **do**
    $L(\langle X, Y, r \rangle) = L(\langle X, Yr \rangle) \setminus \{\pi_X \sigma_y(r) \mid sup(\pi_X \sigma_y(r)) < min\text{-}sup\}$
    if $L(\langle X, Y, r \rangle) \neq \emptyset$ **then**
        $L_{Freq}(r) = L_{Freq}(r) \cup L(\langle X, Y, r \rangle)$
        $L = L \cup \{\langle X, Y, r \rangle\}$
return L **and** $L_{Freq}(r)$

Fig. 6. Scanning the table r

2. In the latter case, $t_i.X$ has to be counted for the support of $q = \pi_X \sigma_{t_i.Y}(r)$. Two cases are then possible:
   (a) If AUX$(r)[i]$ contains a super schema of Y then q has been processed previously, and thus is already associated with $\langle X, Y, r \rangle$. In this case, the support of q is incremented.
   (b) Otherwise, q is processed for the first time, and so, is not associated with $\langle X, Y, r \rangle$. In this case, we check if q can be pruned (see below), and if not, its support is initialized to 1 and q is associated with $\langle X, Y, r \rangle$.

Once these actions are performed, all supports of all classes that have to be computed are known. All classes whose support is greater than or equal to *min-sup* are put in $L_{Freq}(r)$ and the set L of frequent generic classes is output.

In our algorithms, pruning is performed at two distinct levels: for generic classes in Algorithm **mine**, and for classes in Algorithm **scan**. In Algorithm **mine**, a generic class $\langle X, Y, r \rangle$ is pruned if at least one of its predecessors (according to $\preceq$) contains no frequent classes, which entails that no class in $\langle X, Y, r \rangle$ can be frequent. However, if $\langle X, Y, r \rangle$ is not pruned, it may happen that a particular class $\pi_X \sigma_y(r)$ of $\langle X, Y, r \rangle$ can be pruned. This is checked in Algorithm scan (see item 2(b)), according to Algorithm pruneQuery shown in Figure 7.

It is important to note that Proposition 7 of [10] shows that this latter pruning is *partial*, in the sense that not all predecessors of the class $\pi_X \sigma_y(r)$ are tested. We opted for such a partial pruning for efficiency reasons, as processing a complete pruning would damage performance.

Algorithm pruneQuery

Input: A class $q = \pi_X \sigma_y(r)$.
Output: boolean.
Method:
if there exist $A \in Y$ and $a \in dom(A)$ such that
$$q = \pi_X \sigma_{y'}(r) \notin L_{Freq}(r) \text{ and } y = y'a \text{ then}$$
    return true;
return false;

Fig. 7. Class Pruning

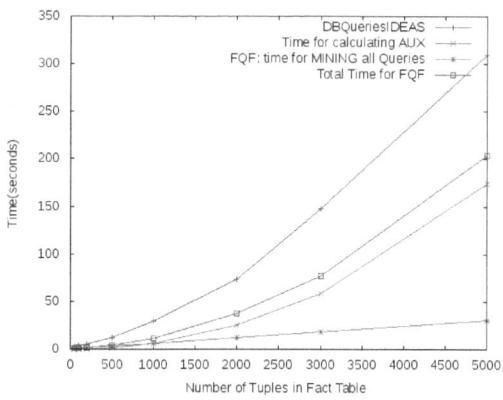

Fig. 8. Runtime over the size of the fact table for FQF and the implementation of [10]

5   Experiments

We performed experiments on an Pentium Duo Core with 2Go main memory running on Ubuntu Linux 2.6. The algorithms are implemented in Java using JDBC to communicate with MySql. Datasets have been generated using our own generator, adapted from the IBM data generator (www.almaden.ibm.com).

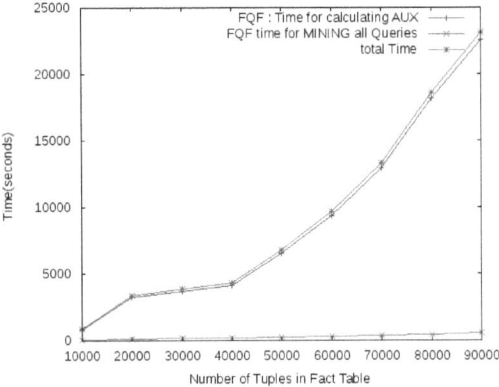

Fig. 9. Runtime over the size of the fact table

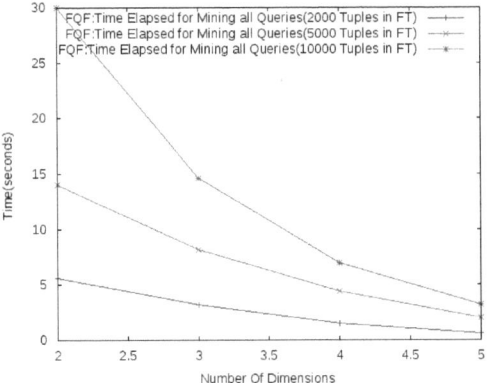

Fig. 10. Runtime over the number of dimensions

The generated databases over star schemas are denoted by dbdDaTtMm where d is the number of dimensions, a is the total number of attributes, t is the number of tuples in the fact table, and m is the number of measure attributes. In all our experiments, except those reported in Figures 11 and 12, the support threshold is set to 0.6 times the number of tuples in the fact table, that is $0.6 \times t$. We also mention that all runtimes reported below include the computation time of the construction of auxiliary tables. In Figures 8 and 9, the runtimes excluding the computation of the auxiliary tables are also shown.

Figure 8 shows the runtimes of FQF compared to those presented in [10] for db2D12TtM1, with t between 50 and 5000. Clearly, FQF outperforms our previous implementation presented in [10]: the reduction of runtime between the implementation in [10] and FQF is always greater than 33%. It should also be noticed from Figure 8 that the runtime for only mining frequent classes is very low, since less than 40 seconds.

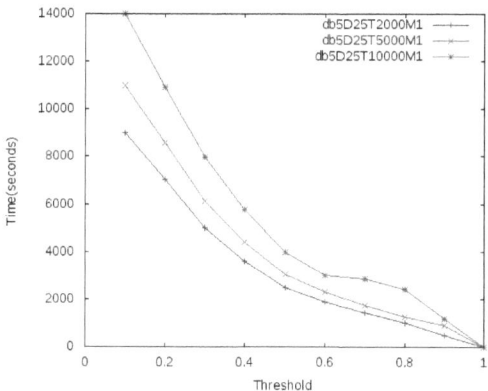

Fig. 11. Runtime over support

Similarly, as shown in Figure 9, the time spent in mining the frequent classes for the databases db2D12TtM1 with t between 10,000 and 90,000 is very low compared to that for calculating the auxiliary tables. Moreover, this runtime increases slowly with the size of the fact table. We also emphasize that, in these experiments, we had no main memory overflow, contrary to what happened with the previous implementation presented in [10], when t exceeds 5000.

Figure 10 reports on runtime over the number of dimensions, for the databases dbdD12TtM1 where d ranges from 2 to 5 and for t equal to 2000, 5000 and 10,000. This figure clearly shows that the time spent for mining frequent classes decreases significantly when the number of dimensions increases. This is so because, given a number of attributes (12 in our case), when d increases, more functional dependencies are available, and so, less classes have to be processed. It is important to note that, according to our previous statement that the number of scans of the database is in $O(N \times |U|)$ (where N is the number of dimensions and $|U|$ the total number of attributes), one would rather expect an *increase* of runtime when N increases. However, what these experiments show is that, although the increase of the number of dimension tables entails more scans, this is compensated by a drastic reduction of the number of generic classes.

Figure 11 shows the runtime over the support threshold (expressed as a ratio of the size of the fact table), for databases db5D25TtM1 with t equal to 2000, 5000 and 10,000. Clearly, runtime decreases rapidly when the support increases.

We recall from Section 2 that the only other work aiming at mining all frequent queries from a relational database is that in [7], and thus, we could compare our algorithm only to the Conqueror algorithm ([7]). This has been done using the IMDB database (http://www.imdb.com), for various support thresholds (expressed in numbers of tuples). To do so, we first transformed the IMDB database into a star schema having 3 dimensions, 6 attributes and no measure. In this experiment, the fact table contains 158,441 tuples.

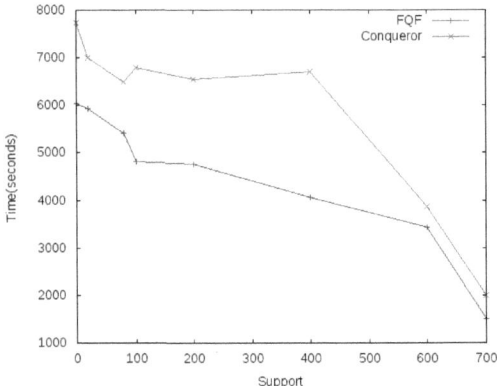

Fig. 12. FQF versus Conqueror

As shown in Figure 12, our algorithm performs better than the Conqueror algorithm. This is so because, in [7], functional and inclusion dependencies are not taken into account, as we do in our approach. However, we recall in this respect that selection conditions of the form $Y = Y'$ (where Y and Y' are attribute sets) are considered in [7], which is not the case in our approach.

We end this section by two important remarks regarding the computation of the auxiliary tables.

1. The computation of auxiliary tables can be seen as a pre-processing, because it has to be computed only *once* for all runs of FQF, provided that, meanwhile, the database has not been updated. This remark is important regarding runtime, because, as shown in Figures 8 and 9, when auxiliary tables are available, the runtime of FQF is very low even for large fact tables.
2. When a database table r is updated, maintaining up to date the associated auxiliary table $AUX(r)$ can be achieved efficiently. Indeed, in the case of insertion of a new tuple t in r, and assuming that t becomes the last tuple of r, a new row is added to $AUX(r)$ and the associated schemas are obtained through one scan of r. If a tuple t_i is deleted from r, then $AUX(r)[i]$ must be deleted from $AUX(r)$, and only the rows $AUX(r)[j]$ such that $j > i$ and $match(t_j, t_i) \in AUX(r)[j]$ have to be updated.

6   Conclusion and Further Work

We presented new algorithms for mining frequent queries in databases over a star schema, based on theoretical results introduced in our previous work [10]. We showed through experiments that, in this particular case, mining frequent conjunctive queries becomes tractable. Our approach relies on the computation of auxiliary tables that can be seen as a pre-processing phase. An important point in this respect is that, assuming these auxiliary tables are available, the time for mining frequent queries becomes very low, as shown in our experiments.

Future work consists in processing further tests and optimizing our algorithms. We plan to generalize our approach to database schemas other than star schemas, and to study the rules that can be obtained based on frequent queries.

References

1. Agrawal, R., Mannila, H., Srikant, R., Toivonen, H., Verkamo, A.: Fast discovery of association rules. In: Advances in Knowledge Discovery and Data Mining. MIT Press, Cambridge (1996)
2. Dehaspe, L., Raedt, L.D.: Mining association rules in multiple relations. In: Džeroski, S., Lavrač, N. (eds.) ILP 1997. LNCS, vol. 1297. Springer, Heidelberg (1997)
3. Diop, C., Giacometti, A., Laurent, D., Spyratos, N.: Composition of mining contexts for efficient extraction of association rules. In: Jensen, C.S., Jeffery, K., Pokorný, J., Šaltenis, S., Bertino, E., Böhm, K., Jarke, M. (eds.) EDBT 2002. LNCS, vol. 2287, p. 106. Springer, Heidelberg (2002)
4. Esposito, R., Meo, R., Botta, M.: Answering constraint-based mining queries on itemsets using previous materialized results. J. of Intelligent Information Systems 26(1) (2006)
5. Goethals, B., den Bussche, J.V.: Relational association rules: getting warmer. In: Hand, D.J., Adams, N.M., Bolton, R.J. (eds.) Pattern Detection and Discovery. LNCS (LNAI), vol. 2447, p. 125. Springer, Heidelberg (2002)
6. Goethals, B., Hoekx, E., den Bussche, J.V.: Mining tree queries in a graph. In: 11th ACM SIGKDD Intl. Conference on Knowledge Discovery and Data Mining, KDD (2005)
7. Goethals, B., Page, W.L., Mannila, H.: Mining association rules of simple conjunctive queries. In: SIAM (2008)
8. Han, J., Fu, Y., Wang, W., Koperski, K., Zaiane, O.: Dmql: A data mining query language for relational databases. In: SIGMOD-DMKD 1996 (1996)
9. Jen, T.-Y., Laurent, D., Spyratos, N.: Mining all frequent selection-projection queries from a relational table. In: EDBT 2008. ACM Press, New York (2008)
10. Jen, T.-Y., Laurent, D., Spyratos, N.: Mining Frequent Queries in Relational Databases. In: IDEAS. ACM Press, New York (2009)
11. Jen, T.-Y., Laurent, D., Spyratos, N., Sy, O.: Towards mining frequent queries in star schemes. In: Bonchi, F., Boulicaut, J.-F. (eds.) KDID 2005. LNCS, vol. 3933, pp. 104–123. Springer, Heidelberg (2006)
12. Meo, R., Psaila, G., Ceri, S.: An extension to sql for mining association rules. Data Mining and Knowledge Discovery 9 (1997)
13. Ullman, J.: Principles of Databases and Knowledge-Base Systems. Comp. Sc. Press, Rockville (1988)

Evaluating Evidences for Keyword Query Disambiguation in Entity Centric Database Search

Elena Demidova[1], Xuan Zhou[2], Irina Oelze[1], and Wolfgang Nejdl[1]

[1] L3S Research Center, Hannover, Germany
{demidova,oelze,nejdl}@L3S.de
[2] CSIRO ICT Centre, Australia
xuan.zhou@CSIRO.au

Abstract. A number of existing approaches attempt to reduce ambiguity of user's keyword queries by translating them to structured database queries. This disambiguation process relies on a proper assessment of whether a structured query represents the intent behind the keyword query. In this paper we systematically analyze a number of intuitive statistical measures that can potentially be used in this disambiguation process. We evaluate the impact of these measures through experiments on real-world data.

Keywords: entity search, keyword query disambiguation, statistical analysis.

1 Introduction

Online databases have been increasingly used to collect and disseminate information about real-world entities, such as people, products, publications and genes. Users of these databases need effective solutions to retrieve the desired entities quickly and accurately [2, 3, 7]. Keyword search has been widely used for this purpose, on account of its usability and efficiency. However, as a keyword search interface may not offer sufficient expressiveness for users to precisely specify their informational needs, it may return a large number of irrelevant results, which prohibit users from retrieving desired entities.

To cope with the limitations of keyword search, some recent work [1, 4, 5, 6, 10] proposed to perform keyword query disambiguation before retrieving entities from databases. The disambiguation process aims at translating a keyword query to a structured database query, which accurately expresses the user's informational need. The structured query can then be executed to retrieve the exact information desired by the user from the database. Due to the ambiguity of a keyword query, there usually exist a large number of structural queries as its possible interpretations. A crucial step of keyword query disambiguation is to assess the likelihood of the possible interpretations and pick the most probable ones to be executed against the database. In practice, a variety of statistics related to keywords, database and query logs can be utilized to make such assessment. Although some of the statistical parameters, such as TF/IDF scores, keyword frequency and length of the join path in a structured query, have been used by existing ranking functions [4, 6, 10] or probabilistic models e.g. [5], they are

P. García Bringas et al. (Eds.): DEXA 2010, Part II, LNCS 6262, pp. 240–247, 2010.
© Springer-Verlag Berlin Heidelberg 2010

far from exhaustive. Many intuitive and relevant parameters remain uninvestigated. Moreover, as the number of parameters increases, exhaustive tuning is required to compose them into an optimal estimation function. There is no systematical study for evaluating and comparing the impact of the various parameters in keyword query disambiguation.

In this paper, we present and evaluate a set of generic statistical parameters for keyword query disambiguation in the context of entity centric search. Our study was conducted over a heterogeneous real-world dataset with 13 entity types and five million data instances. We show the significance of each individual parameter in keyword query disambiguation, as well as the effects of their linear aggregation.

2 Parameters for Keyword Query Disambiguation

We view a database as a set of entities. Each entity is represented as a set of attribute-value pairs, each mapping a set of keywords to the value of an attribute. For example, "name:{Hanks,Tom}" is an attribute-value pair representing that an entity's "name" is "Tom Hanks".

Keyword Query: A keyword query is entity centric, such that each keyword is supposed to occur in an attribute of the desired entity. Some examples of keyword queries are K_1="tom hanks" and K_2="hanks 2001".

Query Interpretation: To construct a structured query from keywords, we first interpret each keyword to an attribute-value pair, and then connect the resulting pairs to build a conjunctive Boolean query, which we call query interpretation. For example, "name:Tom AND name:Hanks" is an interpretation of the keyword query K_1="Tom Hanks", which searches for an entity having both keywords, "Tom" and "Hanks", in the attribute "name". As another example, "actor:Hanks AND year:2001" is an interpretation of K_2="Hanks 2001", which retrieves Hanks' movies in 2001. A keyword query K can have a number of valid interpretations in the database, constituting the **interpretation space** of K.

As an interpretation space of a single keyword query is usually large, the objective of keyword query disambiguation is to quickly identify the most likely interpretation desired by the user. In what follows, we introduce a number of statistics that can potentially be used to assess this likelihood.

2.1 Keyword Specific Parameters

A number of existing approaches considers each keyword in a keyword query independently of other keywords and aims to assess the likelihood of each attribute-value pair given a keyword, i.e. $L(A_j:k_i \mid k_i)$ for every $k_i \in K$ and $A_j \in DB$. A number of statistics can be used to assess this likelihood.

Attribute Specific Keyword Frequency (AKF). This parameter assumes that the formation of a keyword interpretation can be modeled as a random process. For an attribute A_j, this process randomly picks one of its instances a_j and randomly picks a keyword k_i from that instance to form a keyword interpretation $A_j{:}k_i$. Then, the likelihood of a attribute-value pair, is the likelihood that $A_j{:}k_i$ is formed through this random process. This likelihood can be estimated using Equation 1:

$$L(A_j : k_i \mid k_i) \propto AKF = \frac{\left|A_j : k_i \in A_j\right|}{\left|A_j\right|},$$

(1)

where $|A_j|$ is the number of instances of the attribute A_j and $|A_j : k_i \in A_j|$ is the number of instances of A_j containing k_i.

Attribute Specific Keyword Selectivity (AKS). Most probabilistic IR models consider Inverse Document Frequency (IDF) as an important parameter for ranking documents, such that a more selective keyword is usually given a higher weight [8]. To apply the same principle to keyword query disambiguation, we should assign a higher likelihood to an attribute-value pair that has higher selectivity in the database:

$$L(A_j : k_i \mid k_i) \propto AKS = \frac{\left|A_j\right|}{\left|A_j : k_i \in A_j\right|}.$$

(2)

It is obvious that AKF is inversely proportional to the AKS.

Keyword Closeness (KCL). When a user issues a keyword query, the keywords in that query are usually highly correlated. Therefore, it is more likely that these keywords occur in a small number of attributes than being spread over many attributes. We measure KCL as an average number of keywords per attribute.

$$KCL(Q,K) = \frac{|K|}{|A_j \in Q|},$$

(3)

where $|A_j \in Q|$ is the number of distinct attributes in the structured query Q.

Interpretation Completeness (ICP). As we consider each keyword in a keyword query to be meaningful, we prefer a complete query interpretation that includes all the keywords to an incomplete interpretation with only a subset of the user's keywords. We measure completeness of a query interpretation by the proportion of the query terms it includes:

$$ICP(Q,K) = \frac{|k_i \in Q \cap K|}{|K|}.$$

(4)

2.2 Database Specific Parameters

Apart from the properties of keywords, the likelihood of a structured query can also be influenced by the database in use. For example, a person's name can be used more frequently than a person's age in users' queries. We consider such structural patterns as relevant parameters to assess the likelihood of a query interpretation.

Entity Type Popularity (EPL and EPD). If a database possesses a representative query log, given a query Q, we can access the popularity of the target entity type of Q, which is an indicator of the likelihood of Q. This popularity can be measured by:

$$popularity(Q_T) = \frac{|Q_T \in L|}{|L|}, \tag{5}$$

where Q_T is a query to retrieve an entity of type T. If the database maintains a query log, L represents the query log. In this case, this parameter is called EPL. If a query log is not available, we can approximate the popularity of an entity type as its frequency in the database. In this case, L represents the database, and the parameter is called EPD.

Attribute Popularity (APL and APD). Similar to the entity type, popularity of attributes is also an indicator of the relevance of a query interpretation. If a database possesses a representative query log, we can access the popularity of an attribute as the frequency of its usage in the predicates of all logged queries (APL). Alternatively, we can approximate attribute popularity using the frequency of the attribute's occurrences in database entities (APD). Attribute popularity can be calculated as:

$$popularity(A) = \frac{|A \in L|}{|L|}. \tag{6}$$

Average Attribute Cardinality (AAC). Given an entity, some of its attributes have only singular values, such as a movie's title, while some other attributes may have multiple values, such as movie's actors. Average attribute cardinality (AAC) represents the average number of values an attribute can have. Normally, an attribute with a lower cardinality is used more often to identify an entity than an attribute with a higher cardinality. Thus, we can use the cardinality to assess the likelihood of a structured query. AAC of an attribute is calculated as:

$$AAC(A) = \frac{1}{\log(1 + \frac{\sum_{entities} Count(A)}{|Contain(A)|})}, \tag{7}$$

where $Count(A)$ is the number of instances of the attribute A in an entity and $|Contain(A)|$ is the total number of entities containing the attribute A. As the deviation of AAC is usually very large, we use $\log()$ to normalize it.

3 Evaluation

Our evaluation was conducted in two steps. In the first phase, we studied if each ranking parameter can contribute to keyword query disambiguation, in other words, if there is a significant correlation between a parameter and the likelihood of a query interpretation. To this end we used Spearman's rank correlation coefficient [9].

In the second phase of our evaluation, we studied how the ranking parameters can be combined to amplify their effectiveness. A straight-forward approach is to apply a linear combination of the parameters and focus on determining the optimal weight of each parameter. While this approach assumes a linear dependence among the parameters, this is a frequently used model [11]. Basically, the score of the query interpretation is calculated using Formula 8:

$$score(Q_i) = w_1 \cdot f_1 + w_2 \cdot f_2 + \dots + w_n \cdot f_n \quad , \tag{8}$$

where Q_i is a query interpretation, f_i is a normalized value of the ranking parameter i and w_i is a weight of this parameter i. Our experiments applied linear regression [12] to obtain the weights.

3.1   Dataset and Queries

In our experiments we used an entertainment subset of the freebase dataset (www.freebase.com). Freebase is a typical example of an online database with a big number of textual attributes. The entertainment subset of freebase is heterogeneous, including 13 entity types such as film, tv program, music track, book, comic, videogame, opera, play, and artwork, with approximately five million entities in total. The number of attributes for each entity type ranges from 11 to 44 with an average of 24.

As freebase does not provide an associated query log, we extracted 6,800 keyword queries from a query log of the MSN search engine, such that each extracted query has a target URL to a Wikipedia article and this Wikipedia URL identifies an entity in our dataset. The lengths of the keyword queries range from 2 to 6. To identify the intended interpretation (i.e. structured query) for a keyword query we applied a semi-automatic disambiguation procedure. For every keyword query we fetched the corresponding database entity, as identified by its Wikipedia URL. For each keyword, we created its interpretation by mapping it to the attribute where it occurs most frequently. We then aggregated these keyword interpretations in a query interpretation. Finally, we manually evaluated a sample of 200 interpretations and concluded that they all had reasonable semantics.

3.2   Significance of the Disambiguation Parameters

In the first set of experiments, we assessed the correlation between the value of each disambiguation parameter and the relevance of query interpretation. For each keyword query in the query set, we used two of its interpretations. One was the user intended interpretation, which was identified through the procedure introduced in Section 3.1. We assigned this query interpretation the relevance score 1.0, indicating that it is the user intended structured query. The other interpretation was a randomly selected query interpretation that was maximally different from the user intended interpretation (with a totally different set of attribute-value pairs). We assigned this interpretation with the score 0.0, indicating that it is highly unlikely to be intended by user. This resulted in 13K query interpretations as a test set. For every query interpretation in the test set, we computed the score of each disambiguation parameter presented in Section 2. Finally, we computed the correlation coefficient between the score of each parameter and the relevance score of the query interpretation over the 13K interpretations. Table 1 presents the results.

Table 1. Correlation coefficient between the parameter score and score of the intended query interpretation

	AKF	AAC	APL	EPL	ICP	EPD	AKS	APD	KCL
Correlation Coefficient	0.483	0.236	0.724	0.589	-0.015	-0.01	0.043	0.512	0.436

As presented in Table 1, apart from ICP, EPD and AKS, the rest of the query disambiguation parameters are significantly and positively correlated with the relevance scores of query interpretations. There is a strong correlation between the attribute / entity type popularity in query log and the relevance score of the interpretation. This is expected, as the majority of the tested queries targeted on a small number of entity types and attributes, following the Pareto Distribution. Attribute popularity in the database, attribute specific keyword frequency and keyword closeness are important query disambiguation factors too. When attribute specific keyword frequency performs well, it is natural that attribute specific keyword selectivity will be insignificant, as these two parameters are inversely proportional. The insignificance of ICP is due to the fact that we ignored incomplete queries when picking the test examples, such that majority of the target query interpretations in our query set are complete. Entity type popularity in database (EPD) does not seem to be correlated with the score of the interpretation. This shows that the number of existing entities in a database does not necessarily correspond to the frequency of their usage. In contrast, attribute popularity in database exhibits a strong correlation with the relevance score. We believe this reflects the usage of freebase, i.e., the more often an entity is used, the more attributes of it are filled by the users.

3.3 Combination of the Disambiguation Parameters

In the second set of experiments, we studied how to combine the disambiguation parameters to amplify their effects. As stated before, we focused on using linear combination to form a ranking function (Formula 8). We considered three scenarios. In the first scenario, we assume a newly created database without a query log. The attribute popularity in the database also does not reflect its usage. In this case, the best combination of parameters is AKF, AAC and KCL. In the second scenario, we assume a user created database (i.e. freebase) without a representative query log. In this case, the best combination is AKF, AAC, KCL and APD. Finally, we assume a database with a representative query log, for which the best combination is AKF, KCL and APL (AAC and APD are completely dominated by the other parameters). Using the same training set, we performed linear regression on the three combinations of parameters and obtained the ranking functions in Table 2.

Table 2. Three multi-parameter ranking functions

Ranking Functions	Accuracy on Test Sets
Combi_1 = $1.05 \times$ AKF $+ 0.86 \times$ AAC $+ 1.14 \times$ KCL	0.5901
Combi_2 = $2.97 \times$ APD $+ 0.78 \times$ KCL $+ 0.75 \times$ AKF $+ 0.056 \times$ ACC	0.6713
Combi_3 = $2.03 \times$ APL $+ 0.44 \times$ KCL $+ 0.24 \times$ AKF	0.715

246   E. Demidova et al.

3.4 Application to Keyword Query Disambiguation

We applied all the disambiguation parameters as individual ranking functions, together with the three ranking functions in Table 2, to the disambiguation of 6,800 test
keyword queries. Fig. 1 and Fig. 2 show the results.

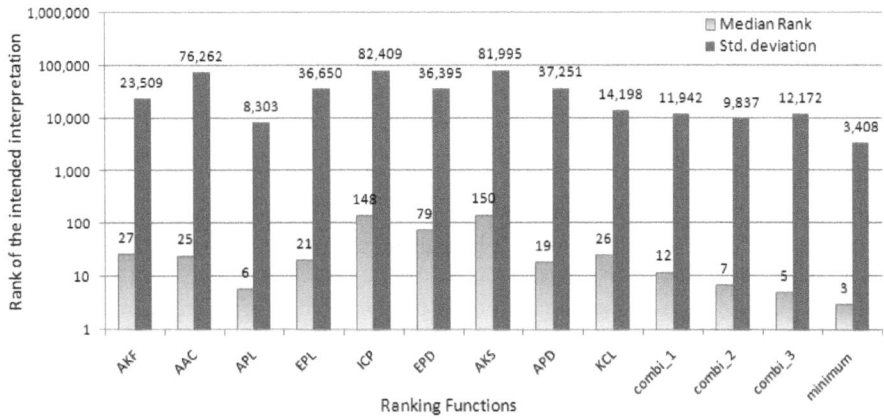

Fig. 1. Median rank & standard deviation of the correct query interpretation

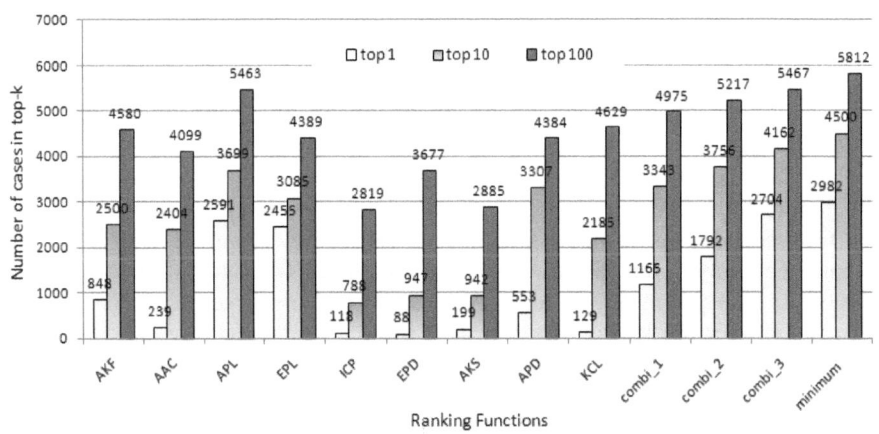

Fig. 2. Disambiguation effectiveness: result within the top-k interpretations

Fig. 1 presents the median rank and standard deviation of the correct query interpretation obtained using each ranking function. For instance, using the APL ranking
function to rank possible query interpretations, the median rank of the first correct
query interpretation over the query set is six with standard deviation of 8,303. Fig. 2
shows the faction of keyword queries whose correct interpretations can be found
within the top-k results of keyword disambiguation. For each keyword query, we
computed the best rank that can be achieved by all the individual factors. We denote

it by minimum and plot it on the right of Fig. 1 and Fig. 2. Both figures (especially Fig. 2) show that the multi-parameter ranking functions outperform each individual parameter. However, the best rank of a query given by one of the individual parameters (represented by minimum) is better than the multi-parameter functions. This indicates that, although each individual parameter is important for the disambiguation, a linear combination might be not the optimal way to aggregate the effects of all the parameters. Further investigation is required to explore more complex relationships between the parameters.

4 Conclusion

In this paper we studied a set of statistical parameters for keyword query disambiguation in database entity search. Our experiments show that some parameters are highly relevant, while the others do not seem to contribute much. We also show that a linear combination of these parameters can achieve improved effectiveness. Nevertheless, keyword query disambiguation is a difficult problem, especially for large databases. Our study was limited to linear models, i.e., correlation coefficient and linear regression. Further investigation is required to explore more complex relationships between the parameters.

Acknowledgments. This work has been partially supported by the FP7 EU Project OKKAM (contract no. ICT-215032).

References

1. Al-Muhammed, M., Embley, D.W.: Ontology-Based Constraint Recognition for Free-Form Service Requests. In: ICDE 2007 (2007)
2. Chakaravarthy, V.T., Gupta, H., Roy, P., Mohania, M.: Efficiently linking text documents with relevant structured information. In: VLDB 2006 (2006)
3. Chaudhuri, S., Das, G.: Keyword Querying and Ranking in Databases. In: VLDB 2009 (2009)
4. Chu, E., Baid, A., Chai, X., Doan, A., Naughton, J.: Combining keyword search and forms for ad hoc querying of databases. In: SIGMOD 2009 (2009)
5. Demidova, E., Zhou, X., Nejdl, W.: IQ^P: Incremental Query Construction, a Probabilistic Approach. In: Proc. of ICDE 2010 (2010)
6. Kandogan, E., Krishnamurthy, R., Raghavan, S., Vaithyanathan, S., Zhu, H.: Avatar semantic search: a database approach to information retrieval. In: SIGMOD 2006 (2006)
7. Kumar, R., Tomkins, A.: A Characterization of Online Search Behavior. IEEE Data Eng. Bull. (2009)
8. Manning, C.D., Raghavan, P., Schütze, H.: Introduction to Information Retrieval. Cambridge University Press, Cambridge (2008)
9. Myers, J.L., Well, A.D.: Research Design and Statistical Analysis, 2nd edn., p. 508. Lawrence Erlbaum, Mahwah (2003), ISBN 0805840370.
10. Tata, S., Lohman, G.M.: SQAK: doing more with keywords. In: SIGMOD 2008 (2008)
11. Taylor, M., Zaragoza, H., Craswell, N., Robertson, S., Burges, C.: Optimisation methods for ranking functions with multiple parameters. In: CIKM 2006 (2006)
12. Witten, I.H., Frank, E.: Data mining: practical machine learning tools and techniques. Morgan Kaufmann, San Francisco (2005), ISBN 0-12-088407-0

Typicality Ranking of Images Using the Aspect Model

Taro Tezuka and Akira Maeda

College of Information Science and Engineering
Ritsumeikan University
{tezuka,amaeda}@media.ritsumei.ac.jp

Abstract. Searching images from the World Wide Web in order to know what an object looks like is a very common task. The best response for such a task is to present the most typical image of the object. Existing web-based image search engines, however, return many results that are not typical. In this paper, we propose a method for obtaining typical images through estimating parameters of a generative model. Specifically, we assume that typicality is represented by combinations of symbolic features, and express it using the aspect model, which is a generative model with discrete latent and observable variables. Symbolic features used in our implementation are the existences of specific colors in the object region of the image. The estimated latent variables are filtered and the one that best expresses typicality is selected. Based on the proposed method, we implemented a system that ranks the images in the order of typicality. Experiments showed the effectiveness of our method.

Keywords: Image retrieval, Typicality, Bag-of-features, Generative model.

1 Introduction

One important use of web-based image search is to know the visual characteristics of an object. In such a case, what the user wants is the most "typical" look of the object. In existing web image search engines, however, the set of high ranked search results contain images that are not typical. The goal of this paper is to propose and evaluate a method that extracts typical images from the result of web image search by applying a generative model, a type of probabilistic model. Although typicality is a difficult concept to capture, but in this paper we define it as follows:

Definition: *An image I is a typical image for query Q if the word Q is an appropriate label for I, given that the evaluator has enough knowledge on the object referred by Q.*

Our proposed method estimates "aspects" expressed in a set of images, and select an aspect assumed to express typicality. We then rank images using conditional probability. One of the characteristics of our method is that it expresses typicality using discrete probabilistic variable. Many models for classification and dimension reduction use continuous variables, including k-means and PCA (principal component analysis). Our model consists of discrete variables only. In this sense it is an intrinsically symbolic approach. The method can be used to obtain a large set of images with labels. The set

P. García Bringas et al. (Eds.): DEXA 2010, Part II, LNCS 6262, pp. 248–257, 2010.
© Springer-Verlag Berlin Heidelberg 2010

has a wide range of applications. For example, it can be used to create a general use visual encyclopedia. It can also be utilized in a car navigation system by providing the user with the exterior image of the destination.

One of the future goals is to build a general task image recognition engine, which gives the name of an object when an arbitrary image is given. Attaching correct labels to a large set of images obtained from the Web would contribute in building such system. Based on our proposed method, we implemented a system and named it "Typi" after "typicality". The paper consists of the following sections. Section 2 gives related work. Section 3 describes our method in detail. Section 4 illustrates implementation, and section 5 describes the result of evaluation. Section 6 is the conclusion.

2   Related Work

There are a number of web-based image search engines available now, for example Google Image Search, Yahoo! Image Search, and Bing Images. There have also been works of applying object identification to images on the Web, for example WebSeek by Smith and Chang[1].

Recently, bag-of-features approach, originated from bag-of-words approach used in text information retrieval, is gaining much attention. Vogel and Schiele used combination of local features to represent higher order concepts such as objects, and evaluated precisions of image retrieval methods[2]. Fei-Fei and Perona used a generative model used in text analysis to classify natural scene images[3].

There has been some research on finding typical images of objects. Kennedy and Naaman proposed a method of extracting typical images of landmarks[4]. In addition to using visual features, their method used geotags, location metadata attached to images contributed on image sharing sites. In contrast, our method relies on visual features alone. Wu and Yang proposed a system for finding street landmarks such as signs based on extracting object fingerprints from images[5].

3   Method

In this section, we describe our proposed method. Figure 1 shows the flow of the system.

Our previous paper describes the methods of extracting object regions in more details[6]. The rest of the section mainly describes the method of selecting the "top aspect" assumed to express typicality.

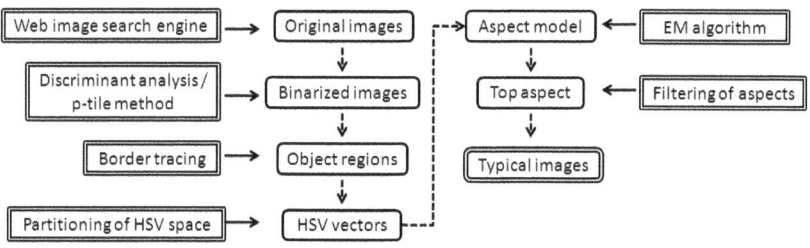

Fig. 1. System flow for extracting typical images

3.1 Aspect Model

The result of web image search usually contains various objects that are relevant to the query. For example, for a query "iris", search results would contain a type of wildflowers and a part of eyeball. Also, there are irises with various colors, ranging from purple to yellow. There are also images that show a field with iris or a part of iris. Such variety can be well expressed using a mixture model, which is a part of probabilistic models.

We use the aspect model, a model consisting of discrete observable and latent variables. The reason for applying this model is that we assume typicality to be expressed by a mixture of typical features. In the case of iris, for example, object region can be purple, yellow, or a mixture of red and white.

The aspect model[7] is a model that assumes an observed pair of discrete features (x, y) (dyadic data) is conditionally independent under a discrete latent variable z. Its graphical model is illustrated in Figure 2.

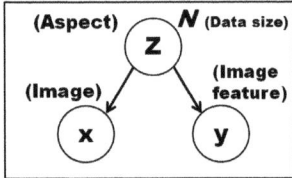

Fig. 2. Graphical model for aspect model

x and y are observable variables, and z is a latent variable. Both are discrete. Observed data consists of pairs (x, y), which is called dyadic data. N is the size of data. Based on the graphical model, we obtain the following conditional independence. It can be considered as the generative model of (x, y).

$$p(x, y, z) = p(x|z)p(y|z)p(z) \tag{1}$$

An aspect z is the first variable to be generated in the model. x and y are generated under conditioned probability $p(x|z)$ and $p(y|z)$.

If we express the observed frequency of the pair (x, y) by $n(x, y)$, the log-likelihood of the dyadic data $L(x, y)$ is expressed as $\ln \prod_{x,y} p(x, y)^{n(x,y)}$. Based on the above mentioned conditional independence, we can transform the equation as follows. We maximize $L(x, y)$ using the EM algorithm.

$$L(x, y) = \sum_{x,y} n(x, y) \ln p(x, y) = \sum_{x,y} n(x, y) \ln \sum_{z} p(x|z)p(y|z)p(z) \tag{2}$$

One practical example where the aspect model is used is on analyzing a set of documents covering different topics. When a term t appears on a document m, it is considered as a dyadic data (t, m). The data is generated from the latent variable z, which expresses the topics.

In this paper, x indicates an image and y indicates and image feature described in the next subsection.

3.2   Image Features

In our proposed method, we model typicality as a mixture of image features. In this paper, we focus on color features, but in the future work it would be extended to other features such as textures and shapes. In order to express features as a vector, we divide color space into color regions. From now on, we refer to each color region as a "color". Similar colors are grouped into a component of the vector. Hue is divided into finer details, since it is usually intrinsic to the object, while brightness and saturation varies depending on lighting. Colors with low brightness is considered as black, and those with low saturation is considered as either dark gray, light gray, or white.

For each color, the system counts how many pixels exist with that color, and create an *HSV vector* having the numbers as its components. The fact that an image feature y appeared in the object region of an image x is considered to be an observed datum (x, y). The object region is extracted using border tracing.

3.3   Extracting an Aspect Expressing Typicality

An aspect z with higher $p(z)$ is more likely to be observed, therefore considered to be more important in the image set. $p(y|z)$ indicates the probability that an image feature y is generated from an aspect z.

If the aspect captures the typical characteristics of an object, it is likely to consist of several colors, rather than of a single color. In a pre-experiment, we observed that aspects with high probability on a single color are less likely to express typicality. On the other hand, if it generates all colors equivalently, it does not have any characteristics, and is not appropriate as an expression of typicality, even if it has high $p(z)$. We therefore introduce entropy $H[p(y|z)]$ for filtering such inappropriate color. Aspects are sorted by $H[p(y|z)]$, and those that comes below or over threshold ranks are removed. This filtering can be expressed as follows.

$$\alpha|Z| < rank(H[p(y|z)]) < \beta|Z| \tag{3}$$

$|Z|$ is the number of the aspects, $rank(H[p(y|z)])$ is the rank of an aspect z when sorted by the decreasing order of entropy $H[p(y|z)]$. $0 \leq \alpha, \beta \leq 1$ are the coefficients for determining the range that the aspects are used.

From the set of aspects that fulfilled the condition on entropy indicated by Expression 3, we select the aspect with the highest $p(z)$ as the "top aspect". Using this aspect, the "typicality" of an image is calculated as follows.

$$typicality(x, m) = \sum_{z_m} p(x|z_m)p(z_m) \tag{4}$$

$p(x|z)$ indicates the probability that an image x is created from an aspect z. $p(z)$ is the probability that the aspect z appears. z_m refers to the m-th aspect when sorted in the decreasing order of $p(z)$. $p(y|z)$ indicates the probability that an image feature y occurs from an aspect z.

4   Implementation

In this section, we describe implementation of "Typi", a typical image retrieval system based on our proposed method.

4.1   System Structure

Typi was implemented using C#. It consists of modules for image collection, feature extraction, parameter estimation, and evaluation. API for Google Image Search[8] is used for collecting images. Since the API provides the search engine's ranking on images, we use it to compare with our method in the evaluation section.

4.2   Parameters

We have used the following parameters for implementation. Colors *black*, *dark gray*, *light gray*, and *white* is defined using brightness V and saturation S. If the brightness is below 0.2, it is considered *black*. In the region that the saturation is below 0.2, if the brightness is between 0.2 and 0.6, it is *dark gray*. If the brightness is between 0.6 and 0.8, it is *light gray*. If the brightness is over 0.8, it is *white*. For the remaining region, brightness is divided into 3, saturation into 3, and hue into 18. The resulting HSV vector has $(18 \times 3 \times 3) + 4 = 166$ components. We have chosen these parameters after empirical tests on various possibilities.

For filtering aspects using entropy, we used $\alpha = \frac{1}{2}$ and $\beta = \frac{4}{5}$. The number of values that the aspect z can take is 10. This is set based on a pre-experiment indicating that when 100 images obtained as search results were clustered, the number of groups consisting of more than 2 elements are usually less than 10.

Criteria for judging convergence are that the difference is below 10^{-5}, or repeated the process over 300 times. Since the aspect model has local maxima, we do 5 trials starting from different initial values. We use the set of parameters with the highest log-likelihood.

4.3   Interface

Figure 3 is a system snapshot. Results of search engine's ranking and of our method are presented. Figure 4 is the mode for evaluation. The evaluator can click on the images and classify them into correct and incorrect ones, enabling evaluation with fewer loads.

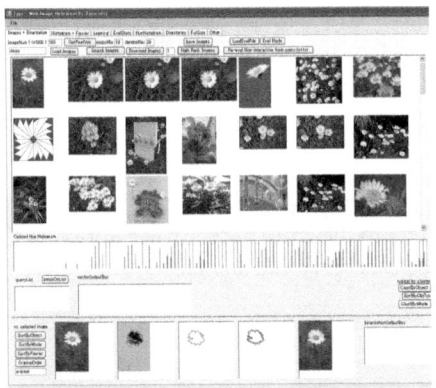 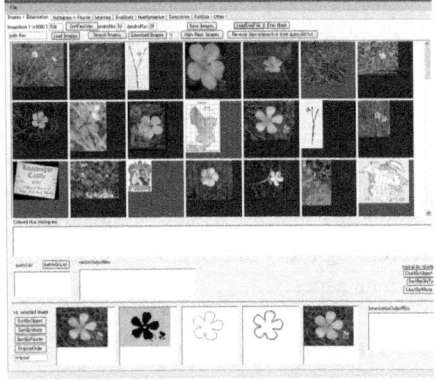

Fig. 3. Snapshot of "Typi" interface   **Fig. 4.** Interface for evaluation

5 Evaluation

5.1 Evaluation Method and Target

For performing experiments, we used 20 queries from a category "wildflowers". 100 images were collected for each query, resulting in 2,000 images in total. We evaluated averaged and individual top-k precisions. Table 1 illustrates the queries used in the experiment. In case of query consisting of two or more words (such as "scarlet pimpernel"), the query was put into double quotes, enabling phrase search.

Table 1. Queries used for experiments

dandelion, daisy, buttercup, iris, water arum, hawkweed, calliopsis, columbine, searocket, pale flax, harebell, wild radish, scarlet pimpernel, lady's slipper, baby blue eyes, chinese houses, ice plant, franciscan wallflower, clematis, forget me nots

Since objects referred by these queries have typical shapes and colors, it is easy to judge correct and incorrect images in the evaluation. In order to reduce processing time, the system uses scaled down images provided by the search engine, rather than the original images available on the Web. Therefore the maximum size of the images is 150×150 pixels.

5.2 Evaluation Criteria

In order to judge whether an image is typical or not, we need a unified criterion. As mentioned in Section 1, we defined "a typical image I of a query Q" by "the term Q is an appropriate label for the image I". Therefore, it is a necessary condition that the object specified by the query appears in the image.

If an image contains more than one object and it is not sufficient in knowing the object's visual characteristics, we considered it to be incorrect image. For example, in case of a query "dandelion", the image should have a dandelion of large enough size that the evaluator can identify it as a dandelion. In evaluation, we assume that the evaluator has enough knowledge on the object being queries.

5.3 Ranking Example

In this subsection, we exemplify the top-ranked images by the rankings of a search engine and the top aspect. In the figures 5-8, images are ordered from top-left, going right and then down. We added "x" beside images that were judged to be incorrect under our evaluation scheme.

Figures 5-6 are the results for a query "iris". Since iris also means a part of an eyeball, top ranked images by search engines contained these pictures (10th and 15th images). This is one weakness of the text-based ranking mechanism used by the image search engine. The images obtained by our method contain more correct images. The precision is higher than that of the search engine.

Fig. 5. Search engine ranking for "iris"

Fig. 6. Top aspect ranking for "iris"

We explain evaluation scheme described in subsection 5.2 using this example. The 4th image in Figure 5 is a collection of flowers. Attaching a label "iris" to it is not appropriate, so we consider it as an incorrect image. The 10th image has a flower bed, but since the label "iris" is not appropriate for it either, so we judge it to be an incorrect image.

Figures 7-8 are the results for a query "chinese houses". Since "chinese houses" can refer to Chinese buildings, the result of search engine contains many such images. On the other hand, the ranking by the top aspect contains flowers mainly, resulting in a high precision.

5.4   Top-k Precision by Categories

We performed experiments using 100 images obtained by an existing web image search engine. Using the search engine's ranking and our proposed method, we evaluated the top-k precisions. We used the result of Google Image Search for obtaining the search engine's ranking [9]. The result is illustrated in Figure 9.

Since the original set consists of 100 images for each query, top-100 precision is the ratio of correct images to the whole set. In this case, the value is 0.58. While the top-5 precision of the search engine's ranking is 70%, the ranking by the top aspect has 79%. For top-30 precision, the search engine's ranking has 66% and our method has 76%.

Fig. 7. Search engine ranking for "chinese houses"

Fig. 8. Top aspect ranking for "chinese houses"

The dotted line in Figure 9 indicates the top-k precision of the ranking obtained from the aspect with highest $p(z)$. The graph also shows that the proposed method has higher precision than the method that does not filter aspects by entropy in the way indicated by Expression 3.

One strong point of our approach is that it is based on the probability theory. For example, since all values are actually parameters of distributions, threshold values can be set with a probabilistic basis.

5.5 Processing Time

We have measured processing time necessary for the extraction of object region and the construction of feature vectors. We used Intel Core2 Duo 2.00GHz 2GB RAM for the experiment. Figures 10 - 11 are histograms indicating how many images required a certain amount of time for processing. The unit is in milliseconds. For the object region extraction, the mode is at around 40 milliseconds. There are some outliers, but they fall within triple the time of the mode. For the construction of feature vectors, the mode is at around 25 milliseconds. Outliers fall within double the time of the mode. The average time required for learning the aspect model for 100 images was 71.1 milliseconds.

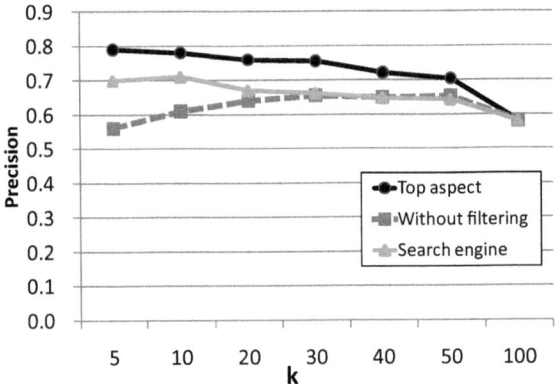

Fig. 9. Average top-k precisions for 20 queries

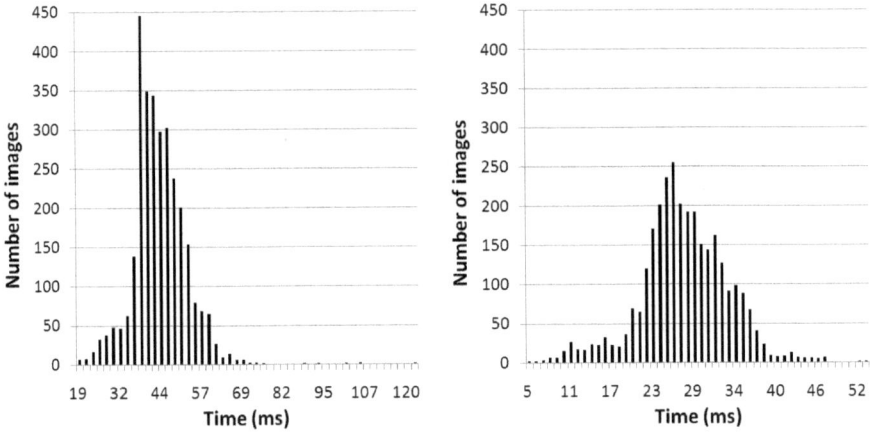

Fig. 10. Time for region extraction   **Fig. 11.** Time for feature vector construction

6   Conclusion

In this paper, we proposed a method that ranks web image search result in the order of typicality, by extracting the top aspect. Our present implementation uses color features only, but we plan to use more complex image features in future work.

In the evaluation, we used wildflowers, which are objects that have strong color characteristics. There are also objects that do not have strong color features. In the future work, we plan to use shapes and textures in addition to color, to deal with such objects.

Acknowledgments

This work was supported in part by a MEXT Grant-in-Aid for Strategic Formation of Research Infrastructure for Private University "Sharing of Research Resources by

Digitization and Utilization of Art and Cultural Materials" (Grant Number: S0991041) and MEXT Grant-in-Aid for Young Scientists (B) "Object Identification System using Web Image Collection and Machine Learning" (Leader: Taro Tezuka, Grant Number: 21700121).

References

1. Smith, J.R., Chang, S.: Visually searching the Web for content. IEEE Multimedia 4(3), 12–20 (1997)
2. Vogel, J., Schiele, B.: On Performance Characterization and Optimization for Image Retrieval. In: Proceedings of the 7th European Conference on Computer Vision, Copenhagen, Denmark, pp. 51–55 (2002)
3. Fei-Fei, L., Perona, P.: A Bayesian Hierarchical Model for Learning Natural Scene Categories. In: Proceedings of the IEEE International Conference on Computer Vision and Pattern Recognition 2005, San Diego, California, pp. 524–531 (2005)
4. Kennedy, L., Naaman, M.: Generating diverse and representative image search results for landmarks. In: Proceedings of the 17th International World Wide Web Conference, Beijing, China, pp. 297–306 (2008)
5. Wu, W., Yang, J.: Object fingerprints for content analysis with applications to street landmark localization. In: Proceedings of the ACM International Conference on Multimedia 2008, Vancouver, Canada, pp. 169–178 (2008)
6. Tezuka, T., Maeda, A.: A Hierarchical Model Approach for Measuring Typicality of Images. In: Proceedings of the 4th International Conference on Ubiquitous Information Management and Communication (ICUIMC 2010), Suwon, Korea (January 2010)
7. Hoffman, T., Puzicha, J., Jordan, M.I.: Learning from Dyadic Data. Advances in Neural Information Processing Systems 11, 466–472 (1999)
8. API for Google Image Search, http://www.codeproject.com/KB/IP/google_image_search_api.aspx
9. Google Image Search, http://images.google.com

Plus One or Minus One: A Method to Browse from an Object to Another Object by Adding or Deleting an Element

Kosetsu Tsukuda, Takehiro Yamamoto,
Satoshi Nakamura, and Katsumi Tanaka

Department of Social Informatics, Graduate School of Informatics,
Kyoto University
Yoshida-Honmachi, Sakyo, Kyoto 606-8501, Japan
{tsukuda,tyamamot,nakamura,tanaka}@dl.kuis.kyoto-u.ac.jp

Abstract. Recently, users can find various kinds of information in the
Web. When a user browses information, he/she sometimes wants to
browse more desirable information by adding/deleting few more. However, there is no service to browse such desirable information from current
information. We focused on such users' browse/search intention. Here,
each information consists of some elements such as ingredients, persons,
and places. We call the information "collective Web object". In this work,
we propose a method to enable users to browse from current collective
Web object to desirable collective Web object by adding one element into
it or deleting one element from it. In addition, we introduce the concept
of structural stability of collective Web object based on constructing
elements and apply our method to recipe search. We implemented a prototype system and performed experiments to evaluate the usefulness and
the applicability of our method.

Keywords: Collective Web object, structure, stability, typicality.

1 Introduction

Recently, the amount of Web information and the number of Web services have
been increasing rapidly. People can obtain various kinds of information, such as
recipes about pot-au-feu, sightseeing places in Spain, or information of Liga Es-
panola, by searching and browsing Web pages. When a user browses information,
he/she sometimes wants to browse more desirable information as follows.

- A user plans to cook a pot-au-feu and the user reached a recipe of simple
  pot-au-feu. However, the user was not satisfied with the ingredients used in
  it because this recipe is very typical and simple. So, the user wants to know
  what the user should add one ingredient to it and such desirable recipe.
- A user plans to visit Spain for sightseeing and reached a Web page that
  introduces a tour which contains Madrid, Barcelona, Bilbao and Andalusia.
  The user is interested in this tour and all places. However, the user does not
  have enough time to visit all places, and it is difficult for the user to decide

P. García Bringas et al. (Eds.): DEXA 2010, Part II, LNCS 6262, pp. 258–266, 2010.
© Springer-Verlag Berlin Heidelberg 2010

to remove one place from them. Then, the user wants to be recommended one place to remove by a system and to browse a page which contains all but the removed place.

There are many other situations except for the ones described above. There is, however, no conventional Web service that enables users to navigate from the current information to another desirable one by adding an element to the information that the user is interested in or deleting an element from the information that the user is not interested in. In addition, there is no service to support to find more typical information or more atypical information from current information. In this paper, we focused on such user's browsing/searching intention.

Here, the information consists of some elements such as ingredients, sightseeing places, and so on. In addition, each information is characterized by such elements. We define such information as "collective Web object". The objective of our work is to enable users to obtain a desired collective Web object by changing constructing elements.

In order to realize such system, we introduce the concept of structural stability of collective Web object and propose a method to calculate the stability by using the combination of constructing elements. The stability of a collective Web object is higher if it is typical. The stability of a collective object is lower if it is not typical. The user will be able to reach desirable Web object by checking the stability by using our system. There are many researches about recommendation of Web pages[2][3] and ingredients[4]. But in these researches, a user can not obtain eccentric items. Yoshida et al.[5] proposed a method to enable users to add or delete queries when they browse. However, they did not take care of the structure and its stability. In this study, we apply our method to recipe search. In the recipe search, a recipe is a collective Web object, and an ingredient is an element. We implemented the prototype system and showed the usefulness of our system by evaluation tests.

2   Collective Web Object

When a user searches about "sightseeing in Spain", there are countless sightseeing courses such as "visiting major cities in Spain", "visiting world heritages in Spain", "visiting little-known sightseeing spots in Spain". Regarding many sightseeing spots in Spain like "Madrid", "Barcelona", "Bilbao", and "Andalusia" as elements, sightseeing courses which are based on each viewpoint are changed by the combination of the elements. Moreover, the combination of each sightseeing spot is included in the same category of "sightseeing in Spain". Even if they belong to the same category, the nature of them changes by the combination of sightseeing spots. In the same way, about the recipes of "pot-au-feu", there are variations such as "Japanese style pot-au-feu", "Italian pot-au-feu", or "typical pot-au-feu", and each of them changes by the combination of ingredients such as carrot, potato, wiener, onion, tomato, or lotus root (Figure 1).

As mentioned above, an *object* that a user finally obtains such as "a typical sightseeing course in Spain" or "Japanese style pot-au-feu" belongs to a *category*

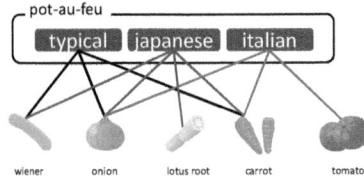
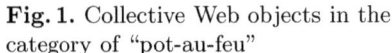

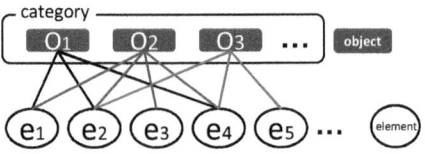

Fig. 1. Collective Web objects in the category of "pot-au-feu"

Fig. 2. Model of general collective Web objects

like "sightseeing in Spain" or "pot-au-feu", and changes its nature by the combination of *elements* like "Madrid" or "potato". We define such an object as a *collective Web object*. We show the model of this in Figure 2. That is, if we set a category, there are countless collective Web objects in the category, and each of them is constituted by some elements. Furthermore, the nature of a collective Web object is defined by the elements.

A collective Web object consists of two kinds of elements: Essential elements included in almost all collective Web objects in a category and not-essential elements included in some objects. For example, in the category of "pot-au-feu", there are many kinds of pot-au-feu such as "Japanese style pot-au-feu", "Italian pot-au-feu" or "typical pot-au-feu", but ingredients like "carrot" or "onion" are essential in any kind of pot-au-feu.

In this study, therefore, we define that a collective Web object has two layers which are constituted by elements. The two layers are a base layer and an upper layer of the base. We regard that essential elements constitute the base layer of collective Web objects and not-essential elements constitute the upper layer of them. In the category of pot-au-feu, for example, "carrot" or "onion" is the base elements because they are essential in any kinds of pot-au-feu, and "lotus root", "tomato", or "bacon" is those which constitute the upper layer because they are not necessary in all kinds of pot-au-feu.

We can say that a collective Web object that includes all base elements has the stable structure and a collective Web object that includes partial base elements has the unstable structure. Here, we classify the structure of objects into the following four types (see Figure 3).

(a) A collective Web object whose base and upper layer are stable: This object includes all base elements and typical elements of the upper layer. The structure is the most stable. In the example of pot-au-feu, "typical pot-au-feu" corresponds to this type.

(b) A collective Web object whose base is stable but upper layer is unstable: This object includes all base elements and atypical elements of the upper layer. In the example of pot-au-feu, "Japanese style pot-au-feu" and "Italian pot-au-feu" correspond to this type.

(c) A collective Web object whose base is unstable but upper layer is stable: This object includes few base elements but includes typical elements of

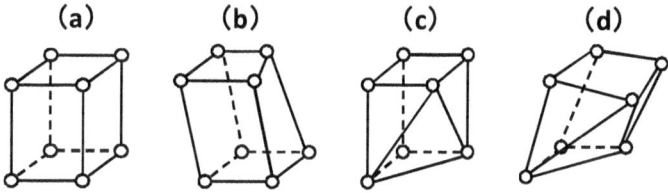

Fig. 3. Various structures of collective Web objects

the upper layer. In the example of pot-au-feu, "pot-au-feu of my home" that does not include an onion corresponds to this type.

(d) A collective Web object whose base and upper layer are unstable: This object includes few base elements and includes atypical elements of the upper layer. The structure is the most unstable. In the example of pot-au-feu, "creative Italian pot-au-feu" that does not include carrot or onion but includes a tomato and an apple corresponds to this type.

In this study, we seek base and upper elements and calculate structual stability by collecting collective Web objects in a category on a large scale.

When a user adds an element to a collective Web object or delete an element from it, the stability of the structure changes. So, in this study, we calculate the structure and the stability of an object a user is interested, and add/delete an element focusing on the stability after adding/deleting it. The user can change a collective Web object to a more typical or atypical one.

3   Constructing Structure and Calculating Stability of Collective Web Object Based on Typicality

The system recommends an addition/deletion element to/from a collective Web object a user is interested. The flow is as follows:

(1) The system seeks a category that includes the collective Web object the user is interested in.
(2) The system collects the set of collective Web objects included in the category seeked in (1).
(3) The system extracts elements that constitute each collective Web object.
(4) Based on (2) and (3), the system judges whether each element constitutes the base or upper layer, and seeks the most stable structure.
(5) Based on (2), (3), and (4), the system calculates the structure and stability of the collective Web object the user is interested.
(6) The system recommends an addition/deletion element to/from the object.

In this study, we especially focus on (4), (5), and (6), and propose a method.

3.1   Constituting Structure of Collective Web Object

In order to analyze the structure of a collective Web object, first we have to divide each element in a category into two groups: elements that constitute the base layer and elements that constitute the upper layer. Then we find the elements that constitute the most typical collective Web object in the category to calculate the stability of the collective Web object the user is interested.

To detect the base elements, the system chooses the element which is included in the most collective Web objects in the category, and classifies it to the set of base elements B. Then the system finds the element which has the highest co-occurrence with B, and adds it to B when the co-occurence is higher than a threshold. The system repeats this step until the co-occurence becomes less than the threshold. As for the most typical set of elements T in the category, the system finds them in the same way.

3.2   Calculating Stability of Collective Web Object

In this study, we use the difference between the set of elements included in the target collective Web object and those which constitute the typical collective Web object in the category, in order to calculate the stability of a collective Web object. We define the stability of the most typical one as 1. That is, the nearer to 1 the stability is, the more stable the structure is, and the smaller the stability is, the more unstable it is. We define the set of elements included a collective Web object o as $E_o = \{e_1, e_2, ..., e_l\}$ (l means the number of elements included in the target collective Web object). We define $Stability(E_o)$ as follows:

$$Stability(E_o) = affinity(E_o) - \delta \cdot difference(E_o) \tag{1}$$

where $affinity(E_o)$ means the value of affinity between all elements in E_o and $difference(E_o)$ means the value of difference between E_o and T. The way to calculate these values is as follows.

The value of $affinity(E_o)$ is that of co-occurrence between all elements in E_o, and can be calculated in the following equation:

$$affinity(E_o) = \frac{1}{|E_o|C_2} \sum_{e_i, e_j \in E_o} co_1(e_i, e_j) \tag{2}$$

where $co_1(e_i, e_j)$ is defined as follows:

$$co_1(e_i, e_j) = \begin{cases} 1 & \frac{|R(e_i) \cap R(e_j)|}{min(|R(e_i)|, |R(e_j)|)} > \theta \\ 0 & otherwise \end{cases} \tag{3}$$

That is, the more the number of elements with bad affinity in E_o is, the smaller the value of $affinity(E_o)$ becomes. where $R(e)$ means the set of collective Web objects that include e.

The value of $difference(E_o)$ can be calculated in the following equation:

$$difference(E_o) = \frac{1}{2} \left\{ (1 - \mu) \sum_{e_i \in E_o - T} \frac{1 - R'(e_i)}{|E_o - T|} + \mu \frac{\sum_{e_i \in T - E_o} R'(e_i)}{\sum_{e_i \in T} R'(e_i)} \right\} \tag{4}$$

where $R'(e)$ is defined as $R'(e) = \frac{|R(e)|}{|R(e_{max})|}$. In equation (4), the first member has the value from 0 to 1 when there are elements which are included in E_o but not in T. The value becomes large as there are many elements with low occurrence rate in $E_o - T$. That is, the more there are rare elements in E_o, the larger the gap with T becomes. The second member has the value from 0 to 1 when there are elements which are included in T but not in E_o. The value becomes large as there are many essential elements in $E_o - T$. That is, the fewer there are essential elements in E_o, the larger the gap with T becomes.

3.3 Extracting Addition/Deletion Element

We recommend addition elements that can change the current collective Web object into more typical or atypical one. That is, in the category which includes the collective Web object a user is interested, we recommend elements each of which changes a collective Web object so that it has the maximum or minimum stability after adding it. However, we remove the elements which change the stability to too low from additional candidate elements.

In addition, we recommend deletion elements from the collective Web object a user is interested. Each of elements is not the base elements and changes the collective Web object so that it has the maximum or minimum stability after deleting it.

4   Experiments and Discussions

When we regard a recipe as a collective Web object, the category name corresponds to the general name of the dish, and the elements correspond to ingredients. Therefore, in order to evaluate the utility of our method, we evaluated whether a user can change a recipe to more typical or atypical one by adding/deleting ingredients recommended by the system. In the experiments, we chose six categories from COOKPAD[1]: "carbonara", "neapolitan", "pork miso soup", "minestrone", "tomato salad", and "tuna salad". The number of recipes included in each category was 73, 59, 141, 76, 80, and 85.

4.1   Evaluating Addition/Deletion Ingredients

To evaluate the appropriateness of addition/deletion ingredients recommended by the system, we inspected to what extent they were accepted by users. First we extracted twenty recipes at random from each category. Then we showed subjects three recipes in each category. About following four items, the subjects chose a more suitable ingredient in each item recommended by the proposed method and the baseline method.

(1) An ingredient that puts close typical recipe by adding it.
(2) An ingredient that puts close atypical recipe by adding it.
(3) An ingredient that puts close typical recipe by deleting it.
(4) An ingredient that puts close atypical recipe by deleting it.

Table 1. Percentage of ingredients chosen by subjects in each category

		addition typical	addition atypical	deletion typical	deletion atypical
carbonara	proposed method	100.0%	55.6%	66.7%	0.0%
	baseline	100.0%	33.3%	44.4%	0.0%
	unchosen	0.0%	11.1%	11.1%	100.0%
neapolitan	proposed method	100.0%	44.4%	55.6%	44.4%
	baseline	88.9%	44.4%	55.6%	55.6%
	unchosen	0.0%	11.1%	22.2%	33.3%
pork miso soup	proposed method	88.9%	55.6%	77.8%	44.4%
	baseline	77.8%	44.4%	55.6%	55.6%
	unchosen	0.0%	0.0%	11.1%	44.4%
minestrone	proposed method	88.9%	77.8%	77.8%	22.2%
	baseline	88.9%	22.2%	22.2%	0.0%
	unchosen	0.0%	0.0%	0.0%	77.8%
tomato salad	proposed method	66.7%	44.4%	44.4%	22.2%
	baseline	55.6%	55.6%	55.6%	33.3%
	unchosen	0.0%	0.0%	22.2%	66.7%
tuna salad	proposed method	55.6%	33.3%	55.6%	22.2%
	baseline	77.8%	66.7%	66.7%	22.2%
	unchosen	0.0%	0.0%	33.3%	66.7%
average	proposed method	83.4%	51.9%	63.0%	25.9%
	baseline	81.5%	44.4%	50.0%	27.8%
	unchosen	0.0%	3.7%	16.7%	64.8%

To recommend ingredients by proposed method in each item, we first extracted the ingredients that can be added to/deleted from a recipe by the method mentioned in section 3.4. Then we recommended the ingredient which had the highest/lowest stability in the recipe after adding in the additional candidates to (1)/(2). Similarly, we recommended the ingredient which had the highest/lowest stability in the recipe after deleting in the deletion candidates to (3)/(4).

As a baseline method, we first extracted the ingredients that the number of recipes using was more than a threshold in the category which included the recipe to evaluate, and removed the ingredients used in the target recipe from them. Then, from the candidate ingredients, we recommended the ingredient that was used by the most recipes to (1), and the fewest to (2). Moreover, from the set of ingredients which were not the base ingredients in the category and were used in the target recipe, we recommended the ingredient that was used by the fewest recipes to (3), and the most to (4). When there were some candidates, we chose one ingredient at random and recommended it.

When the same ingredients were recommended by both the proposed method and the baseline method, a subject chose both of them if he/she thought they were appropriate, and did not choose when he/she thought they were not. The result of the experiment conducted by three subjects is shown in Table 1.

Overall, it often occurred that the same ingredients are recommended by both the proposed method and the baseline method in (1) and (4). This is because when there is an ingredient that is used in many recipes in the candidate ingredients to add/delete, the stability often became the highest/lowest by adding/deleting it. The ingredients which were recommended to put close an atypical recipe by deleting them were seldom chosen because the subjects thought that the recipe was not formed as a dish if they deleted the ingredient. Moreover, it is one of the reasons that generally users do not think they want to put close an atypical recipe by deleting an ingredient. In the category of "tomato salad" and "tuna salad", the different ingredients were often recommended as those which can put close a typical recipe by adding it in two methods, but there was not difference between two methods. That is because the degree of freedom of ingredients was high and the ingredients which can put close a typical recipe were greatly different between subjects.

In the category of "minestrone", there was large difference between two methods about the ingredient that can put close an atypical/typical recipe by adding/deleting it. The reason is that, in the category of minestrone, not only some kinds of meats and vegetables but also seasonings such as sugar, soy sauce, and ketchup are used. So, seasonings were recommended more often than pork miso soup. When it comes to seasonings, the balance of a recipe sometimes collapses by the combination of them, so it can be thought that the ingredients recommended by the proposed method were chosen more often.

5   Conclusion

In this paper, we proposed a method to browse from a collective Web object to another one by adding or deleting an element. To achieve it, we defined the structual stability of a collective Web object based on the typicality of constructing elements. We especially focused on a recipe as a collective Web object and enabled a user to obtain a desired recipe (Web object) by adding/deleting an ingredient. By experiments, we found that the proposed method works well in the category in which there is affinity between elements.

We plan to visualize a structure of a collective Web object. In this paper, we targeted a recipe because we can easily collect a category, Web objects, and elements from the Web. We are planning to target a general Web object in the future. To realize this, the system must extract elements a user is interested in the Web page and estimate which category the topic of the page belongs to. Moreover, the system also must collect the Web objects which belong to the category. This is very difficult to solve but very interesting issue.

Acknowledgement

This work was supported in part by the "Informatics Education and Research Center for Knowledge-Circulating Society" (Project Leader: Katsumi Tanaka, MEXT Global COE Program, Kyoto University), the MEXT Grant-in-Aid for Scientific Research on Priority Areas: "Cyber Infrastructure for the Information Explosion Era", "Contents Fusion and Seamless Search for Information Explosion" (Project Leader: Katsumi Tanaka, A01-00-02, Grant #:18049041), and the Grant-in-Aid for challenging Exploratory Research: "Research on Mobile Collaborative Search" (Satosh Nakamura, Grant #:22650018).

References

1. COOKPAD, `http://cookpad.com`
2. Niwa, S., Doi, T., Honiden, S.: Web Page Recommender System based on Folksonomy Mining. In: ITNG 2006, pp. 388–393 (2006)
3. Zhu, T., Greiner, R., Haeubl, G., Price, B., Jewell, K.: A Trustable Recommender System for Web Content. In: IUI 2005, pp. 83–88 (2005)
4. Shidochi, Y., Takahashi, T., Ide, I., Murase, H.: Finding replaceable materials in cooking recipe texts considering characteristic cooking actions. In: CEA 2009, pp. 9–14 (2009)
5. Yoshida, T., Nakamura, S., Tanaka, K.: WeBrowSearch: Toward Web Browser with Autonomous Search. In: Benatallah, B., Casati, F., Georgakopoulos, D., Bartolini, C., Sadiq, W., Godart, C. (eds.) WISE 2007. LNCS, vol. 4831, pp. 135–146. Springer, Heidelberg (2007)

Using Transactional Data from ERP Systems for Expert Finding

Lars K. Schunk[1] and Gao Cong[2]

[1] Dynaway A/S, Alfred Nobels Vej 21E, 9220 Aalborg Øst, Denmark
[2] School of Computer Engineering, Nanyang Technological University, Singapore
schunk@gmail.com, gaocong@ntu.edu.sg

Abstract. During the past decade, information retrieval techniques have been augmented in order to search for experts and not just documents. This is done by searching document collections for both query topics and associated experts. A typical approach assumes that expert candidates are authors of intranet documents, or that they engage in social writing activities on blogs or online forums. However, in many organizations, the actual experts, i.e., the people who work on problems in their day-to-day work, rarely engage in such writing activities. As an alternative, we turn to structured corporate data—transactions of working hours provided by an organization's ERP system—as a source of evidence for ranking experts. We design an expert finding system for such an enterprise and conclude that it is possible to utilize such transactional data, which is a result of required daily business processes, to provide a solid source of evidence for expert finding.

Keywords: Expert Finding, Information Retrieval, Enterprise Resource Planning, ERP.

1 Introduction

In many information-intensive organizations, one of the most prominent organizational challenges is the management of knowledge, and the ability to locate the appropriate experts for any given information need is essential. In small organizations, locating an expert may be a simple matter of asking around. However, in large organizations with several specialized departments, which may even be geographically scattered, this approach becomes infeasible.

A traditional solution is to maintain a database of employees and skills where each employee fills in his or her experience, skills, and fields of specialization [7]. This approach has some rather demotivating disadvantages. First, it requires a great deal of resources to maintain. Each employee will often be responsible for updating his or her profile in order to keep it current, which requires a very dedicated organization staff. Second, because of this human factor, the system is subject to imprecision, partly due to employees' over- or underrating of their skills, and partly due to a mismatch in granularity between profile descriptions, which tend to be mostly general, and queries, which tend to be specific.

P. García Bringas et al. (Eds.): DEXA 2010, Part II, LNCS 6262, pp. 267–276, 2010.

During the past decade, *expert finding systems* have emerged. The purpose of such a system is to automate the process of associating people with topics by analyzing information that is published within the organization, such as task descriptions, reports, and emails. From a user point of view, it typically is a variant of a traditional search engine; the user inputs a query topic, but instead of retrieving a set of relevant documents, it retrieves a set of relevant people— supposedly experts on the topic suggested by the query.

An important aspect of expert finding systems is the association between documents and expert candidates. Existing expert finding approaches typically assume variations on the following points: 1) candidates write textual content such as papers or forum posts; 2) if the name, email address, or other identifier of a candidate appears in such a document, then that document is related to the candidate. In short, they assume that expert candidates are creators of information.

However, such assumptions can be problematic. E.g., the overview of workshop [2] states, among other things, that when "looking at the chain of emails in which a request for expertise is passed from one person to another, it is also clear that mere candidate mentions do not necessarily imply expertise."

Furthermore, in many organizations, the actual experts, i.e., the people who work on problems in their day-to-day work, are too busy to engage in such writing activities. In these settings, expert finding approaches such as those mentioned above are of limited use. In [8], for instance, it is noted that less than 10% of a workforce studied were engaged in writing blogs. Though this figure has been increasing, it likely has a natural limit far below 100%. Nevertheless, employees are expert candidates even if they are not active creators of textual information, and it would be useful to capture their expertise to facilitate expert finding. In this paper, we therefore disregard the assumptions above, noting that certain types of documents are written without direct candidate annotation. Instead, we turn to other means of forming associations between documents and candidates.

We explore the potential of enterprise resource planning (ERP) systems, such as Microsoft Dynamics AX, as a source of expertise evidence. Many organizations maintain enormous amounts of transactional data in such ERP systems. To each record of transactional data it is possible to attach textual documents, but such documents often do not contain candidate information within their textual contents. Reasons for this include: 1) context is captured by the structured data that surrounds the documents in the ERP system; 2) documents are initially written without any specific people in mind, and then later different people are associated with the documents through the organization's various business processes.

Our enterprise setting is that of thy:data,[1] a Danish software consulting and development company within the area of business solutions primarily based on Microsoft's ERP system Dynamics AX.[2] thy:data employs Dynamics AX for project management, and this system acts as our case in this paper. We believe

---

[1] http://www.thydata.dk
[2] http://www.microsoft.com/dynamics/ax

that thy:data is representative of many companies dealing with similar consulting and development services.

We focus on transactions of hours worked on projects by candidates as our evidence of expertise. To our knowledge, this is the first work to exploit transactional data from ERP systems as expertise evidence for expert finding. We employ two methods for ranking candidates; one based on the classic TF-IDF ranking approach, and the other based on language modeling approaches [3].

One of the major benefits of our approach is that we do not depend on active knowledge-sharing from the employees because we leverage information that is created by required daily business processes within the organization, namely registration of working hours spent on activities. This is in contrast to other sources of evidence, such as corporate blogs and discussion forums, which require that expert candidates engage in knowledge-producing activities of a more voluntary nature, as is used in most previous work (e.g., [1,3,8,13]).

From a more pragmatic point of view, our work in this paper shows good opportunities for implementing expert finding systems that integrate directly with modern ERP systems. This could be done by developing an expert finder as an integral module of an ERP system. Such an effort would provide several benefits including: 1) direct access to all ERP data for establishing expertise evidence; 2) easy integration with human resource modules, task management modules, etc.; 3) easy access to expert finding for all daily users of the ERP system, thus boosting their productivity with minimal extra effort.

2   Related Work

The expert finding task introduced in the TREC[3] Enterprise Track [5] in 2005 has generated a lot of interest in expert finding, and a number of approaches have been developed. One of the central issues is how to establish a connection between candidates and topics. Usually, expertise evidence is found by analyzing documents that somehow relate to the expert candidates.

The P@NOPTIC system [6] presents a simple approach in which this connection is established by building an *expert index*, which consists of *employee documents*—one document is created for each employee. The employee document representing a given employee is the concatenated text of all intranet documents in which that employee's name occurs. With the employee documents in place, the system can match queries against the expert index using any standard information retrieval technique, and retrieve in ranked order the employee documents that match. With the one-to-one correspondence between employee documents and employees, it is easy to go from matching employee document to relevant employee.

Nearly all systems that took part in the 2005 and 2006 editions of the expert finding task at TREC adopted a language modeling approach (e.g., [9]), first introduced by Balog et al. [3]. Based on the idea in [10] of applying language

---

[3] Text REtrieval Conference: http://trec.nist.gov

modeling to information retrieval, Balog et al. rank candidates by their probabilities of generating a given query.

The association between candidates and documents can be refined in various ways. E.g., instead of capturing associations at document level, they may be estimated at snippet level around occurrences of candidate identifiers. Such use of proximity-based evidence has been found to achieve better precision in general. However, Balog et al. [4] note that "the mere co-occurrence of a person with a topic need not be an indication of expertise of that person on the topic." E.g., the name of a contact person may be mentioned in many documents and thus frequently co-occur with many topics, but the contact person is not necessarily an expert on the topics.

Recently, "Web 2.0 data," such as that provided by blogs and discussion forums, has been incorporated into expert finding systems based on the assumption that documents that generate much Web 2.0 data due to user activity are more interesting than documents that spawn only little activity, and that the users who exhibit activity around a document are related to the document [1, 8, 13]. As is also pointed out in [1], these systems cannot retrieve candidate experts who have never blogged or commented on another user's activity.

Apparently, there has not been much research that disregards the presumption that experts—in one way or another—are authors of document content. Furthermore, to our knowledge, neither has transactional context provided by ERP systems been the subject of expert finding research in the past. In this paper, we aim to capture the expertise of people who do not directly produce textual content by utilizing structured data from the organization's ERP system.

3 Corporate Transactional Data

We want to utilize transactional data to form document-candidate associations, and we will completely disregard the fact that candidate information may exist within the documents. Thus, document content is only used for matching query topics, just as in a traditional information retrieval system. When finding related experts, we want to rely entirely on transactional data from an ERP system.

Many companies maintain structured information about their employees, including how many hours they have spent working on different activities. If hours worked on activities can be allocated to documents, then this would seem a good starting point for establishing document-candidate associations necessary for expert search. Let us consider what it means when an employee has worked a large number of hours on some task. We can interpret this fact in at least two ways: 1) The task requires much work, and this employee has developed a valuable degree of expertise within the topics of the task. Thus, we assume that people who work on a task become knowledgeable on topics relevant to the task. This way we capture expertise even if the experts are too busy to write and publish their knowledge. We use this assumption in our basic models. 2) The employee has had difficulty completing the task because he is not an expert on the topic. The employee still may be a relevant person because he has spent time on the task and may have some valuable insight on the problems. However, if much of our evidence falls into this category, we may have to adjust the models. Therefore, after having presented the basic models, we propose an extension to take this potential shortcoming into account.

3.1 Enterprise Data Setting

At thy:data, they store *task descriptions* in a Dynamics AX-based task management system. These documents contain specifications for desired software functionality. Typically, a document describes one well-defined function of a larger system and represents a single unit of work that usually can be completed by one or two employees. The task descriptions are written by consultants, and afterwards a software developer must be assigned to the task. The employees who work on a task register their work hours in the system. Thus, the operational task management system contains both structured data (hours worked) and unstructured data (task descriptions).

All of thy:data's activities are organized into a *project hierarchy*. There is a number of top-level projects, and each project contains a number of sub-projects, which in turn can contain sub-projects themselves, etc. Each project can have a number of *activities* associated with it. An activity usually represents a certain well-defined task such as a software development task. The activities have various textual data associated with them. This includes descriptions of the task that the activity represents, as well as notes written by the people who have worked on the activity. The textual data can be stored directly in dedicated fields in the database or in elaborate documents outside of the database. Besides textual data, activities have transactions associated with them. Such a transaction represents a number of working hours that a certain employee has spent on the activity.

3.2 A Model of the Data

We can view the activities as constituting a central entity that ties together employees and documents. The employees are connected to the activities via transactions, and the activities are connected to the documents. We can view the *hours worked* measure on the transactions as an indicator of how strongly a given employee is associated with a given document.

These entities can be modeled as a weighted bipartite graph with two disjoint sets of vertices: a set of documents and a set of employees. Weighted edges between the two sets are derived from the activity and transaction entities. An example of this is shown in Figure 1. Here we see that document d_1 is associated with employee e_1 because e_1 has worked 60 hours on the activity to which d_1 is attached.

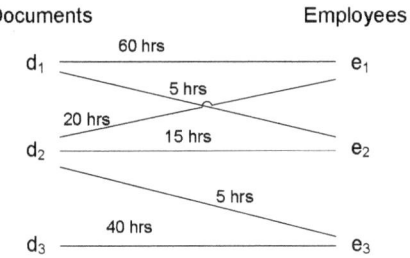

Fig. 1. Hours worked as a weighted bipartite graph with documents and employees

4   Method Design

We propose two basic models for ranking candidates based on the hours worked measure described in Section 3. The two models are variations on two well-known approaches, namely the TF-IDF approach and the document language modeling approach. Both of our models rely on document-candidate associations, so this aspect will be discussed first.

4.1   Document-Candidate Associations

A central part of the document language model described in [3] are the document-candidate associations, which provide a measure of how strongly a candidate is associated with a document. Given a collection of documents D and a collection of candidates C, to each pair (d, ca), where $d \in D$ and $ca \in C$, a non-negative association score $a(d, ca)$ must be assigned such that $a(d, ca_1) > a(d, ca_2)$ if candidate ca_1 is more strongly associated with document d than candidate ca_2.

Balog et al. provide an $a(d, ca)$ measure by using a named entity (NE) extraction procedure that matches identification information of candidate ca with document d. E.g., if ca's email address occurs in d, then $a(d, ca) > 0$.

We want to replace this $a(d, ca)$ measure with one that takes hours worked into account instead of using NE extraction. To do this, we formalize the model of the hours worked that was developed in an intuitive manner in Section 3.2. Let D be the set of documents, and C the set of candidates. Let $G = (V, E)$ be a bipartite graph where $V = D \cup C$ is the set of vertices and $E = \{\{d, ca\} \mid d \in D$ and $ca \in C\}$ is the set of edges. To each edge $\{d, ca\} \in E$ we assign weight $w(d, ca)$ as follows:

$$w(d, ca) = \text{total number of hours worked on } d \text{ by } ca \qquad (1)$$

Now we can introduce a simple document-candidate association measure $a(d, ca)$ in place of the one that was presented in [3]:

$$a(d, ca) = \begin{cases} w(d, ca) & \text{if } \{d, ca\} \in E \\ 0 & \text{otherwise} \end{cases} \qquad (2)$$

4.2   Modifying the TF-IDF Approach

In the classic TF-IDF approach used in traditional information retrieval, we calculate the relevance $r(d, Q)$ of a document d to a query Q as follows:

$$r(d, Q) = \sum_{t \in Q} TF(d, t) IDF(t) \tag{3}$$

where $TF(d, t)$ is the term frequency of term t in document d and $IDF(t)$ is the inverse document frequency of term t [11].

Now we modify this measure so that we can rank candidates. We want to establish a measure of the relevance $r(ca, Q)$ of a candidate ca to a query Q, much like the measure $r(d, Q)$ above. Suppose we have found the *one and only* document d that is relevant to the query Q. Then we can define the relevance of candidate ca to Q like this:

$$r(ca, Q) = a(d, ca) \tag{4}$$

The candidate who has worked the most hours on document d will be the top ranked candidate in terms of relevance to Q. However, many documents may be relevant to Q, some more than others. If we use the relevance of documents $r(d, Q)$ as weights on the document-candidate associations, we have the following definition of $r(ca, Q)$:

$$r(ca, Q) = \sum_{d \in D} r(d, Q) a(d, ca) \tag{5}$$

where D is the set of all documents in the collection. The more relevant documents that a candidate has worked on, the more likely it is that he is a relevant candidate, which we take into account by summing over all documents.

4.3   Modifying the Document Language Modeling Approach

In the document language modeling approach introduced in [3], the ranking of a candidate is calculated as the probability of that candidate generating a query. This is expressed as follows:

$$P(Q|ca) = \sum_{d \in D_Q} P(Q|d) P(d|ca) \tag{6}$$

where

$$P(Q|d) = \prod_{t \in Q} ((1 - \lambda) P_{mle}(t|d) + \lambda P(t|D)) \tag{7}$$

is the probability of the query Q given document d's language model by employing the Jelinek-Mercer smoothing method [12], and

$$P(d|ca) = \frac{a(d, ca)}{\sum_{d' \in D} a(d', ca)} \tag{8}$$

is the probability of document d given candidate ca. Put simply, given candidate ca, the document d with highest probability $P(d|ca)$ will be the document with which ca is most strongly associated.

We can tailor this approach to the present setting by simply replacing the document-candidate association measure $a(d, ca)$ with another one that relies on hours worked instead of the rule-based method using NE extraction. Having provided such a substitute in Section 4.1, it is straightforward to plug this into the document language model.

4.4   Extending the Basic Models

Now that the basic models are in place, we will consider some possible extensions. One could imagine some adjustments to the document-candidate association measure presented in Section 4.1. Consider the following scenario. Suppose that, given a query Q, the set D_Q are deemed relevant documents. Furthermore, candidates ca_1 and ca_2 are deemed relevant candidates. ca_1 has worked hundreds of hours on just one relevant document while ca_2 has worked moderate numbers of hours, say 10–20, on several relevant documents. Which candidate is more relevant? By Equation 5, ca_1 is likely to score higher because the hundreds of hours worked on one document will boost his score significantly. But the work on this *single* document may represent an exception. In contrast, if someone has worked moderate numbers of hours on *several* relevant documents, this may reflect the fact that he actually is an expert who completes his tasks quickly.

To take this into account, we introduce another measure, *document count*, denoted by $dc(ca, D_Q)$, which is the number of relevant documents in D_Q to which candidate ca is associated:

$$dc(ca, D_Q) = |\{d \mid d \in D_Q \text{ and } a(d, ca) > 0\}| \tag{9}$$

We can extend Equation 5 with the document count measure:

$$r(ca, Q) = dc(ca, D_Q) \sum_{d \in D_Q} r(d, Q) a(d, ca) \tag{10}$$

Likewise, we can extend Equation 6:

$$P(Q|ca) = \frac{dc(ca, D_Q)}{|D_Q|} \sum_{d \in D_Q} P(Q|d) P(d|ca) \tag{11}$$

where the document count has been converted to a probability.

Applying these to the example scenario above would boost ca_2's relevance score to reflect the fact that he has worked on several relevant tasks even if the total hours worked are less than those of ca_1.

5   Evaluation

As identified in workshop [2], a major challenge in expert finding research is obtaining real data for evaluation purposes. Currently, existing data sets are built from publicly accessible pages of organizational intranets [2]. These collections do not contain alternative non-textual sources of evidence such as transactional data from ERP systems used in our work, so they are not applicable to our evaluation purposes. The lack of both annotated resources and relevant queries renders the evaluation of our approaches particularly challenging.

For now, our prototype works on a document collection from the Aalborg department of thy:data, which consists of 1319 documents and 28 employees. To get an idea of the effectiveness of the system, we interviewed some key employees, posed a set of ten expertise queries, and noted the corresponding expert employees. We fed the queries to the system and observed how well the candidate experts ranked at different cutoff points. The results are shown in Table 1 for P@1, P@3, and P@5 (precision at rank 1, 3, and 5, respectively). For the language modeling approach, we set $\lambda = 0.5$ by following previous work [3]. We tested the TF-IDF approach both with and without the document count (DC) extension described in Section 4.4.

Generally, these results indicate that the system is fairly accurate with most results being above 70%. The table also shows a slight improvement when we apply the document count extension to the TF-IDF approach. It may seem surprising that the TF-IDF approach performs slightly better than the language modeling approach because the latter is generally considered superior. We note that this is not a full-scale evaluation, which to this point has not been feasible for this project. However, the primary objective of this work was to facilitate expert finding when expertise evidence is *not* available within documents. This objective has been fulfilled. The approaches taken are based on previous results that have performed well in full-scale empirical studies. This constitute the "subjective" aspect of this work. By augmenting these approaches to take hours worked into account, we have added an "objective" aspect. Given the assumption that hours worked are correlated with level of expertise, we can safely incorporate this measure when ranking the employees.

Table 1. P@1, P@3, and P@5 for TF-IDF and language modeling approaches

Approach	P@1	P@3	P@5
Lang. Model without DC	0.700	0.733	0.665
TF-IDF without DC	0.900	0.833	0.670
TF-IDF with DC	1.000	0.867	0.710

6 Conclusion and Future Work

We proposed an approach for capturing expertise of people when we cannot rely on the assumption that expert candidates are creators of information. Companies often maintain structured data that may indicate associations between documents and candidates. We utilized one such type of structured data, namely hours worked, so that we no longer need to rely on the assumption that candidate information exists *within* documents, an assumption that may not always be warranted. Our basic models are simple and apparently effective. Because registration of hours worked is a required daily business process, it becomes a solid source of evidence, which even captures expert candidates who may never have written a single document.

We discussed potential shortcomings of using hours worked as expertise evidence, and we proposed one extension to account for these. We can think of more ways to improve precision of expert finding by using structured ERP data. E.g., measures such as employee seniority or salary class may be used as weights on the hours worked, assuming that hours worked by experienced employees are more indicative of expertise than hours worked by newcomers. Furthermore, we may consider the recency of hours worked, assuming that recent transactions imply recently applied expertise, whereas very old transactions may be ignored. Finally, the approach presented here could be combined with traditional approaches in order to increase general expert finding effectiveness.

References

1. Amitay, E., Carmel, D., Golbandi, N., Har'El, N., Ofek-Koifman, S., Yogev, S.: Finding people and documents, using web 2.0 data. In: Proceedings of SIGIR Workshop: Future Challenges in Expertise Retrieval, pp. 1–5 (2008)
2. Balog, K.: The sigir 2008 workshop on future challenges in expertise retrieval (fcher). SIGIR Forum 42(2), 46–52 (2008)
3. Balog, K., Azzopardi, L., de Rijke, M.: Formal models for expert finding in enterprise corpora. In: Proceedings of ACM SIGIR, pp. 43–50 (2006)
4. Balog, K., de Rijke, M.: Non-local evidence for expert finding. In: Proceedings of CIKM, pp. 489–498 (2008)
5. Craswell, N., de Vries, A., Soboroff, I.: Overview of the trec 2005 enterprise track. In: The 14th Text REtrieval Conference Proceedings, TREC 2005 (2006)
6. Craswell, N., Hawking, D., Vercoustre, A.-M., Wilkins, P.: P@noptic expert: Searching for experts not just for documents. In: Poster Proceedings of AusWeb (2001)
7. Davenport, T., Prusak, L.: Working Knowledge: How Organizations Manage What They Know. Harvard Business School Press, Boston (1998)
8. Kolari, P., Finn, T., Lyons, K., Yesha, Y.: Expert search using internal corporate blogs. In: Proceedings of SIGIR Workshop: Future Challenges in Expertise Retrieval, pp. 7–10 (2008)
9. Petkova, D., Croft, W.B.: Hierarchical language models for expert finding in enterprise corpora. In: Proceedings of ICTAI, pp. 599–608 (2006)
10. Ponte, J., Croft, W.: A language modeling approach to information retrieval. In: Proceedings of ACM SIGIR, pp. 275–281 (1998)
11. Silberschatz, A., Korth, H., Sudarshan, S.: Database System Concepts, 5th edn. McGraw-Hill, New York (2006)
12. Zhai, C., Lafferty, J.: A study of smoothing methods for language models applied to ad hoc information retrieval. In: Proceedings of ACM SIGIR, pp. 334–342 (2001)
13. Zhou, Y., Cong, G., Cui, B., Jensen, C., Yao, J.: Routing questions to the right users in online communities. In: Proceedings of ICDE, pp. 700–711 (2009)

A Retrieval Method for Earth Science Data Based on Integrated Use of Wikipedia and Domain Ontology

Masashi Tatedoko[1], Toshiyuki Shimizu[1], Akinori Saito[2], and Masatoshi Yoshikawa[1]

[1] Graduate School of Informatics, Kyoto University
m.tatedoko@db.soc.i.kyoto-u.ac.jp, {tshimizu,yoshikawa}@i.kyoto-u.ac.jp
[2] Graduate School of Science, Kyoto University
saitoua@kugi.kyoto-u.ac.jp

Abstract. Due to the recent advancement in observation technologies and progress in information technologies, the total amount of earth science data has increased at an explosive pace. However, it is not easy to search and discover earth science data because earth science requires high degree of expertness. In this paper, we propose a retrieval method for earth science data which can be used by non-experts such as scientists from other field, or students interested in earth science. In order to retrieve relevant data sets from a query, which may not include technical terminologies, supplementing terms are extracted by utilizing knowledge bases; Wikipedia and domain ontology. We evaluated our method using actual earth science data. The data, the queries, and the relevance assessments for our experiments were made by the researchers of earth science. The results of our experiments show that our method has achieved good recall and precision.

1 Introduction

Due to the recent advancement in observation technologies and progress in information technologies, the total amount of data related to earth science has increased at an explosive pace. Furthermore, with advanced methods of data visualization and data publication, such data are expected to be shared and utilized among different disciplines and enhance earth science education.

There are two major demands in using earth science data. The first is the demand for a more sophisticated method of outreach by data providers. The second is the demand for better methods of utilizing data from different fields. However, these demands have not been satisfied, and only a few scientists has managed to utilize multi-discipline data sets. This is because each of the fields within earth science requires high degree of expertness, and even between the nearest fields, there exists a huge gap. This gap between each field invokes problems in data discovery and data utilization.

To overcome these problems, there are several approaches to create metadata for earth science data. Metadata portals, such as the GCMD(Global Change

P. García Bringas et al. (Eds.): DEXA 2010, Part II, LNCS 6262, pp. 277–284, 2010.

Master Directory)[1], are attempting to solve the data discovery problem by collecting metadata of diverse data sets, written in a standard form. However, data retrieval against such metadata requires knowledge of technical terminology, and it is difficult to retrieve data from another field of study. Some other approaches to solve the data utilization problem is the annotation of metadata against data formats. These approaches, such as netCDF(network Common Data Form)[2] and GrADS(Grid Analysis and Display System)[3], can define what a portion of data describes. However, it is difficult to utilize them among several data sets in an integrated manner since vocabularies are not uniformed.

So far, the data discovery problem, and the data utilization problem remain unresolved. In our research, however, we will address only the data discovery problem by proposing a data retrieval method which can be used by non-experts such as scientists from other fields, or students interested in earth science.

The goal of our research is to design a portal for retrieval of earth science data where non-experts can access to the desired data. We considered keyword searches are suitable for non-experts. Previous works such as GeoNetwork[4] and Gfdnavi[1] are assumed to be used by experts, thus are designed to achieve high precision against the queries. However, our data retrieval method is assumed to be used by non-experts, and is aimed to achieve high recall by supplementing expertise in queries by utilizing knowledge bases. Therefore, even if a technical terminology cannot be specified in a query, relevant data set can be discovered from keywords, such as "Aurora", "Great Hanshin Earthquake" etc.

There are three elements to specify a portion of earth science data: data set(*what*), spatial condition(*where*), and temporal condition(*when*). We think it is necessary to support specifying the data set, that is supporting *what* is most important for the promotion of utilization of the earth science data. Specifying the time and space where the phenomenon is likely to occur also requires expertise, but we think the necessity of such expertise is relatively low, because many visualization method can be used to overcome the difficulty. By utilizing our proposed method to support specifying the data sets, users can query by using limited vocabularies. For example, input query "Aurora" will return not only satellite images of aurora but also geomagnetism data.

When our proposal system receives a query q, it returns the subset D' of the target data sets D, which is associated with the query q. The query process starts from extracting the article corresponding to q using Wikipedia. The articles in Wikipedia covers wide variety of concepts. Therefore some bridge connecting common concept to academic concept are necessary to obtain the earth scientific data set d_i from D. We chose the domain ontology as the bridge data. The domain ontology is a kind of ontology which is specialized to an academic area. In this research, we used SWEET(Semantic Web for Earth and Environmental Terminology) as the domain ontology, which is a kind of geosciences ontology

---

[1] http://gcmd.nasa.gov/
[2] http://www.unidata.ucar.edu/software/netcdf/
[3] http://www.iges.org/grads/
[4] http://www.fao.org/geonetwork/

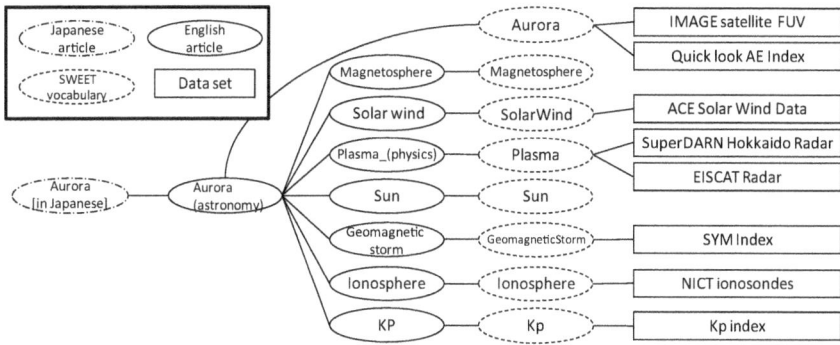

Fig. 1. The graph made from the query "Aurora"

made by NASA. We connect Wikipedia to the domain ontology so that the process is able to obtain the data set d_i associated with q. Finally, the extracted data sets are ranked by the relevance scores which are calculated using relevance degrees between articles of Wikipedia.

For example, Figure 1 shows the graph outputted as the result of query "Aurora". The graph is made based on Wikipedia and SWEET ontology. The start point of the graph is the article in Wikipedia. Our system chooses the article by comparing titles of Wikipedia articles with the given query "Aurora". The system firstly obtains the article "Aurora (astronomy)" by using language links of Japanese article "Aurora". Then it gathers articles linked by and high related to the article "Aurora (astronomy)". From these articles, it obtains vocabulary which is also in SWEET ontology and related with the query. The association of SWEET vocabulary and data sets are made by experts in advance. Therefore the system is able to obtain the data sets related to the given query. Finally, our system ranks the retrieved data sets by the relevance degrees from the start point.

2   Retrieval Method for Earth Science Data

We obtain data sets D_s relevant to an input query q by the following procedure:

1. Search for a Wikipedia article W_s corresponding to q. If the retrieved article is not in English, we retrieve the English edition of the article by using the interlanguage link of Wikipedia.
2. Retrieve articles $W_1...W_n$ relevant to W_s, and measure the relevance degrees between W_i and W_s for $i = 1...n$.
3. Do the string matching between article titles of $W_1...W_n$, W_s and the SWEET ontology. If the title of an article matches any term within the SWEET ontology, the article will be assumed to be an article related to earth science.
4. For every article related to earth science, we obtain the data sets D_s associated to the term in the SWEET ontology.

When the input query consists of multiple terms, the system collects relevant data sets for all terms. In the following subsections, we describe each steps of the procedure for obtaining D_s in detail.

2.1   Retrieval of a Source Wikipedia Article

We obtain a Wikipedia article W_s relevant to q by the following procedure. First, we compare q with every titles of Wikipedia articles by the string matching. If multiple results are found, we assume that the first article retrieved is the relevant article W_s. At this point, the system can also show the candidates to users, and users can choose some other article as W_s. If the chosen W_s is not an English article, we obtain the English edition of the article by using the interlanguage link.

Most earth science data and scientific articles are available in English. However, non-experts may want to search data by using their mother language, and our system supports other languages than English. Since the English edition of Wikipedia has the largest number of articles among all language editions of Wikipedia, we can consider that the English edition of Wikipedia comprehend large number of articles related to earth science. In fact, within the 2995 terms included in the SWEET ontologies, 2130 terms are available as articles of the English edition of Wikipedia. Within the 2130 articles, only 812 of them are available in Japanese. For these reasons, we translate any non-English query into English by using the interlanguage link, and search for relevant articles from the English edition of Wikipedia.

2.2   Retrieval of Relevant Wikipedia Articles

In order to retrieve articles $W_1...W_n$ relevant to W_s, we used $pfibf$[2], which is a method to calculate relevance between articles, proposed by Nakayama et al. We used the relevance degrees between articles as the factor to decide which data set is more relevant. We limited the maximum number of relevant articles to 300, and the relevance degrees take values between 0 and 1. In our implementation, we used Wikipedia API[5] which is external system to obtain relevance degrees between Wikipedia articles.

Because quantifying the strength of relevance between Wikipedia articles is a classical and active research area in Wikipedia mining, there are some researches other than $pfibf$ and it is possible to use them alternatively. In the study by Ito at el. [3], a method calculating relevance between concepts based co-occurrence of links in articles was proposed. The accuracy is as high as $pfibf$ and the algorithm complexity is lower than $pfibf$. Ollivier at el. [4] experimented for Wikipedia using *GreenMeasure* based on Markov chain.

In order to improve the recall, it is possible to retrieve further articles relevant to $W_1...W_n$. In our preliminary experiments, we found that retrieving relevant articles within two steps is effective.

---

[5] http://wikipedia-lab.org/en/index.php/Wikipedia_API

2.3   Integration of Wikipedia and Domain Ontology

Wikipedia has various articles that cover various domains from general event to specific domain. On the other hand, domain ontology is specialized in discipline. In this paper, we use SWEET ontology as domain ontology to extract articles related to earth science from Wikipedia. The matching between articles in Wikipedia and vocabularies in SWEET is detected by string matching.

We extract earth scientific articles from English version of Wikipedia. There are 2,995 terms registered in SWEET. Though the coverage of SWEET is so wide as to include mathematical terms as well as earth scientific terms, we consider these terms as earth scientific terms. We can get 2,130 articles from 6,723,158 articles in English version of Wikipedia as a counterpart of SWEET ontology.

2.4   Retrieval and Ranking of Relevant Data Sets

Every available data sets are associated to one terminology in the SWEET ontologies in advance. This association will be made by experts such as the data providers, and we assume that this association is legitimated. By using this association, we can associate data sets with q. According to the characteristics of the target data sets, we can choose any other domain ontologies to associate with the target data sets. In this research, we used Dagik(DAta-showcase system for Geoscience In Kml)[6] [5] data sets as the target data sets, which covers wide range of fields in earth science. Therefore, we use SWEET as the domain ontology due to its high generality in earth science.

We used the relevance degrees between Wikipedia articles to rank the retrieved data sets for q. If a data set can be retrieved via multiple articles, we summarize every relevance degrees on the route and used it as the score for the data set. The score S_{Di} of a data set D_i is calculated as below.

$$S_{Di} = S_{t_{Di}} + S_{w_{Di}} + S_{h1_{Di}} + S_{h2_{Di}}$$

1. When the title of D_i matches with q, $S_{t_{Di}} = k$. Otherwise, $S_{t_{Di}} = 0$.
2. When D_i can be obtained from the associated term of the domain ontology, which corresponds to an Wikipedia article W_s relevant to q, $S_{w_{Di}} = 1$. Otherwise, $S_{w_{Di}} = 0$.
3. When D_i can be obtained from the associated terms of the domain ontology, which corresponds to Wikipedia articles $W_1...W_i...W_n$, which are relevant to the article W_s by the relevance degrees $r_1...r_i...r_n$, and is accessible with one step, $S_{h1_{Di}} = \sum r_i$. Otherwise, $S_{h1_{Di}} = 0$.
4. When D_i can be obtained from the associated terms of the domain ontology, which corresponds to Wikipedia articles $W_1...W_j...W_m$, which are relevant to the article W_s by the relevance degrees $r_1...r_j...r_m$, and is accessible with two steps, $S_{h2_{Di}} = \sum r_j$. Otherwise, $S_{h2_{Di}} = 0$.

---

[6] http://dagik.org/index.html.en

3 Experiments

In this section, we evaluate our retrieval method. Our approach is a recall-oriented approach, which retrieves relevant data sets as much as possible. On the other hand, existing retrieval methods for earth science data are designed to be used by experts, and they are precision-oriented approaches.

In this experiment, we evaluated our approach using 58 data sets of Dagik[5]. Also, we used SWEET as a domain ontology. The mapping between data sets of Dagik and SWEET ontologies is many-to-one reference.

We show the retrieval results of the following typical 4 queries; q_1: Aurora, q_2: Aurora $\land$ Radar, q_3: Geomagnetism $\land$ Earthquake, q_4: JAXA. These queries and relevance assessments for each query are created by an expert of earth science. In the following, we describe the results and discussion for each query. Tables 1 to 4 show top 10 results obtained by our method for each query. The values in the Score columns are calculated by the procedure in Section 2.4. In the Rel. columns, we show the assessments by the expert; ⊚ means highly relevant, ○ means moderately relevant, and no symbol means irrelevant. Note that we used short names for some of the data set names.

q_1 **Aurora.** The intention of q_1 is to find data sets concerning Aurora which is the general term. For q_1, there are 3 highly relevant data sets and 16 moderately relevant data sets in Dagik. Highly relevant data sets are image data of aurora and data of "REIMEI" which is an aurora observation satellite. Solar wind, plasma, and electric density in the atmosphere are related to occurrence of an aurora closely. As a result, the data sets observing those are moderately relevant.

We obtained 30 data sets using our method. Among those 30 data sets, the top 10 results are shown in Table 1. We obtained data sets such as image data of aurora and solar wind, which we can not obtain without expert knowledge. Thus, our approach satisfies a request of query. Our approach was unable to obtain data sets of geomagnetism and ion density, which are moderately relevant. In this experiments, we used Wikipedia API to obtain the relevance degrees between Wikipedia articles. However, Wikipedia API does not cover whole Wikipedia articles, and sometimes it can not obtain the relevant articles. This problem will be resolved by implementing $pfibf$ in our local database.

q_2 **Aurora $\land$ Radar.** This query is intended to find out data sets about aurora which are observed by radar. For q_2, there are 2 highly relevant data sets and 6 moderately relevant data sets in Dagik. Highly relevant data sets are ground radars to observe ionospheric plasma. The occurrence of the aurora and ionospheric plasma is closely related. Moderately relevant data sets are ground-based observations and satellite data sets. We could obtain all relevant data sets using our method as shown in Table 2.

q_3 **geomagnetism $\land$ Earthquake.** This query is intended to retrieve the geomagnetic disturbance caused by the earthquake based on the expert knowledge. For q_3, there are 8 highly relevant data sets and 4 moderately relevant data sets in Dagik. There are some data sets related to aurora in the highly relevant data

Table 1. Top 10 results for q_1

Data set name	Score	Rel.
IMAGE satellite FUV	23.61	◎
Quick look AE index	23.61	○
REIMEI MAC	23.61	◎
THEMIS	23.61	◎
ACE solar wind	11.55	○
Solar wind: OMNI-2	11.55	○
Solar wind: OMNI	11.55	○
HINODE XRT	9.48	
SOHO EIT 19.5nm image	9.48	
SOHO LASCO C3 image	9.48	

Table 2. Top 10 results for q_2

Data set name	Score	Rel.
ESR radar	55.61	○
Ground Scatter	55.61	◎
Ionospheric Scatter	55.61	◎
KST UHF radar	55.61	○
NICT ionosondes	55.61	○
IMAGE satellite FUV	23.88	○
Quick look AE index	23.88	
REIMEI MAC	23.88	○
THEMIS	23.88	○
ACE solar wind	12.02	

Table 3. Top 10 results for q_3

Data set name	Score	Rel.
Earthquake by USGS	49.01	◎
Plate motion	9.76	○
HINODE XRT	5.36	
SOHO EIT 19.5nm image	5.36	
SOHO LASCO C3 image	5.36	
YOHKOH SXT Daily image	5.36	
ACE solar wind	3.63	
Solar wind: OMNI	3.63	
Solar wind: OMNI-2	3.63	
IMAGE satellite FUV	3.48	

Table 4. Top 10 results for q_4

Data set name	Score	Rel.
HINODE XRT	3.71	
SOHO EIT 19.5nm image	3.71	
SOHO LASCO C3 image	3.71	
YOHKOH SXT Daily image	3.71	
ACE solar wind	1.72	
Solar wind: OMNI	1.72	
Solar wind: OMNI-2	1.72	
ESR radar	0.85	
Ground Scatter	0.85	
Ionospheric Scatter	0.85	

sets because the occurrence of the earthquake leads to the occurrence of aurora. Moderately relevant data sets are geomagnetism and plate motion data sets.

We obtained 24 data sets using our method. Table 3 shows top 10 results obtained by our method. The data sets related to earthquake such as "Earthquake by USGS" and "Plate motion" obtained high score. We obtained only about 30% of relevant data sets. This is because we could not obtain geomagnetism data sets from the articles related to earthquake. This problem depends on the completeness of Wikipedia articles, and will be resolved by appending references to the geomagnetic field in the articles related to earthquakes.

q_4 **JAXA.** JAXA is an institute name, which is short for Japan Aerospace Exploration Agency, and this query is intended to find out data sets provided by JAXA. For q_4, there are 2 highly relevant data sets in Dagik. "REIMEI" and "GEOTAIL" are satellites managed by JAXA.

We obtained 30 data sets using our method. Table 4 shows top 10 results obtained by our method. We could not obtain relevant data sets in the top 10 results. The Wikipedia article of JAXA covers various topics and our method did not work well on such an article with diversity. In such a particular case that we want to search data sets using a condition about data provider, retrieval based on metadata will be more useful than our approach.

Our approach depends on the description of Wikipedia articles. There are about 6,700,000 articles in the English edition of Wikipedia, and there is a diversity in the quantity and the quality of the description for each article. As a solution for this problem, it will be possible to integrate Wikipedia and more specialized knowledge bases such as scientific papers.

4   Conclusions

In this paper, we proposed a retrieval method for earth science data. There were difficulties in discovery of earth science data because earth science requires high degree of expertness. To solve this problem, we proposed a scheme for retrieval of earth science data where non-experts can access to the desired data. Our data retrieval method is assumed to be used by non-experts, and is aimed to achieve high recall by supplementing expertise in queries by utilizing knowledge bases. We used SWEET as domain ontology to extract the articles related to earth science from Wikipedia. We implemented our approach and evaluated by using actual earth science data sets of Dagik. The experimental results show that our method achieved good recall and precision.

We are now planning to implement our method in the portal web site of Dagik. Developing user interfaces is one of our future works. Because the system can be used by non-experts, it can be used for educational purpose. It will also facilitate active research collaborations between different fields of earth science.

Acknowledgments. This work was partially supported by KAKENHI (21700107).

References

1. Isamoto, Y., Watanabe, C., Horinouchi, T., Nishizawa, S.: A cross-search mechanism using faceted navigation for Gfdnavi. IEICE Technical Report 109(186), 21–26 (2009) (in Japanese)
2. Nakayama, K., Hara, T., Nishio, S.: Wikipedia mining for an association web thesaurus construction. In: Benatallah, B., Casati, F., Georgakopoulos, D., Bartolini, C., Sadiq, W., Godart, C. (eds.) WISE 2007. LNCS, vol. 4831, pp. 322–334. Springer, Heidelberg (2007)
3. Ito, M., Nakayama, K., Hara, T., Nishio, S.: Association thesaurus construction methods based on link co-occurrence analysis for wikipedia. In: Proceedings of the 17th ACM Conference on Information and Knowledge Management, Napa Valley, California, USA, October 2008, pp. 817–826 (2008)
4. Ollivier, Y., Senellart, P.: Finding related pages using green measures: An illustration with wikipedia. In: Proceedings of the Twenty-Second AAAI Conference on Artificial Intelligence, Vancouver, British Columbia, Canada, July 2007, pp. 1427–1433 (2007)
5. Saito, A., Yoshida, D.: Dagik: A data-showcase system for the geospace. Data Science Journal 8, S92–S95 (2009)

Consistent Answers to Boolean Aggregate Queries under Aggregate Constraints

Sergio Flesca, Filippo Furfaro, and Francesco Parisi

DEIS - Università della Calabria
Via Bucci - 87036 Rende (CS) Italy
{flesca,furfaro,fparisi}@deis.unical.it

Abstract. A framework for computing consistent answers to boolean aggregate queries in numerical databases violating a given set of aggregate constraints is introduced. Both aggregate constraints and queries are aggregation expressions consisting of linear inequalities on aggregate-sum functions. In particular, our approach works for a specific but expressive form of aggregation expressions (called *steady aggregation expressions*) and computes consistent answers by solving Integer Linear Programming (ILP) problem instances.

1   Introduction

A great deal of attention has been recently devoted to the problem of extracting reliable information from data inconsistent w.r.t. integrity constraints. Most of the work dealing with this problem is based on the notions of *repair* and *consistent query answer* (CQA) introduced in [1]. A repair of an inconsistent database is a new database instance, on the same scheme as the original database, satisfying the given integrity constraints and which is "minimally" different from the original database instance (the minimality criterion aims at preserving the information in the original database as much as possible). Thus, an answer of a given query posed on an inconsistent database is said to be consistent if the same answer is obtained from every possible repair of the database. Based on the notion of CQA, several works investigated the problem of querying inconsistent data considering different classes of queries and constraints. Most of these works deal with "classical" integrity constraints (such as keys, foreign keys, functional dependencies). Indeed, these kinds of constraint often do not suffice to manage data consistency, as they cannot be used to define algebraic relations between stored values. In fact, this issue frequently occurs in several scenarios, such as scientific databases, statistical databases, and data warehouses, where numerical values in some tuples result from aggregating values in other tuples. In our previous work [13], we introduced a new form of integrity constraint, namely *aggregate constraint*, which enables conditions to be expressed on aggregate values extracted from the database. In that work, we characterized the computational complexity of the CQA problem for atomic ground queries in the presence of aggregate constraints. In this paper, we consider a more expressive form of queries (namely, *boolean aggregate queries*), consisting of linear inequalities on aggregate-sum functions. For these queries, we devise a strategy for computing consistent answers. Before presenting our contribution in detail, we provide an example

P. García Bringas et al. (Eds.): DEXA 2010, Part II, LNCS 6262, pp. 285–299, 2010.
© Springer-Verlag Berlin Heidelberg 2010

Table 1. A cash budget

	Year	Section	Subsection	Type	Value			Year	Section	Subsection	Type	Value
t_1	2008	Receipts	beginning cash	drv	50		t_{11}	2009	Receipts	beginning cash	drv	80
t_2	2008	Receipts	cash sales	det	100		t_{12}	2009	Receipts	cash sales	det	100
t_3	2008	Receipts	receivables	det	120		t_{13}	2009	Receipts	receivables	det	100
t_4	2008	Receipts	total cash receipts	aggr	250		t_{14}	2009	Receipts	total cash receipts	aggr	200
t_5	2008	Disbursements	payment of accounts	det	120		t_{15}	2009	Disbursements	payment of accounts	det	130
t_6	2008	Disbursements	capital expenditure	det	20		t_{16}	2009	Disbursements	capital expenditure	det	40
t_7	2008	Disbursements	long-term financing	det	80		t_{17}	2009	Disbursements	long-term financing	det	20
t_8	2008	Disbursements	total disbursements	aggr	220		t_{18}	2009	Disbursements	total disbursements	aggr	120
t_9	2008	Balance	net cash inflow	drv	30		t_{19}	2009	Balance	net cash inflow	drv	10
t_{10}	2008	Balance	ending cash balance	drv	80		t_{20}	2009	Balance	ending cash balance	drv	90
...		...		...	...							

describing the application scenario of our work, and make the reader acquainted with the notions of aggregate constraint and aggregate query.

Example 1. Table 1 represents a two-year *cash budget* of a company, that is a summary of cash flows (receipts, disbursements, etc.). Values '*det*', '*aggr*' and '*drv*' in column *Type* stand for *detail, aggregate* and *derived*, respectively. Specifically, an item is *aggregate* if it is obtained by aggregating items of type *detail* of the same section, whereas a *derived* item is an item whose value can be computed using the values of other items of any type and belonging to any section.

A cash budget must satisfy the following integrity constraints:

κ_1 : for each section and year, the sum of the values of all *detail* items must be equal to the value of the *aggregate* item of the same section and year;

κ_2 : for each year, the *net cash inflow* must be equal to the difference between *total cash receipts* and *total disbursements*;

κ_3 : for each year, the *ending cash balance* must be equal to the sum of the *beginning cash* and the *net cash inflow*.

Table 1 was acquired by means of an OCR (*Optical Character Recognition*) tool from a paper document. The original cash budget was consistent, but some symbol recognition errors occurred during the digitizing phase, as constraints κ_1, κ_2 and κ_3 are not satisfied on the acquired data:

i) for year 2008, in section *Receipts*, the aggregate value of *total cash receipts* is not equal to the sum of detail values of the same section: $100 + 120 \neq 250$;

ii) for year 2009, in section *Disbursements*, the aggregate value of *total disbursements* is not equal to the sum of detail values of the same section: $130 + 40 + 20 \neq 120$;

iii) for year 2009, the value of *net cash inflow* is not equal to the difference between *total cash receipts* and *total disbursements*: $10 \neq 200 - 120$.

The automatic acquisition of cash budget data from paper documents is often performed as the preliminary phase of the decision making process, as it yields data prone to be analyzed by suitable tools for discovering information of interest. In fact, the analysis of company cash budgets is extremely important for both stock and bond investors, since it allows potential liquidity problems to be detected, thus determining the company financial reliability as well as its ability to satisfy financial obligations. Examples of queries which can support this kind of analysis are:

q_1 : for each year, is the value of *net cash inflow* greater than a given threshold, say 20?

q_2 : for years 2008 and 2009, is the sum of *receivables* greater than *payment of accounts*?

q_3 : is the sum of incomings in *cash sales* for both years 2008 and 2009 sufficient to cover the expenses for *long-term financing* of year 2009?

Obviously, since the available data are inconsistent, the mere evaluation of these queries on the data may yield a wrong picture of the real world. Hence, in order to support any analysis task, it is mandatory to retrieve consistent answers to these queries even if the data are inconsistent. □

Besides the typical scenario of numerical inconsistencies due to OCR recognition errors, the problem of extracting reliable aggregate information from data inconsistent w.r.t. the same kind of constraints used in Example 1 arises in several scenarios, such as sensor networks, where errors in the collected data can be due to wrong sensor readings.

In this context, our contribution is a technique for computing consistent answers to *boolean aggregate queries* (such as queries q_1, q_2, q_3 of the example above) on data which are not consistent w.r.t. *aggregate constraints* (such as κ_1, κ_2, κ_3 in the same example). Our work builds on the strategy proposed in [13] for repairing data inconsistent w.r.t. a given set of aggregate constraints. According to this approach, reasonable repairs (namely, *card*-minimal repairs) are those obtained through sets of updates making the database consistent and having minimum cardinality. Correspondingly, consistent answers are those that can be obtained from every possible *card*-minimal repair. Specifically, our contribution consists in showing that consistent answers of boolean aggregate queries can be evaluated without computing every possible *card*-minimal repair, but only solving a pair of Integer Linear Programming (ILP) instances. Thus, our approach enables the computation of consistent query answers by means of well-known techniques for solving ILP problems.

Related Work. The notion of CQA adopted in this paper was introduced in [1]. In that paper, a query rewriting technique was proposed which enables the evaluation of consistent answers of quantifier-free conjunctive queries under binary universal constraints. This approach was extended in [16,15] to work on particular conjunctive queries with existential quantification under key constraints, and these results were further generalized in [21].

Starting from the notion of CQA of [1], several works investigated the problem of querying inconsistent data in the presence of more expressive classes of queries and constraints. The computational complexity of the CQA problem was investigated for union of conjunctive queries under functional and inclusion dependencies in [8], and for several classes of queries under denial constraints and inclusion dependencies in [10]. Several works [2,4,5,17] exploited logic-based frameworks for investigating the problem of computing repairs and evaluating consistent query answers. In [9] a framework for computing consistent query answers was presented, which supports projection-free relational algebra queries in the presence of denial constraints. The CQA problem for aggregate queries was first studied in [3] in the presence of functional dependencies, and further investigated in [15] for aggregate queries with grouping under key constraints.

All the above-cited approaches assume that tuple insertions and deletions are the basic primitives for repairing inconsistent data. In [6,7,14,20], repairs consisting of also value-update operations were considered. In particular, [20] was the first investigating the complexity of the CQA problem in a setting where the basic primitive for repairing data is the attribute-value update.

However, none of these works investigated the problem of computing consistent answers to aggregate queries in the presence of aggregate constraints. The first work addressing aggregate constraints on numerical data is [19], where the consistency problem of very general forms of aggregation was considered, but no issue related to data-repairing was investigated. The form of aggregate constraints considered in this paper was introduced in [13], where the complexity was characterized of several problems regarding the extraction of reliable information from inconsistent numerical data (i.e. repair existence, minimal repair checking, as well as consistent query answer for atomic ground queries). In [11], the architecture of a tool for acquiring and repairing numerical data inconsistent w.r.t. a restricted form of aggregate constraints was presented, along with a strategy for computing reasonable repairs, whereas in [12] the problem of computing reasonable repairs w.r.t. a set of both strong and weak aggregate constraints was addressed.

2   Preliminaries

We assume classical notions of database scheme, relation scheme, and relation instances. Relation schemes will be represented by means of sorted predicates of the form $R(A_1 : \Delta_1, \ldots, A_n : \Delta_n)$, where R is said to be the name of the relation scheme, $A_1, \ldots, A_n$ are attribute names (composing the set denoted as $\mathcal{A}_R$), and $\Delta_1, \ldots, \Delta_n$ are the corresponding domains. Each Δ_i can be either $\mathbb{S}$ (strings) or $\mathbb{Z}$ (signed integers). Attributes [resp. constants] defined over $\mathbb{Z}$ will be said to be *numerical attributes* [resp. *constants*]. The assumption that the numerical domain is $\mathbb{Z}$ yields no loss of generality, as our framework can be easily extended to the case of the rational attributes.

A tuple over a relation scheme $R(A_1 : \Delta_1, \ldots, A_n : \Delta_n)$ is a member of $\Delta_1 \times \cdots \times \Delta_n$. A relation instance of R is a set r of tuples over R. A database scheme $\mathcal{D}$ is a set of relation schemes, and a database instance D of $\mathcal{D}$ is a set of relation instances of the relation schemes of $\mathcal{D}$. Given a tuple t, the value of attribute A of t will be denoted as $t[A]$.

On each relation scheme R, a key constraint is assumed. Specifically, we denote as $\mathcal{K}_R$ the subset of $\mathcal{A}_R$ consisting of the names of the attributes which are a key for R. For instance, in "Cash budget" example, $\mathcal{K}_R = \{Year, Subsection\}$. Given a relation scheme R, we will denote the set of its numerical attributes representing measure data as $\mathcal{M}_R$ (namely, *Measure attributes*). That is, $\mathcal{M}_R$ specifies the set of attributes representing measure values, such as weights, lengths, prices, etc. For instance, in "Cash budget" example, $\mathcal{M}_R$ consists of attribute *Value* only. Throughout this paper, we assume that $\mathcal{K}_R \cap \mathcal{M}_R = \emptyset$, i.e., measure attributes of a relation scheme R are not used to identify tuples belonging to instances of R. Although this assumption leads to a loss of generality, it is acceptable from a practical point of view, since the situations excluded by this assumption are unlikely to occur often in real-life scenarios. As a matter of fact, it was used in [6,13] and it holds in our "Cash budget" example.

Given a boolean formula β consisting of comparison atoms of the form $X \diamond Y$, where X, Y are either attributes of a relation scheme R or constants, and $\diamond$ is a comparison operator in $\{=, \neq, \leq, \geq, <, >\}$, we say that a tuple t over R satisfies β (denoted as $t \models \beta$) if replacing the occurrences of each attribute A in β with $t[A]$ makes β true.

2.1 Aggregation Expressions

Aggregation expressions consist of linear inequalities on aggregate-sum functions defined on numerical databases, and will be used to express both aggregate constraints and boolean aggregate queries.

Given a relation scheme R, an *attribute expression* e on R is either a constant or a numerical attribute of R. Given an attribute expression e on R and a tuple t over R, we denote as $e(t)$ the value e, if e is a constant, or the value $t[e]$, if e is an attribute.

Given a relation scheme R and a sequence $\boldsymbol{y}$ of variables, an *aggregation function* $\chi(\boldsymbol{y})$ on R is a triplet $\langle R, e, \alpha(\boldsymbol{y}) \rangle$, where e is an *attribute expression* on R and $\alpha(\boldsymbol{y})$ is a (possibly empty) boolean combination of atomic comparisons of the form $X \diamond Y$, where X and Y are constants, attributes of R, or variables in $\boldsymbol{y}$, and $\diamond$ is a comparison operator in $\{=, \neq, \leq, \geq, <, >\}$. When empty, α will be denoted as $\bot$.

Given an aggregation function $\chi(\boldsymbol{y}) = \langle R, e, \alpha(\boldsymbol{y}) \rangle$ and a sequence $\boldsymbol{a}$ of constants with $|\boldsymbol{a}| = |\boldsymbol{y}|$, $\chi(\boldsymbol{a})$ maps every instance r of R to $\sum_{t \in r \wedge t \models \alpha(\boldsymbol{a})} e(t)$, where $\alpha(\boldsymbol{a})$ is the (ground) boolean combination of atomic comparisons obtained from $\alpha(\boldsymbol{y})$ by replacing each variable in $\boldsymbol{y}$ with the corresponding value in $\boldsymbol{a}$. We assume that, in the case that the set of tuples selected by the evaluation of an aggregation function χ is empty, χ evaluates to 0.

Example 2. The following aggregation functions are defined on the relational scheme *CashBudget(Year, Section, Subsection, Type, Value)* of Example 1:

$$\chi_1(x, y, z) = \langle CashBudget, Value, (Section=x \wedge Year=y \wedge Type=z \rangle$$
$$\chi_2(x, y) = \langle CashBudget, Value, (Subsection=x) \wedge Year=y \rangle$$
$$\chi_3(x) = \langle CashBudget, Value, (Subsection=x) \rangle$$

Basically, these aggregation functions are equivalent to the following SQL queries:

$\chi_1(x, y, z) =$	$\chi_2(x, y) =$	$\chi_3(x) =$
SELECT sum(Value)	SELECT sum(Value)	SELECT sum(Value)
FROM CashBudget	FROM CashBudget	FROM CashBudget
WHERE Section=x	WHERE Subsection=x	WHERE Subsection=x
AND Year=y	AND Year=y	
AND Type=z		

Function χ_1 returns the sum of *Value* of all the tuples having *Subection* x, *Year* y, and *Type* z. For instance, χ_1('Disbursements', 2008, 'det') returns $120 + 20 + 80 = 220$, and χ_1('Receipts', 2009, 'aggr') returns 200. In our running example, as *Year, Subsection* is a key for *CashBudget*, the sum returned by χ_2 is an attribute value of a single tuple. For instance, $\chi_2(2008, \text{'cash sales'}) = 100$, and $\chi_2(2008, \text{'receivables'}) = 120$. Aggregation function χ_3 returns the sum of *Value* of all the tuples having *Subection* x. Hence, $\chi_3(\text{'cash sales'})$ returns $100 + 100 = 200$.   □

Definition 1 (Aggregation expression). *Given a database scheme* $\mathcal{D}$*, an aggregation expression on* $\mathcal{D}$ *is of the form:* $\forall \boldsymbol{x} \ (\phi(\boldsymbol{x}) \implies \sum_{i=1}^n c_i \cdot \chi_i(\boldsymbol{y}_i) \leq K)$*, where:*

1. *n is a positive integer, and $c_1, \ldots, c_n, K$ are constants in $\mathbb{Z}$;*
2. *$\phi(\boldsymbol{x})$ is a (possibly empty) conjunction of atoms constructed from relation names, constants, and all the variables in $\boldsymbol{x}$;*
3. *each $\chi_i(\boldsymbol{y}_i)$ is an aggregation function, where $\boldsymbol{y}_i$ is a list of variables and constants, and every variable that occurs in $\boldsymbol{y}_i$ also occurs in $\boldsymbol{x}$.*

The assumption that the constants in aggregation expressions are integers yields no loss of generality, as the case of rational constants is easily reducible to this case. A database D satisfies an aggregation expression ae, denoted $D \models ae$, if, for all the substitutions θ of the variables in $\boldsymbol{x}$ with constants making $\phi(\theta(\boldsymbol{x}))$ *true*, the inequality $\sum_{i=1}^n c_i \cdot \chi_i(\theta(\boldsymbol{y}_i)) \leq K$ holds on D. For a set of aggregation expressions $\mathcal{E}$, D satisfies $\mathcal{E}$ (denoted as $D \models \mathcal{E}$) if $D \models ae$ for each $ae \in \mathcal{E}$.

Example 3. The following aggregation expression is defined on the relational scheme *CashBudget* of Example 1 and exploits aggregation function χ_1 defined in Example 2:

$$\forall y, x, z, v \quad CashBudget(y, x, z, v) \implies \chi_1(x, y, \text{'det'}) - \chi_1(x, y, \text{'aggr'}) \leq 0$$

An instance of the *CashBudget* scheme satisfies the condition expressed by this aggregation expression if, for each section and year, the sum of the values of all *detail* items is less than the value of the *aggregate* item of the same section and year.   □

We now introduce a restricted but expressive form of aggregation expressions, namely *steady aggregation expressions*. Let $R(A_1, \ldots, A_n)$ be a relation scheme and $R(x_1, \ldots, x_n)$ an atom, where each x_j is either a variable or a constant. For each $j \in [1..n]$, we say that the term x_j is *associated* with the attribute A_j. Moreover, we say that a variable x_i is a *measure variable* if it is associated with a measure attribute.

Definition 2 (Steady aggregation expression). *An aggregation expression ae on a given database scheme $\mathcal{D}$ is* steady *if:*
1. *for every aggregation function $\langle R, e, \alpha \rangle$ on the right-hand side of ae, no measure attribute occurs in α;*
2. *measure variables occur at most once in ae;*
3. *no constant occurring in the conjunction of atoms ϕ on the left-hand side of ae is associated with a measure attribute.*

Example 4. It is easy to see that the aggregation expression of Example 3 is steady. In fact: (i) the formula α of the aggregation function χ_1 on the right-hand side of the constraint contains no measure attribute; (ii) the unique measure variable v does not occur as argument of χ_1 and it does not appear in any other conjunct on the left-hand side of the constraint; (iii) no constant is associated with a measure attribute on the left-hand side of the constraint.   □

Observe that aggregate constraints enable equalities to be expressed as well, since an equality can be viewed as a pair of inequalities.

In the following, for the sake of brevity, equalities will be written explicitly. Moreover, universal quantification will be omitted and variables in ϕ which do not occur in any aggregation function will be replaced with the symbol '_'.

2.2 Aggregate Constraints

Aggregate constraints are defined by aggregate expressions as follows.

Definition 3 (Aggregate constraint). *An aggregate constraint on a database scheme $\mathcal{D}$ is an aggregation expression on $\mathcal{D}$, and it is* steady *if defined by a steady aggregation expression.*

Given a database D and a set of aggregate constraints $\mathcal{AC}$, we say that D is consistent [resp. inconsistent] w.r.t. $\mathcal{AC}$ if $D \models \mathcal{AC}$ [resp. $D \not\models \mathcal{AC}$].

Example 5. Constraints κ_1, κ_2 and κ_3 of Example 1 can be expressed as follows:

$\kappa_1 : CashBudget(y, x, \_, \_) \implies \chi_1(x, y, \text{'det'}) - \chi_1(x, y, \text{'aggr'}) = 0$

$\kappa_2 : CashBudget(y, \_, \_, \_) \implies \chi_2(\text{'net cash inflow'}, y) -$
$\qquad\qquad (\chi_2(\text{'total cash receipts'}, y) - \chi_2(\text{'total disbursements'}, y)) = 0$

$\kappa_3 : CashBudget(y, \_, \_, \_) \implies \chi_2(\text{'ending cash balance'}, y) -$
$\qquad\qquad (\chi_2(\text{'beginning cash'}, y) + \chi_2(\text{'net cash balance'}, y)) = 0$

Reasoning as in Example 4, it is easy to see that also κ_2 and κ_3 are steady. $\square$

2.3 Repairing Inconsistent Databases

Updates at attribute-level will be used as the basic primitives for repairing data.

Definition 4 (Atomic update). *Let $t = R(v_1, \ldots, v_n)$ be a tuple on the relation scheme $R(A_1 : \Delta_1, \ldots, A_n : \Delta_n)$. An atomic update on t is a triplet $< t, A_i, v_i' >$, where $A_i \in \mathcal{M}_R$ and v_i' is a value in Δ_i and $v_i' \neq v_i$.*

Update $u = < t, A_i, v_i' >$ replaces $t[A_i]$ with v_i', thus yielding the tuple $u(t) = R(v_1, \ldots, v_{i-1}, v_i', v_{i+1}, \ldots, v_n)$. We denote the pair $< tuple, attribute >$ updated by an atomic update u as $\lambda(u)$. For instance, performing $u = < t_2, Value, 130 >$ in the case of our running example, results in the tuple: $u(t_2) = CashBudget(\text{'Receipts'}, \text{'cash sales'}, \text{'det'}, 130)$, and $\lambda(u) = < t_2, Value >$.

Definition 5 (Consistent database update). *Let D be a database and $U = \{u_1, \ldots, u_n\}$ be a set of atomic updates on tuples of D. The set U is said to be a consistent database update iff $\forall j, k \in [1..n]$ if $j \neq k$ then $\lambda(u_j) \neq \lambda(u_k)$.*

Informally, a set of atomic updates U is a consistent database update iff for each pair of updates $u_1, u_2 \in U$, either u_1 and u_2 do not work on the same tuple, or they change different attributes of the same tuple. The set of pairs $< tuple, attribute >$ updated by a consistent database update U will be denoted as $\lambda(U) = \cup_{u_i \in U} \{\lambda(u_i)\}$. We will denote as $U(D)$ the database resulting from performing all the atomic updates in U on the database D.

Definition 6 (Repair). *Let $\mathcal{D}$ be a database scheme, $\mathcal{AC}$ a set of aggregate constraints on $\mathcal{D}$, and D an instance of $\mathcal{D}$ such that $D \not\models \mathcal{AC}$. A repair ρ for D is a consistent database update such that $\rho(D) \models \mathcal{AC}$.*

Example 6. A repair ρ_1 for *CashBudget* w.r.t. $\mathcal{AC} = \{\kappa_1, \kappa_2, \kappa_3\}$ consists of increasing attribute *Value* in the tuples t_2 and t_{18} up to 130 and 190 respectively, that is, $\rho_1 = \{< t_2, Value, 130 >, < t_{18}, Value, 190 >\}$.   □

In general, given a database D inconsistent w.r.t. a set of aggregate constraints $\mathcal{AC}$, different repairs can be performed on D yielding a new consistent database. Indeed, they may not be considered "reasonable" the same. For instance, if a repair exists for D changing only one value in one tuple of D, any repair updating all the values in the tuples of D can be reasonably disregarded. To evaluate whether a repair should be considered "relevant" or not, we use the ordering criterion stating that a repair ρ_1 precedes a repair ρ_2 if the number of changes issued by ρ_1 is less than ρ_2.

Example 7. Another repair for *CashBudget* is: $\rho' = \{< t_2, Value, 130 >, < t_{15}, Value, 120 >, < t_{16}, Value, 50 >, < t_{18}, Value, 190 >\}$. However ρ' consists of more atomic updates than ρ_1, where ρ_1 is the repair defined in Example 6.   □

Definition 7 (*Card*-minimal repair). *Let $\mathcal{D}$ be a database scheme, $\mathcal{AC}$ a set of aggregate constraints on $\mathcal{D}$, and D an instance of $\mathcal{D}$. A repair ρ for D w.r.t. $\mathcal{AC}$ is a card-minimal repair iff there is no repair ρ' for D w.r.t. $\mathcal{AC}$ such that $|\lambda(\rho')| < |\lambda(\rho)|$.*

Example 8. In our running example, the set of *card*-minimal repairs is $\{\rho_1, \rho_2\}$, where ρ_1 is the repair defined in Example 6 and $\rho_2 = \{ < t_3, Value, 150 >, < t_{18}, Value, 190 >\}$.   □

2.4   Boolean Aggregate Queries

Aggregate queries are defined starting from aggregate expressions as follows.

Definition 8 (Boolean Aggregate Query). *A boolean aggregate query q on a database scheme $\mathcal{D}$ is an aggregation expression on $\mathcal{D}$, and it is* steady *if it is defined by a steady aggregation expression. The answer of q over a database D instance of $\mathcal{D}$ is* true *if $D \models q$,* false *otherwise.*

Given an aggregate query q, we define $\neg q$ as the aggregation expression $\forall \boldsymbol{x} \; (\phi(\boldsymbol{x}) \implies \sum_{i=1}^{n} c_i \cdot \chi_i(\boldsymbol{y}_i) \geq K + 1)$.

Example 9. Queries q_1, q_2 and q_3 defined in Example 1 can be expressed as follows:

$q_1 : CashBudget(y, \_, \_, \_) \implies \chi_2(\text{'net cash inflow'}, y) \geq 20$
$q_2 : CashBudget(\_, \_, \_, \_) \implies \chi_3(\text{'receivables'}) \geq \chi_3(\text{'payment of accounts'})$
$q_3 : CashBudget(\_, \_, \_, \_) \implies \chi_3(\text{'cash sales'}) \geq \chi_2(\text{'long-term financing'}, 2009).$

It is easy to see that all the above queries are steady.   □

We adapt the notion of consistent query answer introduced in [1] to our setting.

Definition 9 (Consistent query answer). *Let $\mathcal{D}$ be a database scheme, D an instance of $\mathcal{D}$, $\mathcal{AC}$ a set of aggregate constraints on $\mathcal{D}$ and q an aggregate query over $\mathcal{D}$. The consistent query answer to q on D w.r.t. $\mathcal{AC}$, denoted as $\mathrm{CQA}_{\mathcal{D},\mathcal{AC},q}(D)$, is* true *iff, for each card-minimal repair ρ for D w.r.t. $\mathcal{AC}$, it holds that $\rho(D) \models q$.*

3 Query Answering

In this section, we define a strategy for computing consistent answers of boolean steady aggregate queries in the presence of steady aggregate constraints. Before describing our approach, we characterize the complexity of the CQA problem.

Theorem 1. *Let $\mathcal{D}$ be a fixed database scheme, $\mathcal{AC}$ a fixed set of aggregate constraints on $\mathcal{D}$, q a fixed aggregate query over D, and D an instance of $\mathcal{D}$. Deciding whether* $\text{CQA}_{\mathcal{D},\mathcal{AC},q}(D)$ *is* true *is $\Delta_2^p[\log n]$-complete, even if both $\mathcal{AC}$ and q are steady.*

Although steady aggregation expressions are less expressive than (general) aggregation expressions, Theorem 1 states that the consistent query answer problem is hard also when both the aggregate constraints and the query are steady. From a practical standpoint, the loss in expressiveness is not dramatic, as steady aggregate constraints (resp., steady aggregate queries) are expressive enough to model conditions ensuring data consistency (resp., to check conditions on aggregate data) in several real-life contexts. In fact, all the constraints and queries used in our running example are steady.

Our technique for computing consistent answers of steady aggregate queries under steady aggregate constraints is based on a translation of the CQA problem into the Integer Linear Programming (ILP) problem [18], thus allowing us to exploit well-known techniques for solving ILP problems to compute consistent query answers. Our technique exploits the restrictions imposed on steady aggregation expressions. As explained later, this approach does not work for constraints and queries defined by (general) aggregation expressions.

3.1 Expressing Steady Aggregation Expressions as a Set of Inequalities

Given a database scheme $\mathcal{D}$, a set of steady aggregation expressions $\mathcal{E}$ on $\mathcal{D}$, and an instance D of $\mathcal{D}$, we show how the triplet $\langle \mathcal{D}, \mathcal{E}, D \rangle$ can be translated into a set of linear inequalities $\mathcal{S}(\mathcal{D}, \mathcal{E}, D)$ such that every solution of $\mathcal{S}(\mathcal{D}, \mathcal{E}, D)$ corresponds to a database update U such that $U(D) \models \mathcal{E}$.

We first describe the translation for a single steady aggregation expression ae (which has the form: $\forall \boldsymbol{x} \; \phi(\boldsymbol{x}) \implies \sum_{i=1}^n c_i \cdot \chi_i(\boldsymbol{y}_i) \leq K$, where $\forall i \in [1..n]$, $\chi_i(\boldsymbol{y}) = \langle R_i, e_i, \alpha_i(\boldsymbol{y}_i) \rangle$). The translation results from the following three steps (for every relation scheme R_ℓ in $\mathcal{D}$, we will denote its instance in D as r_ℓ):

1) *Associating variables with pairs $\langle$tuple, measure attribute$\rangle$:*
   For each tuple t of a relation instance r_ℓ in D and measure attribute $A_j \in \mathcal{M}_{R_\ell}$, we create the integer variable z_{t,A_j};
2) *Translating each χ_i into sums of variables and constants:*
   Let $\Theta(ae)$ be the set of the ground substitutions of variables in $\boldsymbol{x}$ with constants such that $\forall \theta \in \Theta(ae) \; \phi(\theta \boldsymbol{x})$ is *true* on D. For every ground substitution $\theta \in \Theta(ae)$ and every χ_i, we denote as $T_{\chi_i}(\theta)$ the set of tuples involved in the evaluation of χ_i w.r.t. θ, that is $T_{\chi_i}(\theta) = \{t : t \in r_i \wedge t \models \alpha_i(\theta \boldsymbol{y}_i)\}$, where r_i is the instance in D of the relation scheme R_i in χ_i.

Then, for every ground substitution $\theta \in \Theta(ae)$, we define the translation of χ_i w.r.t. θ as:

$$P(\chi_i, \theta) = \begin{cases} \sum_{t \in T_{\chi_i}(\theta)} z_{t,A_j} & \text{if } e_i \text{ is the measure attribute } A_j; \\ \sum_{t \in T_{\chi_i}(\theta)} e_i(t) & \text{otherwise.} \end{cases}$$

3) *Translating ae into a set of linear inequalities:*
   The expression ae is translated into the set $S(\mathcal{D}, ae, D)$ of linear inequalities containing, for every ground substitution $\theta \in \Theta(ae)$, the inequality $\sum_{i=1}^{n} c_i \cdot P(\chi_i, \theta) \leq K$.

The system of linear inequalities $S(\mathcal{D}, \mathcal{E}, D)$ (which takes into account all the aggregation expressions in $\mathcal{E}$) is then defined as $S(\mathcal{D}, \mathcal{E}, D) = \cup_{ae \in \mathcal{E}} S(\mathcal{D}, ae, D)$.

For the sake of simplicity, in the following we assume that the pairs $\langle t, A_j \rangle$, where A_j is the name of a measure attribute of tuple t, are associated with distinct integer indexes (the set of these indexes will be denoted as $\mathcal{I}$). Therefore, being i the integer associated with the pair $\langle t, A_j \rangle$, the variable z_{t,A_j} will be denoted as z_i.

Example 10. In "Cash budget" example, we associate each pair $\langle t_i, Value \rangle$ with the integer i, thus $\mathcal{I} = \{1, \ldots, 20\}$. The translation of the aggregation expressions of Example 4, which are the constraints κ_1, κ_2, κ_3 of our running example, is the following (we explicitly write equalities instead of inequalities):

$$\begin{cases} z_2 + z_3 = z_4; & z_5 + z_6 + z_7 = z_8; & z_{12} + z_{13} = z_{14}; & z_{15} + z_{16} + z_{17} = z_{18}; \\ z_4 - z_8 = z_9; & z_{14} - z_{18} = z_{19}; & z_1 + z_9 = z_{10}; & z_{11} + z_{19} = z_{20}. \end{cases}$$

□

3.2   Computing Consistent Answers to Boolean Aggregate Queries

Our approach for computing consistent query answers is based on the resolution of specific ILP problems, defined starting from the ILP problem introduced below.

Definition 10 $(\mathcal{ILP}(\mathcal{D}, \mathcal{E}, D))$. *Given a database scheme $\mathcal{D}$, a set $\mathcal{E}$ of steady aggregation expressions on $\mathcal{D}$, and an instance D of $\mathcal{D}$, $\mathcal{ILP}(\mathcal{D}, \mathcal{E}, D)$ is an ILP of the form:*

$$\begin{cases} \mathbf{A} \times \mathbf{z} \leq \mathbf{B}; \\ w_i = z_i - v_i & \forall i \in \mathcal{I}; \\ z_i - M \leq 0; & -z_i - M \leq 0; & \forall i \in \mathcal{I} \\ w_i - M\delta_i \leq 0; & -w_i - M\delta_i \leq 0; \forall i \in \mathcal{I}; \\ z_i, w_i \in \mathbb{Z}; & \delta_i \in \{0, 1\}; & \forall i \in \mathcal{I}; \end{cases}$$

where:
(i) *$\mathbf{A} \times \mathbf{z} \leq \mathbf{B}$ is the set of inequalities $S(\mathcal{D}, \mathcal{E}, D)$ ($\mathbf{z}$ is the vector of variables z_i with $i \in \mathcal{I}$);*
(ii) *for each $i \in \mathcal{I}$, v_i is the database value corresponding to the variable z_i, that is, if z_i is associated with the pair $\langle t, A_j \rangle$, then $v_i = t[A_j]$;*
(iii) *$M = n \cdot (ma)^{2m+1}$, where: a is the maximum among the modules of the coefficients in $\mathbf{A}$ and of the values v_i, and $m = |\mathcal{I}| + r$, and $n = 2 \cdot |\mathcal{I}| + r$, where r is the number of rows of $\mathbf{A}$.*

Basically, for every solution of $\mathcal{ILP}(\mathcal{D}, \mathcal{E}, D)$, the variables z_i are assigned values which satisfy $\mathbf{A} \times z \leq \mathbf{B}$ (that is, $\mathcal{S}(\mathcal{D}, \mathcal{E}, D)$). Hence, the variables z_i of a solution of $\mathcal{ILP}(\mathcal{D}, \mathcal{E}, D)$ take values which, once assigned to the corresponding pairs *tuple, attribute*, make the database satisfy the aggregate expressions $\mathcal{E}$.

In the definition above, each variable w_i represents the difference between the variable z_i associated with a pair $\langle t, A_j \rangle$ and the original database value $v_i = t[A_j]$. The constant M is introduced for a twofold objective: considering solutions of the first two inequalities with polynomial size[1] w.r.t. the database size, and building a mechanism for counting the number of variables z_i which are assigned a value different from the original value of the corresponding pair *tuple, attribute*. The value of M derives from a well-known general result shown in [18] regarding the existence of bounded solutions of systems of linear equalities. In our case, this result implies that, if the first two (in)equalities of $\mathcal{ILP}(\mathcal{D}, \mathcal{E}, D)$ have at least one solution, then they admit at least one solution where (absolute) values are less than M. Hence, the inequalities of $\mathcal{ILP}(\mathcal{D}, \mathcal{E}, D)$ where M occurs entail that:

- $\mathcal{ILP}(\mathcal{D}, \mathcal{E}, D)$ has solution iff the first two inequalities have a solution. In particular, each solution of $\mathcal{ILP}(\mathcal{D}, \mathcal{E}, D)$ can be obtained by taking any solution of the first two inequalities with values less than M and then properly adjusting each δ_i;
- every solution of $\mathcal{ILP}(\mathcal{D}, \mathcal{E}, D)$ is of polynomial size w.r.t. the size of the database. In fact, solutions of the first two inequalities with values larger than M do not correspond to solutions of $\mathcal{ILP}(\mathcal{D}, \mathcal{E}, D)$, as, if $|w_i| > M$, there is no way of choosing δ_i to satisfy both $w_i - M\delta_i \leq 0$ and $-w_i - M\delta_i \leq 0$.
- for every solution of $\mathcal{ILP}(\mathcal{D}, \mathcal{E}, D)$, the sum of the values assigned to variables δ_i is an upper bound on the number of variables z_i different from the corresponding v_i. In fact, if w_i has a value different from 0 (meaning that z_i has a value different from the "original" v_i) then δ_i is assigned 1. It is easy to see that the vice versa does not hold, thus this sum does not represent the exact number of variables z_i different from the original values.

For any solution s of $\mathcal{ILP}(\mathcal{D}, \mathcal{E}, D)$, the value taken by variable z in s will be denoted as $s[z]$. The above-mentioned properties of $\mathcal{ILP}(\mathcal{D}, \mathcal{E}, D)$ are stated in the theorem below.

Theorem 2. *Every solution s of an instance of $\mathcal{ILP}(\mathcal{D}, \mathcal{E}, D)$ one-to-one corresponds to a consistent database update U for D such that:*

(i) *for each z_i associated with the pair $\langle t, A_j \rangle$ and such that $s[z_i] \neq t[A_j]$, U contains the atomic update $\langle t, A_i, s[z_i] \rangle$;*
(ii) *$\forall i \in \mathcal{I}$, it is the case that $-M \leq s[z_i] \leq M$;*
(iii) *$U(D) \models \mathcal{E}$;*
(iv) *$|U| \leq \sum_{i \in \mathcal{I}} s[\delta_i]$.*

Thus, every solution of $\mathcal{ILP}(\mathcal{D}, \mathcal{E}, D)$ corresponds to a consistent database update U making D satisfy $\mathcal{E}$. We point out that removing the steadiness restriction from aggregation expressions may result in breaking this correspondence. Intuitively, this derives

---

[1] Observe that the size of M is polynomial in the size of the database, as it is bounded by $\log n + (2 \cdot m + 1) \cdot \log(ma)$.

from the fact that, in the presence of non-steady aggregation expressions, applying to D the set of updates corresponding to a solution of $\mathcal{ILP}(\mathcal{D}, \mathcal{E}, D)$ may trigger violations of some aggregation expressions in $\mathcal{E}$ which were not encoded in $\mathcal{ILP}(\mathcal{D}, \mathcal{E}, D)$.

It is worth nothing that, since aggregate constraints are aggregation expressions, the above result can be also read as follows: *given a set of aggregate constraints $\mathcal{AC}$ on $\mathcal{D}$ and a database D instance of $\mathcal{D}$, the consistent database update defined by a solution of $\mathcal{ILP}(\mathcal{D}, \mathcal{AC}, D)$ is a repair for D w.r.t. $\mathcal{AC}$ whose cardinality is bounded by $\sum_{i \in \mathcal{I}} \delta_i$ and whose atomic updates assign values bounded by M.* The following corollary strengthens this result, as it states that the minimum of $\sum_{i \in \mathcal{I}} \delta_i$ among all the solutions of $\mathcal{ILP}(\mathcal{D}, \mathcal{AC}, D)$ is the cardinality of any *card*-minimal repair.

Corollary 1. *Let $\mathcal{D}$ be a database scheme, $\mathcal{AC}$ a set of steady aggregate constraints on $\mathcal{D}$, and D an instance of $\mathcal{D}$. A repair for D w.r.t. $\mathcal{AC}$ exists iff $\mathcal{ILP}(\mathcal{D}, \mathcal{AC}, D)$ has at least one solution, and the optimal value of the optimization problem:*

$$\mathcal{OPT}(\mathcal{D}, \mathcal{AC}, D) := \text{minimize} \sum_{i \in \mathcal{I}} \delta_i \text{ subject to } \mathcal{ILP}(\mathcal{D}, \mathcal{AC}, D)$$

coincides with the cardinality of any card-*minimal repair for D w.r.t. $\mathcal{AC}$.*

We now show how the solution of $\mathcal{OPT}(\mathcal{D}, \mathcal{AC}, D)$ can be exploited to compute consistent query answers. Let q be an aggregate query over $\mathcal{D}$. Consider the ILP problem $\mathcal{CQAP}(\mathcal{D}, \mathcal{AC}, q, D)$ obtained by assembling the inequalities in $\mathcal{ILP}(\mathcal{D}, \mathcal{AC} \cup \{\neg q\}, D)$ with the equality $\overline{\lambda} = \sum_{i \in \mathcal{I}} \delta_i$, where $\overline{\lambda}$ is the value returned by $\mathcal{OPT}(\mathcal{D}, \mathcal{AC}, D)$. The following theorem states computing the consistent query answer of q is equivalent to deciding whether $\mathcal{CQAP}(\mathcal{D}, \mathcal{AC}, q, D)$ has solution.

Theorem 3. *Let $\mathcal{D}$ be a database scheme, $\mathcal{AC}$ a set of steady aggregate constraints on $\mathcal{D}$, q a steady aggregate query on $\mathcal{D}$, and D an instance of $\mathcal{D}$. The consistent query answer to q over D w.r.t. $\mathcal{AC}$ is true iff $\mathcal{CQAP}(\mathcal{D}, \mathcal{AC}, q, D)$ has no solution.*

The result above derives from these facts:

- the solutions of $\mathcal{ILP}(\mathcal{D}, \mathcal{AC} \cup \{\neg q\}, D)$ correspond to repairs for D w.r.t. $\mathcal{AC}$ such that q evaluates to *false* on the repaired databases;
- adding the equality $\overline{\lambda} = \sum_{i \in \mathcal{I}} \delta_i$ means considering *card*-minimal repairs only.

Hence, $\mathcal{CQAP}(\mathcal{D}, \mathcal{AC}, q, D)$ has a solution iff there is a *card*-minimal repair for D w.r.t. $\mathcal{AC}$ such that q evaluates to *false* on the repaired databases, i.e., the consistent answer to q is *false*.

Example 11. The ILP problem $\mathcal{CQAP}(\mathcal{D}, \mathcal{AC}, q, D)$ for our running example is shown in Fig. 1. Herein, $\mathcal{AC} = \{\kappa_1, \kappa_2, \kappa_3\}$, q is the aggregate query q_3, $\overline{\lambda} = 2$ (as shown in Example 8), and the inequality $z_2 + z_{12} \leq z_{17}$ encodes $\neg q_3$, whereas the other equalities on variables z_i encode the aggregate constraints κ_1, κ_2, and κ_3. It is easy to see that there is no solution for the problem shown in Fig. 1. Hence, according Theorem 3, the consistent answer of q_3 is *true*.   □

$$\begin{cases} 2 = \sum_{i=\in\mathcal{I}} \delta_i & w_3 = z_3 - 120 & w_{15} = z_{15} - 130 \\ z_2 + z_3 = z_4 & w_4 = z_4 - 250 & w_{16} = z_{16} - 40 \\ z_5 + z_6 + z_7 = z_8 & w_5 = z_5 - 120 & w_{17} = z_{17} - 20 \\ z_{12} + z_{13} = z_{14} & w_6 = z_6 - 20 & w_{18} = z_{18} - 120 \\ z_{15} + z_{16} + z_{17} = z_{18} & w_7 = z_7 - 80 & w_{19} = z_{19} - 10 \\ z_4 - z_8 = z_9 & w_8 = z_8 - 220 & w_{20} = z_{20} - 90 \\ z_{14} - z_{18} = z_{19} & w_9 = z_9 - 30 & w_i - M\delta_i \leq 0 \quad \forall i \in \mathcal{I} \\ z_1 + z_9 = z_{10} & w_{10} = z_{10} - 80 & -w_i - M\delta_i \leq 0 \quad \forall i \in \mathcal{I} \\ z_{11} + z_{19} = z_{20} & w_{11} = z_{11} - 80 & z_i - M \leq 0 \quad \forall i \in \mathcal{I} \\ z_2 + z_{12} \leq z_{17} & w_{12} = z_{12} - 100 & -z_i - M \leq 0 \quad \forall i \in \mathcal{I} \\ w_1 = z_1 - 50 & w_{13} = z_{13} - 100 & z_i, w_i \in \mathbb{Z} \quad \forall i \in \mathcal{I} \\ w_2 = z_2 - 100 & w_{14} = z_{14} - 200 & \delta_i \in \{0, 1\} \quad \forall i \in \mathcal{I} \end{cases}$$

Fig. 1. Instance of $\mathcal{CQAP}(\mathcal{D}, \mathcal{AC}, q, D)$ obtained for the running example

3.3 Reducing the Size of ILPs

We now show that the size of the two ILP problems $\mathcal{OPT}(\mathcal{D}, \mathcal{AC}, D)$ and $\mathcal{CQAP}(\mathcal{D}, \mathcal{AC}, q, D)$ to be solved for computing the consistent answer can be reduced in the number of both variables and (in)equalities, thus improving performance.

We first consider the case of $\mathcal{OPT}(\mathcal{D}, \mathcal{AC}, D)$, which is built starting from $\mathcal{ILP}(\mathcal{D}, \mathcal{AC}, D)$. The latter consists of the inequality $\mathbf{A} \times \mathbf{z} \leq \mathbf{B}$ augmented with further inequalities involving new variables δ_i and w_i. The number of these variables and inequalities depends on the number of variables occurring in $\mathbf{A} \times \mathbf{z} \leq \mathbf{B}$: the presence of a variable z_i entails the presence of a pair of variables w_i, δ_i, as well as 5 inequalities. Hence, a remarkable reduction in size of $\mathcal{ILP}(\mathcal{D}, \mathcal{AC}, D)$ can be obtained by reducing the number of variables occurring in $\mathbf{A} \times \mathbf{z} \leq \mathbf{B}$.

A classical strategy for reducing the size of a set of inequalities consists of removing linearly dependent columns or rows of the coefficient matrix. Although the presence of redundant columns is unlikely to occur on the coefficient matrix of the whole $\mathcal{ILP}(\mathcal{D}, \mathcal{AC}, D)$, the presence of this kind of columns in the matrix $\mathbf{A}$ is a frequent situation in our scenario. For instance, consider the "Cash budget" example: all the variables z_i corresponding to detail items of the cash budget appear in exactly one inequality (i.e., the inequality encoding the constraint that the sum of detail items for each section must be equal to the aggregate item in the same section). It is easy to see that the columns of $\mathbf{A}$ corresponding to these variables are linearly dependent. Removing linearly dependant columns is the same as replacing linear combinations of variables with new variables. For instance, consider the inequality $\mathbf{A} \times \mathbf{z} \leq \mathbf{B}$ in Example 10, which is a subset of the inequalities of the instance of $\mathcal{CQAP}(\mathcal{D}, \mathcal{AC}, q, D)$ of Fig. 1. It is easy to see that the columns corresponding to the variables z_5, z_6, z_7 are linearly dependent, thus these variables can be replaced with the unique variable $z_{5,6,7}$. Reasoning analogously on the other variables, the inequality $\mathbf{A} \times \mathbf{z} \leq \mathbf{B}$ in Example 10 becomes:

$$\begin{cases} z_{2,3} = z_4; & z_{5,6,7} = z_8; & z_{12,13} = z_{14}; & z_{15,16,17} = z_{18}; \\ z_4 - z_8 = z_9; & z_{14} - z_{18} = z_{19}; & z_1 + z_9 = z_{10}; & z_{11} + z_{19} = z_{20}. \end{cases}$$

Correspondingly, the reduced $\mathcal{ILP}(\mathcal{D}, \mathcal{AC}, D)$ consists of $8 + 14 \times 5 = 78$ inequalities on $14 \times 3 = 42$ variables, instead of $8 + 20 \times 5 = 108$ inequalities on $20 \times 3 = 60$ variables.

An analogous reduction can be applied to reduce the size of $\mathcal{CQAP}(\mathcal{D}, \mathcal{AC}, q, D)$: in this case, the inequality $\mathbf{A} \times \mathbf{z} \leq \mathbf{B}$ on which the reduction is applied encodes both the constraints $\mathcal{AC}$ and the negated query $\neg q$. The reduced version of the instance $\mathcal{CQAP}(\mathcal{D}, \mathcal{AC}, q, D)$ of Fig. 1 is reported in Fig. 2. In this case, z_5, z_6, z_7 have been grouped into $z_{5,6,7}$, while z_{15}, z_{16}, z_{17} into $z_{15,16,17}$. In this figure, the set of the indices of the variables z_i occurring in the reduced version of $\mathcal{CQAP}(\mathcal{D}, \mathcal{AC}, q, D)$ is denoted as $\mathcal{I}^{red}$. Observe that $\mathcal{I}^{red} = \mathcal{I} \setminus \overline{\mathcal{I}} \cup \mathcal{I}^{new}$, where $\overline{\mathcal{I}} = \{5, 6, 7, 15, 16, 17\}$ are the indices of variables z_i grouped in some new variables and $\mathcal{I}^{new}$ are the indices of the new variables grouping old ones.

It is worth noting that the elimination of linearly dependent columns yields no reduction of size when applied on the whole coefficient matrixes of $\mathcal{ILP}(\mathcal{D}, \mathcal{E}, D)$ or $\mathcal{CQAP}(\mathcal{D}, \mathcal{AC}, q, D)$: it is easy to see that the inequalities different from $\mathbf{A} \times \mathbf{z} \leq \mathbf{B}$ make all the columns of the coefficient matrixes linearly independent. Hence, it is mandatory that linearly dependent columns in $\mathbf{A} \times \mathbf{z} \leq \mathbf{B}$ are removed before generating $\mathcal{ILP}(\mathcal{D}, \mathcal{E}, D)$ and $\mathcal{CQAP}(\mathcal{D}, \mathcal{AC}, q, D)$.

$$
\begin{cases}
2 = \sum_{i \in \mathcal{I}^{red}} \delta_i \\
z_2 + z_3 = z_4 \\
z_{5,6,7} = z_8 \\
z_{12} + z_{13} = z_{14} \\
z_{15,16,17} = z_{18} \\
z_4 - z_8 = z_9 \\
z_{14} - z_{18} = z_{19} \\
z_1 + z_9 = z_{10} \\
z_{11} + z_{19} = z_{20} \\
z_2 + z_{12} \leq z_{17} \\
w_1 = z_1 - 50
\end{cases}
\quad
\begin{array}{l}
w_2 = z_2 - 100 \\
w_3 = z_3 - 120 \\
w_4 = z_4 - 250 \\
w_{5,6,7} = z_{5,6,7} - 120 - 20 - 80 \\
w_8 = z_8 - 220 \\
w_9 = z_9 - 30 \\
w_{10} = z_{10} - 80 \\
w_{11} = z_{11} - 80 \\
w_{12} = z_{12} - 100 \\
w_{13} = z_{13} - 100 \\
w_{14} = z_{14} - 200 \\
w_{15,16,17} = z_{15,16,17} - 130 - 40 - 20
\end{array}
\quad
\begin{array}{l}
w_{18} = z_{18} - 120 \\
w_{19} = z_{19} - 10 \\
w_{20} = z_{20} - 90 \\
w_i - M\delta_i \leq 0 \quad \forall i \in \mathcal{I}^{red} \\
-w_i - M\delta_i \leq 0 \quad \forall i \in \mathcal{I}^{red} \\
z_i - M \leq 0 \quad \forall i \in \mathcal{I}^{red} \\
-z_i - M \leq 0 \quad \forall i \in \mathcal{I}^{red} \\
z_i, w_i \in \mathbb{Z} \quad \forall i \in \mathcal{I}^{red} \\
\delta_i \in \{0, 1\} \quad \forall i \in \mathcal{I}^{red}
\end{array}
$$

Fig. 2. Reduced-size instance of $\mathcal{CQAP}(\mathcal{D}, \mathcal{AC}, q, D)$ obtained for the running example

4   Conclusions

We have introduced a framework for computing consistent answers to boolean aggregate queries in numerical databases violating a given set of aggregate constraints, which exploits a transformation into integer linear programming (ILP), thus allowing us to exploit well-known techniques for solving ILP problems. Further work will be devoted to devising strategies for computing consistent answers to more expressive forms of queries.

References

1. Arenas, M., Bertossi, L.E., Chomicki, J.: Consistent query answers in inconsistent databases. In: Proc. 18th ACM Symp. on Principles of Database Systems (PODS), pp. 68–79 (1999)
2. Arenas, M., Bertossi, L.E., Chomicki, J.: Answer sets for consistent query answering in inconsistent databases. Theory and Pract. of Logic Program (TPLP) 3(4-5), 393–424 (2003)

3. Arenas, M., Bertossi, L.E., Chomicki, J., He, X., Raghavan, V., Spinrad, J.: Scalar aggregation in inconsistent databases. Theor. Comput. Sci. (TCS) 3(296), 405–434 (2003)
4. Barceló, P., Bertossi, L.E.: Repairing databases with annotated predicate logic. In: Proc. 9th Int. Workshop on Non-Monotonic Reasoning (NMR), pp. 160–170 (2002)
5. Barceló, P., Bertossi, L.E.: Logic programs for querying inconsistent databases. In: Dahl, V., Wadler, P. (eds.) PADL 2003. LNCS, vol. 2562, pp. 208–222. Springer, Heidelberg (2002)
6. Bertossi, L.E., Bravo, L., Franconi, E., Lopatenko, A.: The complexity and approximation of fixing numerical attributes in databases under integrity constraints. Inf. Systems 33(4-5), 407–434 (2008)
7. Bohannon, P., Flaster, M., Fan, W., Rastogi, R.: A cost-based model and effective heuristic for repairing constraints by value modification. In: Proc. Int. Conf. on Management of Data (SIGMOD), pp. 143–154 (2005)
8. Calì, A., Lembo, D., Rosati, R.: On the decidability and complexity of query answering over inconsistent and incomplete databases. In: Proc. 22nd ACM Symp. on Principles of Database Systems (PODS), pp. 260–271 (2003)
9. Chomicki, J., Marcinkowski, J., Staworko, S.: Computing consistent query answers using conflict hypergraphs. In: Proc. 13th Conf. on Information and Knowledge Management (CIKM), pp. 417–426 (2004)
10. Chomicki, J., Marcinkowski, J.: Minimal-change integrity maintenance using tuple deletions. Information and Computation (IC) 197(1-2), 90–121 (2005)
11. Fazzinga, B., Flesca, S., Furfaro, F., Parisi, F.: Dart: A data acquisition and repairing tool. In: Proc. Int. Workshop on Incons. and Incompl. in Databases (IIDB), pp. 297–317 (2006)
12. Flesca, S., Furfaro, F., Parisi, F.: Preferred database repairs under aggregate constraints. In: Prade, H., Subrahmanian, V.S. (eds.) SUM 2007. LNCS (LNAI), vol. 4772, pp. 215–229. Springer, Heidelberg (2007)
13. Flesca, S., Furfaro, F., Parisi, F.: Querying and Repairing Inconsistent Numerical Databases. ACM Transactions on Database Systems (TODS) 35(2) (2010)
14. Franconi, E., Palma, A.L., Leone, N., Perri, S., Scarcello, F.: Census data repair: a challenging application of disjunctive logic programming. In: Nieuwenhuis, R., Voronkov, A. (eds.) LPAR 2001. LNCS (LNAI), vol. 2250, pp. 561–578. Springer, Heidelberg (2001)
15. Fuxman, A., Fazli, E., Miller, R.J.: Conquer: Efficient management of inconsistent databases. In: Proc. ACM SIGMOD Int. Conf. on Management of Data (SIGMOD), pp. 155–166 (2005)
16. Fuxman, A., Miller, R.J.: First-order query rewriting for inconsistent databases. J. Comput. Syst. Sci. 73(4), 610–635 (2007)
17. Greco, G., Greco, S., Zumpano, E.: A logical framework for querying and repairing inconsistent databases. IEEE Trans. on Knowledge and Data Engineering (TKDE) 15(6), 1389–1408 (2003)
18. Papadimitriou, C.H.: On the complexity of integer programming. Journal of the Association for Computing Machinery (JACM) 28(4), 765–768 (1981)
19. Ross, K.A., Srivastava, D., Stuckey, P.J., Sudarshan, S.: Foundations of aggregation constraints. Theorethical Computer Science (TCS) 193(1-2), 149–179 (1998)
20. Wijsen, J.: Database repairing using updates. ACM Transactions on Database Systems (TODS) 30(3), 722–768 (2005)
21. Wijsen, J.: Consistent query answering under primary keys: a characterization of tractable queries. In: Proc. 12th Int. Conf. on Database Theory (ICDT), pp. 42–52 (2009)

Identifying Interesting Instances for Probabilistic Skylines*

Yinian Qi and Mikhail Atallah

Department of Computer Science, Purdue University
{yqi,mja}@cs.purdue.edu

Abstract. Significant research efforts have recently been dedicated to modeling and querying uncertain data. In this paper, we focus on skyline analysis of uncertain data, modeled as uncertain objects with probability distributions over a set of possible values called *instances*. Computing the exact skyline probabilities of instances is expensive, and unnecessary when the user is only interested in instances with skyline probabilities over a certain threshold. We propose two filtering schemes for this case: a preliminary scheme that bounds an instance's skyline probability for filtering, and an elaborate scheme that uses an instance's bounds to filter other instances based on the dominance relationship. We experimentally demonstrate the effectiveness of our filtering schemes on both real and synthetic data sets and show the efficiency of our schemes compared with other algorithms.

1 Introduction

Skyline analysis is widely used in multi-criteria decision making applications where different criteria often conflict with each other [2]. Given a set of points in data space $\mathcal{D}$, we say point p_1 *dominates* p_2 if p_1 is no worse than p_2 in all dimensions and better than p_2 in at least one dimension. The skyline analysis then returns all points that are not dominated by any other point in the data set, known as *"skyline points"* [4].

In applications where uncertainty is inherent, such as sensor networks, data integration, and location based applications, the skyline analysis needs to be performed on uncertain data. It was [16] that first proposed *probabilistic skylines* for uncertain data modeled as uncertain objects with probability distributions over a set of possible values called *instances*. Unlike the traditional skyline analysis, with uncertain data many points can represent mutually exclusive instances of the same object, and each instance is now associated with a *skyline probability*: the probability that the instance (i) is the one that occurs among the mutually exclusive set of instances (i.e., the object) to which it belongs, and (ii) is not dominated by any occurring instance of another object. The skyline probability of an object is the sum of the skyline probabilities of its instances.

---

* Portions of this work were supported by NSF Grants CNS-0627488, CNS-0915436 and CNS-0913875, by AFOSR Grant FA9550-09-1-0223, and by CERIAS sponsors at Purdue University.

P. García Bringas et al. (Eds.): DEXA 2010, Part II, LNCS 6262, pp. 300–314, 2010.

While skyline results in [16] are objects with skyline probabilities over a given threshold, [2] computes skyline probabilities for all instances by motivating instance-level probabilistic skylines with no threshold. Following [2], in this paper, we study the instance-level probabilistic skylines when thresholds are available. Our goal is to leverage the threshold to efficiently identify instances whose skyline probabilities meet the threshold by designing a scheme that quickly filters instances out while minimizing the number of the expensive skyline probability computations. The key idea of filtering is that the determination of "above threshold" or "below threshold" of an instance's skyline probability is done using computations that are less expensive than the full-blown computation of its exact skyline probability. These easier-to-compute values fall into two categories: upper bounds on the exact skyline probabilities, and lower bounds. If an instance's upper bound is shown to be below the threshold, its exact skyline probability is guaranteed to be below the threshold, i.e., the instance is *"uninteresting"*; we call this *"negative filtering"*. Conversely, if an instance's lower bound is shown to be above or equal to the threshold, then its exact skyline probability is guaranteed to meet the threshold, i.e., the instance is *"interesting"*; we call this *"positive filtering"*. Our main contributions are summarized as follows:

i) We propose an instance-level probabilistic skyline problem for identifying instances with skyline probabilities over a given threshold.
ii) We present two instance-level filtering schemes:

Preliminary filtering scheme: Techniques for avoiding the expensive computation of exact skyline probabilities by bounding them with easier-to-compute values for comparing to the threshold.

Elaborate filtering scheme: Techniques for massive filtering through inter-instance comparisons that leverage one instance's bounds to filter other instances based on the dominance relationship.

iii) Our experimental results on both real and synthetic data sets show that our algorithm is highly effective in filtering out unqualified instances. In addition, we show the efficiency of this new algorithm over our previous algorithm in [2] by comparing their respective time costs.

The rest of the paper is organized as follows: Section 2 formally defines our problem. Section 3 introduces probabilistic range trees for bounding skyline probabilities. Section 4 and Section 5 present our two filtering schemes as well as the final algorithm for probabilistic skylines. Section 6 discusses our experimental study on both real and synthetic data sets. Section 7 reviews the related work, and Section 8 concludes our work.

2   Problem Definition

Our probabilistic skyline problem uses the same uncertain data model as in [2], where uncertain data is modeled as *uncertain objects* with probability distributions over a set of possible values called *instances*. An example is given in Fig. 1.

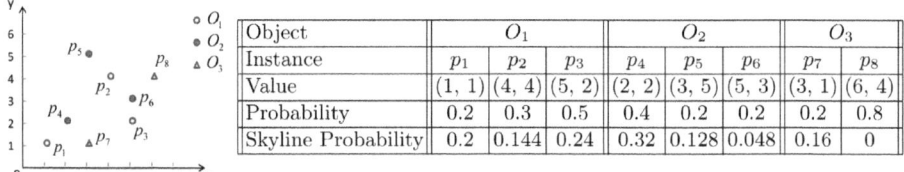

Fig. 1. Probabilistic skylines with three objects and eight instances

Notice that object O_2 does not exist with probability 0.2, since the probabilities of its three instances sum up to $0.8 < 1$.

Generally, we consider each instance as a d-dimensional point in data space $\mathcal{D}$. The dominance relationship "$\prec$" between such points (i.e. instances) is the same as the dominance relationship between points for certain data: Given a set S of n d-dimensional data points: $p_1, \cdots, p_n$ in $\mathcal{D}$ ($\mathcal{D}_1, \cdots, \mathcal{D}_d$), point p_i is said to *dominate* point p_j (i.e., $p_i \prec p_j$) if $\forall k \in [1, d], p_i.\mathcal{D}_k \leq p_j.\mathcal{D}_k$ and $\exists l \in [1, d], p_i.\mathcal{D}_l < p_j.\mathcal{D}_l$. The transitivity of the dominance relationship holds between instances [16], i.e. if $p_1 \prec p_2$, $p_2 \prec p_3$, then $p_1 \prec p_3$.

Definition 1. *The skyline probability of an instance p, i.e., $Pr_{sky}(p)$, is the probability that p exists and no instance of other uncertain objects that dominates p exists. Let m be the total number of uncertain objects and let $p \in O_k$, we have:*

$$Pr_{sky}(p) = Pr(p) \cdot \prod_{i=1, i\neq k}^{m} \left(1 - \sum_{q \in O_i, q \prec p} Pr(q)\right) \tag{1}$$

The skyline probability of an uncertain object is the sum of the skyline probabilities of all its instances.

We denote the upper bound of $Pr_{sky}(p)$ as $Pr_{sky}^+(p)$, and the lower bound as $Pr_{sky}^-(p)$. In Fig. 1, for p_6 to be a skyline point, none of the instances: p_1, p_3, p_4, p_7 should exist. Since p_6 and p_4 both belong to O_2, the existence of p_6 guarantees that p_4 does not exist (instances of the same uncertain object are mutually exclusive). Hence $Pr_{sky}(p_6) = Pr(p_6) * (1 - Pr(p_1) - Pr(p_3)) * (1 - Pr(p_7)) = 0.2 * 0.3 * 0.8 = 0.048$.

The probabilistic skyline problem we study in this paper is defined as follows:

Definition 2. *Given a data set S of n instances that belong to m uncertain objects and a probability threshold θ, the instance-level probabilistic skyline analysis returns all instances with skyline probabilities at least θ, i.e., return the skyline set S_{sky} such that: $S_{sky} = \{p \in S | Pr_{sky}(p) \geq \theta\}$.*

3   Probabilistic Range Trees

We propose two indexing structures based on the range tree [21] to facilitate bounding and computing skyline probabilities. We augment the original range

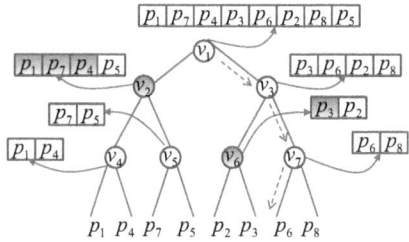

Fig. 2. A two-dimensional probabilistic range tree

trees with additional probabilistic information stored at the internal nodes, which can be leveraged when querying the trees to quickly bound the skyline probability of a given instance p (the query instance). We call such trees *probabilistic range trees* (PRT). Section 3.2 introduces a general PRT built upon all n instances with probabilistic information. A similar indexing structure is described in Section 3.3, which is built for every uncertain object and has different probabilistic information. Our algorithms for the preliminary filtering use both trees, as we will see later in Section 4.

3.1   Overview

We explain the construction of a PRT on n two-dimensional points (representing all instances in the data set S): A complete binary tree T is built on top of the points sorted according to dimension D_1. Each internal node v of T points to an *info-list* L_v that contains the points at the leaves of the subtree of T rooted at v, sorted according to their D_2 dimension. Therefore, if v has children u and w, then L_v is the merge of L_u and L_w; we assume that every element of L_v stores its rank in each of the lists L_u and L_w (which implies that once a search item's position has been located in L_v it can be located in L_u and L_w in constant time). The space is obviously $O(n \log n)$. We also assume a derived probability (defined later in Section 3.2 and Section 3.3) is associated with every element of L_v.

Fig. 2 illustrates a two-dimensional PRT built on top of the eight instances in Fig. 1. The leaves of the PRT are the instances sorted by the first dimension. Each of the internal nodes v_1 to v_7 points to an info-list that contains instances sorted by the second dimension. For example, v_2's info-list (L_{v_2}) has four instances p_1, p_7, p_4, p_5 sorted by the second dimension, which are leaves of the subtree rooted at v_2.

A d-dimensional PRT is built inductively using $d - 1$ dimensional PRTs: A complete tree T is built whose leaves are the n points sorted according to dimension D_1, and each internal node v of T points to a $d - 1$ dimensional PRT that contains the elements at the leaves of the subtree of T rooted at v, organized according to the remaining $d - 1$ dimensions (i.e., ignoring D_1). The space complexity is $O(n(\log n)^{d-1})$. Note that our construction ensures that the points in the info-lists are always sorted according to the last dimension.

3.2   General Probabilistic Range Tree

To bound and compute skyline probabilities, we build the *general probabilistic range tree* (*general-PRT*, denoted as T_g) on all n instances in the data set S.

Definition 3. *Let p be the k-th instance in an info-list L in T_g ($p \in O_i, 1 \leq i \leq m$). let $\hat{L}$ be the list of the first k instances in L, then the probability associated with p (denoted as β_p) in L is defined as:*

$$\beta_p = \prod_{i=1}^{m} (1 - \sum_{q \in \hat{L}, q \in O_i} Pr(q)) \tag{2}$$

In other words, the probability β_p is the probability that no instance in $\hat{L}$ exists, i.e., the probability that p does not exist and no instance before p in L exists. Given a set of instances, we can create an info-list L by adding each instance to L and then sort all instances by their d-th dimension values. Then we compute the probabilistic information β_p for each p in L as follows: We use s_i to record the current probability sum for object O_i that has appeared in L. As we go through the instances in L, we update the corresponding probability sum, and compute β_p based on the β of the instance immediately before p in L (see our technical report [18] for the detailed algorithm). The time needed to compute all $\beta's$ for an info-list L is $O(|L|)$.

3.3   Colored Probabilistic Range Trees

Besides the general-PRT built upon all n instances in S, we also have m specific PRTs, each built upon the instances of the corresponding object. If we render each instance with a color that indicates the source of the instance and match color i to object O_i, then each of these specific PRTs has only one color. Hence we call these trees *colored-PRTs*. For the rest of the paper, whenever we say "instance p of color i", we mean "instance p that belongs to object O_i".

  Unlike the info-lists of the general-PRT, an info-list of a colored-PRT is associated with probability sums for each instance in the list. For the k-th instance p in an info-list L of a colored-PRT, its probability sum $\sigma_p = \sum_{q \in \hat{L}} Pr(q)$ where $\hat{L}$ is the list of the first k instances in L. To compute all σ's, we simply go through L and accumulate the probability sum. The time complexity is $O(|L|)$.

4   A Preliminary Filtering Scheme

Now that we have introduced both the general-PRT and the m colored-PRTs, we can use them to compute the bounds of the skyline probabilities for filtering.

4.1   An Upper Bound

Given a query instance p, we can obtain $Pr_{sky}^+(p)$ by querying the general-PRT T_g as follows: We begin with the base case of $d = 2$, as shown in Fig. 3. Given

Input: the general PRT T_g and a query instance p
Output: an upper bound of $Pr_{sky}(p)$

1. binary search for $p.D_1$ to find the search path $\mathcal{P}$
2. binary search for $p.D_2$ in L_{root}, let the position be k
3. $upperBound = 1$
4. $v = root$ // walk along $\mathcal{P}$ starting from root
5. **while** current node v is not a leaf **do**
6.   **if** the next node $w \in \mathcal{P}$ is $v.rightChild$ **then**
7. $k' = L_v[k].rankL$
8. $u = v.leftChild$
9. $\beta = L_u[k'].beta$ // read β from v's left child
10. $upperBound = upperBound * \beta$
11. $k = L_v[k].rankR$ // locate the position in L_w
12.   **else**   // w is a left child of v
13. $k = L_v[k].rankL$ // locate the position in L_w
14.   **end if**
15. $v = w$ // go one level down
16.   **end while**
17.   **return** $upperBound$

Fig. 3. Compute an upper bound of $Pr_{sky}(p)$

the two-dimensional query $p = (p.D_1, p.D_2)$, we first locate the search path (call it $\mathcal{P}$) in T_g from the root to the position of the value $p.D_1$ among the leaves, then do one binary search for $p.D_2$ in the info-list L_{root} of the root of T_g. We record the position (rank) k of $p.D_2$ in L_{root} and call it the *search position* in L_{root}. We use $L_v[k]$ to denote the k-th instance (let it be q) in the info-list of the node v and $L_v[k].rankL$, $L_v[k].rankR$ to denote the rank of q in the info-lists of v's left child and right child respectively. These ranks are stored so that given the position of q in L_v, we can locate its position in info-lists of v's children in constant time. Starting from the root, we can obtain the search positions in the successive nodes as we walk down the search path $\mathcal{P}$.

We define the *left fringe nodes* of the PRT given the query instance p as the left children of the nodes on the search path $\mathcal{P}$ that are not nodes on $\mathcal{P}$ themselves. For example, in Fig. 2, the search path $\mathcal{P}$ for p_6 is $v_1 \to v_3 \to v_7 \to p_6$. The corresponding left fringe nodes are v_2 and v_6, who are left children of v_1 and v_3 respectively. The leaf p_6 is on $\mathcal{P}$, so it is not a left fringe node despite the fact that it is a left child of v_7 on $\mathcal{P}$.

We use $\hat{L}_v$ to denote the truncated info-list of node v (L_v) with instances up till the search position in L_v. If v is a left fringe node, we call such $\hat{L}_v$ a *qualified info-list* . Fig. 2 highlights two qualified info-lists for the query p_6: One contains the first three instances of L_{v_2} and the other contains the first instance of L_{v_6}.

When we reach the leaf at the end of the query, the variable $upperBound$ in Fig. 3 is the product of all β's we read along $\mathcal{P}$. It is indeed an upper bound of $Pr_{sky}(p)$, as we will see shortly. The time complexity for such a query is $O(\log n)$. In the example of Fig. 2, the upper bound that we get for $Pr_{sky}(p_6)$ is $\beta_{p_4} * \beta_{p_3}$, where p_4 and p_3 are the last instances of the two qualified info-lists.

We can extend the above the algorithm to case $d > 2$. The details can be found in our technical report [18]. The lemma below states that the set of all instances in qualified info-lists ($\hat{L}$'s) is the set of all instances in S that dominate p, which can be easily proved from the search process and the definition of the general-PRT. Note that the notation $\hat{L}_i$ in the lemma is the i-th qualified info-list, not the qualified info-list at node i.

Lemma 1. *Let $\hat{L}_1 \cdots \hat{L}_t$ be the qualified info-lists for query p. Let $S_{\hat{L}} = \cup_{i=1}^t S_{\hat{L}_i}$, where $S_{\hat{L}_i}$ is the set of instances in $\hat{L}_i$. Then we have: 1) $\forall q \in S_{\hat{L}}, q \prec p$; 2) $\forall q' \in S - S_{\hat{L}}, q' \nprec p$.*

For every $\hat{L}_i$, let β_i be the β associated with the last instance in $\hat{L}_i$, i.e., β_i is the probability that none of the instances in $\hat{L}_i$ exists. The theorem below shows that although we cannot compute $Pr_{sky}(p)$ directly from β_i's, we can compute the upper bound $Pr_{sky}^+(p)$ from them.

Theorem 1. *Let β_i be the probability associated with the last instance in $\hat{L}_i (1 \le i \le t)$ where $\hat{L}_i$ is a qualified info-list for query p, then $\prod_{i=1}^t \beta_i \ge Pr_{sky}(p)$.*

The proof is available in our technical report [18]. Theorem 1 shows that $\prod_{i=1}^t \beta_i$ is indeed a $Pr_{sky}^+(p)$. This directly points out a way for filtering the query instance: Given a threshold θ, as soon as we see the current product of β's fall below θ, we can stop and declare that p is not in the skyline, since $Pr_{sky}(p) < \theta$ must also hold.

4.2   A Tighter Upper Bound

While using the general-PRT alone gives us an upper bound of the skyline probability, a tighter upper bound can be achieved by using both the general-PRT and the colored-PRTs. Therefore, we can compute the tighter upper for filtering if the upper bound computed in Section 4.1 fails to filter instances.

Let $p \in O_k$. We prove the following in our technical report [18]:

$$\prod_{i=1}^t \beta_i \ge \frac{\prod_{i=1}^t \beta_i \cdot Pr(p)}{1 - \sum_{q \in O_k, q \prec p} Pr(q)} \ge Pr_{sky}(p)$$

i.e. $\prod_{i=1}^t \beta_i \cdot Pr(p) / \left(1 - \sum_{q \in O_k, q \prec p} Pr(q)\right)$ is a tighter upper bound of $Pr_{sky}(p)$ than $\prod_{i=1}^t \beta_i$. We know $Pr(p)$ and $\prod_{i=1}^t \beta_i$ from querying the general-PRT, to obtain this tighter upper bound, the only part we need to know is $\sum_{q \in O_k, q \prec p} Pr(q)$, which is a probability sum that can be obtained by querying the PRT of color k. The algorithm for computing this sum given a query instance p is the same as computing the upper bound with the general-PRT in Fig. 3 except this time we carry a sum instead of a product along the search path: Whenever a new probability σ is read from a qualified info-list, we add it to the current sum. The final sum is then the sum of all σ's we read as we walk along the path.

The corollary below for querying colored-PRTs can be derived immediately from Lemma 1, which proves that the probability sum returned by querying the PRT of color k is indeed $\sum_{q \in O_k, q \prec p} Pr(q)$:

Corollary 1. *The set of instances in all qualified info-lists obtained by querying the PRT of color k is the set of all instances of color k in S that dominate p.*

4.3   A Lower Bound

We start with $d = 2$. For every instance (x, y), we define $s_{iL}(x)$ (resp., $s_{iB}(y)$) to be the sum of the probabilities of instances of color i that are to the left of x (resp., below y). It is straightforward to preprocess the n instances so that a query that asks for $s_{iL}(x)$ or $s_{iB}(y)$ can be processed in $O(\log n_i)$ time, where n_i is the number of instances of color i: Simply x-sort (resp., y-sort) the instances of color i and store in that sorted list the prefix sums of the probabilities. For each instance p in the list, the prefix sum of p is the sum of probabilities of all instances in the list up till p. Then we process a $s_{iL}(x)$ (resp., $s_{iB}(y)$) query by locating x (resp., y) in that list and reading the relevant prefix sum. Doing such preprocessing for all m colors takes $O(\sum_{i=1}^{m} n_i \log n_i) = O(n \log n)$. Then we can compute in $O(n \log n)$ time for all n instances the following lower bound:

$$Pr(p) \cdot \max \left\{ \prod_{i=1, i \neq k}^{m} (1 - s_{iL}(p.D_1)), \; \prod_{i=1, i \neq k}^{m} (1 - s_{iB}(p.D_2)) \right\}$$

The above lower bound can be easily extended to $d > 2$ by computing the sums of probabilities for each dimension.

5   An Elaborate Filtering Scheme

Recall that the preliminary filtering tries to filter out instances by bounding their respective skyline probabilities. The improved filtering scheme of the present section adds inter-instance comparisons to achieve wholesale filtering (positive or negative), i.e., it considers the impact of one instance's elimination or survival on other instances related to it by the dominance relationship. Therefore, the order in which instances are processed (individually, by bounding skyline probabilities as in the preliminary scheme) becomes crucial.

5.1   Filtering Rationale

Before presenting our elaborate filtering scheme, we first define a ratio called the *"key ratio"* for an instance p:

Definition 4. *For any instance $p \in O_k$, p's key ratio $r_p = \dfrac{Pr(p)}{1 - \sum_{p' \in O_k, p' \prec p} Pr(p')}$. If $r_p \geq \frac{1}{2}$, we call p a "target instance".*

Input: data set S, threshold θ
Output: the candidate list $Cand$ after filtering
1. create the initial $Cand$ from S //Section 5.2
2. **for each** instance p in $Cand$ **do**
3.   compute $Pr^+_{sky}(p)$ // Section 4.1 and 4.2
4.   **if** $Pr^+_{sky}(p) < \theta$ **then**
5.   remove p from $Cand$
6.   get the set of instances in $Cand$ dominated by p
7.   **for each** instance q in the set **do**
8.   **if** p is a target instance **then**
9.   remove q from $Cand$
10.   **else**
11.   **if** p, q are of the same color and $Pr(p) \geq Pr(q)$ **do**
12.   remove q from $Cand$
13.   **end if**
14.   **end if**
15.   **end for each**
16.   **end if**
17. **end for each**
18. **return** $Cand$

Fig. 4. Algorithm for negative filtering in the elaborate filtering scheme

r_p can be easily computed in $O(\log n)$ by querying the PRT of color k to get the probability sum $\sum_{p' \in O_k, p' \prec p} Pr(p')$. The following theorem states the conditions for negative filtering (see our technical report [18] for detailed proof):

Theorem 2. *Given instances $p \prec q$ $(p \in O_k, q \in O_l)$. If $Pr_{sky}(p) < \theta$, then:*
1) $k \neq l$: If p is a target instance, then $Pr_{sky}(q) < \theta$.
2) $k = l$: If p is a target instance or if $Pr(p) \geq Pr(q)$, then $Pr_{sky}(q) < \theta$.

We call instances satisfying the above conditions "killers" – the elimination of themselves causes the massive extinction of others from the skyline result set. In contrast, the corollary below states the conditions for instances to be "saviors" – the survival of themselves causes the survival of others in the final skyline result. The proof of this corollary depends on the proof of $Pr_{sky}(p) \geq Pr_{sky}(q)$, which is exactly the same as the proof for Theorem 2 given in [18].

Corollary 2. *Given instances $p \prec q$ $(p \in O_k, q \in O_l)$. If $Pr_{sky}(q) \geq \theta$, then:*
1) $k \neq l$: If p is a target instance, then $Pr_{sky}(p) \geq \theta$.
2) $k = l$: If p is a target instance or if $Pr(p) \geq Pr(q)$, then $Pr_{sky}(p) \geq \theta$.

5.2   Instance Scheduling and Elaborate Filtering

The theorem and corollary in the previous section together point out a way of filtering instances massively based on a single instance's skyline probability. As we have mentioned earlier, the order in which instances are processed is crucial. The goal of our elaborate filtering scheme is to maximize both negative filtering (*"killing"*) and positive filtering (*"saving"*) as we process the candidate list so that the number of the PRT queries (either for bounding or computing the exact skyline probability) is minimized. We propose the following heuristic for scheduling instances to achieve this goal: Using the standard dominance counting techniques [17], we preprocess all n instances in $O(n \log n)$ time to compute two quantities $count^+(p)$ and $count^-(p)$ for every instance p, where $count^+(p)$ is the number of instances dominated by p and $count^-(p)$ is the number of instances that dominate p. We first sort the instances according to $count^+$ in the descending order. The list then becomes our initial candidate list for computing the skyline results. Suppose that instances $p \prec q$. If p is not a skyline result and is also a target instance, p can kill all instances that it dominates (the number of such instances is $count^+(p)$), i.e., p kills q. On the other hand, if q is a skyline result, q can save all target instances that dominate it (the number of such instances is $\leq count^-(q)$), i.e., q saves p if $r_p \geq \frac{1}{2}$.

The algorithm for the elaborate filtering first does the negative filtering, then the positive filtering. After scheduling all n instances to form the initial candidate list, we process each instance p in the candidate list in order by upper bounding $Pr_{sky}(p)$ (using the techniques in the preliminary filtering scheme). Then we do the negative filtering as shown in Fig. 4. In line 6, we obtain the set of instances that are dominated by p by querying a mirror of our general-PRT (i.e. instead of returning instances that dominate p, it returns instances that are dominated by p). The order that we process instances guarantees that the current instance, if turned out to be a killer, can kill the largest number of instances (because its $count^+$ is the biggest among the unprocessed candidates).

After the candidate list has been exhausted, i.e. all killings have been done, we sort the remaining instances in the list by their $count^-$ in the descending order. We then process each instance q in this new candidate list in order by computing $Pr_{sky}^-(q)$ and compare it with the threshold θ to see whether q survives as a skyline result. If it survives, we move it from the candidate list to the skyline result S_{sky}. The rest of the algorithm is similar to the one in Fig. 4. Notice that we do negative filtering (killing) first, followed by positive filtering (saving). This is because killing and saving are NOT symmetric: A killer p kills all instances dominated by p, whereas a savior q only saves a portion of all instances that dominate q — only the target instances among them can be saved. Hence killing filters more than saving. It should come before saving to minimize the number of instances that need to be processed or further evaluated.

Our final algorithm for probabilistic skylines consists of two stages:

1. **Filtering stage:**

    1) Initialize the skyline result S_{sky} to an empty set
    2) Initialize the candidate list to be all n instances in the data set S
    3) Use the elaborate filtering scheme to reorder the candidate list, eliminate instances with skyline probabilities below the threshold θ, and move those with skyline probabilities at least θ to S_{sky}.

2. **Refining stage:**

1) For each remaining instance p in the candidate list, compute the exact $Pr_{sky}(p)$ by querying the general PRT: From qualified info-lists ($\hat{L}$'s) for query p (see Section 4.1), we obtain all instances in S that dominate p, from which we can compute the exact $Pr_{sky}(p)$ according to Equation 1

2) Add p to S_{sky} if $Pr_{sky}(p) \geq \theta$

3) Return the final S_{sky} as the set of skyline results

6   Experimental Results

We evaluate the effectiveness and the efficiency of our algorithm on both real and synthetic data sets on a MAC with Intel T2500 2GHz CPU and 2GB main memory. All algorithms are implemented in C++.

6.1   Data Sets

In our experiments, we use the real data set: the NBA data set as in [16], kindly provided to us by the authors of [16]. We treat each player as an uncertain object and the records of the player as the instances of the object with three dimensions: number of points, number of assists, and number of rebounds. We assign random probabilities to instances of the same object such that the probabilities sum up to 1. We also use a data generator same as in [2] to generate synthetic data sets.The default values for parameters used in our experiments are: number of uncertain objects $m = 20,000$, number of instances for an object uniformly distributed in the range $[1, 30]$, number of dimensions $d = 3$ and threshold $\theta = 0.01$. Although the absolute value of θ seems small, it is already highly selective for instance-level skyline probabilities. This is because an uncertain object may have many instances, resulting in small occurrence probabilities for these instances to begin with before bounding/computing their skyline probabilities.

6.2   Effectiveness of Filtering

The effectiveness of our schemes is measured by the percentage of instances filtered in the schemes (i.e. filtering percentage).

Effectiveness of the Preliminary Scheme – We evaluate the filtering percentage by upper and lower bounds in our preliminary scheme on the synthetic data set with $m = 2000$. We also evaluate the filtering capabilities of upper bounds (Section 4.1) and the corresponding tighter upper bounds (Section 4.2) respectively. The results are shown in Fig. 5. We can see that the two upper bounds filter much more than the lower bounds, due to small instance skyline probabilities on randomly generated synthetic data sets.

Effectiveness of the Elaborate Scheme – We evaluate the effectiveness of our elaborate filtering on both the real NBA data set and the synthetic data sets. The filtering percentage after "killing" and that after "saving" are shown in Fig. 6

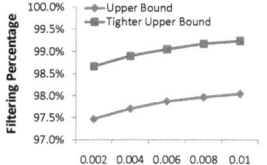 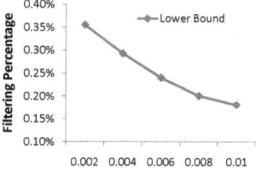 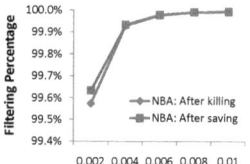

Fig. 5. Filtering percentage by upper and lower bounds **Fig. 6.** Filter for NBA data

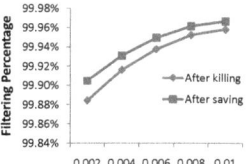 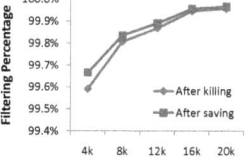 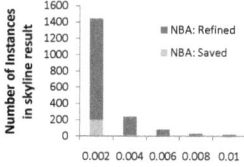

Fig. 7. Effect of threshold and data set size   **Fig. 8.** NBA skyline set

for the NBA data with a varying threshold. The same plot for the synthetic data
is shown in the first chart of Fig. 7. For both data sets, killing filters instances
massively while saving contributes a much smaller portion to the final filtering
percentage. This demonstrates our earlier statement that killing and saving are
not symmetric (Section 5.2). In addition, the filtering percentage on both data
sets increases as the threshold increases. We also plot the filtering percentage
against the data set size (i.e., m, number of objects) on the synthetic data in
Fig. 7. As the number of objects increases, the filtering percentage increases in
general: With more instances present to compete with each other, instances are
likely to be dominated by more, resulting in smaller skyline probabilities.

The final skyline result set consists of two parts: the instances that are filtered
by saving ("saved" ones), and those whose exact skyline probabilities are verified
to meet the threshold ("refined" ones). In Fig. 8 for the skyline set on NBA
data, saved instances are a much smaller portion in the final skyline set than
the refined ones. As the threshold increases, less instances are saved as the lower
bound of an instance's skyline probability is unlikely to be above the threshold.

Fig. 9 shows how filtering and time cost change with respect to dimensions
and the number of instances per object. With increasing dimensions, filtering
percentage decreases because an instance p is less likely to be dominated by
another instance q that has values better than or equal to p's own values in every
dimension. Increasing dimensions also bring increasing overhead in constructing
and querying PRT's, resulting in higher total time cost. We also evaluate the
effect of the number of instances per object in the third chart of Fig. 9. The x-
axis represents the maximum number of instances an uncertain object can have,
e.g., 90 means the instance count per object is generated uniformly between 1
and 90. We fixed the total number of instances to 20,000 while changing the

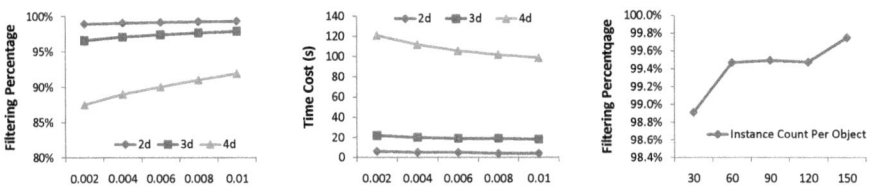

Fig. 9. Effect of dimensions and instance count per object

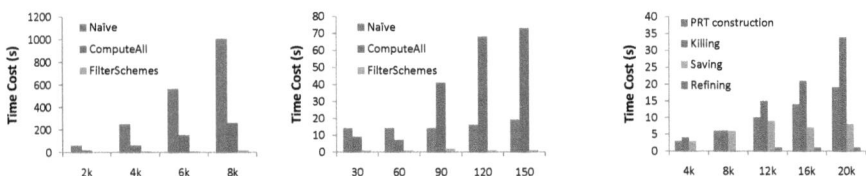

Fig. 10. Comparison between different algorithms   **Fig. 11.** Time cost analysis

range of the instance count per object from [1, 30] to [1, 150]. More instances per object suggests that each instance now has a smaller probability to occur in the first place, resulting in its smaller skyline probability. Therefore, the filtering percentage increases, as the number of interesting instances decreases.

To show that our algorithm (call it `FilterSchemes`) is efficient in leveraging the threshold for filtering, we implement the naïve approach (call it `Naïve`) for probabilistic skylines for benchmarking, which uses a nested loop to compute the exact skyline probability of an instance by looking at all other instances. We also compare `FilterSchemes` with the algorithm we proposed earlier for computing all skyline probabilities [2] (call it `ComputeAll`) to see how our new algorithm with filtering schemes reduces time cost by leveraging the threshold. The time cost of the three algorithms are shown in Fig. 10 for $d = 2$. For the second chart, we fix the total number of instances to be 20,000 while changing the instance count per object. We observe that `FilterSchemes` performs significantly better than `Naïve` and `ComputeAll`, and the advantage of `FilterSchemes` becomes even bigger as the data set size grows, thanks to our effective filtering schemes that exploit the threshold for massive filtering. Moreover, the time cost of both `FilterSchemes` and `Naïve` are insensitive to the number of instances per object, while the performance of `ComputeAll` is greatly affected: the more instances objects have, the slower `ComputeAll` runs. The reason for this is that `ComputeAll` combines weighted dominance counting (WDC) algorithm with the grid method and the decision on which algorithm to use is based on whether an object is a "frequent" object [2]. With more instances per object, there are more frequent objects, suggesting that the expensive WDC algorithm has to be run more times.

Fig. 11 provides a detailed view on how the time cost of our algorithm breaks down to three parts: the time cost for constructing PRT's (the general-PRT and the colored-PRT's), and the time cost for killing, saving and refining. We fix the instance count per object to be 30 while varying the number of total instances. In Fig. 11, although refining takes little time, it is only because there is a small percentage of instances left to be further evaluated – most of instances have already been filtered either by killing or saving.

Since our algorithms are designed specifically for the instance-level filtering with a more general uncertain model than the one in [16] (but same as the one in [2]), while [16] focuses on the object-level filtering for probabilistic skylines, we do not think a comparison of the two will yield convincing results when either their algorithms or our algorithms have to be specifically modified and optimized in order to suit the other's case and become comparable.

7   Related Work

Many studies have been conducted to design efficient skyline algorithms for large data sets [4,15,9,13]. Various kinds of skyline analysis have been proposed for different settings [20,23,12], but they all focus on certain data. Recently, much research has been done in querying [5,7,3,19,10,8] and indexing uncertain data [6,11,1]. Advanced data analysis with uncertain data such as probabilistic skyline analysis has also been studied [2,14,22]. [16,2] are closest to our work. The goal in [16] was to find all uncertain objects with skyline probabilities greater than or equal to a given threshold. Two algorithms (top-down and bottom-up) were proposed to efficiently compute the skyline results by leveraging upper and lower bounds of the objects' skyline probabilities to avoid the expensive computations of the exact skyline probabilities. In contrast, [2] took a different approach by motivating the problem of computing skyline probabilities for all instances and proposed a sub-quadratic algorithm for doing so. It also removed the two assumptions made in [16] that the instances of the same object has equal probabilities and the probabilities of the instances sum up to 1. Our work in this paper follows [2] in studying instance-level probabilistic skylines. However, we now aim at minimizing the number of exact skyline probability computations by introducing a threshold and designing algorithms to leverage the threshold for filtering. Our filtering schemes enable us to quickly identify interesting instances whose skyline probabilities meet the given threshold.

8   Conclusions

In this paper, we study the problem of computing the probabilistic skylines at the instance level for uncertain objects with multiple instances. We propose two filtering schemes to avoid the expensive skyline probability computations by leveraging a given threshold. In our preliminary filtering scheme, we design indexing structures to facilitate bounding of a query instance's skyline probability. Our more elaborate filtering scheme uses the preliminary scheme, and further explores the dominance relationship between instances for massive filtering. Our experiments show that our algorithm effectively reduces the search space and efficiently identifies instances with skyline probabilities above or equal to the threshold.

References

1. Agarwal, P.K., Cheng, S.-W., Tao, Y., Yi, K.: Indexing uncertain data. In: PODS (2009)
2. Atallah, M., Qi, Y.: Computing all skyline probabilities for uncertain data. In: PODS (2009)
3. Beskales, G., Soliman, M.A., Ilyas, I.F.: Efficient search for the top-k probable nearest neighbors in uncertain databases. In: VLDB (2008)
4. Börzsönyi, S., Kossmann, D., Stocker, K.: The skyline operator. In: ICDE (2001)
5. Cheng, R., Kalashnikov, D.V., Prabhakar, S.: Querying imprecise data in moving object environments. In: TKDE (2004)
6. Cheng, R., Xia, Y., Prabhakar, S., Shah, R., Vitter, J.S.: Efficient indexing methods for probabilistic threshold queries over uncertain data. In: VLDB (2004)
7. Cheng, R., Xia, Y., Prabhakar, S., Shah, R., Vitter, J.S.: Probabilistic verifiers: Evaluating constrained nearest-neighbor queries over uncertain data. In: ICDE (2008)
8. Cormode, G., Li, F., Yi, K.: Semantics of ranking queries for probabilistic data and expected ranks. In: ICDE (2009)
9. Godfrey, P., Shipley, R., Gryz, J.: Maximal vector computation in large data sets. In: VLDB (2005)
10. Hua, M., Pei, J., Zhang, W., Lin, X.: Ranking queries on uncertain data: A probabilistic threshold approach. In: SIGMOD (2008)
11. Kanagal, B., Deshpande, A.: Indexing correlated probabilistic databases. In: SIGMOD (2009)
12. Kossmann, D., Ramsak, F., Rost, S.: An online algorithm for skyline queries. In: VLDB (2002)
13. Lee, K.C.K., Zheng, B., Li, H., Lee, W.-C.: Approaching the skyline in z order. In: VLDB (2007)
14. Lian, X., Chen, L.: Monochromatic and bichromatic reverse skyline search over uncertain databases. In: SIGMOD (2008)
15. Papadias, D., Tao, Y., Fu, G., Seeger, B.: An optimal and progressive algorithm for skyline queries. In: SIGMOD (2003)
16. Pei, J., Jiang, B., Lin, X., Yuan, Y.: Probabilistic skylines on uncertain data. In: VLDB (2007)
17. Preparata, F., Shamos, M.: Computational Geometry: An Introduction. Springer, Heidelberg (1985)
18. Qi, Y., Atallah, M.: Identifying interesting instances for probabilistic skylines (longer version of the present submission). Purdue Technical Report (2009), http://www.cs.purdue.edu/homes/yqi/research.html
19. Soliman, M.A., Ilyas, I.F., Chang, K.C.-C.: Urank: formulation and efficient evaluation of top-k queries in uncertain databases. In: SIGMOD (2007)
20. Vlachou, A., Doulkeridis, C., Kotidis, Y., Vazirgiannis, M.: Skypeer: Efficient subspace skyline computation over distributed data. In: ICDE (2007)
21. Willard, D.E.: New data structures for orthogonal range queries. SIAM J. Comput. (1985)
22. Zhang, W., Lin, X., Zhang, Y., Wang, W., Yu, J.: Probabilistic skyline operator over sliding windows. In: ICDE (2009)
23. Zhu, L., Zhou, S., Guan, J.: Efficient skyline retrieval on peer-to-peer networks. Future Generation Communication and Networking (2007)

GPU-WAH: Applying GPUs to Compressing Bitmap Indexes with Word Aligned Hybrid[*]

Witold Andrzejewski and Robert Wrembel

Poznań University of Technology, Institute of Computing Science, Poznań, Poland
{Witold.Andrzejewski,Robert.Wrembel}@cs.put.poznan.pl

Abstract. Bitmap indexes are one of the basic data structures applied to query optimization in data warehouses. The size of a bitmap index strongly depends on the domain of an indexed attribute, and for wide domains it is too large to be efficiently processed. For this reason, various techniques of compressing bitmap indexes have been proposed. Typically, compressed indexes have to be decompressed before being used by a query optimizer that incurs a CPU overhead and deteriorates the performance of a system. For this reason, we propose to use additional processing power of the GPUs of modern graphics cards for compressing and decompressing bitmap indexes. In this paper we present a modification of the well known WAH compression technique so that it can be executed and parallelized on modern GPUs.

1 Introduction

A data warehouse architecture has been developed for the purpose of integrating data from multiple storage systems within an enterprise. The integrated data are stored in a central database, called a data warehouse. Data stored there are analyzed by OLAP queries, for the purpose of discovering trends, anomalies, hidden dependencies between, and predicting trends. OLAP queries typically access, filter, aggregate large volumes of data. Efficient processing of OLAP queries is often supported by the so-called bitmap indexes [21].

A *bitmap index*, in the simplest form, is composed of the collection of bitmaps (cf. Section 2). A bitmap is a vector of bits. Each bit is mapped to a row in an indexed table. One bitmap B_v is created for one value v of an indexed attribute. If the value of a bit in B_v is equal to 1, then a row corresponding to this bit has value v.

Queries whose predicates involve attributes indexed by bitmap indexes can be answered fast by performing bitwise AND, or OR, or NOT operations on bitmaps, that is a big advantage of bitmap indexes. Unfortunately, the size of a bitmap index increases when the cardinality of an indexed attribute increases. Thus, for attributes of high cardinalities (wide domains) bitmap indexes become very large. As a consequence, they cannot fit in main memory and the

---

[*] This work was supported from the Polish Ministry of Science and Higher Education grant No. N N516 365834.

P. García Bringas et al. (Eds.): DEXA 2010, Part II, LNCS 6262, pp. 315–329, 2010.

efficiency of accessing data with the support of such indexes deteriorates [29]. In order to improve the efficiency of accessing data with the support of bitmap indexes defined on attributes of high cardinalities various bitmap index compression techniques have been proposed in the research literature (cf. Section 3). Typically, compressed indexes have to be decompressed before being used by a query optimizer that incurs a CPU overhead and deteriorates the performance of a query.

Paper Contribution. In this paper we propose an extension to the well-known Word Aligned Hybrid (WAH) bitmap compression technique that has been reported to provide the shortest query execution time [25,26]. Our extension, called GPU-WAH, allows to parallelize compressing and decompressing steps of WAH and execute them on Graphics Processing Units (GPU). In our implementation (cf. Section 5) we take advantage of the fact, that modern GPUs may process up to 240 threads in parallel, to obtain blazingly fast compression and decompression as well as possible massively parallel comparison of multiple bitmaps. In our experiments we compared the performance of standard WAH run on a CPU with the performance of GPU-WAH. The results show (cf. Section 6) that GPU-WAH significantly reduces compression/decompression time.

2   Definitions

A *bitmap* is a vector of bits. Bitmap literals will be denoted as a string of ones and zeros starting with the most significant bit, and finished with letter "b", e.g., $111000b$. Each bit in a bitmap is assigned a unique, consecutive number starting with 0. The i-th bit of bitmap B is denoted as B_i. The number of bits stored in bitmap B is called a bitmap *length* and it is denoted as $\|B\|$. We define operation *concatenation* of two bitmaps, denoted as $+$, that creates a new bitmap such that it contains all bits of the first bitmap, followed by all bits of the second bitmap. Formally, given bitmaps A and B, their concatenation creates new bitmap C such that: $\|C\| = \|A\| + \|B\| \wedge \forall_{i=0...\|A\|-1} C_i = A_i \wedge \forall_{i=\|A\|...\|C\|-1} C_i = B_{i-\|A\|}$ (e.g., $01b + 10b = 1001b$; this stems from the fact, that the second operand of operator $+$ consists of the bits that will be more significant in the result).

A *subbitmap* of B is any subvector of B that may be created by removing some of the bits from the beginning and from the ending of B. A subbitmap of bitmap B, such that it contains bits from i to j is denoted as $B_{i \rightarrow j}$. Formally, for a given bitmap B, subbitmap $C = B_{i \rightarrow j}$ must satisfy the condition: $j < \|B\| \wedge \forall_{k=i...j} B_k = C_{k-i}$.

Substitution is an operation that replaces a subbitmap of a given bitmap with another bitmap. Given bitmaps B and C, substituting subbitmap $B_{i \rightarrow j}$ with C is denoted as $B_{i \rightarrow j} \leftarrow C$, and is formally defined as: $B \leftarrow B_{0 \rightarrow i-1} + C + B_{j+1 \rightarrow \|B\|-1}$.

We distinguish two special bitmaps: 1_x and 0_x which are composed of x ones or x zeros respectively. We assume all bitmaps to be divided into 32bit subbitmaps called *words*. Given bitmap B, we denote the i-th word by $B(i)$ (0 based). Formally, $B(i) \equiv B_{i*32 \rightarrow i*32+31}$. In case, where the length of B is not the multiplication of 32, we assume the missing trailing bits to be 0. We distinguish several classes of words. Any word whose 31 less significant bits equal 1 is called a *pre-fill full word*. Any word whose 31 less significant bits equal 0 is called a *pre-fill empty word*. Any word D such, that $D_{30\to31} = 10b$, and the rest of the bits encode a number is called a *fill empty word*. Any word D such, that $D_{30\to31} = 11b$, and the rest of the bits encode a number is called a *fill full word*. Any word D such that $D_{31} = 0b$ and the rest of the bits are zeros and ones, is called a *tail word*.

Given any array A of numbers we define operation *exclusive scan* that creates array SA of the same size as A, such that $\forall_{k>0} SA[k] = \sum_{i=0}^{k-1} A[i] \wedge SA[0] = 0$.

A graphics card hardware (GPU, memory) will be called a *device*. The computer hardware (CPU, memory, motherboard), which sends tasks and data to the device, will be called a *host*. A function which is run concurrently in many threads on a device will be called a *kernel*. The subset of data stored in a database will be called a *query*. The process of finding data specified in a query by means of compressed bitmap indexes will be called *query processing*. During a query execution, bitmaps have to be decompressed and processed by bitwise operations.

3   Related Work

Multiple bitmap compression techniques have been proposed in the research literature. Some of them are based on the the run-length encoding, e.g., BBC [2], WAH [25,27,28], PLWAH [7], and some combine the run-length encoding with the Huffman compression, e.g., RL-Huffman [18], RLH [24]. The run-length encoding consists in representing a continuous vector of bits having the same value (either "0" or "1") as: (1) the common value of all bits in the vector and (2) the length of the vector. A bitmap is divided into words before being encoded. Words that include all ones or all zeros are compressed (they are called fills). Words that include intermixed zeros and ones cannot be compressed (they are called tails). Words are organized into *runs* that typically include a fill and a tail.

BBC divides bit vectors into 8-bit words, WAH and PLWAH divide them into 31-bit words, whereas RLH uses words of a parameterized length. PLWAH is the modification of WAH. PLWAH improves compression if tail T that follows fill F differs from F on few bits only. In such a case, the fill word encodes the difference between T and F on some dedicated bits. Moreover, BBC uses four different types of runs, depending on the length of a fill and the structure of a tail. WAH, PLWAH, and RLH use only one type of a run.

The compression techniques proposed in [18] and [24] additionally apply the Huffman compression [16] to the run-length encoded bitmaps. The main differences between [18] and [24] are as follows. First, in [18] only some bits in a bit vector are of interest, the others, called "don't cares" can be replaced either by zeros or ones, depending on the values of neighbor bits. In RLH all bits are of interest and have their exact values. Second, in [18] the lengths of homogeneous subvectors of bits are counted and become the symbols that are encoded by the Huffman compression. RLH uses run-length encoding for representing distances between bits having value 1. Next, the distances are encoded by the Huffman compression.

Utilizing GPUs in database applications is as of yet not a very well researched field of computer science. Most of the research is being focused on such areas, as: advanced rendering, image and volume processing as well as scientific computations (e.g., numerical algorithms and simulation). The application of GPUs to the compression of images has been presented in [9]. A few papers on increasing the processing power of typical database operations by means GPUs have been proposed so far. They mainly focus on efficient sorting (cf. [11,13,5]), evaluation of query predicates and computing aggregates (cf. [12]), query execution with the support of GPUs and processing indexes (R-trees [17], hash [10], inverted lists [8]). Some approaches to accelerating data mining techniques on GPUs have also been proposed (cf. [3,4,1,23]). None of the aforementioned approaches applies GPUs to compressing and decompressing bitmap indexes.

4   Algorithms

In this section we present three algorithms: an algorithm for extending an input bitmap (denoted as Algorithm 1), an algorithm for compressing an extended input bitmap (Algorithm 2) and an algorithm for decompressing of a compressed bitmap to its extended version (Algorithm 3). We also present several suggestions on query processing scheme using bitmap indexes and a device.

As the first step of compression, bitmaps are extended by Algorithm 1, which appends a single 0 bit after each consecutive 31bit subbitmap. The algorithm starts with appending zeros to the end of input bitmap B, so that its length is a multiplication of 31 (line 1). Next, the number n of 31bit subbitmaps of B is calculated. Once the value n is calculated, it is possible to find the size of the output bitmap E $(32*n)$ and allocate it (line 3). Given E, the algorithm, obtains subbitmaps of B, appends a 0 bit and stores the results in the appropriate words of E. Notice that each operation of storing refers to a different word of the output bitmap, and therefore each word may be computed in parallel.

Algorithm 1. Parallel extension of data

Require: : Input bitmap B
Ensure: : Extended input bitmap E
1: $B \leftarrow B + 0_{31-\|B\| \bmod 31}$
2: $n \leftarrow \|B\|/31$
3: Create bitmap E filled with zeros, such that $\|E\| = 32 * n$
4: **for** $i \leftarrow 0$ to $n-1$ **in parallel do**
5:     $E(i) \leftarrow B_{i*31 \rightarrow (i+1)*31-1} + 0_1$
6: **end for**
7: E contains the extended bitmap B

Extended bitmaps are compressed by Algorithm 2. It is composed of five stages. Executions of stages must be sequential, however each of these stages is composed of operations that may be executed in parallel.

The first stage (lines 2–7) determines classes of each of the words in the input bitmap E (whether each of the words is either a tail word or a pre-fill word). The most significant bit in each of the words is utilized to store the word class information. If the word is is a pre-fill word, we store 1 in the most significant bit. If the word is a tail word, we leave it without change as the most significant bit is zero by default (cf. Algorithm 1). To distinguish between a full and empty pre-fill word, one just needs to check the second most significant bit. Let us notice, that tail words already have the final form, consistent with the WAH algorithm. Pre-fill words also have correct two the most significant bits, but they are not yet rolled into a single word and their 30 less significant bits do not encode counters (i.e., they are not yet fill words). This will be achieved in the subsequent stages. Notice, that each word in this stage is processed independently and therefore all of the words may be calculated in parallel.

The second stage (lines 8–15) divides the input bitmap into blocks of words of a single class, where each block will be compressed (in the subsequent stages) into a single fill or tail word. To store the information about ending positions of the aforementioned blocks we use array F. F has the size equal to the number of words in the input bitmap E. The array stores 1 at position i if the corresponding word $E(i)$ in the bitmap is the last word of the block. Otherwise, the array stores 0. Word number i is the last word of the block if is a tail word (the most significant bit is equal to zero), or the word number $i + 1$ a pre-fill word of a different class (the words differ on the two most significant bits). This stage may also be easily parallelized, as each of the positions in array F may be calculated independently.

The third stage (line 16) performs an exclusive scan on array F and stores the result int array SF. The result of this operation is directly tied to storing compression results. As each block found in the previous stage will result in a single word in the compressed bitmap, we know that the algorithm will output as many words, as there are ones in array F. It is easy to notice, that for consecutive indexes i such that $F[i] = 1$, values $SF[i]$ will be consecutive natural numbers starting with zero. Such values, may therefore be used as the output indexes into the output compressed bitmap. Moreover, it is possible to obtain the number of ones in array F by summing the last value stored in array SF with last value in array F (notice, that the last value in F is always equal to 1). Efficient, parallel algorithms for performing the scan operation have been proposed in the research literature (cf. [15,22]).

The fourth stage (lines 18–23) prepares array T of the size equal to the number of words in the output bitmap. For each word $E(i)$, for which value $F[i]$ is equal to 1 (last words of the blocks) the algorithm stores, in array T at the position $SF[i]$, the number of words in all of the blocks up to, and including the considered word. The aforementioned number of words is equal to $i + 1$. Values stored in T are used by the last compression stage for calculating the numbers of words in blocks as well as they allow to retrieve words from the input bitmap E. This stage may be easily parallelized, as all of the writes are independent and may be performed in any order.

The last, fifth stage (lines 24–36) generates the final compressed bitmap. Computations performed in previous stages, allow to compute each word of the compressed bitmap in parallel and independently on the other words. The stage starts with obtaining the preprocessed words from bitmap E, from the positions, where array F stores ones (using the indexes stored in array T). Each of these words is a representative for its block. If the retrieved word is a tail word, it is stored in the output bitmap C in its original state. If the retrieved word is a pre-fill word, two operations are performed. First, the number of words in the corresponding block is calculated using the data stored in array T. The number of words in block i is equal to $T[i] - T[i-1]$, except for $i = 0$ where it is equal to $T[0]$. Second, the calculated number of words is encoded on 30 less significant bits of the retrieved pre-fill word (which creates a new fill word). Regardless of the class of the obtained word, it is stored in the output, i.e., compressed bitmap C. Once these operations are finished, bitmap C contains the compressed result. This stage may be easily parallelized as well, as all of the output words are computed independently.

Let us now analyze Algorithm 3 that implements decompression. It is composed of several stages, each of which must be completed, before the next one is started, however each stage may process input data in parallel.

The first stage (lines 1–9) creates array S of the size equal to the number of words in the compressed bitmap C. For every word $C(i)$ in the compressed bitmap, the algorithm calculates the number of words that should be generated in the output decompressed bitmap, based on the data contained in word $C(i)$, and store the calculated value in array S at position i. This stage is just a prerequisite for the next stage. Notice that this stage may be easily parallelized, as each value of S may be calculated independently.

The second stage (line 10) performs an exclusive scan on array S and stores the result in array SS. The result of this operation is directly tied to storing decompression results. Notice that after exclusive scan, for each word $C(i)$, array SS at position i stores the number of the word in the output decompressed bitmap at which decompression of the considered word should start. Based on the results of the exclusive scan one may also calculate the size of the output decompressed bitmap. This size is equal to the sum of the last values in arrays S and SS.

The third stage (lines 11–15) creates array F, whose size is equal to the number of words in the output decompressed bitmap. The array initially contains only zeros. Next, for each position $SS[i]$ stored in array SS we store 1 in array F at position $SS[i] - 1$. We omit position stored in $SS[0]$ as it is always equal to 0, and there are no entries of negative positions. The aim of this stage is to create an array, where 1 marks the end of the block into which some fill or tail word is extracted. Each assignment in this stage may be executed in parallel, as each assignment targets a different entry in array F.

The fourth stage (line 16) performs an exclusive scan on array F and stores the result in array SF. Once this stage is completed, array SF contains at each

---

Algorithm 2. Parallel compression of extended data

Require: : Extended input bitmap E
Ensure: : Compressed Bitmap C
 1: $n \leftarrow \|E\|/32$
 2: Create an array F of size n {0 based indexing}
 3: **for** $i \leftarrow 0$ to $n - 1$ **in parallel do**
 4:     **if** $E(i) = 0_{32}$ **or** $E(i) = 1_{31} + 0_1$ **then**
 5:         $E(i)_{31} \leftarrow 1b$
 6:     **end if**
 7: **end for**
 8: **for** $i \leftarrow 0$ to $n - 2$ **in parallel do**
 9:     **if** $E(i)_{30 \rightarrow 31} \neq E(i+1)_{30 \rightarrow 31}$ **or** $E(i)_{31} = 0$ **then**
10:         $F[i] \leftarrow 1$
11:     **else**
12:         $F[i] \leftarrow 0$
13:     **end if**
14: **end for**
15: $F[n - 1] \leftarrow 1$
16: $SF \leftarrow$ exclusive scan on the array F
17: $m \leftarrow F[n - 1] + SF[n - 1]$ {m is the number of words in the compresed bitmap}
18: Create an array T of size m {0 based indexing}
19: **for** $i \leftarrow 0$ to $n - 1$ **in parallel do**
20:     **if** $F[i] = 1$ **then**
21:         $T[SF[i]] \leftarrow i + 1$
22:     **end if**
23: **end for**
24: Create a bitmap C such, that $\|C\| = m * 32$
25: **for** $i \leftarrow 0$ to $m - 1$ **in parallel do**
26:     $j \leftarrow T[i] - 1$
27:     $X \leftarrow E_{(j)}$
28:     **if** $X_{31} = 1b$ **then**
29:         $count \leftarrow j + 1$
30:         **if** $i \neq 0$ **then**
31:             $count \leftarrow count - T[i - 1]$
32:         **end if**
33:         $X \leftarrow$ 30bit representation of $count + X_{30 \rightarrow 31}$
34:     **end if**
35:     $C(i) \leftarrow X$
36: **end for**
37: C contains the compressed bitmap E

---

position i the number of the word in the input compressed bitmap C, which should be used to generate output word $E(i)$.

Fifth stage (lines 17–29) performs the final decompression. For each word $E(i)$ in the output bitmap, the algorithm performs the following tasks. First, the number of the word in compressed bitmap C which should be used to generate word $E(i)$ is retrieved from array SF, from position i. Second, the word of the retrieved number is read from compressed bitmap C, and based on its type, value $E(i)$ is derived. If the retrieved word is a tail word, it is inserted into $E(i)$ without any further processing. If the retrieved word is a fill word, depending on whether it is an empty or full word, 0_{32} or $1_{31} + 0_1$ is inserted into $E(i)$, respectively. Once the last stage is finished, E contains the decompressed bitmap. As all of the output words are calculated independently, calculation of each word may be run in parallel.

Compressing/decompressing bitmaps using a GPU requires data transfer between the host memory and the device memory. This transfer is done by means of the PCI-Express x16 bus. The transfer is very slow as compared to the internal

---

Algorithm 3. Parallel decompression of compressed data

Require: : Compressed Bitmap C
Ensure: : Extended input bitmap E
 1: $m \leftarrow \|C\|/32$
 2: Create an array S of size m
 3: **for** $i \leftarrow 0$ to $m - 1$ **in parallel do**
 4:    **if** $C(i)_{31} = 0b$ **then**
 5:       $S[i] \leftarrow 1$
 6:    **else**
 7:       $S[i] \leftarrow$ the value of *count* encoded on bits $C(i)_{0\rightarrow30}$
 8:    **end if**
 9: **end for**
10: $SS \leftarrow$ exclusive scan on the array S
11: $n \leftarrow SS[m - 1] + S[m - 1]$ {n contains the number of words in a decompressed bitmap}
12: Create an array F of size n filled with zeroes {0 based indexing}
13: **for** $i \leftarrow 1$ to $m - 1$ **in parallel do**
14:    $F[SS[i] - 1] \leftarrow 1$
15: **end for**
16: $SF \leftarrow$ exclusive scan on the array F
17: Create a bitmap E of length $\|E\| = n * 32$
18: **for** $i \leftarrow 0$ to $n - 1$ **in parallel do**
19:    $D \leftarrow C(SF[i])$
20:    **if** $D_{31} = 0b$ **then**
21:       $E(i) \leftarrow D$
22:    **else**
23:       **if** $D_{30} = 0b$ **then**
24:          $E(i) \leftarrow 0_{32}$
25:       **else**
26:          $E(i) \leftarrow 1_{31} + 0_1$
27:       **end if**
28:    **end if**
29: **end for**
30: E contains a decompressed bitmap C

Algorithm 4. Extension and checking of classes of words

```
1: start=i*31;
2: off=start&31; // %32
3: start>>=5; // /32
4: result=(B[start]>>off)&BM31;
5: result|=(B[start+1]<<(32-off))&BM31;
6: test=(result==0 || result==BM31);
7: result=result | (-test & BMMSB);
```

Algorithm 5. Calculating of the number of words to be extracted from a compressed word

---
```
1: bool test=(data&BMMSB)!=0;
2: res=(!test)|((data&BM30)&(-test));
```

Algorithm 6. Deriving of the output word based on the compressed word

---
```
1: bool testCase1=((data&BM2MSB)!=BMMSB);
2: bool testCase2=((data&BM2MSB)==BM2MSB);
3: res=(data&(-testCase1))|(-testCase2);
4: res=res&BM31
```

---

device memory bandwidth [20]. This problem can be partially eliminated by processing all of the query on the device. There are several benefits of such an approach: (1) there is no need to download decompressed bitmaps from the device, (2) only compressed bitmaps need to be uploaded that are small and the transfer may be performed in parallel with computations performed on the device, (3) the computing power of the device can be used in order to perform bitwise operations. The only task of the host during the query processing should be to initiate data transfers when needed and to start kernels on the device. Other than that, the host is free to do any other tasks. The device should decompress the received bitmaps and perform bitwise operations on them. Once all of the calculations are finished, the final bitmap should be transfered from the device to the host. This last stage is unfortunately very slow as the device→host transfers are the slowest. Moreover, the resulting bitmap is decompressed, and therefore very large. Nonetheless it is beneficial, as we only need to transfer one such bitmap (we would have to download every decompressed bitmap if the query was performed on the host).

5   Implementation

The algorithms presented in the previous section were implemented in C++ and C for CUDA using the NVIDIA CUDA platform [6]. In this section we outline their implementations and focus on the implementation details that allowed us to remove almost all of branching (alternative flows of control, e.g. if-then-else structures) from the kernel code.

In our implementation we used a straightforward storage model for bitmaps. Each bitmap is represented as an array of 32bit unsigned integers. The same representation is used on the device and the host. For the exclusive scan operation on the CUDA platform we use a very efficient implementation, which is a part of the CUDPP library [14].

The code fragments, presented in the following paragraphs, utilize several constants: BM31, BMMSB, and BM30. All of these constants are 32bit unsigned integers and contain values 0x7fffffff, 0x80000000, and 0x3fffffff respectively.

In our implementation we have created two compression procedures, where one of them incorporates the extension algorithm and the other does not. The first of these two implementations integrates the extension algorithm with the first stage of compression. Let us consider the aforementioned, integrated portion of the source code (cf. Algorithm 4). This code performs two operations: it extracts 31 bit subbitmap from the input bitmap and appends either 0 or 1 depending on whether the retrieved 31 bit subbitmap contains only the same bits (a pre-fill word) or both values of bits (a tail word).

The next source code is a fragment of the implementation of the decompression algorithm (cf. Algorithm 5). This fragment calculates the number of words that should be generated from the given fill or tail word stored in variable data, and it roughly corresponds to the lines 4–8 of the algorithm 3. It does not require any if-the-else structures.

The last fragment of the source code (cf. Algorithm 6) also represents the implementation part of the decompression algorithm. This code derives, based on the given fill or tail word (stored in `data`), a word that should be stored into the output decompressed bitmap, and it roughly corresponds to the lines 20–28 of the algorithm 3. It does not require any if-the-else structures as well.

6   Experiments

Experiments were performed on an Core i7 2.8GHz CPU and NVIDIA Geforce 285 GTX graphics card. Their aim was to measure:

- compression, decompression, and recompression (applied to an extended bitmap, cf. Algorithm 1) time using CPU,
- compression, decompression, and recompression time using GPU,
- time of uploading the input bitmap to the graphics cards memory (for each type of operation separately),
- time of downloading the output bitmap from the graphics cards memory (for each type of operation separately).

Each of the tested input bitmaps was composed of $96 * 10^7$ bits. In the experiments we used bitmaps with their densities varying from 0.5 to $1/65536$ ($1/2^i$ where $i = 1, 2, \ldots, 16$). The bits whose values were set to 1 were randomly selected. We generated 10 instances of each of the bitmaps for each experiment. The execution times discussed below represents averages of 10 experiments.

The results of the experiments for compression and recompression are presented in Figures 1 and 2, respectively. Both of these figures are very similar, as essentially both of them present results from measuring the same algorithm. The only difference is that the compression includes the extension algorithm and the recompression does not. While comparing these two charts one may notice, that the extension algorithm requires about 70ms on host for the tested bitmaps, and indeed the difference between compression and recompression time is about 57ms in the best case and 115ms and the worst case. The same difference on the device is much smaller: 0.85ms in the best case and 1,12ms in the worst case.

While analyzing Figures 1 and 2 one may also notice that the compression is about 16–24 times faster on the device, than it is on the host. Unfortunately, if we include the transfer times, the difference in speed is reduced to 3.6–5.8 times faster than host. The similar numbers calculated for recompression are 12.4–21 times faster on device without data transfers and 2.6–4 times faster on device with data transfers. One may also notice, that the device→host transfer times monotonically depend on the bitmap density as the less dense the bitmap is, the smaller the compression (recompression) result.

Let us now consider Figure 3 that presents comparison of the decompression times on the device and the host. The decompression on the device is 4.75–10 times faster than on the host. Unfortunately, after decompression, one must transfer the large decompressed bitmap from the device to the host memory over the slow PCI-Express x16 bus. Moreover, graphics cards are designed to optimize the host→device transfers rather then back. This results in large transfer

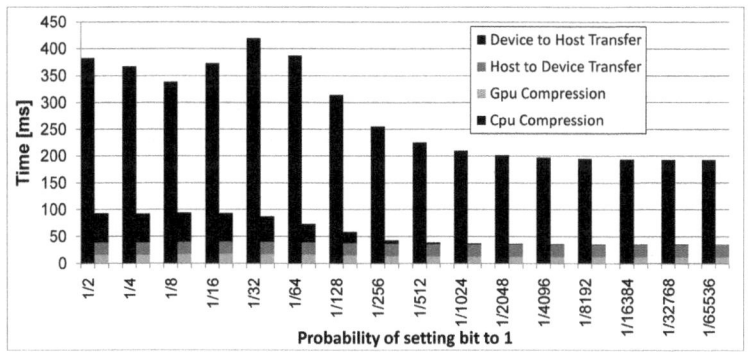

Fig. 1. Compression and data transfer times for bitmaps of varying density

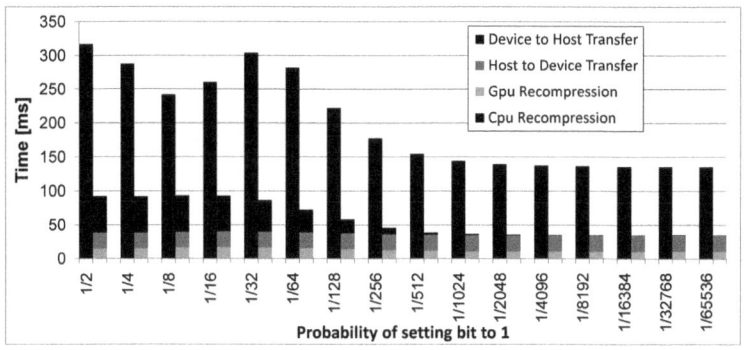

Fig. 2. Recompression and data transfer times for bitmaps of varying density

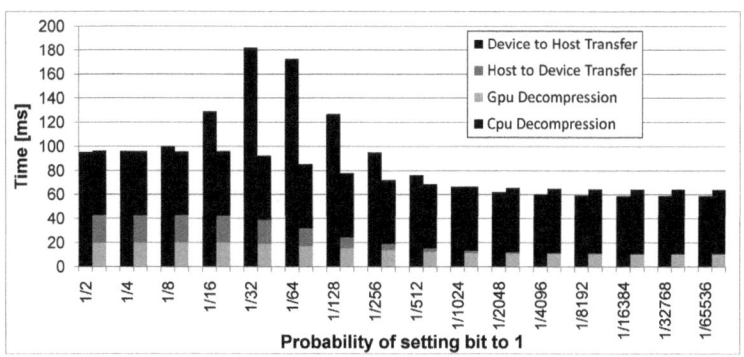

Fig. 3. Decompression and data transfer times for bitmaps of varying density

times that consume all of the benefit from performing the decompression on the device. From Figure 3 we can observe that the whole process of transferring and decompressing a bitmap may be even slower on the device, than it is on the host. We may also notice, that host→device transfer times depend monotonically on the bitmap density, similarly as observed for the device→host transfers during compression/recompression of data. In this case, the same data is sent back to the device in order to be decompressed.

Let us consider the observed problem of large data transfer times between the device and the host. Let us notice, that even when the total decompression time is a bit slower on the device we still benefit from the fact, that the host is free and may in the same time perform other tasks. Moreover, we may also utilize the fact, that data transfers and computations on the device may be performed in parallel. However, as was suggested in the section 4, by performing the whole query on the device, we may be able to accelerate query processing, while still freeing the host from most of the work and removing the need for the costly device→host transfers.

As an example, let us consider an attribute whose domain cardinality is equal to 1024. As shown in Figure 3 total decompression time on the device and host are almost equal. The decompression on the host requires 66.42ms, whereas the decompression on the device requires 66.64ms, where the host→device transfer took 1.64ms, the transfer device→host took 53.46ms and the decompression itself took 11.54ms. To analyze the query performance time we also need the times of performing bitwise operations on bitmaps. The time of performing a bitwise operation on two bitmaps of size $96 * 10^7$ bits on the device consumes about 2.79ms. The same time on the host consumes about 62.67ms. Let us also assume, that the buffer for the compressed bitmaps on the device may contain at most two compressed bitmaps. Let us now consider a hypothetical query which requires 3 bitmaps to be decompressed and processed.

Figure 4 illustrates the hypothetical query processing in the Gannt chart. The query processing starts with the DMA data transfer of the first compressed bitmap $C1$ from the host to the device. Once the transfer is finished, the device may start decompression of the transfered data (the result is stored in bitmap $E1$). Meanwhile, the transfer of the second compressed bitmap $C2$ is started. Once the bitmap $C1$ is decompressed, it may be discarded and the transfer of the last compressed bitmap $C3$ may start. While bitmap $C3$ is transfered, the device may decompress bitmap $C2$ and store the decompression result in $E2$. Once bitmaps $C1$ and $C2$ are decompressed, the device may perform an in-place bitwise operation on their decompressed versions ($E1$ and $E2$), and store the result in $E1$. Bitmap $E2$ may now be discarded, and the decompression of the compressed bitmap $C3$ may start (the result is stored in $E3$). Once bitmap $C3$ is decompressed, the device may perform an in-place bitwise operation between bitmaps $E1$ and $E3$, storing the result in $E1$. $E1$ now stores a decompressed bitmap with the results of the query and may therefore be sent to the host. The whole query processing time, including transfers, takes 95.61ms. Notice, that the query is performed only on the device, and data transfers are performed by

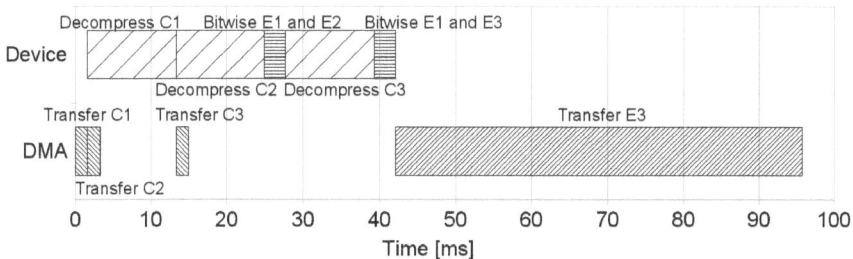

Fig. 4. Gannt chart showing the query processing on the GPU

the DMA. The only work performed by the host is the initialization of DMA transfers and starting of kernels.

Let us now consider the same query, but performed entirely on the host. We need to perform 3 decompressions and 2 bitwise operations on the decompressed bitmaps, which gives the total time equal to 3*66.42ms+2*62.67ms=324.6ms. Notice, that for such a simple case, we have processed a query over 3 times faster on the device, than on the host. The more bitmaps are used, the higher the speedup is. Moreover, while the device processes a query, the host is free to do any other tasks.

We would also like to address the strange, non-monotonic compression and decompression times dependency on the bitmap density. They are probably caused by an increased number of misses in the branch prediction. Notice, that the worst processing time appears for bitmaps of density 1/32 and similar. For such bitmaps, there is a high probability of interleaving of fill and tail words. In our implementation, different parts of code are used for each type of word. We count the words sequentially, and therefore some optimizations presented in the previous sections do not apply here. As there is a high probability that the next generated word will be a different than the previous one, it may cause the branch prediction algorithm to give inaccurate results and lead to the decreased efficiency of data processing.

7   Summary

In this paper we presented an extension, called GPU-WAH, of the WAH compression/decompression technique. GPU-WAH allows to parallelize compressing and decompressing steps and to execute them on the graphics processing units on the CUDA platform. We also discussed the implementation of GPU-WAH and presented its experimental comparison to the standard CPU-based WAH. As the experiments showed, GPU-WAH performs 3.6-5.8 times faster than the CPU-based WAH. We concluded that the decompression is several times faster on a GPU than on the CPU, but the data transfer between GPU memory and computer memory is a bottleneck. Nonetheless, we have presented a query processing scheme which may reduce query processing time by several times.

Future work will focus on: (1) implementing on the CUDA platform other compression techniques including BBC, PLWAH, RLH-n, and comparing their efficiency, (2) implementing query processing technique according to the proposed scheme and test whether its performance is consistent with the theoretical results, (3) applying several optimizations dedicated to given graphics cards computing capabilities, including the upcoming FERMI architecture[19].

References

1. Andrzejewski, W.: Fast K-Medoids Clustering on PCs. In: ADMKD Workshop (2007)
2. Antoshenkov, G., Ziauddin, M.: Query processing and optimization in Oracle RDB. VLDB Journal 5(4), 229–237 (1996)
3. Böhm, C., Noll, R., Plant, C., Wackersreuther, B.: Density-based clustering using graphics processors. In: Proc. of ACM Conference on Information and Knowledge Management (CIKM), pp. 661–670 (2009)
4. Cao, F., Tung, A.K.H., Zhou, A.: Scalable Clustering using graphics processors. In: Yu, J.X., Kitsuregawa, M., Leong, H.-V. (eds.) WAIM 2006. LNCS, vol. 4016, pp. 372–384. Springer, Heidelberg (2006)
5. Chen, S., Zhao, J., Qin, J., Xie, Y., Heng, P.-A.: An efficient sorting algorithm with CUDA. Journal of the Chinese Institute of Engineers 32(7), 915–921 (2009)
6. CUDA. What is CUDA?,
   http://www.nvidia.com/object/what_is_cuda_new.html
7. Deliège, F.: Concepts and Techniques for Flexible and Effective Music Data Management. PhD thesis, Aalborg University, Denmark (2009)
8. Ding, S., He, J., Yan, H., Suel, T.: Using graphics processors for high-performance ir query processing. In: Proc. of Int. Conf. on World Wide Web, pp. 1213–1214 (2008)
9. Erra, U.: Toward real time fractal image compression using graphics hardware. In: Bebis, G., Boyle, R., Koracin, D., Parvin, B. (eds.) ISVC 2005. LNCS, vol. 3804, pp. 723–728. Springer, Heidelberg (2005)
10. Gosink, L.J., Wu, K., Bethel, E.W., Owens, J.D., Joy, K.I.: Bin-hash indexing: A parallel method for fast query processing. Research report, Lawrence Berkeley National Laboratory (2008)
11. Govindaraju, N., Gray, J., Kumar, R., Manocha, D.: GPUTeraSort: high performance graphics co-processor sorting for large database management. In: Proc. of ACM SIGMOD Int. Conf. on Management of Data, pp. 325–336 (2006)
12. Govindaraju, N.K., Lloyd, B., Wang, W., Lin, M., Manocha, D.: Fast computation of database operations using graphics processors. In: Proc. of ACM SIGMOD Int. Conference on Management of Data, pp. 215–226 (2004)
13. Greß, A., Zachmann, G.: GPU-ABiSort: Optimal Parallel Sorting on Stream Architectures. In: IEEE International Parallel and Distributed Processing Symposium (IPDPS), p. 45 (2006)
14. Harris, M., Owens, J.D., Sengupta, S., Tseng, S., Zhang, Y., Davidson, A., Satish, N.: CUDA Data Parallel Primitives Library (CUDPP)
15. Harris, M., Sengupta, S., Owens, J.D.: Parallel prefix sum (scan) with cuda. In: GPU Gems 3. Addison-Wesley, Reading (2007)
16. Huffman, D.A.: A method for the construction of minimum-redundancy codes. In: Proc. of the Institute of Radio Engineers, pp. 1098–1101 (1952)

17. Kunjir, M., Manthramurthy, A.: Using graphics processing in spatial indexing algorithms. Research report, Indian Institute of Science, Database Systems Lab. (2009)
18. Nourani, M., Tehranipour, M.H.: Rl-huffman encoding for test compression and power reduction in scan applications. ACM Transactions on Design Automation of Electronic Systems 10(1), 91–115 (2005)
19. NVIDIA. NVIDIA's Next Generation CUDA Compute Architecture: Fermi. White Paper, NVIDIA
20. NVIDIA CUDA Toolkit 2.3. NVIDIA CUDA C Programming Best Practices Guide
21. O'Neil, P., Quass, D.: Improved query performance with variant indexes. In: Proc. of ACM SIGMOD Int. Conference on Management of Data, pp. 38–49 (1997)
22. Sengupta, S., Harris, M., Zhang, Y., Owens, J.D.: Scan primitives for gpu computing. In: Graphics Hardware 2007, pp. 97–106. ACM, New York (2007)
23. Shalom, S.A.A., Dash, M., Minh, T.: Efficient K-Means Clustering Using Accelerated Graphics Processors. In: Song, I.-Y., Eder, J., Nguyen, T.M. (eds.) DaWaK 2008. LNCS, vol. 5182, pp. 166–175. Springer, Heidelberg (2008)
24. Stabno, M., Wrembel, R.: RLH: Bitmap compression technique based on run-length and Huffman encoding. Information Systems 34(4-5), 400–414 (2009)
25. Stockinger, K., Wu, K.: Bitmap indices for data warehouses. In: Wrembel, R., Koncilia, C. (eds.) Data Warehouses and OLAP: Concepts, Architectures and Solutions, pp. 157–178. Idea Group Inc., USA (2007) ISBN 1-59904-364-5
26. Wu, K., Otoo, E.J., Shoshani, A.: Compressing bitmap indexes for faster search operations. In: Proc. of Int. Conference on Scientific and Statistical Database Management (SSDBM), pp. 99–108 (2002)
27. Wu, K., Otoo, E.J., Shoshani, A.: On the performance of bitmap indices for high cardinality attributes. In: Proc. of Int. Conference on Very Large Data Bases (VLDB), pp. 24–35 (2004)
28. Wu, K., Otoo, E.J., Shoshani, A.: Optimizing bitmap indices with efficient compression. ACM Transactions on Database Systems (TODS) 31(1), 1–38 (2006)
29. Wu, M., Buchmann, A.: Encoded bitmap indexing for data warehouses. In: Proc. of Int. Conference on Data Engineering (ICDE), pp. 220–230 (1998)

Containment of Conjunctive Queries with Negation: Algorithms and Experiments

Khalil Ben Mohamed, Michel Leclère, and Marie-Laure Mugnier

LIRMM (CNRS - University of Montpellier), France
{benmohamed,leclere,mugnier}@lirmm.fr

Abstract. We consider the containment problem for conjunctive queries with atomic negation. Firstly, we refine an existing algorithm based on homomorphism checks, which itself improves other known algorithms in databases, and analyze it experimentally. Secondly, we present a new algorithm based on the translation of the containment problem into the problem of checking the unsatisfiability of a propositional logical formula, which allows us to use a SAT solver, and we experimentally compare both algorithms.

1 Introduction

The query containment problem is a fundamental problem in databases. It takes two queries q_1 and q_2 as input, and asks if q_1 is contained in q_2 (noted $q_1 \sqsubseteq q_2$), i.e. if the set of answers to q_1 is included in the set of answers to q_2 for all databases (e.g. [AHV95]). Algorithms based on query containment can be used to solve various problems, such as query evaluation and optimization [CM77][ASU79], rewriting queries using views [Hal01], detecting independence of queries from database updates [LS93], etc. In this paper, we consider the problem of deciding on containment for conjunctive queries with atomic negation (denoted $CQC^\neg$ hereafter). The so-called (positive) conjunctive queries form a class of natural and frequently used queries and are considered as the basic database queries [CM77]. Conjunctive queries with negation extend this class with negation on atoms. Note that $CQC^\neg$ is equivalent to important problems in artificial intelligence, such that: checking entailment / deduction between two first-order logic clauses (without function); query answering with boolean conjunctive queries with negation on a knowledge base composed of a set of positive and negative factual assertions (while making the open-world assumption).

When only positive conjunctive queries are considered, query containment checking is NP-complete [AHV95]. When atomic negation is considered, the problem becomes much more complex: it is π_2^P-complete[1] [FNTU07][CM09] and very few algorithms for solving it can be found in the literature.

This paper is devoted to refining and proposing algorithms solving $CQC^\neg$ and testing them experimentally. An algorithm scheme was introduced in [LM07], which itself improves the previous proposals in [Ull97] and [WL03]. All three algorithms use homomorphism as a core notion. We first compare experimentally several heuristics, which

---

[1] $\pi_2^P = (co\text{-}NP)^{NP}$.

P. García Bringas et al. (Eds.): DEXA 2010, Part II, LNCS 6262, pp. 330–345, 2010.

allows to refine this algorithm scheme. We then propose another approach, which consists of building a propositional logical formula from the queries q_1 and q_2, such that q_1 is contained in q_2 if and only if this formula is valid, i.e. always true. Equivalently, the negation of this formula is unsatisfiable, which allows to use a SAT solver (SAT is the problem of deciding whether a given propositional formula is satisfiable) and benefit from practical improvements achieved in this domain [Sai08]. However, the translation of the queries into the propositional formula is generally exponential in the size of the queries. Thus the question is whether –or when– the second algorithm can be better than the first one. We provide first experimental answers to this question.

Due to the lack of benchmarks or real-world data available for CQC¬, we built a random generator. We analyzed the influence of several parameter values on the problem instance difficulty in order to define difficult instances, on which the algorithms were run. In databases, conjunctive queries with negation are generally imposed to be *safe*, i.e. all variables in the query must occur in at least one positive subgoal. Hence, even if all of our results hold for general conjunctive queries with negation, we restrict the experiments to safe queries in this paper.

Paper layout. Section 2 recalls the framework of [LM07] and expresses previously proposed algorithm schemes in this framework. In Section 3, we present our experimental methodology and use it to compare several heuristics, which leads to propose a refined algorithm. The second approach is presented and compared to the refined algorithm in 4. Section 5 outlines the prospects of this work.

2   Framework

We recall here some basic definitions about queries with Datalog-like notations. More details can be found in [AHV95] for instance. A *conjunctive query with negation (CQ¬)* is of the form: $q = ans(x_1 \ldots x_q) \leftarrow p_1, \ldots, p_n, n_1, \ldots, n_m$, where each p_i (resp. n_i) is a positive (resp. negative) *subgoal*, $1 \leq n + m$, and *ans* is a special relation (which defines the answer part of the query). The left part of the query is called its *head* and the right part is its *body*. Each subgoal is of form $r(t_1, \ldots, t_k)$ (positive subgoal) or $\neg r(t_1, \ldots, t_k)$ (negative subgoal) where r is a relation and $t_1, \ldots, t_k$ is a tuple of terms (i.e. variables or constants). All variables $x_1 \ldots x_q$ occur at least once in the body of the query. Without loss of generality, we assume that the same subgoal does not appear twice in the body of the query. A $CQ¬$ is *boolean* if it has no variable in its head (we note *ans()*). A $CQ¬$ is *positive* if it has no negative subgoal ($m = 0$). A $CQ¬$ is *safe* if each variable occurring in a negative subgoal also occurs in a positive one.

In the following, we will focus on boolean queries because having a non-empty *ans* part can only make the query containment problem easier. For the same reason, we can consider that queries contain no constants. Note however that the framework and all results hold for general $CQ¬$.

In [LM07], $CQ¬$ are seen as labeled graphs. This allows to rely on graph notions that have no simple equivalent in logic (such as pure subgraphs, see later). More precisely, a $CQ¬$ q is represented as a bipartite, undirected and labeled graph Q, called *polarized graph (PG)*, with two kinds of nodes: term nodes and relation nodes. Each term of the query becomes a term node, that is unlabeled if it is a variable, otherwise it is labeled by the constant itself. A positive (resp. negative) subgoal with relation r becomes a relation node labeled $+r$ (resp. $-r$) and it is linked to the nodes assigned to its terms. The labels on edges correspond to the position of each term in the subgoal (see Figure 1 for an example). For simplicity, the subgraph corresponding to a subgoal, i.e. induced by a relation node and its neighbors, is also called a *subgoal*. We note it $+r(t_1,\ldots,t_k)$ (resp. $-r(t_1,\ldots,t_k)$) if the relation node has label $+r$ (resp. $-r$) and list of neighbors $t_1,\ldots,t_k$. We note $\sim r(t_1,\ldots,t_k)$ a subgoal that can be positive or negative, i.e. $\sim \in \{+,-\}$. Subgoals $+r(t_1,\ldots,t_k)$ and $-r(u_1,\ldots,u_n)$ with the same relation but different signs are said to be *opposite*. Opposite Subgoals $+r(t_1,\ldots,t_k)$ and $-r(t_1,\ldots,t_k)$ with the same list of arguments are said to be *contradictory*. Given a relation node label (resp. subgoal) l, $\bar{l}$ denotes the complementary relation node label (resp. subgoal) of l, i.e. it is obtained from l by reversing its sign. Queries are denoted by small letters (q_1 and q_2) and the associated graphs by the corresponding capital letters (Q_1 and Q_2). We note $Q_1 \sqsubseteq Q_2$ iff $q_1 \sqsubseteq q_2$. A PG is *consistent* if it does not contain two contradictory subgoals.

Homomorphism is a core notion in this work. A *homomorphism* h from a PG Q_2 to a PG Q_1 is a mapping from nodes of Q_2 to nodes of Q_1, which preserves bipartition (the image of a term -resp relation- node is a term -resp. relation- node), preserves edges (if rt is an edge with label i in Q_2 then $h(r)h(t)$ is an edge with label i in Q_1), preserves relation node labels (a relation node and its image have the same label) and can instantiate term node labels (if a term node is labeled by a constant, its image has the same label, otherwise there is no constraint on the label of its image). Note that this notion corresponds exactly to the well-known *query homomorphism* defined on positive conjunctive queries; it can be seen as an extension of query homomorphism to negative subgoals.

When there is a homomorphism h from Q_2 to Q_1, we say that Q_2 *maps to* Q_1 by h. Q_2 is called the *source* graph and Q_1 the *target* graph. If Q_2 and Q_1 have only positive subgoals, $Q_1 \sqsubseteq Q_2$ iff Q_2 maps to Q_1. When we also consider negative subgoals, only one side of this property remains true: if Q_2 maps to Q_1 then $Q_1 \sqsubseteq Q_2$; the converse is false, as shown in Example 1.

Example 1. See Figure 1: Q_2 does not map to Q_1 but $Q_1 \sqsubseteq Q_2$. Indeed, if we complete q_1 w.r.t. relation p, we obtain the union of four queries $q_{1,1} = ans() \leftarrow p(t) \wedge s(t,u) \wedge s(u,v) \wedge s(v,w) \wedge \neg p(w) \wedge p(u) \wedge p(v)$, $q_{1,2} = ans() \leftarrow p(t) \wedge s(t,u) \wedge s(u,v) \wedge s(v,w) \wedge \neg p(w) \wedge \neg p(u) \wedge p(v)$, $q_{1,3} = ans() \leftarrow p(t) \wedge s(t,u) \wedge s(u,v) \wedge s(v,w) \wedge \neg p(w) \wedge p(u) \wedge \neg p(v)$ and $q_{1,4} = ans() \leftarrow p(t) \wedge s(t,u) \wedge s(u,v) \wedge s(v,w) \wedge \neg p(w) \wedge \neg p(u) \wedge \neg p(v)$. Each of the queries is a way of completing q_1 w.r.t. p. Q_2 maps to each of the graphs associated with them. Thus q_1 is contained in q_2.

One way to solve CQC$^\neg$ is therefore to generate all "complete" PGs obtained from Q_1 using relations appearing in Q_1, and then to test if Q_2 maps to each of these graphs.

Definition 1 (Complete graph and completion). *Let Q be a consistent PG. It is complete w.r.t. a set of relations $\mathcal{P}$, if for each $p \in \mathcal{P}$ with arity k, for each k-tuple of term nodes (not necessarily distinct) $t_1,\ldots,t_k$ in Q, it contains $+p(t_1,\ldots,t_k)$ or $-p(t_1,\ldots,t_k)$. A completion Q' of Q is a PG obtained from Q by repeatedly adding*

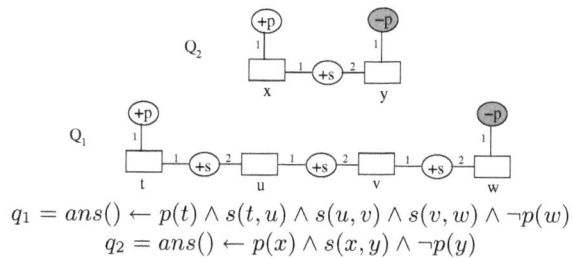

$$q_1 = ans() \leftarrow p(t) \land s(t,u) \land s(u,v) \land s(v,w) \land \neg p(w)$$
$$q_2 = ans() \leftarrow p(x) \land s(x,y) \land \neg p(y)$$

Fig. 1. Polarized graphs associated with q_1 and q_2

new relation nodes (on term nodes present in Q) without yielding inconsistency. Each addition is a completion step. *A completion of Q is called* total *if it is a complete graph w.r.t. the set of relations considered, otherwise it is called* partial.

Theorem 1. *[LM07] Let Q_1 and Q_2 be two PGs (Q_1 consistent), $Q_1 \sqsubseteq Q_2$ iff Q_2 maps to all total completions of Q_1 w.r.t. the set of relations appearing in Q_1.*

We can further restrict the set of relations considered to those appearing in opposite subgoals both in Q_2 and in Q_1 [LM07]. In the sequel, this set is called the *completion vocabulary* of Q_2 and Q_1 and denoted $\mathcal{V}$.

Let us outline previous algorithmic proposals for checking containment of $CQ^\neg$ queries in this framework. Although it is expressed in a different framework, the first proposal [Ull97] can be recast as follows (see [LM07] for more details): it consists of computing the set of total completions of Q_1 and checking the existence of a homomorphism from Q_2 to each of them (say Q_1^c). The complexity of this algorithm is prohibitive: $\mathcal{O}(2^{(n_{Q_1})^k \times |\mathcal{V}|} \times hom(Q_2, Q_1^c))$, where n_{Q_1} is the number of term nodes in Q_1, k is the maximum arity of a relation, $\mathcal{V}$ is the completion vocabulary and $hom(Q_2, Q_1^c)$ is the complexity of checking the existence of a homomorphism[2] from Q_2 to Q_1^c.

Two kinds of improvements are defined in [WL03] and [LM07]: first, some necessary conditions for containment are exhibited, which can be used to tentatively detect a failure before generating completions; secondly, completions can be incrementally built and checked.

In [WL03], the following necessary but not sufficient condition for containment is exhibited (for *safe* queries but it remains true for general $CQ^\neg$): if $Q_1 \sqsubseteq Q_2$ then there must be a homomorphism, say h, from the positive part of Q_2, say Q_2^p, to Q_1; moreover, this homomorphism should not contradict the negative subgoals of Q_2: for all subgoals $-r(t_1, \ldots, t_k)$ in Q_2, Q_1 should not contain $+r(h(t_1), \ldots, h(t_k))$. This property can be used as a filter: if there is no such homomorphism from Q_2^p to Q_1, then $Q_1 \not\sqsubseteq Q_2$. It is generalized in [LM07] with the notion of *pure subgraphs* and *compatible* homomorphism, as detailed below. Then we have: if $Q_1 \sqsubseteq Q_2$ then, for each *pure* subgraph Q_2' of Q_2, there must be a *compatible* homomorphism from Q_2' to

---

[2] Homomorphism checking is NP-complete. A brute-force algorithm solves it in $\mathcal{O}(n_{Q_1}^{n_{Q_2}})$, where n_{Q_2} is the number of term nodes in Q_2.

Q_1 w.r.t. Q_2. In the following definitions, we add some notions and notations that we will use in the sequel of this paper.

Definition 2 (pure subgraph). *A PG is said to be* pure *if it does not contain opposite subgoals (i.e. each relation appears only in one form, positive or negative). A pure sub-graph of Q_2 is a subgraph of Q_2 that contains all term nodes in Q_2 (but not necessarily all relation nodes)[3] and is pure.*

We will use the following notations for pure subgraphs of Q_2:

- Q_2^{max} denotes a pure subgraph that is maximal for inclusion;
- Q_2^+ is the Q_2^{max} with all positive relation nodes in Q_2;
- Q_2^- is the Q_2^{max} with all negative relation nodes in Q_2;
- Q_2^{Max} denotes a Q_2^{max} of maximal cardinality.

Note that if a relation label $+r$ or $-r$ in Q_2 does not appear in Q_1 then $Q_1 \not\sqsubseteq Q_2$. Thus, we assume in the following that all relation labels in Q_2 appear in Q_1, which implies that the subgraph induced by the relations not in $\mathcal{V}$ is pure, and all Q_2^{max} contain it. Hence, we have Q_2^p (the positive part of Q_2) $\subseteq Q_2^+$ but the contrary is generally false.

Example 2. See Figure 1: there are two Q_2^{max} (which are also of maximal cardinality): the first one is Q_2^+, which contains $+p(x)$ and $+s(x,y)$, and the second one is Q_2^-, which contains $-p(x)$ and $+s(x,y)$ (because $\mathcal{V} = \{p\}$, thus $s \notin \mathcal{V}$).

Intuitively, a homomorphism from a pure subgraph of Q_2 to Q_1 is "compatible" if it can be extended to a homomorphism from Q_2 to a total completion of Q_1.

Definition 3 (Border, Compatible homomorphism). *Let Q_2 and Q_1 be two PGs and Q_2' be a pure subgraph of Q_2. The relation nodes of $Q_2 \setminus Q_2'$ are called* border rela-tion nodes *of Q_2' w.r.t. Q_2. A homomorphism h from Q_2' to Q_1 is said to be* compatible *w.r.t. Q_2 if, for each border relation node inducing the subgoal $\sim r(t_1, \ldots, t_k)$, the opposite subgoal $\overline{\sim r}(h(t_1), \ldots, h(t_k))$ is not in Q_1 and for each pair of opposite bor-der relation nodes respectively on $(c_1, \ldots, c_k)$ and $(d_1, \ldots, d_k)$, $(h(c_1), \ldots, h(c_k)) \neq (h(d_1), \ldots, h(d_k))$.[4]*

Now, let us consider the search space leading from Q_1 to its total completions and partially ordered by the relation "subgraph of". In [LM07], this space is explored as a binary tree with Q_1 as root. The children of a node are obtained by adding, to the graph associated with this node (say Q_1'), a relation node in positive and negative form (each of the two new graphs is thus obtained by a completion step from Q_1'). The aim is to find a set of partial completions covering the set of total completions of Q_1, i.e. the question becomes: "is there a set of partial completions $\{Q_{1,1}, \ldots, Q_{1,n}\}$ of Q_1 such that (1) Q_2

---

[3] Note that this subgraph does not necessarily correspond to a set of subgoals because some term nodes may be isolated.

[4] The last condition is necessary to ensure that a compatible homomorphism from Q_2' to Q_1 can be extended to a homomorphism from Q_2 to a total completion of Q_1. However, it is necessarily satisfied if Q_2' is a pure subgraph that is maximal for inclusion. We only need it for the second approach presented in this paper.

maps to each $Q_{1,i}$ for $i = 1 \ldots n$; (2) each total completion Q_1^c of Q_1 is covered by a $Q_{1,i}$ (i.e. $Q_{1,i}$ is a subgraph of Q_1^c) ?". After each completion step, it is checked whether Q_2 maps to the current partial completion: if yes, this completion is one of the sought $Q_{1,i}$, otherwise the exploration continues. Figure 2 illustrates this method on the very easy Example 1. Two graphs $Q_{1,1}$ and $Q_{1,2}$ are built from Q_1, respectively by adding $+p(v)$ and $-p(v)$. Q_2 maps to $Q_{1,1}$, thus there is no need to complete $Q_{1,1}$. Q_2 does not map to $Q_{1,2}$: two graphs $Q_{1,3}$ and $Q_{1,4}$ are built from $Q_{1,2}$, by adding $+p(u)$ and $-p(u)$ to $Q_{1,2}$. Q_2 maps to $Q_{1,3}$ and to $Q_{1,4}$, respectively. Finally, the set proving that Q_1 is included in Q_2 is $\{Q_{1,1}, Q_{1,3}, Q_{1,4}\}$ (and there are four total completions of Q_1 w.r.t. p). Algorithm 1 implements this method (the numbers in the margin are relative to the refinements studied in Section 3.2).

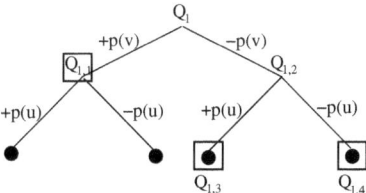

Fig. 2. The search tree of Example 1. Each black dot represents a Q_1^c and each square a $Q_{1,i}$.

Algorithm 1. recCheck(Q_1)

Input: a consistent PG Q_1
Data: Q_2, $\mathcal{V}$
Result: true if $Q_1 \sqsubseteq Q_2$, false otherwise
begin

    if *there is a homomorphism from Q_2 to Q_1* **then return** *true* ;
    if Q_1 *is complete w.r.t. $\mathcal{V}$* **then return** *false* ;

(3)   *** *Filtering step* ***\\
(1)   Choose $r \in \mathcal{V}$ and $t_1, \ldots, t_k$ in Q_1 such that $+r(t_1, \ldots, t_k) \notin Q_1$ and $-r(t_1, \ldots, t_k) \notin Q_1$;
    Let Q_1' be obtained from Q_1 by adding $+r(t_1, \ldots, t_k)$;
    Let Q_1'' be obtained from Q_1 by adding $-r(t_1, \ldots, t_k)$;
(2)   **return** *recCheck(Q_1') AND recCheck(Q_1'')* ;

end

The algorithm proposed in [WL03] can be seen as exploring the same space of graphs but in a radically different way. At each step, it generates all homomorphisms from Q_2^p to the current Q_1. Then, for each compatible homomorphism in this set, say h, and for each negative subgoal $-p(t_1, \ldots, t_k)$ in Q_2 that cannot be mapped to Q_1 by extending h, a new query to test is generated from Q_1 by adding the positive subgoal $+p(h(t_1), \ldots, h(t_k))$; intuitively, the idea is that each total completion of Q_1 contains either $+p(h(t_1), \ldots, h(t_k))$ or $-p(h(t_1), \ldots, h(t_k))$; if $-p(h(t_1), \ldots, h(t_k))$ were present, then h could be extended to $-p(h(t_1), \ldots, h(t_k))$, thus it remains to test the $+p(h(t_1), \ldots, h(t_k))$ case. In Example 1 (see also Figure 1), only one homomorphism can be found at each step: the homomorphism $\{x \mapsto t, y \mapsto u\}$ from Q_2^p to Q_1

leads to Q_1' obtained by adding $+p(u)$; at the next step, $\{x \mapsto u, y \mapsto v\}$ from Q_2^p to Q_1' leads to Q_1'' obtained by adding $+p(v)$; finally, there is a homomorphism from Q_2^p to Q_1'' that can be extended to the negative subgoal, thus no new graph is generated. This algorithm can be seen as developing an *and/or* tree: a homomorphism h leads to success if all queries Q_i' directly generated from it lead to containment; a query Q_i' leads to containment if there is a homomorphism from Q_2^p to $Q_i'^p$ leading to success. The and/or tree is traversed in a breadth-first manner. Contrarily to Algorithm 1, partial completions built by this algorithm do not partition the space (basically because only positive subgoals are generated), which leads to the problem of detecting that a newly generated graph is not the same as a graph already generated (see the discussion in [LM07] for more details). In this paper, we focus in refining the algorithm in [LM07]. The experimental comparison of both ways of exploring the space of graphs remains to be done.

3 Experimental Methodology and Algorithm Refinements

In this section, we briefly present our experimental methodology, then we propose and analyze three refinements of Algorithm 1.

3.1 Methodology

Due to the lack of benchmarks or real-world data available for the studied problem, we built a random generator of polarized graphs. The chosen parameters are as follows:

- the number of *term* nodes (i.e. the number of terms in the associated query)[5];
- the number of distinct *relations*;
- the *arity* of these relations (set at 2 in the following experiments);
- the *density* per relation, which is, for each relation r, the ratio of the number of subgoals with relation r in the graph to the number of subgoals with relation r in a total completion of this graph w.r.t. $\{r\}$.
- the *percentage of negation* per relation, which is, for each relation r, the percentage of *negative* subgoals with relation r among all subgoals with relation r.

An instance of CQC⁻ is obtained by generating a pair (Q_1, Q_2) of PGs corresponding to *safe* queries. In this paper, we chose the same number of term nodes and the same percentage of negation for both graphs. In the sequel we adopt the following notations: *nbT* represents the number of term nodes, *nbR* the number of distinct relations and *SD* (resp. *TD*) the Source (resp. Target) graph Density per relation. The difficulty of the problem led us to restrict the value of *nbT* to between 5 and 8 (5 for the first experiments, 8 after improvement 1 which greatly decreases the running time).

In order to discriminate between different techniques, we first experimentally studied the influence of the parameters on the "difficulty" of instances. We measured the difficulty in three different ways: the running time, the size of the search tree and the number of homomorphism checks (see [BLM10] for more details). Concerning the *negation*

---

[5] We do not generate constants; indeed, constants tend to make the problem easier to solve because there are fewer potential homomorphisms; moreover, this parameter does not influence the studied heuristics.

percentage, we checked that the maximal difficulty is obtained when there are as many negative relation nodes as positive relation nodes. One can expect that increasing the number of relations occurring in graphs increases the difficulty, in terms of running time as well as the size of the searched space. Indeed, the number of completions increases exponentially (there are $(2^{n_{Q_1}^2})^{nbR}$ total completions for *nbR* binary relations). These intuitions are only partially validated by the experiments: see for instance Table 1, which shows, for each number of relations, the density values at the difficulty peak. We observe that the difficulty increases up to a certain number of relations (3 here, with a CPU time of 14809 and a Tree size of 216911) and beyond this value, it continuously decreases. Moreover, the higher the number of relations, the lower the *SD* that entails the greatest difficulty peak, and the higher the difference between *SD* and *TD* at the difficulty peak. In following experiments, we always take the *SD* and *TD* values corresponding to a difficulty peak.

Table 1. Influence of the number of relations (*nbT*=5)

nbR	SD	TD	CPU time (ms)	Tree size
1	0.24	0.24	19	82
2	0.12	0.24	7168	111540
3	0.08	0.4	14809	216911
4	0.08	0.68	12793	119911
5	0.08	0.8	4556	42566

For each value of the varying parameter, we considered 500 instances and computed the mean search cost of the results on these instances (with a timeout set at 5 minutes). The program is written in Java. The experiments were performed on a Sun fire X4100 Server AMD Opteron 252, equipped with a 2.6 GHz Dual-Core CPU and 4G of RAM, under Linux. In the sequel we only show the CPU time when the three difficulty measures are correlated.

3.2   Refinements

We now analyze three refinements of Algorithm 1, which concern the following aspects:

1. the choice of the next subgoal to add;
2. the choice of the child to explore first;
3. dynamic filtering at each node of the search tree.

1. Since the search space is explored in a depth-first manner, the choice of the next subgoal to add, i.e. $\sim r(t_1, \ldots, t_k)$ in Algorithm 1 (Point 1), is crucial. A brutal technique consists of choosing r and $t_1, \ldots, t_k$ randomly. Our proposal is to guide this choice by a compatible homomorphism, say h, from a Q_2^{max} to the current Q_1. More precisely, the border relation nodes $\sim r(e_1, \ldots, e_k)$ w.r.t. this Q_2^{max} can be divided into two categories. In the first category, we have the border nodes s.t. $\sim r(h(e_1), \ldots, h(e_k)) \in Q_1$, which can be used to *extend* h; if all border nodes are in this category, h can be extended to a homomorphism from Q_2 to Q_1. The choice of the subgoal to add is based on a node $\sim r(e_1, \ldots, e_k)$ in the second category: r is its relation symbol and $t_1, \ldots, t_k = h(e_1), \ldots, h(e_k)$ are its neighbors (note that neither $\sim r(h(e_1), \ldots, h(e_k))$ nor $\overline{\sim r}(h(e_1), \ldots, h(e_k))$ is in Q_1 since $\sim r(e_1, \ldots, e_k)$ is in the second category and h is compatible). Intuitively, the idea is to give priority to relation nodes potentially able to transform this compatible homomorphism into a homomorphism from Q_2 to a (partial) completion of Q_1, say Q_1'. If so, all completions including Q_1' are avoided.

Figure 3 shows the results obtained with the following choices:

- *random choice*;
- *random choice + filter*: random choice and Q_2^+ as filter (i.e. at each `recCheck` step a compatible homomorphism from Q_2^+ to Q_1 is looked for: if none exists, the false value is returned);
- *guided choice*: Q_2^+ used both as a filter and as a guide.

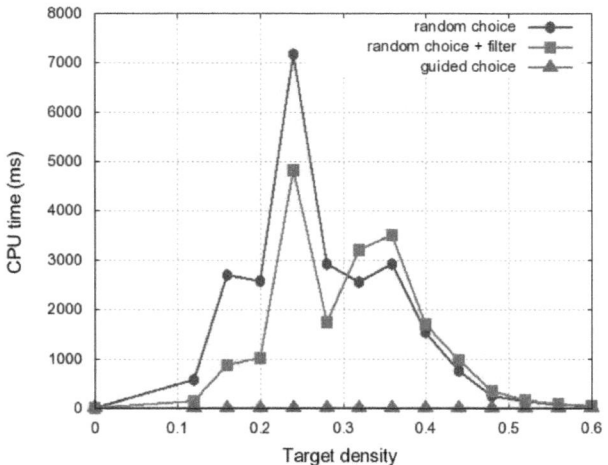

Fig. 3. Influence of the completion choice : nbT=5, nbR=2, SD=0.12

Note that the guided choice comes with an implicit filter: indeed, when a compatible homomorphism from Q_2^+ to a partial completion of Q_1 (say Q_1') is sought, the false value is returned if none exists (since $Q_1' \not\sqsubseteq Q_2$). In order to only discriminate choice heuristics, we also considered a random choice with a filter. As expected, the guided choice is always much better than the random choice (with or without filter).

2. Experiments have shown that the order in which the children of a node, i.e. Q_1' and Q_1'' in Algorithm 1 (Point 2), are explored is important. Assume that Point 1 in Algorithm 1 relies on a guiding subgraph. Consider Figure 4, where Q_2^+ is the guiding subgraph (hence the border is composed of negative relation nodes), "Extension" means "Q_1'' first" and "Contradiction" means the reverse order: we see that it is always better to explore Q_1' before Q_1''. If we take Q_2^- as the guiding subgraph, then the inverse order is better. More generally, let $\sim r(e_1, \ldots, e_k)$ be the border node that defines the subgoal to

add. Let us call h-*extension* (resp. h-*contradiction*) the graph built from Q_1 by adding $\sim r(h(e_1), \ldots, h(e_k))$ (resp. $\overline{\sim r}(h(e_1), \ldots, h(e_k))$). See Example 3. It is better to first explore the child corresponding to the h-*contradiction*. Intuitively, by contradicting the compatible homomorphism found, this gives priority to failure detection.

Example 3. See Figure 1. $Q_2^+ = \{+p(x), +s(x,y)\}$. Let Q_2^+ be the guiding subgraph. The only border node of Q_2^+ w.r.t. Q_2 is $-p(y)$. $h = \{x \mapsto t, y \mapsto u\}$ is the only compatible homomorphism from Q_2^+ to Q_1. The h-extension (resp. h-contradiction) is obtained by adding $+p(u)$ (resp. $-p(u)$).

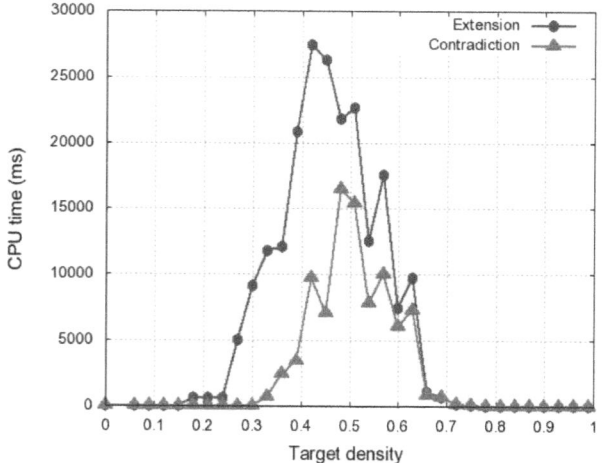

Fig. 4. Influence of the exploration order : *nbT=8, nbR=3, SD=0.06*

3. The last improvement consists of performing dynamic filtering at each node of the search tree. Once again, the aim is to detect a failure sooner. More precisely, we consider a set of Q_2^{max} and check if there is a compatible homomorphism from each element in this set to the newly generated graph. Table 2 shows the obtained results (at a difficulty peak) with the following configurations:

- *Max*: Q_2^{Max} as guide and no filter (other than Q_2^{Max});
- *Max-$\overline{Max}$*: Q_2^{Max} as guide and $Q_2^{\overline{Max}}$ (the subgraph on the relation nodes in $Q_2 \setminus Q_2^{Max}$) as filter;
- *Max-all*: Q_2^{Max} as guide and all Q_2^{max} as filters.

Unsurprisingly, the stronger the dynamic filtering, the smaller the size of the search tree. The CPU time is a bit higher for *Max-all* but this configuration checks much more homomorphisms than the others. Since our current algorithm for (compatible) homomorphism checking can be considerably improved, these results show that *Max-all* is the best choice.

The algorithm finally obtained is called `recCheckPlus` and it is shown in Algorithm 2. It is initially called with $(Q_1, \emptyset)$. The second parameter is used to memorize

Table 2. Influence of the dynamic filtering: $nbT=8$, $nbR=3$, $SD=0.06$, $TD=0.54$

Configuration	CPU time (ms)	Tree size	Hom check
Max	7404	3810	5539
$Max\text{-}Max$	7421	3765	6209
$Max\text{-}all$	8427	2249	13331

the compatible homomorphism found for the father of the current node, in the case where this node is an h-extension of its father (see Q_1'' in the algorithm); otherwise, the compatible homomorphism for its father has been contradicted and a new one has to be computed, which is done in the *chooseCompletionSubgoal* subalgorithm. More precisely, this subalgorithm returns a completion literal as explained in Point 1 and a new compatible homomorphism h if needed.

Algorithm 2. recCheckPlus(Q_1, h)

　　Input: a consistent PG Q_1 and a compatible homomorphism h from the guiding subgraph
　　　　　to the father of Q_1 (empty for the root)
　　Data: $Q_2, \mathcal{V}$
　　Result: true if $Q_1 \sqsubseteq Q_2$, false otherwise
　　begin
　　　　if *there is a homomorphism from Q_2 to Q_1* **then return** *true* ;
　　　　if *Q_1 is complete w.r.t. $\mathcal{V}$* **then return** *false* ;
(3)　　　**if** *dynamicFiltering(Q_1) = failure* **then return** *false* ;
(1)　　　$l, h \leftarrow$ **chooseCompletionSubgoal**(Q_1, h) ;
　　　　Let Q_1' be obtained from Q_1 by adding $\bar{l}$;
　　　　Let Q_1'' be obtained from Q_1 by adding l ;
(2)　　　**return** *recCheckPlus($Q_1', \emptyset$) AND recCheckPlus(Q_1'', h)* ;
　　end

4　Second Approach

In this section, we present the second method, which consists of translating the CQC$^\neg$ problem into the problem of checking unsatisfiability of a propositional formula in conjunctive normal form (*i.e.* a conjunction of disjunctions of propositional literals), called the UNSAT problem. This method is then experimentally compared to recCheckPlus.

4.1　Method

Let us first explain the main ideas of this method. Instead of exploring the space of graphs in a depth-first manner in order to find a set of partial completions that covers all total completions, a *candidate* covering set is built at once; it is built from all compatible homomorphisms from a specific pure subgraph of Q_2 to Q_1; this candidate is indeed a covering set if and only if a formula built from it is valid; this formula is built by considering for each compatible homomorphism the relation nodes that should be added to Q_1 to obtain a homomorphism from Q_2. More specifically, we proceed in three steps:

1. Compute all compatible homomorphisms from Q_2' (the special subgraph of Q_2) to Q_1;
2. Build a propositional formula F_{prop}, that is the disjunction of, for each compatible homomorphism h of Step 1, the conjunction C of missing subgoals in Q_1 for h to be a homomorphism from Q_2 to $Q_1 \wedge C$;
3. Check if $\overline{F_{prop}}$ (an UNSAT instance) is unsatisfiable: if yes, then $Q_1 \sqsubseteq Q_2$, otherwise $Q_1 \not\sqsubseteq Q_2$.

Step 1: Compute the candidate covering set. The method is based on a specific pure subgraph of Q_2, which necessarily maps to the Q_1 part of each total completion of Q_1 when $Q_1 \sqsubseteq Q_2$ (note that this is not true for all pure subgraphs of Q_2).

Definition 4 (Stable subgraph). *The* stable subgraph *of Q_2, denoted Q_2^s, contains all term nodes in Q_2 and all subgoals in Q_2 with a relation that does not appear in the completion vocabulary of Q_1 and Q_2 (note that Q_2^s is included in all Q_2^{max}).*

Since the stable subgraph does not contain relations belonging to subgoals added to Q_1 during completion, we have the desired property:

Property 1. Let Q_1 and Q_2 be two PGs. If h is a homomorphism from Q_2 to a total completion of Q_1, then h is a compatible homomorphism from Q_2^s to Q_1.

Furthermore, we can replace, without incidence on the CQC$^\neg$ problem, each variable in Q_1 by a **new** constant. This modification preserves all homomorphisms to Q_1. In the sequel, we consider that Q_1 contains only constants. In particular, the query q_1 in Example 1 becomes: $q_1 = ans() \leftarrow p(a) \wedge s(a,b) \wedge s(b,c) \wedge s(c,d) \wedge \neg p(d))$. This will allow to obtain a propositional formula in Step 2.

Example 4. Let Q_1 and Q_2 be PGs of Figure 1. $Q_2^s = s(x,y)$. There are 3 compatible homomorphisms from Q_2^s to $Q_1 : h_1 = \{x \mapsto a, y \mapsto b\}$, $h_2 = \{x \mapsto b, y \mapsto c\}$ and $h_3 = \{x \mapsto c, y \mapsto d\}$.

Step 2: Build the propositional formula. With each compatible homomorphism computed at Step 1, we build a conjunction of the "missing" subgoals in Q_1. Each partial completion of Q_1 obtained by adding these subgoals to Q_1 is an element of the candidate covering set.

Definition 5 (Minimal conjunction). *The* minimal conjunction *of Q_1 w.r.t. a compatible homomorphism h, denoted C^m, is the conjunction composed of the atom $\blacksquare$[6] and the subgoals $\sim r(h(t_1), \ldots, h(t_k))$ such that $\sim r(t_1, \ldots, t_k) \in (Q_2 \setminus Q_2^s)$ and $\sim r(h(t_1), \ldots, h(t_k)) \notin Q_1$.*

Example 5. For $h_1 : C_1^m = \neg p(b)$; for $h_2 : C_2^m = p(b) \wedge \neg p(c)$; for $h_3 : C_3^m = p(c)$.

Then we build the entire formula, which is the disjunction of all minimal conjunctions:

---

[6] $\blacksquare$ is the tautology: it is necessary when h is a homomorphism from Q_2 to Q_1, otherwise the conjunction would be empty.

Definition 6 (Disjunction of Stable Minimal Conjunctions ($\mathcal{DSMC}$)). *We call $\mathcal{DSMC}(Q_2, Q_1)$ the disjunction of the atom $\square$[7] and all minimal conjunctions w.r.t. the compatible homomorphisms from Q_2^s to Q_1 (i.e. $\square \vee C_1^m \vee \ldots \vee C_i^m$, denoted $\bigvee C^m$).*

Example 6. $\mathcal{DSMC}(Q_2, Q_1) = \neg p(b) \vee (p(b) \wedge \neg p(c)) \vee p(c)$.

The theorem validating this approach is the following (Q_1 is assumed to be consistent):

Theorem 2. $Q_1 \sqsubseteq Q_2$ *iff* $\mathcal{DSMC}(Q_2, Q_1)$ *is valid.*

The following definitions and lemmas are used to prove the theorem.

Definition 7 (Total conjunction). *Let Q_1^c be a total completion of Q_1. The total conjunction of Q_1^c, denoted C, is the conjunction of all subgoals $\sim r(t_1, \ldots, t_k) \in (Q_1^c \setminus Q_1)$.*

Definition 8 (Disjunction of Total Conjunctions ($\mathcal{DTC}$)). *We call $\mathcal{DTC}(Q_1)$ the disjunction of all total conjunctions of Q_1 (i.e. $C_1 \vee \ldots \vee C_j$).*

Lemma 1. $\mathcal{DTC}(Q_1)$ *is valid.*

Lemma 2. *If $Q_1 \sqsubseteq Q_2$ then for all $C \in \mathcal{DTC}(Q_1)$, there is a $C^m \in \mathcal{DSMC}(Q_2, Q_1)$ such that $C^m \subseteq C$ (i.e. all subgoals in C^m are also in C).*

Property 2. Let C^m be a minimal conjunction of Q_2 w.r.t. a compatible homomorphism from Q_2^s to Q_1. $Q_1 \wedge C^m \sqsubseteq Q_2$.

Proof of Theorem 2:
$\Longleftarrow$ Since $\mathcal{DSMC}(Q_2, Q_1)$ is valid, $Q_1 \equiv Q_1 \wedge \bigvee C^m \equiv \bigvee (Q_1 \wedge C^m)$. According to Property 2, $\bigvee (Q_1 \wedge C^m) \sqsubseteq Q_2$. Thus $Q_1 \sqsubseteq Q_2$.
$\Longrightarrow$ Let $Q_1 \sqsubseteq Q_2$. According to Lemma 2, for all $C \in \mathcal{DTC}(Q_1)$, there is $C^m \in \mathcal{DSMC}(Q_2, Q_1)$ s.t. $C = C^m \wedge C'$ where C' is a conjunction of subgoals. *By absurd:* Assume $\mathcal{DSMC}(Q_2, Q_1)$ is not valid. Then there is an interpretation $\mathcal{I}$ s.t. for all $C^m \in \mathcal{DSMC}(Q_2, Q_1)$, $\mathcal{I} \models \neg C^m$. Thus for all $C \in \mathcal{DTC}(Q_1)$, $\mathcal{I} \models \neg C$. Thus $\mathcal{DTC}(Q_1)$ is not valid, which contradicts Lemma 1. Hence $\mathcal{DSMC}(Q_2, Q_1)$ is valid. $\square$

Step 3: Translate into UNSAT. The negation of $\mathcal{DSMC}(Q_2, Q_1)$ is a propositional conjunctive normal form, which enables us to use a SAT solver.

Example 7. $CF = \overline{\mathcal{DSMC}(Q_2, Q_1)} = p(b) \wedge (\neg p(b) \vee p(c)) \wedge \neg p(c)$. CF is unsatisfiable, thus $\mathcal{DSMC}(Q_2, Q_1)$ is valid and $Q_1 \sqsubseteq Q_2$.

4.2 Algorithm and Experiments

Algorithm 3 implements this method. The UNSAT call uses the well-known *Sat4J* solver[8].

---

[7] $\square$ is the absurd literal: it is necessary when there is no compatible homomorphism from Q_2^s to Q_1, otherwise the disjunction would be empty.

[8] http://www.sat4j.org/

Algorithm 3. UNSATCheck(Q_1)

Input: a consistent PG Q_1
Data: Q_2, Q_2^s
Result: true if $Q_1 \sqsubseteq Q_2$, false otherwise
begin
  $\mathcal{DSMC}(Q_2, Q_1) \leftarrow \square$;
  $h_1, \ldots, h_n \leftarrow$ **findAllCompatibleHomomorphisms**(Q_2, Q_2^s, Q_1) ;
  foreach $h_i, i = 1 \ldots n$ **do**
    foreach *border node* $\sim r(t_1, \ldots, t_k) \in Q_2 \setminus Q_2^s$ **do**
      $C_i^m \leftarrow \blacksquare$;
      if $\sim r(h_i(t_1), \ldots, h_i(t_k)) \notin Q_1$ **then**
        $C_i^m \leftarrow C_i^m \wedge \sim r(h_i(t_1), \ldots, h_i(t_k))$;
    $\mathcal{DSMC}(Q_2, Q_1) \leftarrow \mathcal{DSMC}(Q_2, Q_1) \vee C_i^m$;
  return *UNSAT*($\overline{\mathcal{DSMC}(Q_2, Q_1)}$)
end

To compare UNSATCheck and recCheckPlus, we used the size of the vocabulary completion (directly correlated with the size of the stable subgraph) as the varying parameter. Indeed, this parameter is crucial for UNSATCheck (note that this parameter has also an influence for recCheckPlus): the bigger the stable subgraph, the lower the number of compatible homomorphisms and then the size of the obtained formula. We built 1000 random instances for each value of $|\mathcal{V}|$ (for each relation, the percentage of negation is equal to 0%, 50% or 100%) and compared UNSATCheck and recCheckPlus on them: see Table 3.[9]

Table 3. Detailed comparison of the two algorithms : $nbT=8$, $nbR=3$, $SD=0.06$, $TD=0.51$

Size of $\mathcal{V}$	Size of the stable subgraph	recCheckPlus CPU time	UNSATCheck CPU time			
			Total	1st step	2nd step	3rd step
3	0	6482	27038	10541	15378	1119
2	6	800	444	172	229	43
1	12	5	20	9	2	9
0	18	1	2	2	0	0

We observe that for both algorithms the maximal difficulty is for $|\mathcal{V}| = 3$. Then UNSATCheck is the worst because the stable subgraph contains only term nodes, thus the number of compatible homomorphisms and then the size of the obtained formula are exponential in the sizes of the initial queries. As expected, the increase of the size of the stable graph makes the algorithms better. For $|\mathcal{V}| = 2$, UNSATCheck is a little better than recCheckPlus. When the size of the stable graph is the highest, the results for both algorithms are similar. These results show that the choice of an algorithm rather than another depends on the size of the stable graph. These are only preliminary results, further experiments are needed to refine this choice.

---

[9] Note that 8 term nodes is the highest value that UNSATCheck can deal with (with $|\mathcal{V}| = 3$ and our implementation): beyond 8, the memory space explodes.

344   K. Ben Mohamed, M. Leclère, and M.-L. Mugnier

5 Perspectives

In this paper, we have refined the algorithm proposed in [LM07] and checked experimentally several choices. These refinements heavily rely on special subgraphs, called pure subgraphs. Using pure subgraphs of maximal cardinality as guiding and filtering graphs seems a good choice. However, the size of pure subgraphs is not the "ultimate" criterion, as shown in Figure 5: for each instance, we ran recCheckPlus with all possible Q_2^{Max} (they all have the same size). The Maximum (resp. Minimum) curve is obtained by choosing, for each instance, the worst (resp. best) Q_2^{Max}, i.e. that leads to the highest (resp. lowest) CPU time. We conclude that the choice of the Q_2^{Max} used to guide among all Q_2^{Max} is a determining step. However, finding criteria allowing to better discriminate between pure graphs is an open issue.

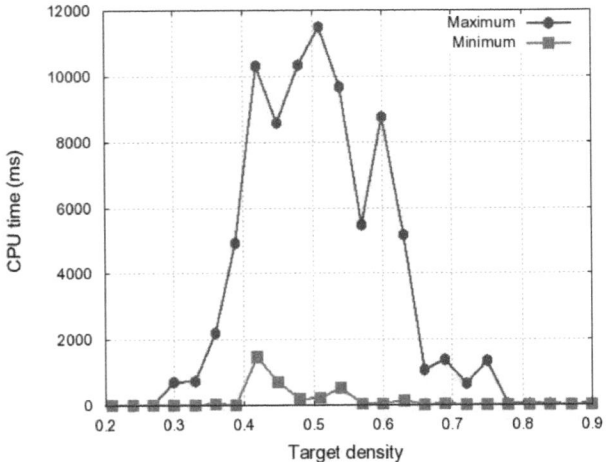

Fig. 5. Comparison between the worst and the best choice of Q_2^{Max}: $nbT=8$, $nbR=3$, $SD=0.06$

In [WL03] another way of exploring the query space is proposed. The associated algorithm is much more complex to follow and implement than recCheck. Furthermore, some parts were not specified (for instance how to avoid generating a query that was already generated). We are currently implementing this algorithm, while integrating the improvements designed for recCheck.

References

[AHV95]   Abiteboul, S., Hull, R., Vianu, V.: Foundations of Databases: The Logical Level. Addison-Wesley, Reading (1995)
[ASU79]   Aho, A.V., Sagiv, Y., Ullman, J.D.: Equivalences among relational expressions. SIAM J. Comput. 8(2), 218–246 (1979)
[BLM10]   Ben Mohamed, K., Leclère, M., Mugnier, M.-L.: Deduction in existential conjunctive first-order logic: an algorithm and experiments. Technical Report RR-10010, LIRMM (March 2010)

[CM77]   Chandra, A.K., Merlin, P.M.: Optimal implementation of conjunctive queries in relational databases. In: 9th ACM Symposium on Theory of Computing, pp. 77–90 (1977)

[CM09]   Chein, M., Mugnier, M.-L.: Graph-based Knowledge Representation and Reasoning—Computational Foundations of Conceptual Graphs. In: Advanced Information and Knowledge Processing. Springer, Heidelberg (2009)

[FNTU07]   Farré, C., Nutt, W., Teniente, E., Urpí, T.: Containment of conjunctive queries over databases with null values. In: Schwentick, T., Suciu, D. (eds.) ICDT 2007. LNCS, vol. 4353, pp. 389–403. Springer, Heidelberg (2006)

[Hal01]   Halevy, A.Y.: Answering queries using views: A survey. VLDB Journal 10(4), 270–294 (2001)

[LM07]   Leclère, M., Mugnier, M.-L.: Some Algorithmic Improvments for the Containment Problem of Conjunctive Queries with Negation. In: Schwentick, T., Suciu, D. (eds.) ICDT 2007. LNCS, vol. 4353, pp. 404–418. Springer, Heidelberg (2006)

[LS93]   Levy, A.Y., Sagiv, Y.: Queries independent of updates. In: VLDB 1993: Proceedings of the 19th International Conference on Very Large Data Bases, pp. 171–181. Morgan Kaufmann Publishers Inc., San Francisco (1993)

[Sai08]   Sais, L.: Problème SAT: progrès et défis. Hermes (2008)

[Ull97]   Ullman, J.D.: Information Integration Using Logical Views. In: Afrati, F.N., Kolaitis, P.G. (eds.) ICDT 1997. LNCS, vol. 1186, pp. 19–40. Springer, Heidelberg (1996)

[WL03]   Wei, F., Lausen, G.: Containment of Conjunctive Queries with Safe Negation. In: Calvanese, D., Lenzerini, M., Motwani, R. (eds.) ICDT 2003. LNCS, vol. 2572, pp. 343–357. Springer, Heidelberg (2002)

Ranking Objects Based on Attribute Value Correlation[*]

Jaehui Park and Sang-goo Lee

School of Computer Science and Engineering, Seoul National,
Seoul 151-742, Republic of Korea
{jaehui,sglee}@europa.snu.ac.kr

Abstract. There has been a great deal of interest in recent years on ranking query results in relational databases. This paper presents a novel method to rank objects (e.g., tuples) by exploiting the correlations among their attribute values. Given a query, each attribute value is assigned a score according to mutual occurrences with the query and its distribution status in the columns of the attribute. These attribute value scores are aggregated to get a final score for an object. Furthermore, a *concept vector* is proposed to provide a synopsis of the attribute value in a given database. A concept vector is utilized to get the similar objects. Experimental results demonstrate the performance of our ranking method, RAVC (Ranking with Attribute Value Correlation), in terms of search quality and efficiency.

Keywords: Ranking function for structured data, attribute value correlation, attribute importance.

1 Introduction

Relational databases, which follow the Boolean retrieval model, have to be queried with conditions that exactly match the information needs of the user. Users have suffered from formulating the search conditions that give satisfactory results for their needs. In this setting, the users may face the following two problems [2].

Empty answers problem. If a query has conditions that are too selective, the answers may be empty or too limited. In this case, the conditions of the original query have to be rewritten into less restrictive ones.

Many answers problem. If the conditions of a query are not discriminative enough in selecting answers, too many answers are retrieved. In this case, it is desirable to order the answers by quantitative measures that present the relevance to the information needs of the user.

Our methods resolve above two problems by proposing a relevance-based retrieval model for relational databases. Our intuition is as follows: data statistics, such as

---

[*] This research was supported by the MKE(The Ministry of Knowledge Economy), Korea, under the ITRC(Information Technology Research Center) support program supervised by the NIPA(National IT Industry Promotion Agency). (grant number NIPA-2009-C1090-0902-0031).

P. García Bringas et al. (Eds.): DEXA 2010, Part II, LNCS 6262, pp. 346–359, 2010.

value correlation (e.g., mutual occurrence) and deviation (e.g., entropy), are useful to estimate the relevance of an object to a given query. Main focuses of this paper are as follows:

1) We introduce a novel ranking method that utilizes the database statistics to improve the retrieval effectiveness. Developing an effective ranking function is crucial to solve the many answers problem. We note that the statistical relationships between attribute values reside in the underlying databases could be useful to rank tuples since a tuple is an instance of generalized binary relations of attribute values. In our method, tuples are ranked based on the aggregated correlations of their attribute values because the correlations reflect the implicit information of statistical relationships for the tuples.This paper speculates on how much an attribute value correlates to the user's query based on the mutual occurrences over the underlying database. Each attribute value in a tuple is assigned corresponding scores from the quantified correlations. Moreover, each attribute for a relation is assigned a weight by analyzing the attribute value distribution status in columns of the attributes. Each score of the attribute value is then aggregated as a final ranking score for each tuple.

2) We propose a *concept vector* to alleviate the empty answers problem. The intuition is similar to the semantics of the ranking function; a synopsis of a concept is summarized by the group of statistically related attribute values. It can be utilized to expand the scope of the Boolean query results. The synopsis of objects specified by the query is summarized as a concept vector, which is the form of a weighted vector of all the common attribute values and their frequencies. Each dimension of the vector is represented as an attribute value, not an attribute. We give illustrative examples in the following section.

Table 1. Examples of used car tuples

	Year	Make	Model	Color	Style
t_1	2008	Toyota	Camry	Silver	Sedan
t_2	2009	Toyota	Camry	Silver	Sedan
t_3	2008	Toyota	Camry	Black	Sedan
t_4	2008	Toyota	Camry	Blue	Sedan
t_5	2008	Toyota	Camry	Gold	Coupe
t_6	2008	Honda	Accord	Silver	Sedan

1.1 Illustrative Examples

Table 1 illustrates a used car database D with a single table in which the car instances are stored as tuples with various attributes such as *Year*, *Make*, *Model*, *Color*, and *Style*. Each tuple t_i in D represents a used car for sale. The set of tuples $t_1 \sim t_5$ represents Boolean query results for the query '*Camry*'. Ranking is not provided by the traditional Boolean retrieval model. Given a tuple t_i, we assign a *Score(t_i)* based on the correlations of its attribute values. In this unordered set of tuples, we find the mutual occurrences of attribute values for the given query '*Camry*'. In the column of the attribute *Make*, every attribute value is *Toyota*. This is one of the simple examples that we realize the correlation between these two attribute values (*Camry* and *Toyota*) represent the strong statistical relationships. Basic intuition for this study is that the correlation is interpreted as a relevance that is measured by the ratio of the mutual occurrences of the attribute values. In the given set of tuples, the attribute value *Toyota* is most relevant to the query, and the attribute value *Silver* is relatively less relevant. This evidence of relevance can be naturally interpreted to the semantics of our ranking function. Analogously, we determine the relevance of attributes as well as the relevance of attribute values. Based on the value distribution within a column of an attribute, we can determine which attribute correlates more to the query. It is measured by the expected value of mutual occurrences. We realize the attribute *Model* correlates to the attribute *Make* and *Style* given query '*Camry*'. For the corresponding attributes, we assign weights representing the importance of the attribute for a given query based on its distribution status. Several approaches [10, 5, 13, 9] have been proposed to assign weights for attributes to relax search conditions based on the data distributions. However, there is no clear idea how to decide how much the attribute is important for a given query. We note that the difference of attribute value deviations within columns represent the importance of an attribute for a query. Intuitively, the attribute *Color* is less important to the query '*Camry*' since the values binding to the attribute *Color* are highly distributed. Our approach is similar to the way of capturing important information about random variables in descriptive statistics.

Assuming the Boolean query results, t_1~t_5, are too small. Our method provides additional results that are similar to the original query results. Although the tuple t_6 does not match the query keyword '*Camry*', the tuple t_6 is added because it is similar to the synopsis of *Camry* which is represented by the tuples t_1~t_5. It is beneficial to identify the similar objects, such as *Accord*, to expand the results set. The synopsis of *Camry* can be summarized by the several properties of tuples t_1~t_5, such as attribute values and their frequencies. The model *Camry* is *Sedan* style car made by *Toyota*. Four cars produced in year of 2008 and one car in year of 2009. The synopsis is modeled as a weighted vector, *concept vector*.

1.2 Contributions

The contributions of this paper are follows:

1) A novel method is presented to rank the tuples in the query results while analyzing the data statistics of the attribute values. We provide an effective ranking function without any prior knowledge, such as user feedbacks or query workloads to learn evidences from user interventions.
2) A *concept vector* is proposed to extend the limited result set by identifying the similar tuples for a query. This can be utilized in various extensions associated to the metrics of object similarity (or relevance) in relational databases.

1.3 Organization

This paper is organized as follows. Section 2 reviews some related works. Section 3 proposes the method for tuple ranking based on the data statistics: data correlation and data distribution status. Section 4 describes how to expand the result set by adding similar tuples. In Section 5, we discuss several research dimensions for the generalization of this work. The experimental results are presented in Section 6. The paper is concluded in Section 7.

2 Related Work

Ranking query results in relational databases has been actively investigated in recent years. Many approaches have been proposed to deal with the *empty answers problem* and the *many answers problem*. Traditional approaches [15, 14] have focused on employing relevance-feedback techniques for developing ranking function. In contrast, our method requires no training data such as query feedbacks. The articles [11, 1, 3, 6] empowering users to access databases using simple keywords to tackle these problems; developing ranking function is an important component in tuple retrieval in relational databases. A major concern of this paper is developing ranking method only based on data statistics. Recent approaches [10, 5, 13, 9] have focused on the ranking tuples based on the data frequency within the specified attributes and the reduction of the search conditions in the original query statement in order to expand the scope of the query. *TF-IDF* concept from Information Retrieval (IR) is extended to a database containing a mix of categorical as well as numerical data in [3, 2]. *Approximate Functional Dependency (AFD)* is used to answer typed keyword queries over web databases in [10]. AFD captures the relationships between attributes of a relation and can be used to determine the degree to which a change in the binding value of an attribute influences other attributes. Authors assume the AFD as a measure of the importance of an attribute value to the query keywords. In [9], the contextual preferences are exploited to rank the query result. It also uses the query workload in the pre-processing step. In [4], the principle of the probabilistic model from the IR, which is the *Probabilistic Information Retrieval (PIR)* model [8], is adapted into structured data ranking while considering the data dependencies. This approach considers the ranking function based on the dependency between the two types of attributes, which are the specified and unspecified attributes in a structured query. However, this ranking approach can potentially lead to unintuitive results, for example, by ranking higher the high-priced items of low-quality because they consider that the tuples with infrequent combinations of values should be ranked higher. Several techniques [6, 4] have been proposed to support top-k query processing in score-based ranking approach. However, efficient top-k query processing is not a main focus of this paper. The *Entropy* measure is adopted from the Information Theory field for the purpose of this study and to show how a novel weighting scheme can be used to measure the importance of an attribute.

The work most related to this study is [13], which assigns a score to a tuple based on the attribute value distribution in an e-commerce context. This paper has differences in the following aspects:

1) The attribute weight assignment only considers the distribution difference between the database and the query results. This study assigns a weight to an attribute based on the attribute value distribution status.

2) Authors assume a domain-specific knowledge involved to derive a ranking function: "Lower price is more desirable than the higher price for selecting items in e-commerce context." Our method only considers the data statistics only, but performs better in several contexts.

3 Ranking Objects

This section defines the problem in ranking database query results (e.g., a set of tuples) and presents the relevance-based ranking function considering the correlation between the attribute values.

3.1 Problems Formulation

Consider a database table D with n tuples $T = \{t_1, t_2, ..., t_n\}$ with attributes $A = \{A_1, A_2, ..., A_m\}$ and a query Q over D with a conjunctive selection condition of the form "*SELECT * FROM D WHERE $A_1 = x_1$ AND $A_2 = x_2$ ... AND $A_s = x_s$*", where each A_i is an attribute that contains the value matching the attribute value x_i, which is the term specified in the parameterized query. The set of attributes $X = \{A_1, A_2, ..., A_s\} \subseteq A$ is known as the set of specified attributes by the query keywords. The query Q is represented by a set of keywords $K = \{x_1, x_2, ..., x_s\}$. The goal is to develop a scoring function $Score(Q,t)$ that captures the strength of the relevance between the tuple t_i and the query Q.

$$Corr(K,t) = \sum_{i}^{s} \sum_{A_u \in A - X} F(E_{A_u} \times t[A_u], x_i).\tag{1}$$

$Corr(K, t)$ indicates the score that captures the extent the attribute values $t[A_u]$ in tuple t correlates to the keywords x_i. Let function $F(x, y)$ return a value indicating the degree to which x is relevant for y. K is a set of query keywords binding to the attributes in X. A_u is an attribute that is not specified in the parameterized query. Let weight E be associated with all values binding attribute A_u based on the value distribution status in the column of attribute A_u.

$$Sim(K,t) = VSim(V_K, V_t).\tag{2}$$

$VSim(x, y)$ returns a value in [0...1] indicating how similar vector x to vector y. V_K and V_t are vector representations of the objects specified by the query K and the tuple t, respectively. The similarity between the tuple t and the synopsis V_K specified by the query K are measured by vector similarity function. The dimensions of the vector V_K is determined by attribute values in the Boolean query results. Each dimension of the vector V_K is assigned weight according to the frequencies of attribute values occurred in the original query results.

Our ranking function for conjunctive query is as follows:

$$Score(Q,t) = Sim(K,t) \times Corr(K,t).\tag{3}$$

The above explanations represent only the simple problem instances. A point query is assumed as a basic setting. In more general settings, a query may contain the range conditions for numerical values, the disjunctive operator, and the NULL values. While our method can be extended to all these general capabilities, the focus of this paper is on ranking tuples of conjunctive keywords on a single categorical table.

3.2 Attribute Value Correlation

The correlation of an attribute value with a given query (which is represented as a set of attribute values) is computed as the ratio of co-occurred tuples to a given tuple set. Table 1 shows the attribute value *Toyota* is most co-occurred for the given query '*Camry*'. On the other hand, the attribute values binding the attribute *Color* are distributed well and it means that all the colors equally correlate to the query, although the quantities are small. That is, database statistics (e.g., correlation) infer that the relevance of the color *Blue* is not significant as that of the make *Toyota* to *Camry*.

$$P(v,q) = log \frac{|D_v \cap D_q|}{|D_q|}. \tag{4}$$

P denotes the measure of mutual occurrence for an attribute value v given a query q. D_i represents a set of tuples contains the attribute value v. This study defines $max(P(v, q))$ as the correlation of the corresponding attribute. The existing approach [10] measure the importance of an attribute based on the most frequent values, $argmax(P(v, q))$ for relaxing query conditions. Our method differs from [10], we focus on the attribute values itself.

For the attribute weight E, we define a measure called *data skewness*. The following example provides a clear exposition of data skewness for weighting an attribute.

Data Skewness. The importance of a tuple can be computed using Equation 4 additively. However, it can be argued that the increase of the relevance is not proportional only to the number of co-occurrences. We need to assign the weight E (in Equation 1) on each attribute A_u based on the expected value of the statistical relationships between associated values. From Table 1, the attribute value *Sedan* is skewed for the query '*Camry*' in a given column; it means the attribute is (almost) dependent to the value *Camry*. Therefore, it is reasonable that the skewed values in the results are considered to be important for the query because the answers in the skewed column of attribute are more predictable. On the other hand, the attribute values for *Colors* are not skewed as much as values for the attribute *Style* or *Make*. As an extreme case, this skewness identifies the soft (functional) dependency between attributes. In this paper, the skewness will be quantified by entropy, which is a well known measure to estimate the asymmetry in data distribution. The relevance of the attribute value *Gold* and *Coupe* is not equal even though they have the same frequencies since the attribute *Style* is more dependent to the query '*Camry*' than the attribute *Color*. This relevance measure for attributes, data skewness, utilizes the semantics of underlying data schema for ranking tuples effectively.

3.3 Entropy-Based Attribute Weight Assignment

There are many methods to assess the data skewness (or deviation) in categorical data. The best approach is the information theoretical method, using the model of *Shannon's entropy* [12]. The entropy is the measure of expected value of the information of the distribution. The entropy is formally defined as follows:

Entropy. A measure of the average information content one is missing when the value of the random variable is unknown. Let V be a discrete random variable on a finite set $V = \{v_1, v_2, ..., v_n\}$, with probability distribution function $p(v) = Pr(V=v)$. The entropy $E(A)$ of V is defined as

$$E(A) = - \sum_{v_i \in t_i[A]}^{n} p(v_i) log_b p(v_i). \tag{5}$$

We assume that the occurrences of an attribute value are independent to other values in the same column. The attribute values can be seen as independent random values of V. The frequencies of each attribute value can vary along with results from the different probabilities of the value occurrences. High entropy means V is from a uniform (or well-distributed) distribution. Low entropy means V have samples that would be more predictable based on the high probability. Therefore, the attribute of a high probability of occurrence will have low entropy. The weight for the attribute value will be higher according to the value of $1/E$.

The final score for attribute value is as follow.

$$Corr(K,t) = \sum_{i}^{s} \sum_{A_u \in A-X} \frac{1}{1 + E(A_u)} \frac{P(t[A_u], K)}{max(P(t[A_u], K)} \tag{6}$$

Intuitively, the reciprocal of function E determines the *predictability* of a certain attribute. Consequently, a ranking score $Corr(K,t)$ is derived for object t given a keyword query K by aggregating all the attribute value scores.

4   Finding Similar Objects

A *concept vector* is proposed to describe a synopsis for a set of objects in a relational database. Similar tuples can be identified by measuring the similarity to the synopsis of a certain topic specified by the query.

4.1   Concept Vector

A synopsis over the database instance is indicated by a set of attribute values that represent a certain topic. We assume that values are related if they occur in the same tuple. A synopsis is created by gathering the related attribute values for a given concept specified by the query. For example, the synopsis of the concept *Camry* is described by pairs of the common attribute value and its frequency in the tuple $t_1 \sim t_5$ in Table 1. For a common attribute value *Toyota*, its frequency is 5. The vector representation of this synopsis for *Camry* is as follows: *[(2008,3), (2009,1), (Toyota,5), (Silver,2), (Black,1), (Blue,1), (Gold,1), (Sedan,4), (Coupe,1)]*. The vector representation summarizes the retrieved objects by the query "*Camry*". This vector is denoted as a *concept vector*, and it will be used for assigning the weight $Sim(K,t)$ for the score $Corr(K,t)$in Equation 3. For any relevant tuples identified by concept vector similarity are assigned a damping factor, which is the weight, $Sim(K,t)<1$. If a tuple t already

Table 2. Synopsis of concept *Camry*

	Freq.	E
2008	3	0.87
2009	1	0.12
Toyota	5	1.0
Silver	2	0.28
Black	1	0.28
Blue	1	0.28
Gold	4	0.28
Sedan	4	0.87
Coupe	1	0.12

resides in the original results, *Sim(K,t)* is 1. Table 2 describes the synopsis of *Camry*. In Table 1, the tuple t_6 of car model *Accord* is added to the original query results $t_1 \sim t_5$ since it has certain similarity for the synopsis of *Camry*.

4.2 Similarity Measure for Concept Vector

Each dimension of the concept vector is weighted by the *E* from the Equation 1 according to a query. The weighted *cosine similarity coefficient* is used to determine the similarity between two vectors. The cosine similarity is defined as the normalized dot-product of the two corresponding vectors. The model is refined by scaling each component with the assigned weight *E* of the corresponding attribute values. Intuitively, the tuples similar to the synopsis specified by the attribute value *c*, which can be represented as an attribute value, will have small differences in attribute values of the least weight. The following is the similarity between the synopsis specified by the attribute value *c* and the tuple *t*.

$$VSim(c, t) = \frac{\sum E[i] \times c[i] \times t[i]}{\sqrt{\sum (E[i] \times c[i])^2 \times \sum (E[i] \times t[i])^2}}. \tag{7}$$

The underlying motivation is to identify tuples that are most similar to the synopsis specified by the attribute value *c*. Tuples are selected by the ratio of similarity. For limited number of exact matches, the similar tuples compensate as a relevant result. The expanded query results contain tuples directly related to the query selection and tuples implicitly related to them in various properties. This relevance model can be useful in many applications and can provide the user more information.

5 Generalizations

In this section, we describe how to extend our methods to deal with several generalizations: joining multiple tuples and discretizing numerical data.

5.1 Joining Multiple Tuples

Given a set of query terms, a relational database returns a set of tuples that contains all the keywords. Identifying tables that contain at least one tuple that matches at least a keyword is a challenging issue in keyword search over relational databases; the tables are connected with primary key to foreign key relationships. The set of tuples may contain inter-connected tuples that have no matches to query keywords. To find optimal set of tuples, we have to deal with the complex graph problem, which is NP-hard. Several approaches to enable keyword search in relational databases are focused on this problem [1, 3, 6]. Our work focuses on a complementary problem: How to rank a join tuples after joining the multiple tuples? A join tuples express the graph as a flat representation. That is, we can simply consider the join tuples as a tuple of many attributes after the joins.

5.2 Dealing with Numerical Values

For more general database instances, we have to consider database of numerical attributes. Recent approaches such as [2] discretize the numerical values to apply same ranking technique. However, ranking heterogeneous type of data is challenging issue. Since numerical values have inherent ordering, we can consider it as search condition. We can provide range search predicate, such as *Mileage >2000*, which is one of the important settings in database query processing. By analyzing numerical values for our correlation based ranking, we can adopt the well studied correlation measures, such as Pearson's correlation coefficient, into our method. In our experiments, we discretized the numeric attribute values into pre-assigned buckets, and then it should be treated as a categorical value.

6 Experiments

This section describes the experimental evaluations for our ranking method. The evaluation results focus on the retrieval quality and the computational efficiency.

6.1 Experimental Setup

For the experimental dataset, a used car database CarDB (Make, Model, Year, Color, Style, Price, Mileage, Location) is set up containing 15955 tuples extracted from *Yahoo! Autos* [16]. In this database, each tuple represents a car for sale, and each column represents an attribute of the car. Since our ranking method only focuses on categorical data, the attributes *Price* and *Mileage* (which are numerical values) are discretized into categorical data (5 buckets: $p_1 \sim p_5$ and $m_1 \sim m_5$). MySQL Server 5.0. RDBMS is used on a system with an AMD Athlon 64 processor 3.2 GHz PC with 2 GB of RAM for the environment of the experiments. Algorithms are implemented in JAVA and connected to the RDBMS through JDBC interface.

Two other ranking methods are implemented to compare with our method, RAVC.

RANDOM. The query results are presented in a random order in this ranking model. This model provides a baseline method.

QRRE. A Query Result Ranking over E-commerce (QRRE) model [13] has been successfully used for ranking objects and addresses the same problems as our method (RAVC). In QRRE, given a tuple t, its ranking score is calculated based on the distribution difference of attributes in query results and in a database using the *Kullback-Leiber distance measure.*

The entire tuple scoring is indexed for up to 3-words query for all instances in the database at preprocessing time because the attribute value distribution analysis time is too long compared to the query processing time. For pragmatic use of this method, many query optimization issues such as a large storage overhead or frequent database updates should be presented; however, the issues are not the focus of this paper. At the query time, the query keywords are joined to find the intersection of the matched tuples.

The search quality evaluation was accomplished using surveys from 14 graduate students from our respective universities and institutions. Users were requested to behave like a used-car dealer customer. We solicited 10 typical queries that represent heterogeneous mix of different profiles of positional car buyers. Since it is not practical to ask participants to select the whole query result for a specific query, we conduct a comparative study for our purpose. We collected the first 10 tuples produced for each implemented ranking methods, and 30 tuples were collected in total. If there is overlap among the resulted tuples from different methods, more tulpes are extracted using the RANDOM algorithm so that 30 unique tuples are collected in total. Next, for each of the queries, each user was asked to rank the top 10 tuples as the relevant tuples that they preferred most from the 30 unique tuples collected for each query. This strategy can evaluate the average precision for each ranking method.

6.2 Implemented Algorithm

The entire object scoring is indexed for every match in the database instances in a preprocessing time because the attribute value distribution analysis time needs high cost of repeatable database scans. At the query time, only the query keywords are joined to find the intersection of the matched objects. We assume that the number of query matches is at least 1 in D for a given query. We built a preliminary index structure, *D-table*, which describes the data statistics in D for efficient ranking process. It is possible to avoid repetitive calculation for deriving a final score for a given query at online stage. D-table maintains the mutual occurrence values and the entropy value for each column in the database D. D-table construction algorithm is as follows.

D-table construction

```
program constructDtable (input:D)
  {tuple score R={(score_i, tuple_i)|i<n}, attribute-value
weight A={(attribute_j, value_j, freq_j, entropy_j|j<m)}};
  for each j in 1,..., m do
    for each i in 1,..., n do
      freq_j := |t_i| / Count(argmax(|t_i|));
      entropy_j := probe t_i to derive (-|t_i|log2|t_i|);
    end
    update A_j : (attribute_j, value_j, freq_j, entropy_j);
  end
```

6.3 Evaluation Results

The quality of query results is evaluated with a two metrics: *average precision* and *average recall*. The Precision metric is used to evaluate how well the user intention is captured by the different ranking models. Precision is the ratio obtained by dividing the number of retrieved tuples that are relevant to the total number of retrieved tuples.

$$Precision = \frac{|relevant\ tuples \cap retrieved\ tuples|}{|retrieved\ tuples|}. \tag{8}$$

The Recall metric is used to evaluate the quality of the relevance-based query result ranking, which utilizes the concept vector. Recall is the ratio obtained by dividing the number of retrieved tuples that are relevant to the total number of relevant tuples.

$$Recall = \frac{|relevant\ tuples \cap retrived\ tuples|}{|relevant\ tuples|}. \tag{9}$$

The overall search quality of the RAVC ranking method is higher than that of QRRE. RAVC effectively distinguish the tuples of the same ranking score in QRRE ranking system. Also, our method clearly outperforms QRRE in the recall metric. By expanding the original query results based on similarity, we progressively improve the recall in the model of relevance-based retrieval in relational databases. Followings are discussions with illustrative figures.

Average Precision. Figure 1 shows the average precision of the different ranking methods for each query. Only the top 10 results are considered to be evaluated in the setting of this study; this measure is precision at 10. The results of the experiments show that RAVC generally produces rankings of higher quality compared to QRRE.

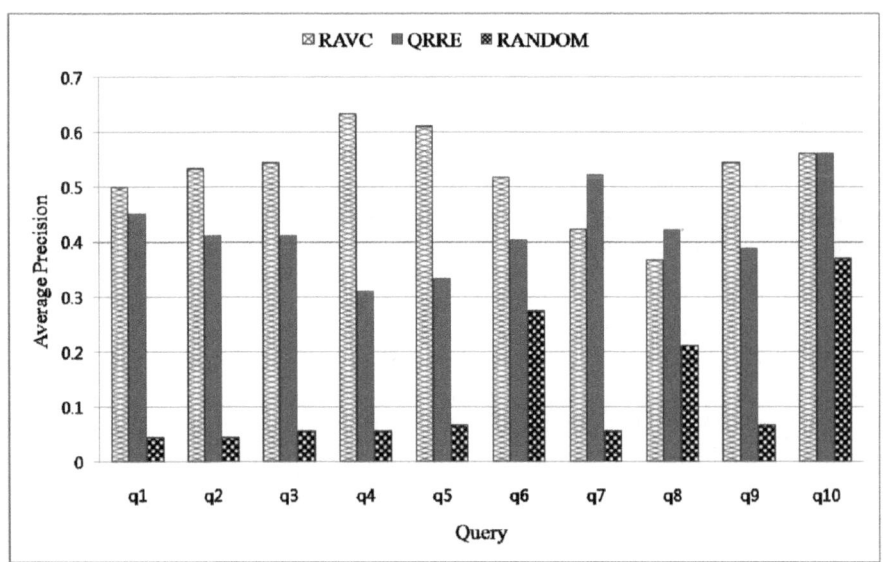

Fig. 1. Average precision

For most of the queries, there is some overlap between the results of RAVC and that of QRRE. The query results for q7 and q8 lost in precision compared to QRRE method. The reason is that the terms that were very commonly surfaced in the query logs but these terms were not popular terms for the car sales databases. Since our method focus on the database statistics only, the vocabularies not appeared in the database cannot be used as supports for ranking tuples. For 8 of 10 queries, the precision of RAVC is 0.15 higher than that of QRRE method on average.

Average Recall. The average recall of answers for each algorithm is shown in Figure 2. The recall of RAVC is consistently higher than QRRE because it expands the result set. Averagely, 14 tuples are expanded for the query set. The average recall of the RAVC is 0.07 higher than that of the QRRE. Recall increases with an increasing number of relevant documents retrieved. By expanding the relevant set of tuples in relational databases, the average precision does not deteriorate at each recall point.

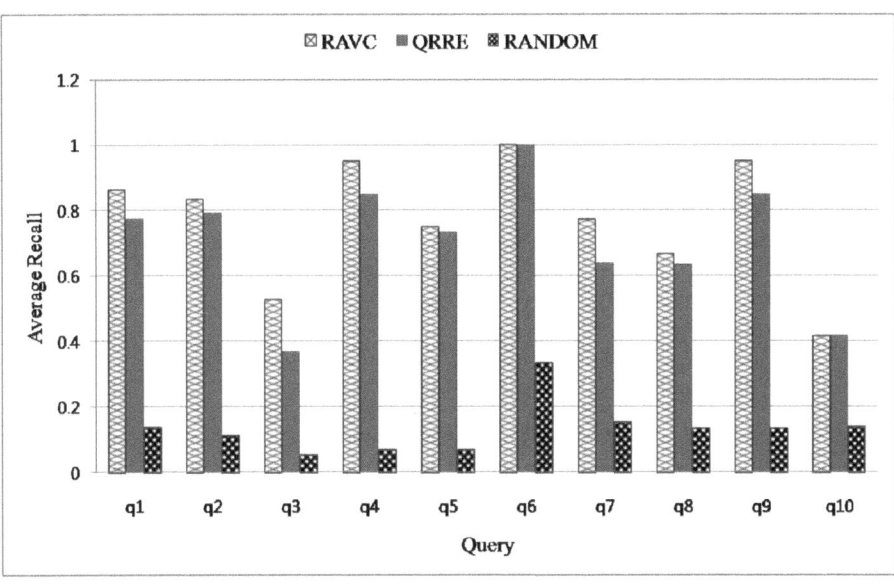

Fig. 2. Average recall

Execution Time. A preprocessed index table, D-table, should be constructed before the query time for efficient processing. The index table contains the data statistics such as data correlation and value distribution status. The construction time usually remains constant over time. The query result ranking in the online processing includes attribute value weight selection, tuple score computation, and the sorting tuples. The three computation modules have a time complexity of $O(n)$ and the sorting algorithm has a time complexity of $O(nlogn)$. The time complexity of our method in the online stage is $O(nlogn)$. Figure 3 shows the online execution time as a function of the number of tuples in the query result. The execution time grows almost linearly with the number of tuples in the query result. This result illustrate our method spends a reasonable amount of time to rank objects even for a large size of the query results.

6.4 Summary

Overall search quality of the RAVC ranking method is higher than that of QRRE. They perform better than the RANDOM model. The RAVC is an effective ranking method for several reasons:

1) Users will have more specifically ordered query results because the RAVC can effectively distinguishes the tuples of same ranking score in QRRE ranking system by assigning a single score to represent the importance of each of the tuples.
2) Our method clearly outperforms QRRE in recall metric. The expanded query results based on the similarity measure will meet the needs of the user in the framework of ranking in relational databases.

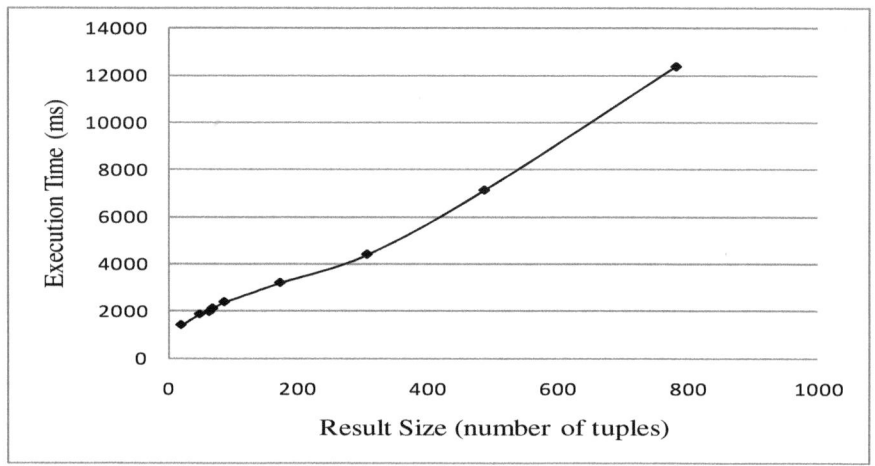

Fig. 3. Execution time

7 Conclusion

We presented a novel method to rank objects in relational databases. Based on the attribute value correlation and its distribution status, a score is assigned to a tuple for ranking. For limited results, similar tuples are added to the Boolean query results based on the concept vector. No domain knowledge or user feedback is required in the whole process. Experimental results showed the search performance and the efficiency of our method, RAVC. The experiments on a real dataset identified that RAVC improves the performance of the existing approach QRRE [13], which is the well-known technique tried to resolve the same problem.

Future plans are to investigate ways to minimize the size of the structure to store the pre-computed data structure by sampling databases. Comprehensive experimental evaluation should be conducted to prove the performance improvement in more principled ways. A large, comprehensive benchmark should be built to extensively evaluate the query result ranking system for future research.

References

1. Agrawal, S., Chaudhuri, S., Das, G.: DBXplorer: A System for Keyword-Based Search over Relational Databases. In: 18th IEEE International Conference on Data Engineering, pp. 5–16. IEEE Press, New York (2002)
2. Agrawal, S., Chaudhuri, S., Das, G., Gionis, A.: Automated Ranking of Database Query Results. In: First Biennial Conference on Innovative Data Systems Research, pp. 888–899. ACM Press, New York (2003)
3. Bhalotia, G., Hulgeri, A., Nakhe, C., Chakrabarti, S., Sudarshan, S.: Keyword Searching and Browsing in Databases using BANKS. In: 18th IEEE International Conference on Data Engineering, pp. 431–440. IEEE Press, New York (2002)
4. Chaudhuri, S., Das, G., Hristidis, V., Weikum, G.: Probabilistic Information Retrieval Approach for Ranking of Database Systems. ACM TODS 31(3), 1134–1168 (2006)
5. Das, G., Hristidis, V., Kapoor, N.S., Sudarshan, S.: Ordering the Attributes of Query Results. In: 26th ACM SIGMOD International Conference on Management of Data, pp. 395–406. ACM Press, New York (2006)
6. Hristidis, V., Papakonstantinou, Y.: DISCOVER: Keyword Search in Relational Databases. In: 28th International Conference of Very Large Data Bases, pp. 670–681. VLDB Endowment, New York (2002)
7. Huhtala, Y., Karkkainen, J., Porkka, P., Toivonen, H.: TANE: An Efficient Algorithm for Discovering Functional and Approximate Dependencies. Comput. J. 42(2), 100–111 (1999)
8. Manning, C.D., Raghavan, P., Schtze, H.: Introduction to Information Retrieval. Cambridge University Press, New York (2008)
9. Meng, X., Ma, Z.M., Yan, L.: Answering Approximate Queries over Autonomous Web Databases. In: 18th International World Wide Web Conference, pp. 1021–1030. ACM Press, New York (2009)
10. Nambiar, U., Kambhampati, S.: Answering Imprecise Queries over Autonomous Web Databases. In: 22th IEEE International Conference on Data Engineering, pp. 45–55. IEEE Press, New York (2006)
11. Binderberge, M.O., Chakrabarti, K., Mehrotra, S.: An Approach to Integrating Query Refinement in SQL. In: Jensen, C.S., Jeffery, K., Pokorný, J., Šaltenis, S., Bertino, E., Böhm, K., Jarke, M. (eds.) EDBT 2002. LNCS, vol. 2287, pp. 15–33. Springer, Heidelberg (2002)
12. Shannon, C.E.: A Mathematical Theory of Communication. SIGMOBILE Mob. Comput. Commun. 5(1), 3–55 (2001)
13. Su, W., Wang, J., Huang, Q., Lochovsky, F.: Query Result Ranking over E-commerce Web Databases. In: 15th ACM CIKM International Conference on Information and Knowledge Management, pp. 575–584. ACM Press, New York (2006)
14. Yong, R., Huang, T.S., Mehrotra, S.: Content-based Image Retrieval with Relevance Feedback in MARS. In: 4th IEEE International Conference on Image Processing, pp. 815–818. ACM Press, New York (1997)
15. Xu, J., Croft, W.B.: Expansion using Local and Global Document Analysis. In: 19th ACM SIGIR International Conference on Research and Development in Information Retrieval, pp. 4–11. ACM Press, New York (1996)
16. Yahoo! Autos, http://autos.yahoo.com

Efficiently Finding Similar Objects on Ontologies Using Earth Mover's Distance

Mala Saraswat

Rustamji Institute of Technology,
Gwalior, India
malasaraswat@gmail.com

Abstract. Ontologies are being progressively used to capture the semantics of information from various sources. They have wide area of
usage ranging from artificial intelligence, natural language processing to
web content and biology. This paper proposes the problem of finding similar objects that have been defined as a set of terms from an ontology.
We consider tree-based ontologies where a node represents a term and an
edge weight defines the distance or dissimilarity between corresponding
terms. For object distance, Earth Mover's Distance (EMD) is used as
it outperforms other distance measures like average and minimum pairwise distance. EMD, however is highly computationally intensive as it
involves solution to linear programming (LP) problem. We propose an
efficient lower bound on computing EMD by aggregating the terms in the
ontology at the first level of the tree. This reduces the number of terms,
thereby decreasing the number of flow variables and making it computationally faster. Range queries that use the lower bound runs faster by up
to a factor of 20, as approximately 97% percentage of database objects
are pruned, thereby saving expensive EMD calculations.

Keywords: Ontologies, Similarity Search, Range queries, Earth Mover's
Distance.

1 Introduction

Ontologies can be defined as explicit formal specification of terms in a particular
domain and relationships among them [6]. It is used to share information by
people, applications and databases. Examples of ontology includes WordNet (an
electronic lexical database) [9], Semantic Web (defined as meaningful web) which
is developed to provide semantics for web resources [10].

Ontologies can be represented by directed acyclic graphs (DAG) since they
comprise a generalization hierarchy. The nodes of the graph represent the concepts and edges represent the relationship (hierarchical) between the concepts.
By DAG representation of the ontology the degree of similarity between two
terms can be estimated. In this paper ontology is assumed as tree.

It is assumed that each object has been labeled with one or more terms (which
form an ontology tree) that define the properties of the object. Terms can be

P. García Bringas et al. (Eds.): DEXA 2010, Part II, LNCS 6262, pp. 360–374, 2010.

mapped to objects as many-to-one mapping. Since ontology as a tree are considered, the distance $d(t_i, t_j)$ between two terms t_i and t_j is defined as length of the shortest path between them on ontology tree. The degree to which two terms are similar depends on location of the terms in hierarchy. The lower in hierarchy the more similar the terms or feature are.

Using similarity search (range search), the distance measure finds objects from the given database set which are within a range from the query object. This paper presents an approach to use ontologies to find similar objects.

In literature many type of distances are used to find dissimilar objects. We compared different distance measures like average pairwise distance, minimum pairwise distance and Earth Mover's distance to find dissimilar objects whose terms form an ontology tree. As EMD is highly computationally expensive *lower-bound* is derived to efficiently find similar objects using the given ontology tree.

The rest of the paper is organized as follows. Section 2 discusses and compares various distance measures where the terms of objects form an ontology tree. Section 3 introduces lower bound approach to speed up similarity search. Section 4 analyzes effect of varying different parameters on range queries for similarity search of objects. Finally, Section 5 concludes the paper.

2 Comparison of Various Distance Measures

This section compares different distance measure (minimum pairwise distance, average pairwise and EMD), given an ontology tree with a set of *terms* as nodes where the terms are defined as attributes or feature of the objects. The edge distances between each term pair is given in the ontology tree which serves as the measure of dissimilarity. It is assumed, with the increase in level of tree the edge distance decreases by half since as we go deeper in hierarchy, the terms are more similar to each other than the terms at upper level. The distance $d(t_i, t_j)$ between two terms t_i and t_j is defined as length of shortest path between them on ontology tree. Since there is only one path between two terms it can be shown from properties of shortest path, that this distance is a metric distance [5]. Different distance measures are defined between the objects using these term distances.

Figure 1 illustrates a particular instance of the problem. An ontology tree is given, which shows the terms and distance between them. The distance decreases exponentially as the level increases. In this example, $O_1 = \{t_1\}$, $O_2 = \{t_2, t_3\}$, $O_3 = \{t_3, t_4, t_5\}$, $O_4 = \{t_3, t_5\}$, $O_5 = \{t_2, t_5\}$, $O_6 = \{t_1, t_6\}$ and query object $Q = \{t_1, t_7\}$.

2.1 Minimum Pairwise Distance

The *minimum pairwise distance* between two objects is defined as minimum pairwise distance between their corresponding set of terms.

The minimum pairwise distance between two objects O_i and O_j is defined as

$$d_{min}(O_i, O_j) = t_i \in O_i, t_j \in O_j min\{d(t_i, t_j)\} \tag{1}$$

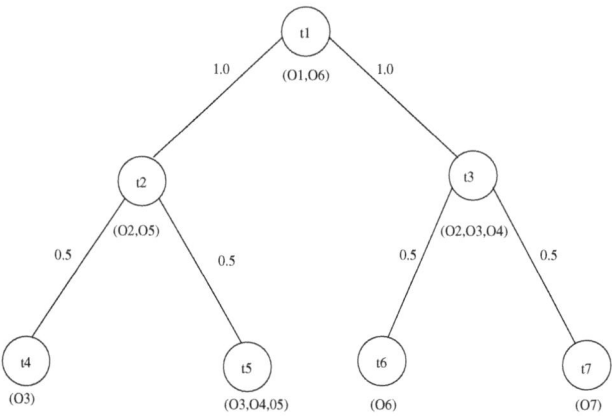

Fig. 1. Running example

The distance is useful when searching objects that have similar terms (concepts). For example, even though a single biological document may contain references to different terms like photoreceptor cells and ganglion cells, it is useful to be able to retrieve it when another document that describes photoreceptor cells is queried.

Table 1. Minimum pairwise distance for the example

	O_1 O_2 O_3 O_4 O_5 O_6 Q
Q	0.0 0.5 0.5 0.5 0.5 1.0 0.0

Table 1 shows the minimum pairwise distances of the query object to other objects in Figure 1. It can be verified, that d_{min} distance is *not* a metric distance, since it does not follow triangular inequality. For example, for the objects O_1, O_3 and O_6 triangular equality is not maintained as $d_{min}(O_1, O_6) + d_{min}(O_6, O_3) = 0.0 + 0.5 < 1.0 = d_{min}(O_1, O_3)$.

2.2   Average Pairwise Distance

The *average pairwise distance* between two objects is defined as average pairwise distance between their corresponding set of terms. The average pairwise distance between two objects O_i and O_j is defined as

$$d_{avg}(O_i, O_j) = \frac{1}{|O_i| \cdot |O_j|} \sum_{t_i \in O_i, t_j \in O_j} d(t_i, t_j) \qquad (2)$$

where $|O_i|$ and $|O_j|$ denote the number of terms describing O_i and O_j respectively. The average is useful in cases where objects are not precisely defined. For example, among genes whose functions are not known with certainty but can be described using some general function terms, the average distance can be used to query all gene pairs with similar function profiles. It can be used when objects are defined using multiple attributes, the similarity among the objects will be based on all its attributes. Table 2 shows the average distance among objects for the example in Figure 1. As can be seen average distance is *symmetric* since

$$\forall i, j \qquad d_{avg}(O_i, O_j) = d_{avg}(O_j, O_i)$$

For example $d_{avg}(Q, O_6) = d_{avg}(O_6, Q) = 1.0$.

Table 2. Average pairwise distance for the example

	O_1	O_2	O_3	O_4	O_5	O_6	Q
Q	0.75	1.25	1.75	1.5	2.0	1.0	0.75

If the term distance is metric, the avg distance follows the triangle inequality property. Assume any three objects A, B and C. We need to prove $d_{avg}(A, B) + d_{avg}(B, C) \geq d_{avg}(C, A)$.

Brameier *et al.* [4], used minimum pairwise distance to group yeast genes (Saccharomyces cerevisiae) according to expression profile and Gene Ontology annotations. Genes with similar expression profiles are more likely to have similar biological function. Minimum pairwise distance $d_{GO}(g_1, g_2)$ was used to compute GO (Gene Ontology) distance between two pair of genes g_1 and g_2 and average pairwise distance $d_{GO}(g, C_i)$ computes gene-cluster distance between gene g and genes that have been assigned to cluster C_i, excluding g if it is in C_i. To explore the relation between gene evolutionary rates and functional similarity, Leonardo *et al.* [8] used average pairwise distance. They performed evaluation of evolutionary rates and functional annotations for the yeast (Saccharomyces cerevisiae). Non-synonymous (dN) and synonymous (dS) substitution rates were calculated for gene sets common to *S.cerevisiae* and other yeast species. Comparison between evolutionary rates between pair of genes ($\triangle dN$ and $\triangle dS$) and functional similarity sGO (measured using Gene Ontology) was made. For any gene pair ij, all term-term similarity values were aggregated at the level of gene products to yield sGO_{ij} using average pairwise distance. It was found $\triangle dN$ and sGO had significant correlation.

2.3   Earth Mover's Distance

The Earth Mover's Distance (EMD) was introduced by Rubner *et al.* [11] to overcome the inconsistencies with perceptual similarity as observed in other dissimilarity measures like Minkowski-form distance, Kullback-Leibler divergence,

Histogram intersection, Jeffrey Divergence, χ^2 statistics. EMD between two objects is computed by adopting the definition as given in [7]. Suppose that objects A and B are annotated with n and m terms respectively. The ground distance $c_{ij} = d(t_i, t_j)$ between two terms t_i in object A and t_j in object B is defined as the length of the shortest path between them on the ontology tree. The feature values indicate the presence or absence of the term and are, hence, 1 for the terms present and 0 otherwise. Normalizing with the number of terms yields $a_i = \frac{1}{n}$ for all $i = 1, \cdots, n$ and $b_j = \frac{1}{m}$ for all $j = 1, \cdots m$. Computing the EMD involves finding a flow matrix $F = \{f_{ij}\}$, where each flow f_{ij} denotes the amount of feature or mass to be moved from term t_i in A to term t_j in B such that object A is transformed into object B. Note that both F and $C = \{c_{ij}\}$ are matrices of size $n \times m$ where n and m are number of feature values of object A and B. The cost of moving mass f_{ij} from term t_i to term t_j is the ground distance of t_i to t_j multiplied by the mass to be moved, or $c_{ij} \times f_{ij}$. The EMD, which is the minimum cost of transforming object A into object B is defined as

$$EMD(A, B) = Fmin \sum_{i=1}^{n} \sum_{j=1}^{m} c_{ij} f_{ij} \text{ subject to} f_{ij} \geq 0, \quad \sum_{j=1}^{m} f_{ij} = a_i \sum_{i=1}^{n} f_{ij} = b_j (3)$$

$$\forall i \in \{1, \cdots, n\}, \qquad \forall j \in \{1, \cdots, m\}$$

The EMD measures the effort to match one object against the other. Similar objects will cause less effort to be transferred to entries in other at minimal cost. EMD is a metric distance. The proof for the same is given in [7].

Table 3. EMD for the example

	O_1	O_2	O_3	O_4	O_5	O_6	Q
Q	0.75	0.75	1.417	1.0	2.0	0.5	0.0

Table 3 shows the EMD between the query object and all other objects for the example in Figure 1.

2.4   Quality Experiment

This section investigates the qualitative properties of the different distance measures, Minimum pairwise distance (MPD), Average pairwise distance (APD) and Earth Mover's distance (EMD) by finding similar documents using nearest neighbor search. Real-life datasets MESH (Medical Subject Heading) ontology and Ohsumed Test Collection are used.

MeSH is the vocabulary thesaurus of National Library of Medicine's (NLM) [2] consisting of sets of medical terms in a hierarchical structure. Experiments were conducted using 49,714 terms from mesh 2009.

The Ohsumed test collection was created to assist information retrieval research [3]. It is a clinically-oriented MEDLINE subset, consisting of large number of references, covering all references from different medical journals over a five-year period (1987-1991). MEDLINE is the National Library of Medicine's (NLM) bibliographic database covering various fields like medicine, dentistry, nursing, veterinary medicine etc.

There are different datafiles associated with Oshumed. There are a set of 63 different Ohsumed case (query-topics) each pertaining to some particular medical case like "30 year old with fever, lymphadenopathy, neurological changes and rash". There is year wise test collection of ohsumed documents from year (88-91) each having various fields like doc-id, abstract, MESH terms, author, source, publication. A file containing truth dataset with fields like the case number (varying from 1-63) and the doc-id, which are relevant/similar to particular case.

The first document for each of the case is considered to be query document. All other documents within the same case are similar, while documents within other case are dissimilar to the query document. We arbitrarily chose 10 cases and 15 documents for each case, which totaled to 150 documents pertaining to various cases. Then 10 query document (1 from each case) are chosen and each is compared with all documents in the collection. This is done by extracting the MESH terms from each of the documents and using MESH ontology for finding the average pairwise, EMD and minimum pairwise distances of all the documents from the selected query document. From the ranked list of (top-10, top-5 and top-2 nearest neighbor) documents which fall into same case as the given query document are counted for each distance measure. For the minimum case if there is a common MESH term, the distance is zero. So there are number of documents which have distance of 0.0 from query object. For this case we counted the number of objects having 0.0 distance from query object. The results are shown and compared in Table 4. It can be seen EMD based similarity measure outperforms other distance measure.

Table 4. Number of similar documents

	Top-10	Top-5	Top-2	Zero distances
APD	2.0	1.0	0.5	0.0
EMD	4.1	2.6	1.4	0.0
MPD	–	–	–	38.8

As can be seen EMD provides more accurate results than the other two. So we chose EMD in our work to find similar objects.

2.5 Computational Complexity

As seen from equation 4, EMD is a large linear programming problem since a flow matrix that will minimize the flows between terms in objects need to be found. To get the flow matrix, GNU Linear Programming Kit (GLPK) is used [1]. The number of flow variables in linear programming increases with the number of terms (features) by which an object is defined. Complexity of simplex is exponential in number of variables. Number of terms by which an object is identified cannot be changed as the object is defined by its terms. This increases the running time for each distance computation.

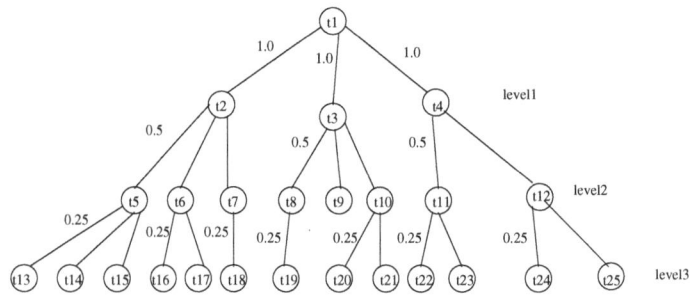

Fig. 2. An ontology tree

3   Using Lower Bound to Speed Up Similarity Search

The running time to solve the linear programming for computing EMD increases with the number of terms in each object. Reducing the number of terms, gives a lower bound bound which will make EMD computation efficient. This section describes how a lower bound for the distance calculation with large number of terms can be computed using lesser number of terms. This changes the object description. Ljosa *et al.* [7] have used lower bound to speed up similarity search for images.

For reducing the number of terms, all terms of a subtree are aggregated to the ancestor at level 1. Aggregating a subtree rooted at node T' at level 1 in the ontology tree implies changing all the nodes in the subtree to T'. Computation of distance between the nodes of subtree is done with respect to T'. This implies that the distance between nodes belonging to same subtree is zero, distance between a node in the subtree and T' is zero and distance between an external node (not belonging to the same subtree) and a node in the subtree is same as distance of external node and node T'. This aggregation of terms slightly decreases the distance between terms but not much difference is there, since the distance at each level goes on decreasing exponentially. General idea of aggregation is depicted through Figure 3 which shows an aggregated ontology tree transformed from the given ontology tree as depicted in Figure 2. The dashed triangle shows all the terms of the subtree which are aggregated to the respective ancestor at

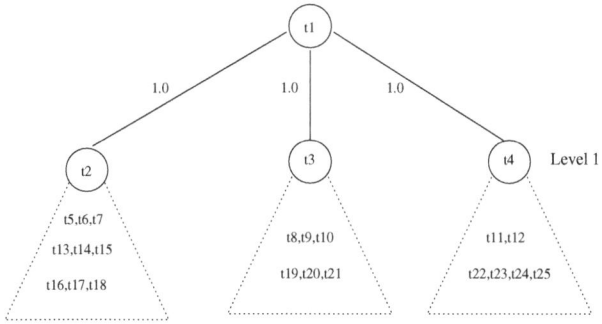

Fig. 3. Aggregating the terms of Ontology tree at level 1

level 1. As seen in Figure 3, t_2 is substituted for all terms in the leftmost subtree, t_3 for all terms in middle subtree and t_4 with all terms in rightmost subtree. Object description changes by introducing the lower bound. For example if object $O = \{t_2, t_5, t_8, t_{15}, t_{16}, t_{18}, t_{20}, t_{22}\}$ has 8 terms, after aggregating the terms at level 1, $O = \{t_2, t_3, t_4\}$ will have 3 terms.

3.1   Computing Lower Bound EMD

Suppose object A has n terms $T_n = \{1, \cdots, n\}$ and $\mathbf{a_i}$ is the feature value of term t_i. Given $n' \leq n$, we change the object description of A to A' by changing T_n (set of n terms) into T_n' (set of n' terms) where $n' \leq n$ by aggregating the terms to their respective ancestors. Given $m' \leq m$, B' is defined similarly for object B.

The feature value $\mathbf{a_i'}$ of object A' will be sum of the feature values of all the terms that are aggregated to the same ancestor at level 1.

$$\mathbf{a_i'} = \sum_{j \in t_i'} \mathbf{a_j} \qquad (4)$$

Feature values for object $O = \{0.125, 0.125, 0.125, 0.125, 0.125, 0.125, 0.125, 0.125\}$ for the above example. After aggregating to their ancestor at level 1 the feature values of object $O = \{0.625, 0.250, 0.125\}$ for terms t_2, t_3, t_4 respectively. Lower bound for EMD between two objects A and B is the EMD between their summaries A' and B' where the ground distance c_{ij} defined on $n \times m$ matrix is modified to another ground distance c_{ij}' defined on $n' \times m'$.

The d_{ij}' distance between two terms t_i' and t_j' (terms at level 1) is never more than the d_{ij} distance between any term corresponding to t_i and t_j. Reduction in the number of terms, reduces the size of ground distance matrix as well as flow matrix. Now the linear programming problem for lower bound EMD can be solved.

$$LEMD(A, B) = F' min \sum_{i=1}^{n'} \sum_{j=1}^{m'} c'_{ij} f'_{ij}$$

$$\text{subject to} f'_{ij} \geq 0, \quad \sum_{j=1}^{m'} f'_{ij} = a'_i \quad \sum_{i=1}^{n'} f'_{ij} = b'_j,$$

$$\forall i \in \{1, \cdots, n'\}, \qquad \forall j \in \{1, \cdots, m'\} \tag{5}$$

The number of variables in LP problem for object A and B is reduced by a factor of $(n/n')(m/m')$ making it less computationally demanding. For instance, as shown in example for Figure 3 if 8 terms are aggregated to 3 terms in both object and Query, the number of variables are reduced by a factor of around 7. For finding similar objects to a given query object (range queries) using lower bound EMD computation to speed up similarity search, first only lower bound EMD is computed. For each object and query, the terms and their feature value is found. These terms are aggregated to the ancestor of the subtree at level 1 and their feature values added. This reduces the number of terms in each object since many terms are substituted by only the terms at level 1. Then distance matrix c'_{ij} is computed. We then find the flow matrix and compute EMD. This is lower bound EMD since we changed the object description by reducing the number of terms. For similarity search lower bound EMD of each object is compared to given range. If lower bound EMD is less then the given range, actual EMD need to be computed. Objects whose lower bound EMD is greater than the given range are pruned since their actual EMD will be more than the lower bound EMD. Again the actual EMD is compared to range, if it is less we get similar the objects within a range to a given query object.

4   Experimental Results

This section presents the results of our proposed lower bound algorithm for computing EMD for similarity search. We compared the performance by introducing lower bound on EMD, with actual EMD calculations based on changing various parameters of the given ontology tree, in order to understand the effects of them.

Experiments were conducted using synthetic datasets and the effect of varying different parameters like number of objects, number of terms per object and branching factor is analyzed. Synthetic datasets are generated randomly with uniform distribution to construct artificial ontology tree. The number of terms and number of objects are varied through order of dimensions, starting from small datasets of size 300 to large datasets of size 10^5. Similarly number of terms per object and branching factor are varied from 3 to 100. For the experiments for similarity search (range search), range is varied from 0.125 to 2.0. An average over different query objects is taken for finding results since the dataset is random.

The terms and objects are randomly generated as per parameters given. We have implemented the code in C, both for actual EMD and lower bound EMD similarity search and used GLPK-4.25 [1], for solving linear programming equations.

4.1   Parameters of Comparison

Results are compared based on following parameters.

1. *Comparison of running time of lower bound EMD with actual EMD for range search*

   In case of similarity search without using lower bound, actual EMD is directly computed for all objects from query object. In case of similarity search using lower bound, first lower bound EMD for all objects from query object is computed. Actual EMD for only those objects whose lower bound EMD is less than the given *range* are computed.

   Equations below show the relation between pruning ratio and the ratio of lower bound and actual EMD time. Let us consider n objects and average time to compute EMD is E for each object from query object. Let g be number of objects left after pruning. Time for computing lower bound is E'.

   Total time for computing EMD for finding similar objects:

   $$T_1 = n \times E$$

   Total time for computing EMD using lower bound:

   $$T_2 = (n \times E') + (g \times E)$$

   If we want incorporating lowerbound to be beneficial then:

   $$T_2 < T_1$$

   or $(n \times E') + (g \times E) < n \times E$

   or $nE' + gE < nE$

   or $gE < n(E - E')$

   or $(n - g)E > nE'$

   or $\frac{(n-g)}{n} > \frac{E'}{E}$

   Theoretically it shows pruning ratio should be greater than the ratio of lower bound EMD to actual EMD for making lower bound to be beneficial. This is because it has to cope with the extra overhead of computing lower bound. Equation 1 signifies the importance of lower bound. It should be such that will prune more objects, reducing the actual EMD computations. Pruning ratio should be more than 25% to get $\frac{E'}{E}$ to 0.25. Practically it is found for value of $E = 27.34$ sec, $E' = 2.73$ sec i.e $\frac{E'}{E} = 0.1$ pruning ratio =77%. So performance is better. For range 2.0 no objects get pruned so pruning ratio is zero, signifying the lower bound is not beneficial.

2. *Comparison of Pruning Ratio*

   It determines the ratio of objects pruned $(n - g)$ to total number of objects (n). The more the ratio, the better is our algorithm for lower bound EMD since less actual EMDs' need to be computed.

3. *Tightness*
Tightness is the value of ratio of lower bound EMD with actual EMD. We
calculated tightness taking average EMD and lower bound EMD of all objects from different query objects. As it is the distance it does not vary with
range.

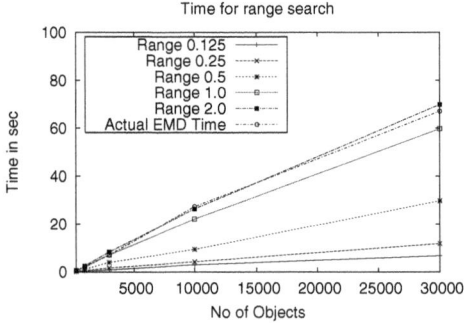

Fig. 4. Comparison of running time over different ranges with number of objects

Variation in number of objects. The figure show the effect of varying the
number of objects keeping other parameters i.e branching factor of ontology tree,
number of terms and number of terms per object constant.

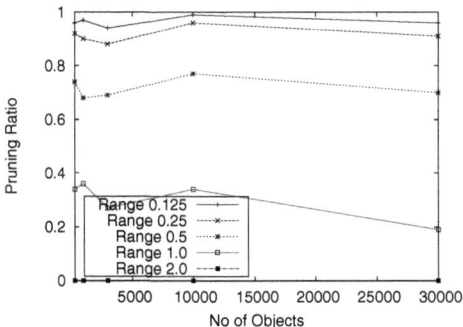

Fig. 5. Comparison of pruning ratio with number of objects

Figure 4 compares the time taken for finding similar objects over different
ranges by varying number of objects keeping other parameters constant. With
the increase in number of objects the time for similarity search also increases,
since more number of objects need to be compared.

The time also varies with the given *range*. As the range increases from 0.125
to 2.0, the time also increases since lower bound becomes less tighter.

With increase in number of objects Pruning ratio remains nearly constant as shown in figure 5. The ratio decreases with range as lowerbound gets less tighter.

Table 5. Comparison of tightness with increase in number of objects

No. of Objects	300	1000	3000	10000	30000
Tightness	0.33	0.32	0.31	0.31	0.30

Table 5 compares the tightness by varying the number of object. It does not vary with increase in the number of object.

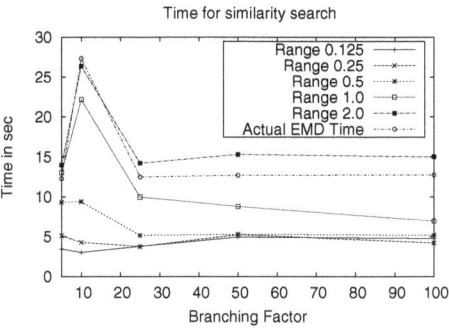

Fig. 6. Comparison of running time over different ranges with branching factor

Variation in branching factor of ontology tree. Figure 6 compares the running time for finding similar objects over different ranges with increase in branching factor of the given ontology tree. With increase in branching factor, the running time increases till 10, since level 1 terms will first increase to 10 (branching factor) which will increase the number of variables for computation. This is because all terms of objects will be aggregated at level 1 while finding lower bound EMD. Running time decreases after that as the height of tree decreases, making lower bound tighter. So more objects get pruned decreasing the running time. After that it is nearly constant as the distance decreases exponentially.

As shown in figure 7, Pruning ratio of object increases with the increase in branching factor of ontology tree since the height of ontology tree decreases and lower bound becomes tighter. As branching factor increases, tightness increases as depicted in Table 6 since the height of tree decreases, decreasing the *ground distance* c_{ij} and thus the actual EMD, increasing the tightness.

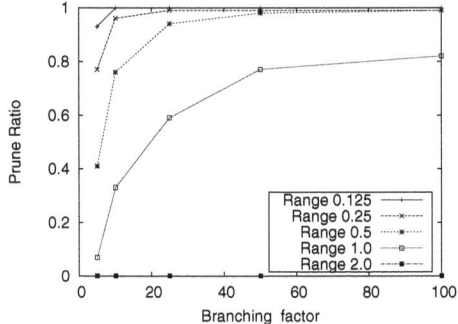

Fig. 7. Comparison of pruning ratio with branching factor

Table 6. Comparison of tightness with increase in branching factor

Branching factor	5	10	25	50	100
Tightness	0.28	0.33	0.43	0.5	0.53

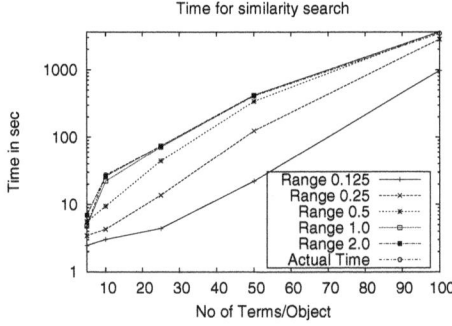

Fig. 8. Comparison of running time over different ranges with number of terms per object

Variation in number of terms per object. Figure 8 shows time taken for finding similar objects over different ranges with increase in the number of terms per object. As the number of terms per object increases, the time also increases since, more are the number of variables to be solved in linear programming, thus increasing the computation time.

As seen from figure 9 with increase in the number of terms per object, more terms aggregates to their ancestors at level 1, decreasing the EMD distance for more number of objects, making the lower bound less tight thus decreasing the pruning ratio.

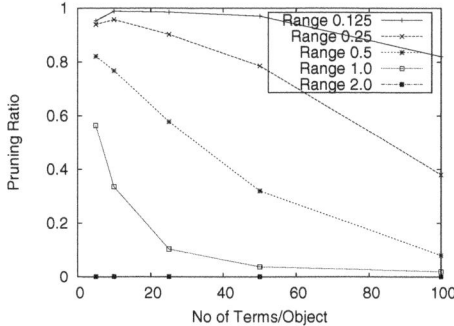

Fig. 9. Comparison of pruning ratio with terms per object

Tightness decreases with increase in the number of terms per object as depicted in Table 7 since more terms will be aggregated to the ancestor at level 1 (where maximum terms are 10), decreasing the number of terms (variables to solve LP) at level 1 and also the ground distance c_{ij} will decrease, making lower bound EMD to be less than actual EMD, hence decreasing tightness.

Table 7. Comparison of tightness with increase in number of terms per object

Terms per object	5	10	25	50	100
Tightness	0.38	0.33	0.28	0.25	0.20

5   Conclusion

The work in the paper was focused on designing and implementing an efficient algorithm that will find similar objects in a given ontology tree. The work considered the problem of speeding up the computation of spatially-sensitive distance measures between objects whose terms come from an ontology tree. Different distance measures are compared using real datasets (average pairwise, minimum pairwise and EMD) and found EMD outperforms other. But since it is computationally hard, we developed a lower bound for computing it. Through a series of experiments by varying different parameters the lower bound approach is justified and results in substantial performance enhancements. Following results through the experiments are justified:-

- EMD outperforms average and minimum pairwise distance as a distance measure for finding similar objects.
- Algorithm proposed provides a low cost approach to similarity search in ontology tree. With less range the algorithm is highly efficient as compared to higher ranges.

- The reduction in running time by incorporating lower bound, is as high as 20 times as approximately 97%objects are pruned. Reduction in cost is a huge gain as this is important for practical problems.

References

1. Gnu linear programming kit, http://www.gnu.org/software/glpk/
2. Medical subject headings,
   http://www.nlm.nih.gov/pubs/factsheets/mesh.html/
3. Ohsumed test collection, http://ir.ohsu.edu/ohsumed/ohsumed.html
4. Brameier, M., Wiuf, C.: Co-clustering and visualization of gene expression data and gene ontology terms for Saccharomyces cerevisiae using self-organizing maps. Journal of biomedical informatics 40(2), 160–173 (2007)
5. Cormen, T., Leiserson, C., Rivest, R., Stein, C.: Introduction to algorithms. MIT Press, Cambridge (1990)
6. Gruber, T.R.: A translation approach to portable ontology specifications. Knowledge Acquisition 5(2), 199–220 (1993)
7. Ljosa, V., Bhattacharya, A., Singh, A.: Indexing spatially sensitive distance measures using multi-resolution lower bounds. In: Ioannidis, Y., Scholl, M.H., Schmidt, J.W., Matthes, F., Hatzopoulos, M., Böhm, K., Kemper, A., Grust, T., Böhm, C. (eds.) EDBT 2006. LNCS, vol. 3896, p. 865. Springer, Heidelberg (2006)
8. Mariño-Ramírez, L., Bodenreider, O., Kantz, N., Jordan, I.: Co-evolutionary rates of functionally related yeast genes. Evolutionary bioinformatics online 2, 295 (2006)
9. Miller, G., Beckwith, R., Fellbaum, C., Gross, D., Miller, K.: Introduction to wordnet: An on-line lexical database. International Journal of Lexicography 3(4), 235–244 (1990)
10. Pease, A., Fikes, R., Hendler, J.: Ontologies and the Semantic Web. In: AAAI Workshop, Edmonton AB (2002)
11. Rubner, Y., Tomasi, C., Guibas, L.: The earth mover's distance as a metric for image retrieval. International Journal of Computer Vision 40(2), 99–121 (2000)

Towards a "More Declarative" XML Query Language*

Xuhui Li[1], Mengchi Liu[1], and Yongfa Zhang[2]

[1] State Key Lab of Software Engineering, Wuhan Univ.,
Wuhan, Hubei, China
lixuhui@whu.edu.cn, mengchi@sklse.org
[2] International School of Software, Wuhan Univ.
Wuhan, Hubei, China
yongfachang@hotmail.com

Abstract. To extract and restructure information in XML documents, various query languages have been proposed in the past decade. These languages take navigational or pattern-based approach to data extraction and often claim to be declarative. However, declarativeness in them is not as prominent as in SQL because they often exhibit a procedural style in handling heterogeneity and presenting tree-like document structure.

In this paper, a new XML query language called XTQ is proposed to address this challenge. XTQ is a pattern-based language which introduces disjunction as well as conjunction operators in composing tree-like patterns named LXT (Logic XML Tree) for data extraction. LXT can expressively handle heterogeneity common in XML queries. Based on a hierarchically structured pattern with considerate restructuring rules, XTQ deploys a flexible hierarchically grouping mechanism in data construction so that complex tree-like structure can be intuitively presented. Examples from common query request show that XTQ can present XML queries more declaratively than existing studies.

1 Introduction

XML documents, a kind of hierarchical documents with user-defined tags, have widely spread over the Internet since the end of last century. All kinds of data are migrated to XML documents to be machine-readable and exchangeable. XML query, which extracts data from and constructs them into XML documents, thus becomes an important and interesting topic in utilizing XML documents.

Many languages have been proposed to query XML documents, such as XPath[1], XQuery[2], XDuce[3], Xcerpt[4] and TQL[5]. To be convenient and user-friendly, most XML query languages claim to be "declarative". Being "declarative" means that the language can enable users to focus on presenting the purpose

* This research is partially supported by the Fundamental Research Funds for the Central Universities of China under contract No.6082010, the Wuhan ChenGuang Youth Sci.&Tech. Project under contract No.200850731369, and the National Science Foundation of China under contract No.60688201.

P. García Bringas et al. (Eds.): DEXA 2010, Part II, LNCS 6262, pp. 375–390, 2010.

of computation and seldom need to consider the procedures to fulfill the purpose, in other words, describes what to do without specifying how to do. Designers of database query languages like SQL usually pursue declarativenss of the language because database users often don't wish to consider the underlying mechanisms in processing query. Unfortunately, this feature is not as prominent in XML query languages as it was in SQL due to the tree-like structure of XML documents.

How to present the purpose of XML query and then to resolve and accomplish it are key to declarativeness of XML query language. We have studied common data manipulation requirements of XML queries and found some issues key to declarativeness in presenting XML query. a) **Separating data extraction and data construction** can enable users to explicitly describe the purpose of query with a global sketch of data extraction and construction. b) **Structurally composing global data extraction based on local ones** can enable users intuitively describe data extraction requests by specifying structural relationship of the data to be extracted. c) **Flexibly presenting data construction and deductively resolving data restructuring process** based on extracted data structure enable users specify various construction requests coherent to extracted data without describing the data transformation procedures. d) **Coherently handling heterogeneity** is of special importance in XML query in that heterogeneous data are often to be gathered and processed homogeneously and homogeneous data are to be separated and processed heterogeneously, because of the semi-structured characteristics of XML documents. Unfortunately, existing XML query languages haven't addressed the above issues as a whole.

In this paper, we propose a new XML query language called XTQ standing for XML Tree Query. XTQ is a pattern-based functional language which meets all the requirements above. XTQ is distinct from existing studies in three aspects: a)XTQ adopts an expressive tree shape pattern named *Logical XML Tree*(LXT) which enables disjunction as well as conjunction operators for composing patterns. The disjunctively composite patterns greatly enhances the flexibility of data extraction and enables declaratively handling heterogeneity in both data extraction and data construction. b) XTQ uses a pattern expression named *variable pattern* together with proper semantic model to coherently specify the hierarchical structure and instance of the extracted data. c) XTQ adopts a flexible and deducible restructuring mechanism to present data restructuring and construction, which enables the declarativeness of data construction.

The rest of the paper is organized as follows. Section 2 introduces the related works in more detail. Section 3 describes the pattern-based mechanisms of data manipulation in XTQ. Section 4 illustrates the outline, the usage and the advantages of XTQ with certain typical examples. Section 5 makes a conclusion.

2 Related Works

The studies of XML query language can be roughly classified based on their data extraction mechanisms as the navigational ones and the positional ones[6].

Early navigational XML query languages such as Lorel[7] use the "." notation denoting the parent-child relationships as major navigation operator, and the

Kleene closure operator is also allowed for expressing navigation through recursively defined data structures to enhance the expressive power. Later studies such as XQL[8] and XPath propose and optimize new navigational mechanisms specifically for XML document query. For example, XQL uses the notations "/" and "//" representing the parent-child and ancestor-descendant relationships for navigation, and it also proposes filtering the navigation results with predicates. Furthermore, XPath deploys the "step" functions instead of notations for navigation and data extraction. The "step" function, in the form of "axis::nodetest[predicate]", is composed of a location function called "axis" and a node test optionally with predicates to filter the results. The "axis", indicating the relative positions such as "child", "descendant" and "parent" can navigate forward or backward along the hierarchical path and access the siblings. To make the path expression more expressive, the logic language XPathLog[9] extends XPath by binding multiple variables to the query terms in predicate, which records the "route" of navigation. The full-fledged navigational language, e.g., XQuery or XSLT, deploys the XPath expression to extract data and uses certain statements like FLWOR expression for result construction.

Navigational languages are easy to understand and to use for most programmers. However, the programs in navigational languages are seldom "declarative" enough, because there is a fundamental gap between data extraction and data construction. Navigation expression like XPath navigates document tree and extracts a flat set of homogeneous data as its navigation target indicates, but XML queries usually are required to output XML document fragments. Therefore, certain control flows like *for* loops, nested queries and sometimes conditional statements are necessary to construct hierarchical document. To iterate the elements from XPath expressions and further fetch the data deeply inside, multiple *for* and *let* statements have to be involved in complex variable binding, which segments the original tree structure and reduces program readability. Therefore, users usually have to write XQuery program in a lengthy and nesting style, and data extraction and construction are often intertwined to describe the procedures of extracting data and constructing results.

The positional or, pattern-based, XML query languages utilize tree- or graph-shaped patterns as sketch of the data of interest. The design of the patterns, especially the composition of the patterns, shows the designers' preference and background. In the database field, studies often use structural composition, that is, complex pattern is composed of simpler ones according to document structure. Early studies such as UnQL[10] and StruQL[11] use simple structural patterns. The patterns in these languages are of simple syntax and semantics and thus trivial in presenting query request, which makes the program lengthy to write and hard to read. More considerate pattern-based languages such as XTreeQuery[12] and Xcerpt propose new structural pattern expression to present the patterns compactly. In comparison with early pattern-based languages, the tree-shape patterns are much more intuitive and compact. However, the composite pattern in them only implicitly describes the conjunctive relationships among fragments, and there is no inherent restructuring mechanism for data transformation.

Therefore, the patterns are often rigid and the languages are not expressive enough to meet the requirements of handling heterogeneity and declaratively presenting data construction.

Another pattern-based approach to XML query deploys type systems for XML documents, e.g. regular expression type system[3], in static typed XML processing languages like XDuce and CDuce[13]. Since the regular expression types for XML data allow various kinds of type composition, e.g., concatenation, union, and recursion, the patterns constructed based on the compound types are powerful and flexible. However, the type-based patterns can only work with conventional functional language core and thus is not convenient for common users who prefer to use query languages. Additionally, type-based pattern is only fit for the XML document fragments with predefined schema, and is not suitable for querying common XML documents.

Logic plays an important role in data query because users tend to consider their query request in a logic way. In data query languages, logic is often used to filter data and deduce results. For example, SQL adopts logic connectives, e.g., "and", "or" and "not", in combining conditions for data filtering, and DataLog, the deduction database language, uses logic rules for data declaration and query deduction. As for XML query languages, XPath and XQuery use logic connectives for data filtering; XML-RL[14], XPathLog and Xcerpt use logic deduction in presenting complex query procedures. However, although logic is popularly used in XML query languages, its use is still limited to the ways in conventional data query. Studies seldom concern the use of logic in presenting query requirements on XML document structure. Existing pattern-based languages, e.g., Xcerpt, often implicitly indicate that the patterns are composed conjunctively, but these languages seldom explore the features of logic in designing patterns syntactically or semantically, and thus lose the valuable expressive power of logic in presenting queries. An exception is TQL which adopts Ambient logic[15], a modal logic describing spatial relations, to form logic-based patterns. TQL inherits many syntactic and semantic features from Ambient logic, which makes it very expressive but relatively complicated and thus hardly accepted by common users having no such backgrounds.

3   Data Manipulation in XTQ

3.1   Logic XML Tree

Logical XML Tree (LXT) is a tree-like pattern expression with logic composition operators. LXT pattern is built based on **atomic pattern**s which carries out basic form of pattern matching. Atomic patterns in LXT include constant values, i.e., XML elements in the form of the pair "*tagname*⇒*content*", variables like "*$x*" prefixed with "$", and expressions substituting one or both sides of the element pair with variable or the wild-card notation "*", e.g., "*title*⇒*$t*", "*$n*⇒*$c*" and "*title*⇒*". Matching an atomic pattern with document fragments can result in variable bindings like "*$x* ↦ *v*" or Boolean values *true* and *false*. For example, matching the pattern "*title*⇒*$t*" with a fragment "*title*⇒*'TCP/IP Illustrated'*" would result in the binding "$\$t \mapsto$ 'TCP/IP Illustrated'" whereas matching the constant value pattern "$title \Rightarrow$ 'Unix'" with the fragment would result in *false*. A special kind of patterns are the ones with the wild-card "*", e.g., "$title \Rightarrow$*". It would match any *title* element and result in *true*, for other elements the result is *false*. Besides, a variable can be directly used as an atomic pattern, indicating that it can be bound to any document fragment under certain context.

Definition 1. An **LXT** (Logical XML Tree) is a pattern expression inductively defined as follows:

1. An **element pattern** composed of a location prefix and an atomic pattern is an LXT pattern. A location prefix can be "/", "//" or "@", which respectively indicates a child, a descendant, or an attribute.

2. A **tree pattern** composed of an element pattern "p_1" and an LXT pattern "p_2", denoted by "p_1 p_2", is an LXT pattern.

3. A **conjunctive pattern** composed of a sequence of LXT patterns, denoted by "$(p_1, p_2, \ldots, p_n)$", is an LXT pattern.

4. A **disjunctive pattern** composed of a sequence of LXT patterns which either all contain variable(s) or all contain no variable(s), denoted by "$(p_1 \parallel p_2 \parallel \ldots \parallel p_n)$", is an LXT pattern.

5. A **negative pattern** composed of an LXT pattern p which contains no variable, denoted by "$not(p)$", is an LXT pattern.

An LXT is often of a tree-like shape because the tree pattern "p_1 p_2" is composed of an element pattern p_1 as root and an arbitrary LXT p_2 as branch. The patterns in p_2 are imposed on the contents of the element matching p_1. This tree-like composition enables LXT pattern to sketch the outline of XML documents as structural pattern does. LXT supports conjunctive, disjunctive and negative compositions which are named after common logic operators since they play similar functions in pattern matching. A typical example is as follows:

```
//book=>*((/author=>$a/email=>$em || /editor=>$e/email=>$em),
          /title=>$t, not(/price=>*))
```

This tree pattern indicates to find all book elements and then bind certain values in the books' content to the variables in the branch pattern. It is composed of the root element pattern "$//book \Rightarrow$*" and the branch pattern consisting of compound compositions of the element patterns, "$/author \Rightarrow \$a/email \Rightarrow \$em$", "$/editor \Rightarrow \$e/email \Rightarrow \$em$", "$title \Rightarrow \$t$", and "$/not (price \Rightarrow$*$)$". The branch pattern is a compound of conjunctive pattern, disjunctive pattern and negative pattern. The negative pattern "$/not(price \Rightarrow$*$)$" indicates that the book element should not contain any price sub-element; the disjunctive composition indicates that the content of the author or the editor sub-element in the book element would be bound to $\$a$ or $\$e$ respectively if any, and further their email sub-elements would be bound to $\$em$; the conjunctive composition indicates that the content of the title sub-element would be bound to $\$t$ and associated with the value of $\$a$ or $\$e$.

LXT is distinct from previous studies in that it introduces the disjunctive pattern composition into data manipulation besides the conventional conjunctive pattern composition. A disjunctive pattern containing no variables are treated as a disjunctive proposition which would yield "*true*" or "*false*" for matching or mismatching in data extraction. A disjunctive pattern containing variables plays an analogous role as the disjoint sum type in type system. It is used to constitute a disjoint union of the data extracted by the subordinate LXT patterns. This disjoint union has a special two-facet feature: on one hand, the elements in the union can be regarded as same kind against the outside conjunctive context; on the other hand, the elements can be treated separately since they are disjoint. That means, through the disjunctive pattern we can either extracted heterogenous data and treat them homogeneously, or separate homogeneous data and treat them heterogeneously. For example, the LXT pattern "*(/author⇒$a || /editor⇒$e)*" indicates that the heterogeneous data, i.e., the content of the authors and the editors, are extracted respectively and can be treated homogeneously with respect to the title; the pattern "*(/author⇒$a || /author⇒$e)*" indicates that the the content of the authors would be extracted and bound to $a and $e respectively, and $a and $e can be restricted and used in different ways. Especially, when we want to directly use heterogenous data homogeneously, we can use the same variable in different patterns composed disjunctively. For example, "*(/author⇒$a || /editor⇒$a)*" would automatically merge the disjunctive union as a single set of bindings of $a.

Negative pattern is used as auxiliary proposition to simplify presentation. It doesn't contain variable because "negative variable" is not intuitively sound and would involve semantic problems when considering restrictions on conjunctive variables. A negative pattern can often be treated as a simplified quantification indicating that every element in current context doesn't satisfy the restriction. For example, "*not(/price⇒*)*" means there is not a price element, and "*not(/author⇒*(not(/email⇒*)))*" means there is not an author without email, i.e., every author has email.

Some syntactic sugars of LXT expressions are provided for programming with XTQ. The parentheses can often be omitted if incurs no misunderstanding. The atomic pattern "*name⇒**" specifying the existence of certain element can be simply abbreviated as "*name*", e.g. "*/author*" or "*@year*". Additionally, the common branches of the tree patterns composed disjunctively can be extracted out. Thus we get the previous example after applying the syntactic sugars as

```
//book((/author=>$a||/editor=>$e)/email=>$em, /title=>$t, not(/price))
```

3.2   Data Extraction

The objective of LXT pattern is to extract and organize the data satisfying the constraints indicated by the pattern when it is bound to a data source. As atomic pattern does, an LXT pattern also indicates a proposition that certain values in the current document context match the pattern, thus matching an LXT with a document fragment is to list the values in the document as proofs satisfying the proposition. Those proofs constitute the data extracted from documents against the LXT pattern, which are often associated with each other and exhibits a coherent structure.

For example, for a simple element pattern "/author⇒$a" all the authors of a book can match it and thus generate a set of proofs, which is named as a **group**, matching $a; for a disjunctive pattern "(/author⇒$a || /editor⇒$e)" all the authors and editors of a book can match it and generate a disjunctive union, which is also named as **enumeration**, consisting of the two groups of $a and $e; for a conjunctive pattern "(/title⇒$t, /author⇒$a)" the proof group is composed of the conjunctive **tuple**s of proofs of "/title⇒$t" and "/author⇒$a" respectively. Actually, we would rather treat this group as a variation of a tuple composed of the groups respectively matching $t and $a.

When we consider complicated situations such as matching the tree-like pattern "//book((/author⇒$a || /editor⇒$e),/title⇒$t)" with a set of books in a bookstore, the groups, enumerations and tuples are naturally hierarchically composed accordingly. To explicitly specify the complicated hierarchical structure of the data extracted with LXT patterns, we introduce variable pattern as described in the rest of this subsection.

Definition 2. A **variable pattern** is an expression defined inductively as follows:

1. Any variable "$x" is a variable pattern. "ϵ" is a variable pattern.
2. For a variable pattern p, a **group** (pattern) denoted by "$\{p\}$" is a variable pattern.
3. For variable patterns p_1, p_2, ..., p_n, a **tuple** (pattern) denoted by "$(p_1, p_2, ..., p_n)$" is a variable, an **option** (pattern) denoted by "$p_1 || p_2 || ... || p_n$" is a variable pattern, and an **enumeration** (pattern) denoted by "$<p_1 || p_2 || ... || p_n >$" is a variable pattern.

Variable pattern is an expression to sketch data structure in XTQ. A variable pattern can be a simple variable or a compound expression using the operators ",", "||", "{ }" and "<>" to compose variables or subordinate variable patterns. We introduce ϵ as an intermediate variable pattern indicating the void pattern which can be omitted in further processing. We directly inherit the "," and "||" operators from LXT to present tuple and option patterns, and introduce new notations of "{ }", "<>" to present group and enumeration.

Based on variable pattern, the structure of extracted data with an LXT pattern is specified by its *raw pattern*.

Definition 3. The **raw pattern** of an LXT l is a variable pattern denoted by $vp(l)$, inductively defined as follows:

1. For an element pattern l, $vp(l) = \{(var(l))\}$ where $(var(l))$ is the tuple of variables in the pattern l.
2. For a tree pattern $l = rp(bp)$, $vp(l) = \{ (var(rp), vp(bp))\}$ or especially $\{vp(bp)\}$ if $vp(rp)$ is ϵ.
3. For a conjunctive pattern $l = (p_1, ..., p_n)$, $vp(l) = (vp(p_1), ..., vp(p_n))$.
4. For a disjunctive pattern $l = (p_1 || ... || p_n)$, $vp(l) = <vp(p_1) || ... || vp(p_n)>$.
5. For a negative pattern l, $vp(l)$ is ϵ.

For the example LXT in the previous subsection, the simple element pattern "/email⇒$em" would yield the raw pattern "{$em}" indicating that the data bound to $em under certain context, e.g., an author or an editor, are naturally grouped together; the tree pattern "/author⇒$a/email⇒$em" would yield the raw pattern "{($a, {$em})}" indicating that the data bound to author, a tuple consisting of an author and the associated group of emails, are grouped together; the disjunctive pattern "(/author⇒$a/email⇒$em || /editor⇒$e/email⇒$em)" would yield the raw pattern "<{($a, {$em})} || {($e, {$em})}>" indicating that the group of authors and the group of editors are gathered in an enumeration; for the whole LXT pattern, the corresponding raw variable pattern is "{(<{($a, {$em})} || {($e, {$em})}>, {$t})}". This pattern actually specifies the structure of extracted data through a hierarchically compound group.

The semantics of data extraction is specified based on a hierarchically structured model named *and-or-set* which is the instance of variable pattern. For the details of and-or-set and the associated operational semantic rules please refer to the full specification of the language[17].

3.3   Data Restructuring and Construction

Variable pattern can not only be used to specify the original data structure but also be flexibly restructured so as to meet various requests of data construction. These requests are presented by restructured patterns which are variants of variable patterns appending certain information on structure manipulation.

Definition 4. A **restructured pattern** is an expression defined as follows:

1. A variable pattern is a restructured pattern.

2. A **restructured group** denoted by "$\{p\}_{ql}$" is a restructured pattern. Here p is a restructured pattern, ql is a variable pattern list named index pattern list in the form of "ql_1 *sep* ... *sep* ql_n" where ql_i is a variable pattern or a index pattern list and *sep* can be ";" "|" or ":". For convenience, group pattern $\{p\}$ is also denoted as restructured group $\{p\}_\epsilon$.

3. For two restructured patterns p and q, the **folded group** denoted by "$\{p\}_{q\%}$" is a restructured pattern.

4. For a restructured group $\{p\}_{ql}$ that is a sub-pattern of a group or a tuple, the **flattened group** denoted by "ˆ$\{p\}_{ql}$" is a restructured pattern.

5. For two restructured pattern p and q, the **hidden tuple** denoted by "p/q" is a restructured pattern. Here p and q satisfy that there is not a variable pattern r which occurs both in p and in q.

6. For two restructured patterns p and q, the **eliminated option** denoted by "$p\backslash q$" is a restructured pattern.

XTQ deploys two term rewriting systems on restructured pattern, defined by the restructuring rules "$p \hookrightarrow q$" and the reduction rules "$p \Rightarrow q$" as shown in Fig.1, to specify data restructuring and construction. The rule "$p \hookrightarrow q$" indicates the restructuring process that the pattern p is to be directly restructured to the pattern q, while the rule "$p \Rightarrow q$" indicates the restructuring result that the data being restructured to the pattern p has a structure indicated by the pattern q.

Tuple-related rules

(p, p')↪ (p', p) (tpl- comm)   ((p, p'), p") ↪ (p, (p', p")) (tpl-assoc)

(p, p')↪ p / p' ⇒ p if var(p)∩ var(p') = φ (tpl-hid)   p ↪ (p, p) (tpl-dupl)

(p, {p'}$_{ql}$) ↪ {(p, p')}$_{p';ql}$ if var(p)∩ var(p') = φ (grp-distr)

($\hat{}${p}$_{ql}$, p$_1$) ⇒ $\hat{}${(p, p$_1$)}$_{ql}$,   (p$_2$,$\hat{}${p}$_{ql}$) ⇒ $\hat{}${(p$_2$,p)}$_{ql}$ (flatten-tpl-red) †

(p, <p' || p">) ↪ <(p, p') || (p, p")> if var(p)∩ (var(p')∪ var(p")) = φ (enum-distr)

(p, p' || p") ↪ (p, p') || (p, p") if var(p)∩ (var(p')∪ var(p")) = φ (opt-distr)

Group-related rules

{p}$_{ql}$ ↪ $\hat{}${p}$_{ql}$ (flatten-intro)   {$\hat{}${p}$_{ql}$}$_{ql'}$ ↪(⇒) {p}$_{ql':ql}$ (flatten-grp-red)

{(p, p')}$_{ql}$ ↪ {{(p, p')}$_{p\%}$}$_{ql}$ ↪ {({(p, p')}, p%)}$_{ql}$ (grp-tpl-fold) ‡

Enumeration&option-related rules

p||p' ↪ p'||p (opt-comm)   <<p||p'> ||p"> ↪ <p|| <p'||p">>> (enum-assoc)

<{p}$_{ql}$ ||{p'}$_{ql'}$> ↪ {p || p'}$_{ql|ql'}$ (grp-merge) {p||p'}$_{ql}$ ↪ <{p}$_{ql}$||{p'}$_{ql}$> (grp-split)

<{p}$_{ql}$ || {p}$_{ql}$ > ⇒ {p}$_{ql}$ (id-grp-merge) † < p || ... || p > ⇒ {p} (id-enum-fold) †

p || p' ↪ p \ p' ⇒ p (opt-elim)   <p> ⇒ p if p is not option (enum-red)

{<p> || p'}$_{ql}$ ⇒ {p || p'}$_{ql}$ p' can be ε (grp-enum-red)

$\dagger$ The rule can be used to preprocess the raw pattern.

$\ddagger$ A folded group {{p}$_{q\%}$}$_{ql}$ cannot be further flattened.

Fig. 1. Restructuring and reduction rules of variable pattern

In Fig.2 we use previous LXT pattern as an example to briefly illustrate concrete pattern restructuring. We start from the raw pattern P0 and illustrate the restructuring process from P1 through P11. P1 hides the pattern "{$em}" in both tuples using the *tpl-hid* rule which projects the tuple to desired sub-patterns and thus hides undesired sub-element in tuples. P2 duplicates the pattern "{$t}" in P1 using the *tpl-dupl* rule which can duplicate the specified pattern in the tuple so as to provide multiple copies of original data for different use. P3 is restructured from P1 with the *enum-distr* rule and be further restructured to

---

Ex.	//book ((/author=>$a || /editor=>$e) /email=>$em, /title=>$t)
P0.	{ (< { ($a, {$em}) } || { ($e, {$em}) } > , {$t}) }
P1.	{ (< { $a / {$em} } || { $e / {$em} } >, {$t}) }
P2.	{ (< { $a / {$em} } || { $e / {$em} } >, {$t}, {$t}) }
P3.	{ < ({ $a / {$em} }, {$t}) || ({ $e / {$em} }, {$t}) >}
P4.	{ < {({ $a / {$em} }, $t)}$_{\$t}$ || { ($e / {$em} , {$t}) }$_{\$e/\{\$em\}}$ >}
P5.	{< {{($a/{$em}, $t)}$_{\$a/\{\$em\}}$}$_{\$t}$ || {{($e/{$em}, $t)}$_{\$t}$}$_{\$e/\{\$em\}}$ >}
P6.	{<{$\hat{}${($a/{$em}, $t)}$_{\$a/\{\$em\}}$}$_{\$t}$ || {$\hat{}${($e/{$em}, $t)}$_{\$t}$}$_{\$e/\{\$em\}}$ >}
P7.	{<{($a/{$em}, $t)}$_{\$t:\$a/\{\$em\}}$ || {$\hat{}${($e/{$em}, $t)}$_{\$t}$}$_{\$e/\{\$em\}}$ >}
P8.	{<{{($a/{$em}, $t)}$_{\$t\%}$}$_{\$t:\$a/\{\$em\}}$ ||{$\hat{}${($e/{$em}, $t)}$_{\$t}$}$_{\$e/\{\$em\}}$>}
P9.	{{{($a/{$em},$t)}$_{\$t\%}$ || $\hat{}${($e/{$em}, $t)}}} $_{(\$t:\$a/\{\$em\})|(\$e/\{\$em\})}$ }
P10.	{<{{($a/{$em},$t)}$_{\$t\%}$}$_{(\$t:\$a/\{\$em\})|(\$e/\{\$em\})}$ ||
	{$\hat{}${($e/{$em}, $t)}$_{\$t}$}$_{(\$t:\$a/\{\$em\})|(\$e/\{\$em\})}$ >}
P11.	{<{{($a/{$em}, $t)}$_{\$t\%}$}$_{\$t:\$a/\{\$em\}}$ \ {$\hat{}${($e/{$em}, $t)}$_{\$t}$}$_{\$e/\{\$em\}}$>}
P12.	{{{($a,$t)}$_{\$t\%}$}$_{\$t:\$a}$}

Fig. 2. Examples of pattern restructuring

P4 and P5 with the *grp-distr* rule. These distributive rules are used to extend the scope of group or enumeration by distributing the elements in group or enumeration to the remainder of the outside tuple. For a tuple "$(p,\{p'\}_q)$", as the *grp-distr* rule shows, the elements in the group "$\{p'\}$" are coupled with the element of the pattern p as new tuples and thus a new group pattern "$\{(p, p')\}_{p';q}$" is formed. Here the index pattern "p'" indicates how the distributive group is formed, and the index pattern list "$p';q$" denotes the provenance of applying the *grp-distr* rule. The *enum-distr* rule is similar to the *grp-distr* rule by replacing the elements in the group with the ones in the enumeration. In P6 the group flatten operator "$\^{}$" is inserted to certain groups. A flattened groups like "$\{\^{}\{p\}\}$" can be reduced to "$\{p\}$", as shown in P7, indicating that the elements of the inner groups are to be directly gathered in the outer group. In P8 the *grp-tpl-fold* rule is used to introduce a new group layer in the group "$\{(\$a/\{\$em\}, \$t)\}_{\$t:\$a/\{\$em\}}$" by classifying the elements on the distinct values of $\$t$, and result in the new folded group "$\{\{(\$a/\{\$em\}, \$t)\}_{\$t\%}\}_{\$t:\$a/\{\$em\}}$". To avoid ambiguity in resolving restructuring process, folded group has a restriction on further restructuring that a folded group cannot be merged, flattened or distributed, and the group containing the folded group cannot be flattened either. That is, none of "$\^{}(\{(p, p')\}_{p\%})$", "$\{(q,p')\}_{p';p\%}$" and "$\^{}\{\{(p, p')\}_{p\%}\}$" is valid. P8 can be restructured to P9 and further to P10 following the *grp-merge* and *grp-split* rules in sequential. The *grp-merge* rule merges the groups disjunctively combined in an enumeration and thus form a larger group, indicating the elements in original groups are to be treated homogeneously. In contrast to group merging, the *grp-split* rule indicates that a group containing the elements of two patterns can be split into an enumeration consisting of two groups each of which contains the elements of an original sub-pattern. P8 can also be restructured to P11 by eliminating a subpattern in the option, following the *opt-elim* rule.

Reduction rules specify a convergent rewriting system of restructured patterns whose normal form is a variable pattern with certain restructuring provenance. For example, P11 can be eventually reduced to P12 in which the major body "$\{\{\{(\$a,\$t)\}\}\}$" is a variable pattern indicating the final restructured data structure and the index patterns "$\$t\%$" and "$\$t:\$a$" indicate the restructuring provenance. Since each restructuring rule maintains the restructuring provenance such as group index pattern and each reduction rule maintains the full information on the parts being of interested, the restructuring process from a raw pattern to a restructured pattern can be figured out unambiguously.

XTQ specifies data construction with construct pattern which is a hybrid of restructured pattern and constant values in a "*tag*⇒*content*" form. The keywords **groupby**, **hid** and **elim** are introduced to signify the group index, the symbol "/" and the symbol "\" respectively, and a keyword **orderby** is introduced to sort the tuple values in a group according to certain element in the tuple. For example, P12 can be embedded the construct pattern "*results*⇒{*bookinfo*⇒ {*pairs*⇒{(*author*⇒$\$a$,*title*⇒$\$t$)} *orderby* $\$a$ *groupby* $\$t\%$ } *groupby* $\$t:\$a$}" is supposed to construct a document fragment in which the root tag is "results", the content is composed of bookinfo elements containing pairs of authors and titles. The pairs are firstly grouped by the distinct values of t and then be ordered by the value of a.

However, a fully specified restructured pattern is often too trivial and complex for users to present data construction easily. XTQ adopts some syntax sugars and guidelines for users to facilitate presenting data construction in practice. These mechanisms are helpful to presenting the queries neatly and intuitively in most cases. Firstly, group flattening is introduced into LXT pattern for preprocessing the raw pattern. An element LXT pattern such as "$/name \Rightarrow \$x$", is allowed to be extended with a flattening prefix "$\hat{ }$", i.e., "$\hat{ }/name \Rightarrow \$x$" which would force the reduction rules to be applied to raw pattern, i.e., reducing "$\hat{ }\{\$x\}$" to "$\$x$", before resolving the restructuring process. Secondly, a construct pattern can be simplified by abbreviating group index pattern lists so that only partial information of group restructuring is maintained for resolving provenance. Thirdly, XTQ parser would deploy an out-left-matching policy to choose a restructuring provenance if multiple possibilities occur due to insufficient provenance information in a simplified. That is, for multiple patterns being able to be restructured to the construct pattern, the one is preferred if it has more group layers outside, e.g., the pattern "$\{\{p\}\}$" corresponds to "$\{\{\hat{ }\{p\}\}\}$" rather than "$\{\hat{ }\{\{p\}\}\}$"; for two patterns with the same outside group layers, the one is preferred if its top level group index pattern occurs closer to the left in the backbone than the other. For example, a construct pattern for the sample LXT, "$results \Rightarrow \{authorsbytitle \Rightarrow (title \Rightarrow \$t\%, \{author \Rightarrow \$a \})\}$", would be resolved as "$results \Rightarrow \{authorsbytitle \Rightarrow (title \Rightarrow \$t\%, \{author \Rightarrow \$a \ elim \ \$t\} \ groupby \ \$t\%)\}_{\epsilon:\$a:\$t}$", which would gather all the authors and then group them by the distinct values of book titles.

4   XTQ Language with Examples

XTQ adopts **QWC** expression, composed of the three clauses *query*, *where* and *construct*, as its major body. To make the program clear and neat, XTQ uses functions instead of nested query. Functions are declared with *QWC* expressions and are invoked in the *where* and the *construct* clauses. To make the language Turing-complete, function can be recursively declared.

The *query* clause uses LXT pattern with data sources to extract data. Flatten operator can be embedded in the LXT patterns as previously mentioned. The data source in the *query* clause is a function to get document from certain URL, e.g., doc(), or a bound variable. In *query* clause multiple data sources with LXT patterns can be combined conjunctively or disjunctively, resulting in an extended variable pattern.

The *where* clause uses conditions to filter extracted data. Conditions are boolean functions using fragments of fully specified construct pattern as arguments. The extracted data matching the argument pattern of a condition would be tested with the condition function and only those enabling the condition would be maintained. The filtered results of conditions are gathered and processed for data construction. The quantifiers "*foreach*" and "*forsome*" are introduced into conditions to form restrictions on groups. Conditions can be combined not only with the common logic connectives "*and*", "*or*" and "*not*" but also with the special connectives "||" and "*with*". The filtering on conditions combined with "and" or "or" are processed respectively and then the filtered results are combined. The connective "||" is special to restrict the disjunctive patterns in parallel. The connective "*with*" is introduced for combining the conditions to be processed sequentially. The syntax and semantics of conditions is much more subtle than it looks like, and details can refer to [17].

The *construct* clause uses construct patterns for data construction based on filtered data.

In this section, we use some typical queries instead of syntax description to illustrate the features of XTQ language. Some queries adopts the W3C XML query use case XMP [16] which is often used as a benchmark by XML query languages, and in [17] we list those queries in XQuery and Xcerpt for a comparison as well as the formal syntax and semantics of XTQ. Other examples are picked out to show the advantages of XTQ in presenting practical queries involving complicated data processing. Readers are encouraged to implement these examples in other languages for a further comparison.

Example 1. The bibliography document "bib.xml" contains book records within the "book" elements. A book record includes a title, one or more authors, one or more editors, a publisher and a price. For each author in the "bib.xml", list the author's name and the titles of all books by that author, grouping inside a "result" element. The result would be ordered with last name and then first name. (XMP Q4)

```
query doc("bib.xml")//book(/title=>$t,/author=>$a)
construct {result=>(author=>$a%,{title=>$t})} orderby $a%/last;$a%/first
```

This simple example shows the convenience of using simplified restructured patterns in declaratively presenting data construction. The parser would resolve the simplified pattern as "$\{\{(\$a\%,\{\$t\})\}_{\epsilon:\$a:\$t}$".

Example 2. For each book with an author, return the book with its title and authors. For each book with an editor, return a reference with the book title and the editor's affiliation. (XMP Q11)

```
query doc("bib.xml")//book(/title=>$t, (/author=>$a || /editor=>$e))
where notnull($a) || notnull($e)
construct {(book=>(title=>$t, {author=>$a})||(reference=>(title=>$t,
                       affiliation=>{$e/affiliation}))}
```

This example shows the basic usage of disjunctive pattern in handling heterogeneity. The disjunctive pattern and the restructuring mechanism facilitates declaratively extracting and constructing heterogeneous data, which make the program clear and compact.

Example 3. A document "reviews.xml" also contains book records within "entry" element. For each book found at both "bib.xml" and "reviews.xml", list the title of the book and its price from each source (XMP Q5).

```
query doc("book.xml")//book(^/title=>$tb, ^/price=>$pb),
     doc("review.xml")//entry(^/title=>$te, ^/price=>$pe)
where $tb = $te
construct {book_with_prices=>(title=>$tb, price_b=>$pb, price_e=>$pe)}
```

This example illustrates the join operation carried out by restrictions on conjunctive pattern. Especially, we use flattened LXT patterns to facilitate resolving pattern restructuring process.

Example 4. For each book that has at least one author, list the title and first two authors, and an empty "et-al" element if the book has additional authors (XMP Q6).

```
query doc("bib.xml")//book(/title=>$t, (/author=>$a || /author=>$b))
where ({$a}.count>0 and {$a}.count<=2) || {$b}.count>2
construct bib=>{book=>(title=>$t,
                       <{author=>$a}||({author=>$b}.[1..2],et-al=>~)>)}
```

This example shows the common approach to heteorgenously handling homogenous data in XTQ. The XTQ program utilizes a disjunctive pattern to separately extract homogeneous data, e.g., "*author*" elements, restrict them simultaneously with different conditions combined disjunctively and then process them heterogeneously. In the query, the group selector function *.[i..j]* fetches the elements in the range specified by i and j, and the symbol "~" signifies an empty content. Other languages often resort to procedural mechanishm, such as the "*if-then-else*" expression in XQuery, to handle such heteorgeneity, which reduces the declarativeness of the program.

Example 5. Gather the names of the authors of each book and label them with a sequence number following their document order.

```
declare addnumber($group, $tag) as (
   query $group/item=>$it, type()/int^/item=>$i
   where $i >= 0 and $i < {$it}.count()
   construct {(number=>$i, $tag=>{$it}.($i))} )
query doc("bib.xml")//book/author=>$a
construct {bookinfo=>addnumber({item=>$a}, "author")}
```

This example shows the usage of functions in XTQ. Here the function addorder deploys a data source "*type()*" to return a virtual XML document comprising the primitive data types with data instances. That is, the query clause "*type()/int^/item⇒$i*" would extract the data from a virtual "int" branch of the virtual document, which denotes the integers, and thus *$i* would indicate the integers corresponding to the number of the group items after proper restriction in the where clause. This example also exemplifies that elements enumeration usually implemented with loops can also be simulated in XTQ in a declarative way.

Example 6. Find the minimal prices of the books in three bookstores. Suppose each bookstore has a document like "bs?.xml" which contains a "storename" element indicating the name and a series of book records extends the ones in "bib.xml" with a balance element indicating the store balance of the book in the store. Suppose each book can be identified by its title. The result should contain the book title, the minimal price and the corresponding bookstore name list which is ordered by the stock balance.

```
query (doc("bsa.xml")||doc("bsb.xml")||doc("bsc.xml"))
        (//storename=>$n,//book(/title=>$t, /price=>$p, /balance=>$b))
construct {(bookinfo=>(title=>$t%,map(key=>{$p}.min(),
                        {mapping=>(key=>$p%,content=>(price=>$p%,
                                    {name=>$n}order by $b))}))}
declare map ($key, $mappings) as
    (query $key/key=>$kf, $mappings/mapping(/key=>$ks, /content=>$c)
    where $kf = $ks
    construct {$c} )
```

This query request is common in practice since various XML documents are often required to be gathered, grouped, analyzed and processed. The *query* clause utilizes disjunctively combined data sources to extract and process heterogeneous data homogenously. The *construct* clause declaratively presents the purpose of data construction, i.e., the data structure for output, and leaves the restructuring process for the parser. As the construct pattern shows, the extracted data are to be grouped in two layers. The outer layer is indexed by book values, i.e., the tuple of book's title and publisher. The inner layer is indexed by book price but only the store information with the minimal price would be output. We also list the definition of the standard XTQ function *map*.

Example 7. Find the books each of which has at least 3 people, including authors and editors, whose email addresses all end with "edu", and return those people.

```
query doc("bib.xml")//book((/author=>$p || /editor=>$p)/email=>$em)
where (foreach $em, endWith($em, "edu")) with count({$p})>2
construct results=>{book=>{people=>$p}}
```

This example shows the usage of the condition connective *"with"*. A compond condition "c_1 *with* c_2" indicates that c_1 is subordinate to c_2 and thus the two conditions are supposed to be processed sequentially. In the example, the first condition restricts the values of the group {*$em*} to be the ones whose elements all end with "edu". The second condition restricts the number of the people to be larger than 2. {*$em*} is subordinate to {*($p,{$em})*} and thus "*with*" is used here to signify that the two conditions be processed one by one. That means, we should count the people only if their email addresses satisfy the first condtion.

5 Conclusion

Existing XML query language often lack a proper declarativeness in presenting complex queries. The navigational languages are often clumsy in constructing tree-structure documents from flat navigation result, and pattern-based languages are not expressive enough due to the simple and rigid pattern structure.

Table 1. Comparison of XML Query Languages

Features / Languages	Separating Data Extr. & Constr.	Structurally Composing Data Extr.	Flexibly Presenting DataConstr.	Deductively Restructuring Data	Coherently Handling Heterogeneity
XTQ	Yes	Strong	Strong	Yes	Yes
XQuery	No	Weak	Strong	No	No
Xcerpt	Yes	Not strong	Weak	No	No
XTreeQuery	Yes	Not strong	Weak	No	No
XDuce	Yes	Strong	Not strong	No	No
TQL	Yes	Strong	Weak	No	No
UnQL	No	Weak	Weak	No	No
XPathLog	Yes	No	Weak	No	No

In this paper, we introduced XTQ, a new XML query language which can declaratively and expressively presenting complex queries. XTQ introduces some coherent mechanisms to meet specific requirements of XML query. XTQ uses LXT pattern, a tree-like pattern expression with conjunctive and disjunctive operators, to present composite data extraction and handle heterogeneity. It adopts a hierarchical pattern structure which coherently reflects the data hierarchy and supports data restructuring. It deploys a set of restructuring rules so that complex data construction can be declaratively presented without specifying the restructured process. These features makes XTQ more declarative than other languages. In the summary Table 1, we list the issues key to declarativeness of XML query language and make a comparison for XTQ and some representative studies.

The full specification of syntax and semantics of XTQ can refer to [17], and the core of XTQ has been implemented based on the operational semantics. Now we are going to extend XTQ for distributed XML data processing. The flexible structure manipulation in XTQ program is promising in specifying uniform views for heterogeneous data in XML integration, and the composition and deduction of the views based on LXT pattern and restructuring mechanism are practical and interesting problems to study. Additionally, fragmenting and distributing big XML documents becomes a practical issue in XML data processing. How to adapt XTQ to fit for manipulating various kinds of distributed document fragments is also an interesting problem we concern.

References

1. Berglund, A., Boag, S., Chamberlin, D., Fernandez, M., Kay, M., Robie, J., Simeon, J.: XML Path Language (XPath) 2.0. Recommendation, World Wide Web Consortium (2007), http://www.w3.org/TR/xpath20/
2. Boag, S., Chamberlin, D., Fernandez, M.F., Florescu, D., Robie, J., Simeon, J.: XQuery 1.0: An XML Query Language. Recommendation, World Wide Web Consortium (2007), http://www.w3.org/TR/xquery/
3. Hosoya, H., Pierce, B.: XDuce: A Typed XML Processing Language. ACM Transactions on Internet Technology 3(2), 117–148 (2003)

4. Schaffert, S., Bry, F.: Querying the Web Reconsidered: A Practical Introduction to Xcerpt. In: Proc. Extreme Markup Languages (August 2004)
5. Cardelli, L., Ghelli, G.: TQL: A Query Language for Semistructured Data Based on the Ambient Logic. Mathematical Structures in Computer Science 14(3), 285–327 (2004)
6. Bailey, J., Bry, F., Furche, T., Schaffert, S.: Web and Semantic Web Query Languages: A Survey. In: Eisinger, N., Małuszyński, J. (eds.) Reasoning Web. LNCS, vol. 3564, pp. 35–133. Springer, Heidelberg (2005)
7. Abiteboul, S., Quass, D., McHugh, J., Widom, J., Wiener, J.: The Lorel Query Language for Semistructured Data. International Journal on Digital Libraries 1(1), 68–88 (1997)
8. Robie, J., Derksen, E., Fankhauser, P., Howland, E., Huck, G., et al.: XQL (XML Query Language) (1999), http://www.ibiblio.org/xql/xql-proposal.html
9. May, W.: XPath-Logic and XPathLog: A Logic-Programming Style XML Data Manipulation Language. Theory and Practice of Logic Programming 3(4), 499–526 (2004)
10. Buneman, P., Fernandez, M., Suciu, D.: UnQL: A Query Language and Algebra for Semistructured Data Based on Structural Recursion. VLDB Journal 9(1), 76–110 (2000)
11. Florescu, D., Levy, A., Fernandez, M., Suciu, D.: A Query Language for a Web-site Management System. SIGMOD Record 26(3), 4–11 (1997)
12. Chen, Z., Ling, T.W., Liu, M., Dobbie, G.: XTree for Declarative XML Querying. In: Lee, Y., Li, J., Whang, K.-Y., Lee, D. (eds.) DASFAA 2004. LNCS, vol. 2973, pp. 100–112. Springer, Heidelberg (2004)
13. Benzaken, V., Castagna, G., Frisch, A.: CDuce: An XML-Centric General-Purpose Language. In: Proc. International Conference on Functional Programming (2003)
14. Liu, M.: A Logical Foundation for XML. In: Pidduck, A.B., Mylopoulos, J., Woo, C.C., Ozsu, M.T. (eds.) CAiSE 2002. LNCS, vol. 2348, p. 568. Springer, Heidelberg (2002)
15. Cardelli, L., Gardner, P., Ghelli, G.: A Spatial Logic for Querying Graphs. In: Widmayer, P., Triguero, F., Morales, R., Hennessy, M., Eidenbenz, S., Conejo, R. (eds.) ICALP 2002. LNCS, vol. 2380, pp. 597–610. Springer, Heidelberg (2002)
16. Chamberlin, D., Fankhauser, P., Florescu, D., et al.: XML Query Use Cases. World Wide Web Consortium (2007), http://www.w3.org/TR/xquery-use-cases
17. Li, X., Liu, M.: XTQ: Syntax, Semantics and Use Cases. Technical Report in SKLSE (2009), http://www.sklse.org:8080/xtq

Reducing Graph Matching to Tree Matching for XML Queries with ID References

Huayu Wu[1], Tok Wang Ling[1], Gillian Dobbie[2], Zhifeng Bao[1], and Liang Xu[1]

[1] School of Computing, National University of Singapore
{wuhuayu,lingtw,baozhife,xuliang}@comp.nus.edu.sg
[2] Department of Computer Science, The University of Auckland, New Zealand
gill@cs.auckland.ac.nz

Abstract. ID/IDREF is an important and widely used feature in XML documents for eliminating data redundancy. Most existing algorithms consider an XML document with ID references as a graph and perform graph matching for queries involving ID references. Graph matching naturally brings higher complexity compared with original tree matching algorithms that process XML queries. In this paper, we make use of semantics of ID/IDREF to reduce graph matching to tree matching to process queries involving ID references. Using our approach, an XML document with ID/IDREF is not treated as a graph, and a general query with ID references will be decomposed and processed using tree pattern matching techniques, which are more efficient than graph matching. Furthermore, our approach is able to handle complex ID references, such as cyclic references and sequential references, which cannot be handled efficiently by existing approaches. The experimental results show that our approach is 20-50% faster than MonetDB, an XQuery engine, and at least 100 times faster than TwigStackD, an existing graph matching algorithm.

1 Introduction

Because XML is an important standard format for data exchange over the Internet, it is important to remove redundant data from documents, which uses unnecessary storage and adds extra cost during data transfer. Consider the example shown in Fig. 1(a). Since both part A and part B are supplied by the same supplier, the information about supplier $s001$ is repeated twice. The most common way to reduce redundancy is to introduce ID and IDREF attributes [20]. ID and IDREF can be likened to primary key and foreign key constraints in relational databases. Using ID/IDREF, each object is stored once under the document root with a unique ID. A new structure for the document tree in Fig. 1(a) with data redundancies removed is shown in Fig. 1(b). The dotted arrows represent the references from IDREF value to the referenced object which will have the same value as its ID.

Despite the importance of ID/IDREF for good XML design, existing algorithms that process queries involving ID references in XML are still not efficient. To the best of our knowledge, all the existing algorithms consider both XML

P. García Bringas et al. (Eds.): DEXA 2010, Part II, LNCS 6262, pp. 391–406, 2010.
© Springer-Verlag Berlin Heidelberg 2010

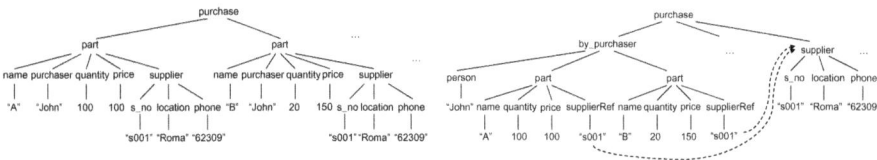

(a) XML tree with data redundancies   (b) XML tree after reducing redundancies

Fig. 1. Example XML document in two schemas, with and without data redundancies

documents with ID/IDREF and XML queries with ID references as digraphs, and perform graph matching to process queries. It is true that an XML document is modeled as a digraph if we consider the ID references as directed edges. However, transforming tree pattern matching to graph pattern matching naturally brings much higher complexity, because graph matching is more costly than tree matching with the same size input [12]. A simple question is whether we have to abandon many efficient tree pattern matching approaches (for XML and queries without ID references), and invent new, but less efficient graph pattern matching algorithms to process such queries with ID references. Fortunately, the answer is no. Unlike the graph model for social networks or other graph databases, ID reference in an XML document is not a random link between nodes. It has strong semantics, which always starts at an IDREF value and references an object with a same ID value. Surprisingly, no existing algorithm captures this semantics during query processing. They normally focus on how to enhance the efficiency of graph matching, but ignore the fact that using the semantic information of ID/IDREF, graph matching can be reduced to a less complex tree matching.

This paper focuses on incorporating semantics of ID/IDREF to reduce graph matching to tree matching to process XML queries with ID references. Besides significantly reducing the pattern matching complexity, our approach also makes all existing efficient tree pattern matching algorithms feasible for queries with ID references. The rest of the paper is organized as follows. We revisit some related work in XML query processing in Section 2. Our semantic approach to processing queries with ID references is presented in Section 3. Section 4 discusses how to handle special references in documents and queries. We present experimental results in Section 5 and conclude our paper in Section 6.

2   Related Work

XML query processing has been studied for many years. Since in most XML query languages (e.g. [4][5]) queries are expressed as *twig* patterns, finding all occurrences of a twig pattern in an XML document is a core operation for XML query processing. In the early stage, a lot of work focused on storing and querying XML data using mature relational database systems [10][21][27][26]. Generally they shred XML data into relational tables, and convert XML queries into SQL to query the database. The advantage of these relational approaches is that they can manage and operate on values efficiently, e.g. performing range search for predicates, and they can make use of existing relational query optimizers to optimize SQL-style XML queries. However, the drawback of the relational approaches is also obvious. A twig pattern XML query may involve many table joins, which are costly. Sometimes it is not easy to decide what tables are to be joined and how many times to join them particularly for queries with "//"-axis (ancestor-descendant axis). Later how to process twig pattern queries natively without using relational databases became a hot topic. The structural join based approach is the most efficient native approach accepted by researchers. In particular, *TwigStack* [6] and subsequent work [9][15][19][8] bring in the idea of holistic structural join, which makes structural join very efficient. Finally, [25] complements structural join based approaches by introducing relational tables to process content search and content extraction.

When XML documents and queries involve ID references, most twig pattern matching based algorithms cannot handle the references between nodes. Many works focus on how to efficiently perform graph matching for such documents and queries that are modeled as graphs. Some traditional approaches [22] generate all possible mappings between each pair of nodes in two graphs and check for correctness. However, this sort of graph matching problem is NP-complete generally [12]. Moreover, these graph matching algorithms can hardly support "//"-axis queries. Later, [23] and [7] consider the structure of XML documents with ID/IDREF as a directed acyclic graph (DAG) and proposes algorithms to process queries on DAGs. However, XML documents with ID/IDREF may be a cyclic graph [13]. Recently, [17] and [16] extend twig pattern query to support queries involving ID references, and propose techniques to solve the extended twig. [24] proposes a new labeling scheme for document graph so that parent-child and ancestor-descendent relationships can be identified and thus queries can be processed. However, all these attempts consider ID references as a random link between nodes and match random graph queries to the document graph. As mentioned in Section 1, graph matching is normally more expensive than tree matching, thus this paper proposes a method to reduce graph matching to much less complex tree matching for XML queries.

3   Semantic Approach for Queries with ID References

3.1   Reference Pattern Query

Twig pattern, or tree pattern, is considered the core XML query pattern when ID reference is ignored in documents and queries. We first extend the twig pattern expression to express ID references in a query, and propose a semantic approach based on this approach. Our extension mainly includes two parts: explicitly marking output nodes and introducing ID reference edges.

Definition 1. *(Output node) Output nodes in a twig pattern query are defined as a group of nodes in the twig pattern such that the query aims to find values for them, based on conditions on other nodes.*

Every query must contain at least one output nodes. For example, in the query shown in Fig. 2(a), *s_no* is an output node since we aim to find the value for c for each supplier that is located in 'Roma'. Besides noting output node, we also note the ID reference within a query. For example a query to find the value for *quantity* of part *B* which is purchased by John and supplied by some supplier located in 'Roma' is issued on the document shown in Fig. 1(b). From the document, we can see that full information about a certain *supplier* is stored separately from the object *part* and that an IDREF property *supplierRef* is used to reference the corresponding *supplier*. By considering both output node and ID reference, we propose the notion of *reference pattern query*. The above query can be issued as a reference pattern query as shown in Fig. 2(b).

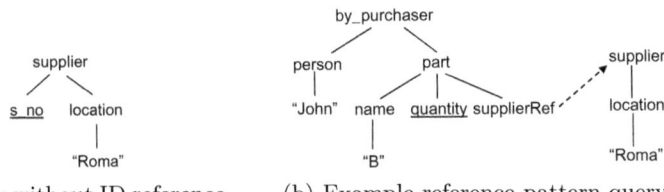

(a) Query without ID reference   (b) Example reference pattern query

Fig. 2. Example twig pattern query and reference pattern query

Definition 2. *(Reference pattern query)* Reference pattern query *generalizes twig pattern query to express ID references between twigs using a dotted referencing arrow. In a reference pattern query, the main body where the referencing arrow starts is called the* referencing part, *and the part to which the referencing arrow points is called the* referenced part. *The referenced part normally corresponds to an object with an ID value. The output nodes in a reference pattern query are marked by underlining them.*

Note that a query issuer is expected to have some schematic information of the underlying XML document. Otherwise, she has no way to compose a structured query expression. Similarly, a query processor is also aware of the document structure. Although sometimes there is no formal schema available for a document, by parsing the document a program can easily summarize its structure. With such requirements, we assume that a user should be able to issue a reference pattern query, and the system can interpret the query, even if the query contains implicit ID references across a "//" relationship. For example, to process the query shown in Fig. 3(a) which finds all locations where John purchases things, the system will identify the ID reference across the "//" edge by consulting the document structural summary, and rewrite the query to be a reference pattern query as shown in Fig. 3(b). Sometimes, an object class may have a recursive reference, e.g. a *paper* cites another *paper*. This case is similar to element recursion in DTDs. The work in [18][11] discussed how to translate queries involving recursive elements into SQL. Since in our work, we use SQL to handle ID reference, these works can be adopted to solve recursive ID references.

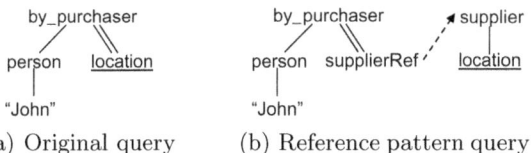

(a) Original query   (b) Reference pattern query

Fig. 3. Case that "//" relationship contains ID reference

3.2   Parsing XML Document with ID References

The hierarchical structure of XML data is normally modeled as a tree. However, when there are references from tree node to tree node, the data model is considered as a graph. Most existing algorithms handle XML queries over documents with ID reference in a graph matching manner, e.g. using an extra index to record ID references between nodes or inventinga new labeling scheme for graphs. As mentioned in Section 1, compared with tree matching, graph matching naturally brings higher complexity. ID reference in an XML document is not a random link between nodes. It always starts from an IDREF attribute and points to an object with the same ID value. If we use the semantics to avoid treating a document as a random graph, the performance could be improved. First, we present how our approach parses an XML document with ID references.

Most XML query processing algorithms assign a label to each document node, so that the parent-child or ancestor-descendant relationship between each pair of document nodes can be easily determined by their labels during query processing. In our approach, we ignore the ID references, and only label property nodes, object nodes and other internal nodes, but not value nodes. IDREF is an internal node, but its value is a value node (as shown in Fig. 4).

Definition 3. *(**Object, property**) In an XML document, the parent node (either an attribute or an element) of each value is a* property. *We consider the parent node of a property node (including IDREF node) as an object*[1].

In the document in Fig. 1(b), the nodes *name*, *quantity* and *supplierRef* are all properties as they have value children, and each *supplier* and *part* are objects as they are parents of certain properties. When we ignore the ID reference and the values, the label assignment of the document in Fig. 1(b) is shown in Fig. 4.

The labeled document nodes are organized as inverted lists based on different tags, which is the same as other approaches; whereas, the values that are not labeled are stored in object-oriented relational tables. In particular, for each object class there is a table, whose schema includes a label field to store the label of each object in this class, and a set of property fields to store the values of each property for a certain object. During this process, both ID value and IDREF value are treated in the same way as other properties. The reference between them is ignored, thus these indexes only keep the tree like nature of the document. The example object tables for *supplier* and *part* are shown in Fig. 5, in which the ID and IDREF attributes are stored in the same way as other

---

[1] It may not be semantically true for all cases, but it will not affect the correctness of query processing.

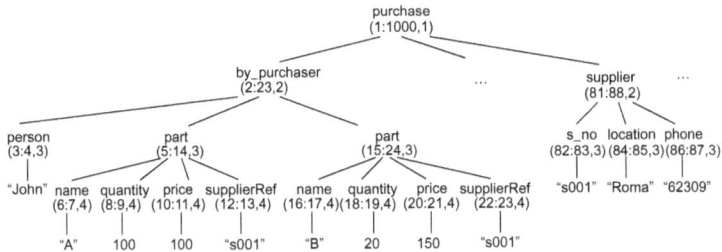

Fig. 4. The purchase document with internal nodes labeled

$R_{supplier}$

Label	S_no	Location	Phone
(81:88,2)	s001	Roma	62309
...	...	...	...

R_{part}

Label	Name	Quantity	Price	SupplierRef
(3:12,3)	A	100	100	s001
(13:22,3)	B	20	150	s001
...	...	...	...	...

(a) *Supplier* table   (b) *Part* table

Fig. 5. Example object tables during document parsing

properties. Note that the IDREF attribute *supplierRef* in the XML document in Fig. 4 is represented directly with the same value in the attribute *supplierRef* in the *part* table. There is no redundancy or duplicated data in the tables.

The implementation details of object table construction and the solution to some potential problems of object tables, e.g., how to store multi-valued property, are discussed in our previous report [25].

3.3   Query Processing with Tree Matching

The ID reference in a reference pattern query is reflected by a dotted arrow. Such a dotted arrow always corresponds to an ID reference in the document, which makes the document a graph structure. During pattern matching, most algorithms consider ID references in both documents and queries as a normal edge, thus they have to perform graph matching. In our approach, we try to reduce the graph matching to a simple tree matching, to improve pattern matching performance. To do this, we treat a document as a tree, as mentioned in the previous section, and also ignore the dotted arrow in a reference pattern query when we perform pattern matching. However, the ID reference is a part of the query constraint which cannot be ignored. Our solution transforms the ID reference in a query to a table join, because (1) the semantics of ID/IDREF are such that all the ID references must be between ID and IDREF attributes, and they are not a random link, and (2) both the corresponding ID value and IDREF value are stored in relevant object tables and the reference between them is the same as the equi-join of the two tables.

The general idea of our approach is to decompose a reference pattern query into a referencing part and a referenced part (the two parts are defined in Definition

2). The query is processed by a tree matching of the referencing part, with any existing twig pattern matching algorithms, and a join between the referencing part and the referenced part. In more detail, the join between the referencing part and the referenced part is eventually performed by a table join between two object tables with ID and IDREF attributes. In fact, a tree matching is a series of structural joins between each adjacent query node. In our heuristic we try to perform the join between the referencing part and the referenced part first, as this operation normally results in high selectivity due to the constraints in the referenced part. However, if the referencing part has no output node, the whole referencing part becomes a predicate, then we match the referencing part first before joining with the referenced part where output nodes are involved. The detailed query processing algorithm is presented in Algorithm 1, in which we suppose o and obj are two objects in the referencing part and referenced part of a reference pattern query respectively. First, we only consider the basic reference pattern query with only one referencing part and one referenced part.

Algorithm 1. Query processor

1: **if** there is no output node in the referencing part **then**
2:    match the referencing part to the document tree, to find the set S of distinct values of the IDREF attribute
3:    join S with R_{obj} for the referenced part, based on the condition $S.IDREF=R_{obj}.ID$
4:    select values for the output node in the joined result, based on other constraints under obj.
5: **else**
6:    Join R_{obj} and R_o based on the condition $R_{obj}.ID=R_o.IDREF$, and select the labels of o based on the predicates under obj and o
7:    create a new inverted list $T_{o'}$ for o and put the selected labels into $T_{o'}$
8:    rewrite the referencing part by changing o to o', which corresponds to $T_{o'}$
9:    match the rewritten referencing part to the document tree with $T_{o'}$ for o' to find values of the output nodes in the referencing part
10:    **if** there are output nodes in the referenced part **then**
11:       find the set S of distinct values of the IDREF attribute from the tree matching result
12:       join S with R_{obj} for the referenced part, based on the condition $S.IDREF=R_{obj}.ID$
13:       select values for the output node in the joined result, based on constraints under obj.
14:    **end if**
15: **end if**

There are three cases of reference pattern query, with respect to the position of output nodes: Case (1) output nodes reside in the referencing part, Case (2) output nodes reside in the referenced part and Case (3) output nodes reside in both the referencing part and referenced part. Now we use examples to illustrate our approach for the three cases.

Example 1. Consider the Case (1) query shown in Fig. 2(b). The query asks for the quantity of part B which is purchased by John and supplied by some supplier in 'Roma'. In this query, only the referencing part contains an output node, *quantity*. The two objects involved in the ID reference are *part* and *supplier*. Using the algorithm, we join R_{part} and $R_{supplier}$ (as shown in Fig. 5) based on s_no=SupplierRef, filter the results by the predicate part.name='B' and supplier.location='Roma', and select the labels for *part*. These labels are put into a new inverted list for *part* and the referencing part is rewritten as shown in Fig. 6(a). In the new query, the subscript of node *part'* explains that this node corresponds to the new inverted list of parts, whose name is 'B' and has a supplier

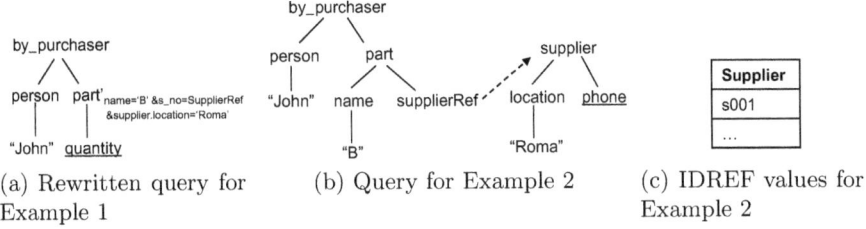

(a) Rewritten query for Example 1

(b) Query for Example 2

(c) IDREF values for Example 2

Fig. 6. Figures for Example 1 and 2

located in 'Roma'. The final step is to match the tree pattern referencing part with the new inverted list for *part'* to the document tree.

Example 2. Consider the Case (2) query shown in Fig. 6(b), in which only the referenced part has an output node, *phone*. By the algorithm, we first match the referencing part to the document tree, to find the labels of each matched *part* object. Then using these labels we can find the distinct values of *supplierRef* in R_{part}. The result is shown in Fig. 6(c). We join these tuples with table $R_{supplier}$ and select *phone* in $R_{supplier}$ based on *location*='Roma'.

Example 3. Consider the Case (3) query in Fig. 7(a). In this query both the referencing part and the object contain output nodes. We first join tables R_{part} and $R_{supplier}$ based on the equality of *supplierRef* and *s_no*, and select *part* labels based on the conditions that part's name is 'B' and supplier is in 'Roma'. Then a new inverted list $T_{part'}$ for *part* is constructed with the selected labels. The referencing part is rewritten by renaming the node *part* to be *part'* so that the new inverted list will take effect (shown in Fig. 7(b)). Using any tree matching algorithm to process the rewritten referencing part, we can find the labels for matched *part*. Then in R_{part} we can extract values for *supplierRef* and the output node *quantity*. To find the value for the other output node *phone*, we join the distinct values of *supplierRef* with the table $R_{supplier}$, and select the *phone* value based on the condition that *location* equals 'Roma'.

During query processing, any reference pattern query requiring graph matching on the document is eventually processed by tree pattern matching and table joins. Furthermore, the tree pattern to be matched is normally much simpler than the original query pattern with references. Since most relational systems can perform selection and join very efficiently with B^+ tree indexes, the overhead

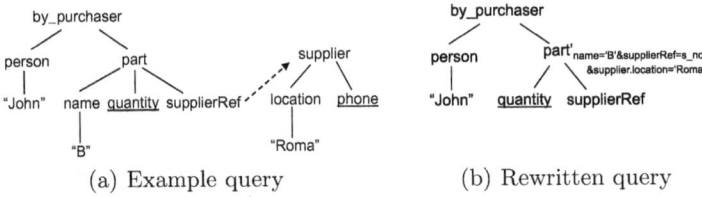

(a) Example query

(b) Rewritten query

Fig. 7. Figures for Example 3

on table operations will not affect the benefit of reducing graph matching to tree matching. Our experiments also prove this. In particular, when the referenced part of a reference pattern query is in a complex pattern, e.g. enclosing other objects, we can perform pattern matching on the referenced part, before joining with the referencing part.

3.4   Correctness

The basic query processing idea in our approach is to replace the structural join using ID references, which is used in other algorithms, with a table join. Actually, an ID reference means the involved IDREF attribute has the same value as the ID attribute. On the one hand, we can visualize such a reference using a graph edge and perform a structural join; on the other hand, we can push the equality of IDREF and ID values to a table join. In this regard, both structural join and table join have the same effect of solving the constraints of ID references.

4   Special References

ID/IDREF in XML documents may lead to very complex patterns. In this section, we introduce two special cases of ID/IDREF and explain how our algorithm handles queries involving these cases.

4.1   Cyclic Reference

ID references in an XML document may cause cycles if we consider parent-child relationships and ID references as directed edges in a document graph. Consider a document which contains such cycles, as shown in Fig. 8. In this document, Roy chooses Lisa as his first partner, while Lisa chooses Roy as her second partner. Then the references between *member* Roy and Lisa generate a cycle. When we process a query containing a reference cycle, e.g. the query shown in Fig. 9(b), many DAG-based graph matching algorithms, e.g. [23][7], are no longer effective. In our approach, two query objects involved in a reference cycle, i.e. two member nodes in this case, play both a referencing part and a referenced part. Thus we ignore the constraints under both of them during pattern matching, and handle these constraints by table joins. For the query in Fig. 9(b), we first match a rewritten query as shown

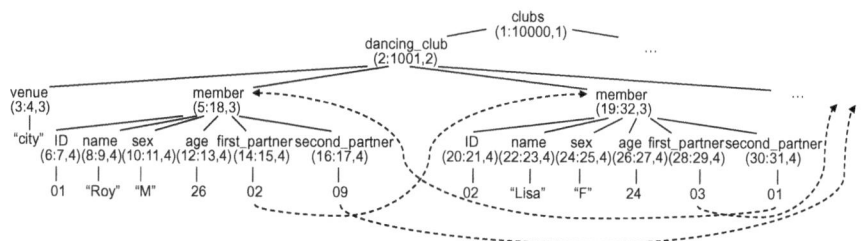

Fig. 8. XML document with cyclic reference

R_member

Label	ID	Name	Sex	Age	First_partner	Second_partner
(5:18,3)	01	Roy	M	26	02	09
(19:32,3)	02	Lisa	F	24	03	01
...	...	...	...	...	...	...

(a) Table for *member*

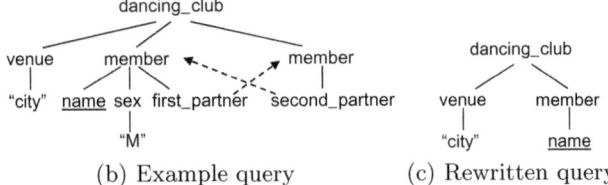

(b) Example query   (c) Rewritten query

Fig. 9. Example query involving cyclic reference

in Fig. 9(c) to the document tree to get the labels of all satisfied members. By joining the result with itself through the member table (shown in Fig. 9(a)) twice, we can easily handle the cyclic situation, and output the desired values.

4.2 Sequential Reference

Sequential references happen when one object references another object, while that object also references a third object. One example document with sequential references is shown in Fig. 10. In this research community document, each *seminar* has a *chair* whose detailed information is stored in some other part of the document; and each *people's affiliation* is also stored in detail separately. The ID/IDREFs between *seminar*, *people* and *affiliation* form a set of sequential references. A reference pattern query with a sequential reference is shown in Fig. 11(b). In this query, we try to find the *topic* of a *seminar* in the 'database' *research area*, which is chaired by a *people* from 'NUS'.

Processing queries involving sequential references in an XML document increases the complexity in many traditional subgraph matching algorithms. Sequential references lead to more cycles if we do not consider the directions of each reference. As we know, in traditional approaches, subgraph matching is done by generating all possible maps between nodes in two graphs and then filtering out

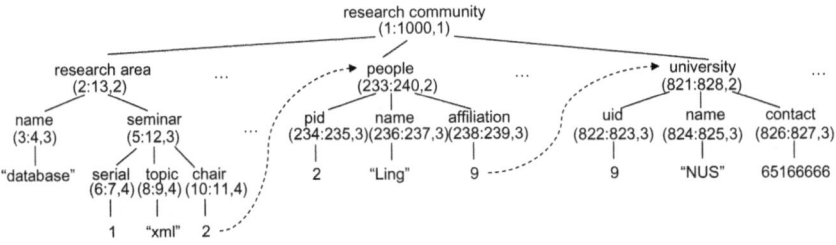

Fig. 10. XML document with sequential reference

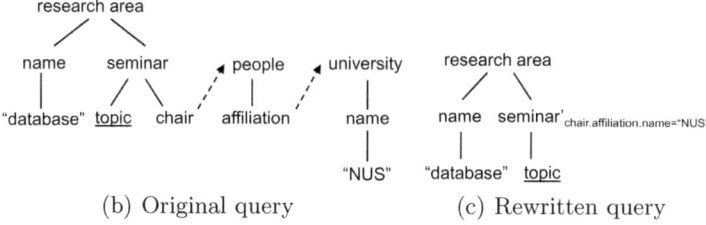

(a) Tables involved

(b) Original query   (c) Rewritten query

Fig. 11. Example query involving sequential reference

incorrect answers. With more cycles, incorrect mappings cannot be pruned as early as that in graphs with fewer cycles. In our approach, we do not need to consider this aspect. For the query in Fig. 11(b), we just perform selection and join between tables for *seminar*, *people* and *university* (shown in Fig. 11(a)).

The selected *label* values for *seminar* will be used to construct a new inverted list for query node *seminar*. Then the original query is rewritten to a new query as shown in Fig. 11(c) by removing references and some other query nodes.

4.3   Complex Reference

Theoretically, a general reference pattern query can be very complex with cyclic and sequential ID references. The last two sections show that table join is powerful to handle both kinds of references between two twig parts. When we deal with a reference pattern query with complex references, we simply decompose the query into twig parts, each of which contains an object referencing or being referenced by other parts. The reference between different parts is solved by table join, and when a twig part is complex itself, we can perform a pattern matching to solve it. We will further research on query optimization on queries with complex reference.

5   Experiments

In this section, we present experimental results, comparing our approach with an XQuery engine [2] and *TwigStackD* [7], which is a stack based approach for XML query processing involving ID/IDREF, which has proven more efficient than traditional graph matching methods. For convenience, we name our table based method as *TBM*. Another recent work on XML graph matching [24] may not be correct when it models a graph pattern query. Thus we do not do a comparison with it[2].

---

[2] More explanations are available at
http://www.comp.nus.edu.sg/∼wuhuayu/problem.pdf

5.1   Experimental Settings

Implementation: We implemented all algorithms in Java. The experiments were performed on a 3.0GHz Pentium 4 processor with 1G RAM.

XML Data Sets: We used three XML data sets for our experiments: Gene ontology data, purchase data and XMark data. Gene ontology data is a 70MB real-life data set, which is taken from a Gene Ontology Project [1]. Purchase data is a 12MB synthetic data set generated by our data generator. The schema of this document is similar to the schema of our example document shown in Fig. 1(a). The characteristics of this document is a large number of ID references, as every part has a supplier reference. We also use 9 XMark benchmark [3] documents with the size varying from 11MB to 111MB to compare document parsing time, and use one of them (23MB) to test execution time. XMark documents contain multiple types of ID references.

Queries: We randomly selected five meaningful queries with ID references for each data set. The queries are shown in Fig. 12. The last element in each query expression is the output node. We only consider the first two cases where the output node resides in either the referencing part or the referenced part in each query. The third case is just a combination of the first two cases. ID references in queries are denoted by '→', and some queries (Q1, Q5, Q11, Q15) contain sequential references.

Query	Data Set	Path Expression
Q1	Gene	//term[n_associations=0][isa/resource→term/isa/resource →term/name='molecular_function']/accession
Q2	Gene	//term[accession='GO0016329']/isa/resource →term/association/evidence/evidence_code
Q3	Gene	//term[name='anticoagulant']/isa/resource →term/dbxref[reference]/database_symbol
Q4	Gene	//term[accession='GO0016172'][name='antifreeze'][//resource →term/dbxref[reference]]/about
Q5	Gene	//term[n_associations=0][isa/resource→term/isa/resource →iterm/isa/resource→iterm/dbxref[reference]]/accession
Q6	Purchase	//part[name='phone'][supplierRef →supplier/location='Sydney']/price
Q7	Purchase	//department[part[name='PC']/supplierRef →supplier/phone='345']/head
Q8	Purchase	//department[head='Fione']/part[name='sofa']/supplierRef →supplier/location
Q9	Purchase	//department[head]/part/supplierRef→supplier[location='London']/phone
Q10	Purchase	//department[name='R&D']//supplierRef→supplier/location
Q11	XMark	//open_auction[itemref →item[location='United States']/incategory →category/name='Seeming mingle teach']/current
Q12	XMark	//bidder[date='11/13/2001']/personref →person[address/city='Birmingham'][profile/gender='male']/phone
Q13	XMark	//person[profile[education='High School']/age=38][watches/watch →open_auction/initial=71.36]/name
Q14	XMark	//closed_auction[buyer/person→person[address/province='Haban']][seller/ person→person[address/city='Lisbon']]/price
Q15	XMark	//bidder[increase='21'][personref→person[age='35'][//watch →open_auction/reserve=50.84]]/date

Fig. 12. Experimental queries

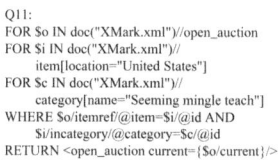

Q11:
FOR $o IN doc("XMark.xml")//open_auction
FOR $i IN doc("XMark.xml")//
    item[location="United States"]
FOR $c IN doc("XMark.xml")//
    category[name="Seeming mingle teach"]
WHERE $o/itemref/@item=$i/@id AND
    $i/incategory/@category=$c/@id
RETURN <open_auction current={$o/current}/>

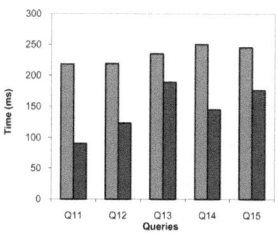

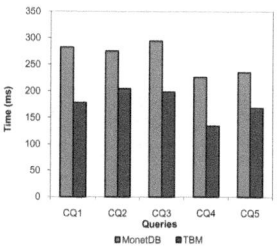

(a) Q11 in XQuery   (b) Result for Q11-Q15   (c) Result for cyclic cases

Fig. 13. Query performance comparison between *MonetDB* and *TBM*

5.2 Experimental Results and Analysis

Comparison with XQuery engine. This experiment is done using *MonetDB* [2], which is a well known memory-based XQuery engine, and a relatively small XMark document (11MB) so that all the processing can be done in memory. MonetDB uses an optimized node-based relational approach [14] to process XML queries. First using *MonetDB* and *TBM*, we process the queries Q11-Q15 which include sequential reference cases. The XQuery expression of Q11 is shown in Fig. 13(a). Other queries can also be expressed as XQuery expressions in a similar way. Due to the space limitations, we do not show the XQuery expression of every query. The experimental result is shown in Fig. 13(b). In the second step, we test queries with cyclic references. In XMark data, we observe that each *person* has several *watches*, each *watch* contains an *open_auction*, each *open_auction* has *bidders*, and each *bidder* is a *person*. We randomly compose five queries within this cycle and the execution time for the two methods is shown in Fig. 13(c).

TBM outperforms *MonetDB* for all the queries by 20-50%. The reason is that XQuery cannot express queries involving ID references using a single path. Instead, XQuery has to do a multiple retrieval for the referencing part and the referenced part of the query, and then do a join between the retrieved results. However, using our method, we handle the reference separately, solving the reference constraint, and also simplifying the query structure and search space.

Comparison with TwigStackD. Our experiments mainly compare the query processing time and document parsing time between *TwigStackD* and our table based methods (*TBM*). We used B^+ trees to organize inverted lists for both approaches to ensure high performance of inverted list accessing. The execution time for *TBM* includes the time of processing ID reference with table joins and the time of structural joins during tree pattern matching. The results of execution time are shown in Fig. 14 (the Y-axis is in logarithmic scale). From the results, we can see for all queries the performance of *TBM* is more than 100 times faster than *TwigStackD*. The reason why *TwigStackD* is worse is that their approach uses graph matching rather than tree matching. As a result, in *TwigStackD* they maintain an index to store position relationships between nodes. Each time they process a query, partial solutions are expanded based on the index. This stage is very costly and seriously affects the performance.

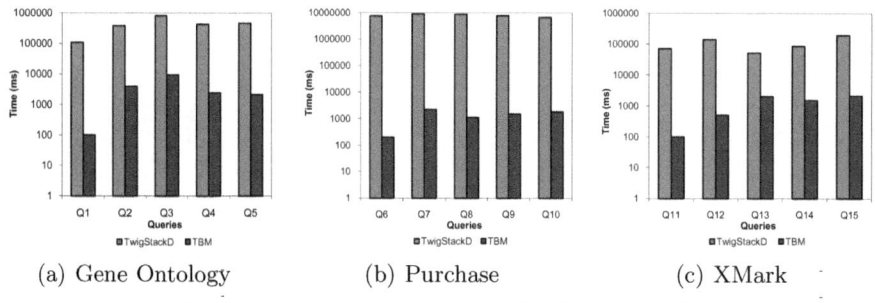

(a) Gene Ontology   (b) Purchase   (c) XMark

Fig. 14. Execution time by *TwigStackD* and *TBM*

In the purchase document, the schema is simple but contains lots of references between *part* and *supplier*. When *TwigStackD* expands partial solutions in a pool, the pool size and amount of checking is very large. That is why the execution time of *TwigStackD* in purchase document is much greater than that in the other two documents where fewer references are involved. The XMark document contains four types of ID/IDREF on objects *item*, *category*, *person* and *open_auction*. Due to the complex references, the index in *TwigStackD* is very large (nearly 3 times greater than the original document size) and it takes quite a long time to build such an index.

Finally we conducted experiments on document parsing time between *TBM* and *TwigStackD*. This parsing time includes node labeling, inverted list constructing and other index building. All these operations are required for structural join based native XML query processing algorithms. In our experiment, we took 12 XMark documents with different numbers of nodes. The results given in Fig. 15, show that when the size of the document grows, the parsing time for *TwigStackD* increases quickly, while using our approach the parsing time is more acceptable for different document sizes.

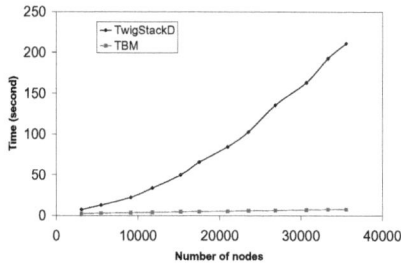

Fig. 15. Document parsing time comparison

6 Conclusions and Further Work

In this paper we analyze the drawbacks of existing work for query processing in XML documents with ID/IDREF. In particular, most existing algorithms consider XML documents and queries as graphs, and perform graph matching to process queries. However, graph matching is generally believed to be less efficient than tree matching, which is a widely accepted approach to process XML queries without considering ID references. Motivated by this finding, we propose a table based semantic approach to reduce graph matching to tree matching to process XML queries involving ID references. When we parse an XML document, we only consider the native hierarchical structure, and do not treat it as a graph with ID references. We build relational tables for each object that may contain ID or IDREF attributes. During query processing, we decompose a reference pattern query into a referencing part and a referenced part using the ID reference involved. Now the referencing part will be a simple tree structure that can be matched to the document tree. The reference between the referencing part and the referenced part is eventually transformed to a table join between the two parts. The experimental results show that our approach is 20-50% more efficient than MonetDB and more than 100 times faster than TwigStackD, a structural join based graph matching algorithm. Furthermore, our approach can also handle complex ID/IDREF relationships such as cyclic references and sequential references, which are bottlenecks for many existing works.

For further work, we will further investigate how to generate a better query plan when dealing with both tree pattern matching and table joins.

References

1. http://www.geneontology.org/
2. MonetDB, http://monetdb.cwi.nl/
3. XMark. An XML benchmark project (2001), http://www.xml-benchmark.org
4. Berglund, A., Chamberlin, D., Fernandez, M.F., Kay, M., Robie, J., Simeon, J.: XML path language XPath 2.0. W3C Working Draft (2003)
5. Boag, S., Chamberlin, D., Fernandez, M.F., Florescu, D., Robie, J., Simeon, J.: XQuery 1.0: An XML query. W3C Working Draft (2003)
6. Bruno, N., Koudas, N., Srivastava, D.: Holistic twig joins: optimal XML pattern matching. In: SIGMOD, pp. 310–321 (2002)
7. Chen, L., Gupta, A., Kurul, M.E.: Stack-based algorithms for pattern matching on dags. In: VLDB, pp. 493–504 (2005)
8. Chen, S., Li, H.-G., Tatemura, J., Hsiung, W.-P., Agrawal, D., Candan, K.S.: Twig^{2}stack: Bottom-up processing of generalized-tree-pattern queries over XML documents. In: VLDB, pp. 283–294 (2006)
9. Chen, T., Lu, J., Ling, T.W.: On boosting holism in XML twig pattern matching using structural indexing techniques. In: SIGMOD, pp. 455–466 (2005)
10. Deutsch, A., Fernandez, M.F., Suciu, D.: Storing semistructured data with STORED. In: SIGMOD Conference, pp. 431–442 (1999)
11. Fan, W., Yu, J.X., Lu, H., Lu, J., Rastogi, R.: Query translation from XPath to SQL in the presence of recursive DTDs. In: VLDB, pp. 337–348 (2005)
12. Garey, M.R., Johnson, D.S.: Computers and Intractability: A Guide to the Theory of NP-Completeness. W.H. Freeman, New York (1979)
13. Gou, G., Chirkova, R.: Efficiently querying large XML data repositories: A survey. IEEE Trans. Knowl. Data Eng. 19(10), 1381–1403 (2007)
14. Grust, T., Keulen, M.V., Teubner, J.: Accelerating XPath evaluation in any RDBMS. ACM Trans. Database Syst. 29(1), 91–131 (2004)

15. Jiang, H., Lu, H., Wang, W.: Efficient processing of twig queries with or-predicates. In: SIGMOD, pp. 59–70 (2004)
16. Jiang, M.: Querying XML data: efficiency and security issues. Ph.D. Thesis. The Chinese University of Hong Kong
17. Kimelfeld, B., Sagiv, Y.: Twig patterns: from XML trees to graphs. In: WebDB, pp. 26–31 (2006)
18. Krishnamurthy, R., Chakaravarthy, V.T., Kaushik, R., Naughton, J.F.: Recursive XML schemas, recursive XML queries, and relational storage: XML-to-SQL query translation. In: ICDE, pp. 42–53 (2004)
19. Lu, J., Ling, T.W., Chan, C.Y., Chen, T.: From region encoding to extended dewey: On efficient processing of XML twig pattern matching. In: VLDB, pp. 193–204 (2005)
20. Morgenthal, J., Evdemon, J.: Eliminating redundancy in XML using ID/IDREF. XML Journal 1(4) (2000)
21. Shanmugasundaram, J., Tufte, K., Zhang, C., He, G., DeWitt, D.J., Naughton, J.F.: Relational databases for querying XML documents: Limitations and opportunities. In: VLDB, pp. 302–314 (1999)
22. Shasha, D., Wang, J.T.-L., Giugno, R.: Algorithmics and applications of tree and graph searching. In: PODS, pp. 39–52 (2002)
23. Vagena, Z., Moro, M.M., Tsotras, V.J.: Twig query processing over graph-structured XML data. In: WebDB, pp. 43–48 (2004)
24. Wang, H., Li, J., Luo, J., Gao, H.: Hash-based subgraph query processing method for fraph-structured XML documents. In: VLDB, pp. 478–489 (2008)
25. Wu, H., Ling, T.W., Chen, B.: VERT: A semantic approach for content search and content extraction in XML query processing. In: Parent, C., Schewe, K.-D., Storey, V.C., Thalheim, B. (eds.) ER 2007. LNCS, vol. 4801, pp. 534–549. Springer, Heidelberg (2007)
26. Yoshikawa, M., Amagasa, T., Shimura, T., Uemura, S.: XRel: a path-based approach to storage and retrieval of XML documents using relational databases. ACM Trans. Internet Techn. 1(1), 110–141 (2001)
27. Zhang, C., Naughton, J.F., DeWitt, D.J., Luo, Q., Lohman, G.M.: On supporting containment queries in relational database management systems. In: SIGMOD Conference, pp. 425–436 (2001)

Improving Alternative Text Clustering Quality in the Avoiding Bias Task with Spectral and Flat Partition Algorithms

M. Eduardo Ares, Javier Parapar, and Álvaro Barreiro

IRLab, Department of Computer Science, University of A Coruña, Spain
{maresb,javierparapar,barreiro}@udc.es

Abstract. The problems of finding alternative clusterings and avoiding bias have gained popularity over the last years. In this paper we put the focus on the quality of these alternative clusterings, proposing two approaches based in the use of negative constraints in conjunction with spectral clustering techniques. The first approach tries to introduce these constraints in the core of the constrained normalised cut clustering, while the second one combines spectral clustering and soft constrained k-means. The experiments performed in textual collections showed that the first method does not yield good results, whereas the second one attains large increments on the quality of the results of the clustering while keeping low similarity with the avoided grouping.

1 Introduction

Data analysis plays nowadays a central role in several fields of science, industry and business. With the ever-growing size of the data collections being compiled and used by public institutions and private firms alike a great need for automatic data analysis tools has arisen, in order to provide a way to exploit those collections in an effective and timely manner.

Clustering is the most popular non-supervised automatic data analysis tool. Given a data collection, the clustering algorithms try to form a meaningful grouping of the data, categorising the data instances (text documents, in our case) in various groups (clusters), such that the instances in the same cluster bear high similarity between them and low similarity with the instances that have been put in the other clusters.

Unfortunately, the concepts of "meaningful grouping" and "high" and "low" similarity are very subjective. Sometimes, and even though the grouping of the data found by a certain clustering algorithm can make sense from a purely mathematical point of view, it might be completely useless or even meaningless to the user. Gondek and Hofmann illustrate in [1] several examples this situation, such as the clustering of news corpora which have been already annotated by a certain criterion (such as region) or the clustering of users' data with gender or income information. The outcome of the algorithm might reflect a grouping of the data

P. García Bringas et al. (Eds.): DEXA 2010, Part II, LNCS 6262, pp. 407–421, 2010.

which is well known, or which would be easy to find with a manual examination. Consequently, it will be of little use to the user of the data analysis tool.

Thus, sometimes mechanisms are needed to find alternative clusterings to the one proposed by the clustering algorithm. If we are trying to avoid the tendency (bias) of the clustering algorithm to fall in a certain grouping of the data that is being clustered the task is called Avoiding Bias. This problem has been tackled by several authors in the last years, which have proposed a wide range of approaches, ranging from distance learning [2] to using constraints [3]. However, it should be underlined that avoiding bias is still a clustering process, where the main focus is providing the user with a meaningful grouping of the data. For instance, the easiest way to find a very different grouping from the one given would be assigning randomly documents to clusters, which would be obviously a very bad solution in terms of clustering quality. Thus, a compromise has to be reached between the quality of the clustering and the distance to the avoided grouping when devising an avoiding bias algorithm.

In this paper we study various ways to obtain an alternative clustering with high quality while keeping the objective of avoiding the known clustering. Concretely, we test two different approaches which use a strategy similar to the one in [3] (using negative constraints to steer the clustering process away from the known clustering), making use of spectral clustering techniques to try to attain that high quality. The first one is introducing negative constraints in the constrained normalised clustering approach proposed by Ji et al. in [4]. The second one is introducing the soft constrained k-means algorithm proposed by Ares et al. in [3], which has been shown to have good results, in the second phase of a normalised cut clustering algorithm [5]. The experiments carried out with these approaches showed that, while the first approach does not yield good results, the combined one (normalised cut plus soft constrained k Means) outperforms soft constrained k-means in terms of quality of the results while keeping a good avoidance of the known clustering.

This paper is organised as follows: in Section 2 the clustering algorithms on top of which the proposed approaches are built are introduced. In Section 3 we tackle the problem of introducing negative constraints in normalised cut, while in Section 4 we introduce the experiments which were carried out and their results. Finally, Sections 5 and 6 are respectively devoted to the related work and the conclusion and future works.

2   Clustering Algorithms

In this section we describe the clustering approaches which we have used in the methods proposed in this paper. Firstly, we survey normalised cut, a very effective spectral clustering algorithm introduced by Shi and Malik in [5], and its constrained counterpart, constrained normalised cut, introduced by Ji et al. in [4]. Afterwards, we outline soft constrained k-means, a constrained clustering algorithm based on k-means introduced by Ares et al. in [3].

2.1   Normalised Cut

The spectral clustering algorithms [6,7] are a family of algorithms which use results from graph spectral theory to perform the clustering of data. Concretely, normalised cut tries to tackle a clustering problem by transforming it into a graph cut problem.

The first step is creating a graph $G = (V, E, W)$ in which the documents to be clustered are the vertices ($V = \{v_1, v_2, ..., v_n\}$), and the weights ($W = \{w_{1,1}, w_{1,2}, ..., w_{n,n}\}$) of the edges ($E$) are related to the similarity between the documents joined by each edge, such that the more similar the documents are, the higher the weight of the edge. Hence, the aim of the clustering process, creating groups of documents such that the documents in the same cluster are very similar and documents in different clusters have low similarity, can be re-formulated as cutting this new graph G in connected components in a way that the weights of the edges which join vertices in different connected components are low and the ones of the edges which join vertices in the same connected component are high.

To measure this, Shi and Malik introduced the normalised cut (NCut) value of a cut of a graph in [5]. For a graph $G = (V, E, W)$ and a cut $\{A_1, A_2, ...A_k\}$ of that graph, NCut is defined as:

$$\text{NCut}(A_1, ...A_k) = \sum_{i=1}^{k} \frac{\text{cut}(A_i, \bar{A}_i)}{\text{vol}(A_i)} \tag{1}$$

$$\text{cut}(A, B) = \sum_{i \in A, j \in B} w_{ij} \tag{2}$$

$$\text{vol}(A) = \sum_{i \in A} \sum_{j=1}^{n} w_{ij} \tag{3}$$

where w_{ij} is the weight of the edge that joins vertices i and j, and $\bar{A}_i$ are the vertices which are not included in A_i (i.e., $\bar{A}_i = V \setminus A_i$).

As it follows from (1), the NCut of a graph cut is minimised when the sum of the weights of the edges joining documents in different connected components are low, while keeping the sizes of the different connected components, which are measured using their volume, as high as possible. This last condition tries to ensure a certain balance between the connected components, to avoid trivial solutions with connected components comprising only very few vertices. Thus, a graph cut with a low NCut value would fulfil the requisites of a good clustering.

Finding a cut $\{A_1, A_2, ..., A_k\}$ of a certain graph G which minimises the NCut value can be transformed [7] into a trace minimisation problem (4), where H is a $n \times k$ matrix (where n is the number of documents to be clustered) which encodes the membership of vertices to connected components as indicated in (5), $D = (d_{ij})$, is a diagonal matrix with $d_{ii} = \text{degree}(v_i)$ and L is the Laplacian matrix ($L = D - W$) of G.

$$\min_{A_1, ...A_k} \text{Tr}(H^T L H) \text{ subject to } H^T D H = I \tag{4}$$

$$H = (h_{ij}) = \begin{cases} \frac{1}{\sqrt{vol(A_j)}} & \text{if vertex i} \in A_j \\ 0 & \text{else} \end{cases} \tag{5}$$

Unfortunately, the condition imposed by (5) on the values of H makes the minimisation in (4) NP-hard. If that discreteness condition is dropped and a simple variable substitution is performed ($Y = D^{\frac{1}{2}}H$), the minimisation can be rewritten in the standard form of a trace minimisation problem (6):

$$\min_{Y \in \mathbb{R}^{n \times k}} \mathrm{Tr}(Y^T \left[D^{-\frac{1}{2}}LD^{-\frac{1}{2}} \right] Y) \text{ subject to } Y^T Y = I \tag{6}$$

It can be shown that (6) is minimised by the matrix Y which contains as columns the eigenvectors corresponding to the smallest eigenvalues of $D^{-\frac{1}{2}}LD^{-\frac{1}{2}}$. However, as the values of Y are not constrained, this matrix is no longer composed of indicator vectors for the connected components. Instead, each of the documents has been projected into $\mathbb{R}^k$, and a further step has to be taken (such as applying a clustering algorithm like k-means) in order to find a discrete segmentation of the points in that space. Once this segmentation has been found, we can transpose it to the original documents, providing a clustering of the original collection.

2.2 Constrained Normalised Cut

Based on the same principles of normalised cut, Ji et al. proposed in [4] a constrained clustering algorithm which makes some changes in the function to be minimised in order to introduce *a priori* knowledge in the clustering process, specifically which pairs of documents the user wants to be grouped by the clustering algorithm into the same cluster.

To achieve this, they introduced a new matrix U with n columns and a row for each constraint used in the algorithm. Thus, a constraint which establishes that data points i and j should be in the same cluster will be encoded as a row of zeroes with the exception of positions i and j, which will be set to 1 and -1 (or vice-versa, as these constraints are non-directional). If membership to connected components is encoded in a matrix H as in (5), the Frobenius norm of the product of matrices U and H will be smaller as more constraints are respected in the clustering, with a minimum of zero when none of them is disregarded. Thus, a new minimisation problem can be written involving both NCut and the supplied constraints:

$$\min_{A_1,...A_k}(\mathrm{NCut}(A_1, ..., A_k) + ||\beta U H||^2) \tag{7}$$

where $\beta > 0$ is a parameter which controls the degree of enforcement of the constraints. The higher that β is, the tighter the enforcement of the constraints is. This minimisation problem, following a derivation similar to the one used in the non constrained case, can be written as:

$$\min_{Y \in \mathbb{R}} \mathrm{Tr}(Y^T \left[D^{-\frac{1}{2}}(L + \beta U^T U)D^{-\frac{1}{2}} \right] Y) \tag{8}$$

again subject to $Y^TY = I$. As this problem is in the standard form of a trace minimisation problem, the same theoretical result used in the unconstrained case can be used here. Thus, this equation is minimised by a matrix Y which contains as columns the eigenvectors which correspond to the smallest eigenvalues of matrix $D^{-\frac{1}{2}}(L + \beta U^TU)D^{-\frac{1}{2}}$. Again, these columns are not proper indicator vectors, so a segmentation of the projected data points has to be performed in order to produce a clustering of the data.

2.3   Soft Constrained k-Means

Batch k-means [8] is one of the most popular flat clustering algorithms. The first step of the algorithm is the initialisation, where some points in the representation space are taken as seeds of the clustering process. Typically these seeds are chosen randomly between the documents to be clustered. Afterwards, the main core of the algorithm is a loop in which documents are assigned to clusters depending on its similarity with clusters' centroids. Once all of them have been assigned the centroids are recalculated and the process starts again. This loop is repeated until a given convergence condition is met (typically when the change in the centroids between a iteration and the next is very small).

Based on batch k-means skeleton, Wagstaff et al. introduced in [9] a constrained clustering algorithm which enables the use of domain knowledge in the clustering process. This domain knowledge can be introduced in the form of two kinds of instance level pairwise constraints: Must-links, which indicate that two documents must be in the same cluster, and Cannot-links, to indicate that two documents must be in different clusters. To honour these constraints they modified the cluster assignment policy, assigning the documents to the closest (most similar) centroid such that this assignment does not violate any constraints. That is, if a document with which the document being assigned has a Must-link constraint has been assigned to a cluster in the current iteration, the document will be assigned directly to that cluster. Otherwise, the document is assigned to the cluster with the closest centroid, excluding those containing documents with which the document being assigned has a Cannot-Link constraint, in order to enforce that kind of constraints. In that paper, the authors show that these constraints can effectively affect the clustering process, leading it towards a better solution. On the other hand, the authors admit as well that the absolute nature of the proposed constraints can make sometimes the presence of this constraints harmful. For instance, Cannot links can lead the clustering process to a dead end, if a document has a Cannot link with at least one document in each cluster.

In order to address these limitations, Ares et al. introduced in [3] two kinds of non absolute constraints: May-Links and May-Not-Links, which indicate that two documents are, respectively, likely or not likely to be in the same cluster. The implementation of these constraints alters again the assignment process of the documents. After the absolute constraints introduced by Wagstaff et al. are accounted for, each cluster is given a score which is initialised with the similarity between the document and its centroid. Then, the score of a given cluster will be increased in a certain factor w for each document with which that document has a

May-Link and was last assigned to that cluster. Conversely, the score of a cluster will be decreased by the same factor for each document with which the document has a May-Not-Link and was last assigned to the cluster. The authors claim that these new constraints overcome the drawbacks of the absolute constraints, while maintaining good effectiveness. Namely, the May-Not-Links are shown to be effectively better than their absolute counterparts (Cannot-links), because their efficacy seems to be similar and the May-Not-Links are not affected by the dead end problem, as it is always possible to find a suitable cluster for all the data points. Anyway, the algorithm proposed in the paper allows as well the introduction of domain knowledge in form of absolute constraints, following the same strategy proposed by Wagstaff et al.

3   Negative Constraints in Normalised Cut

As it was previously explained, Ji et al. proposed in [4] an addition to normalised cut which allowed introducing domain knowledge in the clustering process. However, the method that they propose only allows the introduction of *positive* information, i.e., pairs of documents that the user thinks that they ought to be in the same cluster. But this is not the only kind of information that a user might have available about the documents to be clustered. For instance, it is also very likely that the user has some intuition about which pairs of documents might not (or must not) be in the same cluster (this is what we will call *negative* information). Actually, this negative information is less informative to the clustering algorithm than the positive constraints, as with the positive information we are actually providing the algorithm with fragments of the desired final grouping (or at least we hope to be doing so). However, is precisely this lesser informativeness (and the less restrictions that they impose on the algorithm) which makes the negative constraints more likely to be elicited from the domain knowledge, or even the only information that can be provided, in cases where the nature of the task being tackled does not allow the obtaining of positive information at all. For instance, this is the case of the Avoiding Bias task, which is the main focus of this paper.

In the Avoiding Bias task, the only information available is the grouping of the documents that we are trying to avoid. We can not obtain any positive clues from it, as neither the fact that two documents are in the same cluster, nor the fact that they are in different ones gives us any positive evidence about if they should be in the same cluster in an alternative grouping.

However, if two documents are in the same cluster in the grouping that we are trying to avoid, it is sensible to make some indication (using non absolute negative constraints) to the clustering algorithm that these documents might not be in the same cluster in other grouping, expecting that the distorsion induced by these constraints is on the one hand enough to break the bias of the algorithm to fall in the avoided clustering and on the other hand not strong enough to break completely the structure of the similarities between documents, so that the final clustering of the data is still meaningful. This is precisely the intuition that sustains Ares et al. Avoiding Bias approach in [3].

Obviously, the same point could be made about using positive non absolute constraints on documents which are not in the same cluster in the avoided grouping. However, bringing closer these documents will not have the effect of avoiding the bias of the clustering algorithm to fall in the given grouping. To do so, these constraints should be very strong, but this will likely compress the representation space too much, providing clusters of bad quality.

In this section we will tackle the problem of introducing the negative constraints into the normalised cut clustering algorithm.

3.1 Negative Constraints in Constrained Normalised Cut

In Sect. 2.2 we have explained the approach used by Ji et al. [4] to transform the classic normalised cut algorithm into a constrained clustering one, allowing the use of positive constraints. Intuitively, a similar scheme could be used to try to introduce negative information as well.

In their paper, the authors introduce a matrix U which encodes the positive constraints, such that the Frobenius norm of the product of that matrix and the indicator matrix is in inverse proportion with the number of constraints which are respected by the clustering represented by the indicator matrix, having a minimum of zero when all of them are honoured. Thus, introducing this factor into the function minimised at the core of the normalised cut algorithm (7,8) causes a change in the nature of the solution, now having to find a clustering of good quality (minimising NCut) while respecting as well the constraints (minimising the new term). The influence of the constraints is controlled by a parameter (β), being the enforcement of the constraints greater as the value of β increases, with a minimum in $\beta = 0$, where the the constraints are not taken into account at all.

With that in mind, an apparently easy and intuitive way to introduce the negative constraints would be using a new matrix U_N, which would encode the negative constraints in the same way as the positive ones were encoded in U. Again, the Frobenius norm of the product of U_N with the indicator matrix will be lower as more of the pairs of documents linked by a constraint are in the same cluster, and, vice versa, higher as more of them are not in the same cluster, which is precisely the objective of the negative information. In order to introduce this new term in the minimisation a new parameter (β_N) is needed to control the enforcement of the negative constraints. As this new factor is in direct proportion to the number of negative constraints which are respected in the clustering, it must be introduced in the formula with a minus sign (9,10). Again, the value of β_N is equal or greater than 0, with a harder enforcement of the constraints as its value increases.

$$\min_{A_1,...A_k}(\text{NCut}(A_1, ..., A_k) - ||\beta_N U_N H||^2) \tag{9}$$

$$\min_{Y \in \mathbb{R}} \text{Tr}(Y^T \left[D^{-\frac{1}{2}}(L - \beta_N U_N^T U_N) D^{-\frac{1}{2}} \right] Y) \tag{10}$$

Even though this approach seems theoretically sound, it does not yield good results in the Avoiding Bias task. Our explanation about why this happens is given in Sect. 4.5.

3.2 Combining Soft Constrained k-Means and Normalised Cut

As it has been previously explained (Subsect. 2.1), the normalised cut algorithm is based on transforming the clustering problem into a graph cut problem. The aim of the process is finding a cut of the graph which minimises its normalised cut value. Being this a NP-hard problem, a certain relaxation of the conditions imposed on the solution has to be performed in order to reduce its complexity and make it computationally accessible. Thus, the outcome of this minimisation is a projection of the data points into $\mathbb{R}^k$, instead of the grouping itself, and a last step should be performed to reach the final clustering of the data. In order to perform this last phase, Shi and Malik propose using k-means on the projected data points.

Our proposal in this paper is using the soft constrained k-means algorithm proposed by Ares et al. instead of batch k-means, enabling the introduction of domain knowledge in form of absolute (Must and Cannot-Link) and non-absolute (May and May-Not-Link) constraints. Even though they would be defined over the initial documents, the one to one correspondence between them and the projected documents (the document which was represented by the vertex v_i of the graph is now encoded in the ith row of matrix Y) enables us to apply these same instance level constraints over the corresponding projected documents.

From the point of view of soft constrained k-means, the normalised cut acts as a kind of document preprocessing phase, where the documents are transformed from the chosen document representation to a representation in $\mathbb{R}^k$ based on the normalised cut criterion. The effect of this "preprocessing" is twofold: not only we are benefiting from the increment of cluster quality caused by using the normalised cut algorithm, but also we are likely to experiment an increase in the effect of the pairwise constraints. As documents which are close to constrained ones are affected as well by the changes in the destination of the later ones induced by the constraints, our intuition is that the effectiveness of the constraints in this new data space is increased, as similar documents (over which the same constraints tend to be true) are brought together and dissimilar ones are separated (thus avoiding some non desired "interferences" of the constraints over non related documents).

In terms of performance, the computational cost of this combined approach is the same of that of the normalised cut algorithm, as the cost of the soft constrained k-means and of batch k-means is the same. Consequently, being the costliest operation of the whole algorithm still by a wide margin the calculation of eigenvectors, the total cost will depend on the method chosen to perform that calculus. This cost can be kept fairly moderated if a standard algorithm is used. For instance, using Lanczos algorithm, the time complexity would be $O(kN_{Lanczos}nnz(M))$, where k is the desired number of clusters (i.e. of eigenvectors), $N_{Lanczos}$ is the number of iteration steps of the algorithm and $nnz(M)$

is the number of non zero elements of the matrix $D^{-\frac{1}{2}}LD^{-\frac{1}{2}}$ (see Sect. 2.1), whose eigenvectors are being calculated.

4 Experiments

4.1 Methodology

In order to test the practical behaviour of the algorithms we have set an avoiding bias experiment following the standard methodology of the papers on that subject. Thus, we will use text document collections in which documents have been categorised according to two different criteria. Using the standard methodology of avoiding bias experiments, we will assume alternatively that one of them is the known grouping and we will try to avoid it, evaluating the results of the process comparing the resulting grouping with both the known one (to assess the avoidance that has been achieved) and the "unknown" one (as a way to measure the quality of the results).

We have used as baseline the original soft constrained k-means approach to Avoiding Bias introduced by Ares et al. in [3], where the authors show that it improves an algorithm specially tailored for Avoiding Bias such as Conditional Information Bottleneck [1]. Thus, we have replicated the same experimental conditions used in that paper. The set of constraints was created introducing a constraint for each pair of documents which are in the same cluster in the known grouping of the data (the only a priori information available). In the case of the baseline and of the combined (NC+SCKM) approach, which support bidirectional and unidirectional constraints, we have used the bidirectional ones. Moreover, we will assume that the number of clusters is known, setting it to the number of clusters of the non avoided grouping of the data. Finally, as the clustering seeds were also chosen randomly from the documents, and the outcome of the processes is really dependant on the quality of the initial seeds, several repetitions of the clustering process have to be performed in order to have a faithful representation of the performance of the algorithms. We report the average of these initialisations.

Following this approach, the only parameters which should be initially set are w, the strength of the constraints in the baseline and in the approach based on the combination of normalised cut and soft constrained k-means and β_N, the tightness of the observance of the negative constraints in the approach based on constrained normalised cut. Besides, in our experiments we have detected that the clustering algorithms yielded better results when the number of dimensions of the projection of the documents performed in the spectral phase is greater than the wanted number of clusters. Typically, the best performance was obtained when the number of eigenvectors ranged from 10 to 20 (in opposition to the number of desired clusters, which ranges from 2 to 5), a fact that is likely caused by the combination of two circumstances. Firstly, the high topicality of the collection compared with the number of expected clusters, and, secondly, this relatively small number of desired clusters, which would cause a great loss of information in the projection if we take the same number of eigenvectors. However, taking too many dimensions could result in adding noise to the documents, which would worsen the quality of the clustering. Thus, after some preliminary tests, we have used to create the projection of the documents the first 15 eigenvectors, a value which we have found that performs well in all collections.

4.2 Datasets

To perform the experiments we have used the two datasets used in the baseline experiments, which were originally defined in [1].

Dataset (i) was created from WebKB's Universities Dataset, which was made collecting webpages from the websites of different U.S. universities (Cornell, Texas, Washington, Wisconsin and others). These webpages have been manually tagged according to two aspects: university and topic ("course", "department", "faculty", "project", "staff", "student" and "other"). The dataset used in the experiments is created taking the documents from the Universities of Cornell, Texas, Washington and Wisconsin which were as well tagged as "course", "faculty", "project" "staff", "student", which yields a total of 1087 documents.

Dataset (ii) was created from Reuters RCV-1, a huge document collection composed of about 810,000 news stories from Reuters, one of the most important news agencies. These documents have been manually tagged according to three aspects: topic, geographical area and industry. The dataset used in the experiments is created taking the documents with have been labelled with respectively only one topic and region label and whose topic is "MCAT" or "GCAT" and whose region is "UK" or "INDIA". This yields a total of 1600 documents.

4.3 Document Representation

As in the baseline experiments, we have used Mutual Information as the original representation of the documents (i.e., the one used to build the graph G), as it has been shown to perform consistently better than other $tf \cdot idf$ approaches [10]. Thus, the representation of a document d in a collection of m terms and d documents is a vector (11) where the components are the mi values of the terms (12),(13), calculated used the frequency of the each term t in the document d ($tf(d, f)$).

$$mi(d) = [mi(d, t_1); mi(d, t_2); \ldots; mi(d, t_m)] \qquad (11)$$

$$mi(d, t) = \log\left(1 + \frac{\frac{tf(d,t)}{N}}{\frac{\sum_i^D tf(d_i,t)}{N} \times \frac{\sum_j^m tf(d,t_j)}{N}}\right) \qquad (12)$$

$$N = \sum_i \sum_j tf(d_i, t_j) \qquad (13)$$

The similarity between two documents d_1 and d_2 was computed using the cosine distance between their vectors, which was also also the distance function used to compare the projected documents after the spectral phase.

4.4   Metrics

In order to evaluate the results of our tests we have used two different metrics, which compare the clustering of a collection of n documents yielded by the algorithm $\Omega = \{\omega_1, \omega_2, ...\omega_k\}$ with a certain ground truth $\mathbb{C} = \{c_1, c_2, ...c_j\}$.

Purity (P) [11] measures how well the clustering outcome matches the target split in average. Higher Purity values mean more similarity between Ω and $\mathbb{C}$.

$$P(\Omega, \mathbb{C}) = \frac{1}{n} \sum_k \max_j |\omega_k \cap c_j| \tag{14}$$

On the other hand, Mutual Information (MI) [12] measures how much information about a grouping is conveyed by another. Again, higher values of Mutual Information mean more agreement between Ω and $\mathbb{C}$.

$$MI(\Omega; \mathbb{C}) = \sum_k \sum_j \frac{|\omega_k \cap c_j|}{n} \log \frac{N|\omega_k \cap c_j|}{|\omega_k||c_j|} \tag{15}$$

4.5   Results

In order to set the parameters of the algorithms we have used a crossvalidation strategy. This strategy involved tuning the value of these parameters in one of the avoiding bias problems, specifically in collection (i) avoiding the grouping by "Topic", and using that value in the other problems. The value w chosen for the baseline (soft constrained k-means) was 0.0025, the value which obtained the best compromise between quality and avoidance. In the combined approach (NC+SCKM), as the focus of this paper is improving the quality of the grouping, the value ($w = 0.05$) was chosen as the one which yielded the best similarity (MI) with the non avoided grouping of the documents ("University") while maintaining a similarity with the avoided grouping ("Topic") less or equal to the one achieved by the baseline, which was itself quite low. As for the constrained normalised cut with negative constraints, the tuning process showed poor quality values and a great instability of the algorithm with respect to the values of β_N. Our explanation about why this happens is given at the end of this section.

The results of the performed experiments are shown in Table 1. As in the experiments in [3] and in [1], for each dataset and avoided grouping we report the values of Mutual Information (MI) with the avoided and the non-avoided groupings, to see to which of them the outcome of the clustering process is mostly leaning, and Purity (P) with the non-avoided grouping, to measure the quality of the clustering. Hence, a good result would have high values of MI and P with the non-avoided grouping and a low value of MI with the avoided one. The results reported are the average of the ten different initialisations of seeds and document inspection order tested in each combination of dataset and avoided grouping.

As a preliminary note, it is worth remarking that the results show the expected increase in the quality of clustering of normalised cut with respect to batch k-means. Moreover, they also point out a tendency in the non constrained

Table 1. Results for the avoiding bias experiment with the defined datasets for batch k-means, soft constrained k-means (SCKM), normalised cut and the combined approach (NC+SCKM)

Dataset (i)	Avoiding Topic (k=4)			Avoiding University (k=5)		
	MI(Topic)	MI(Univ.)	P(Univ.)	MI(Univ.)	MI(Topic)	P(Topic)
Batch k-means	0.5069	0.2304	0.4364	0.2972	0.5682	0.6874
SCKM ($w = 0.0025$)	0.0052	0.2789	0.4772	0.0031	0.4499	0.6484
Normalised cut	0.4801	0.4097	0.4994	0.5822	0.5606	0.6794
NC+SCKM ($w = 0.05$)	0.0032	0.9340	0.7684	0.0011	0.6569	0.7163

Dataset (ii)	Avoiding Topic (k=2)			Avoiding Region (k=2)		
	MI(Topic)	MI(Region)	P(Region)	MI(Region)	MI(Topic)	P(Topic)
Batch k-means	0.0075	0.0874	0.8253	0.1400	0.0093	0.9838
SCKM ($w = 0.0025$)	0.0003	0.1194	0.8253	0.0004	0.0075	0.9838
Normalised cut	0.0075	0.1510	0.8253	0.1862	0.0106	0.9838
NC+SCKM ($w = 0.05$)	<0.0001	0.1643	0.8253	<0.0001	0.0164	0.9838

algorithm (in our case, normalised cut) to fall in one of the two groupings of the collections, even though this tendency is sometimes less clear than in the case of the batch k-means.

The similarity of the outcome of the proposed algorithm (NC+SCKM) with the non avoided clustering (which, as it has been said before, is used as a indication of the quality of the clustering) is in all cases greatly increased over the soft constrained k-means results. Moreover, the results show how the introduction of this constrained phase has not any detrimental effect over the quality of the normalised cut results, and in fact improves them in all cases. As for the avoided grouping, the similarity of the results of our technique is still reduced, keeping it in values equal or less than those of the baseline, which were already low.

It should be also noted that the reason for the repeated values of P for the four methods in dataset (ii) is the structure of the dataset, where in each of the possible groupings one of the clusters is much bigger than the other (still, the MI values for that dataset attest the improvements attained using the combined method). Finally, it is also worth remarking that further tests on the training collection have shown that the parameter w of this combined approach is quite stable. This can be seen in Fig. 1(a), which shows that the MI with the avoided and non-avoided groupings are not affected to a greater extent by wide variations around the chosen value of 0.05.

The results of the tests performed with the approach introduced in Sect. 3.1 (which introduces the negative constraints in the core of the constrained normalised cut algorithm) are not included in Table 1 as the quality values achieved were poor and the value of the parameter β_N was very unstable. This is shown in Fig. 1(b): for almost all values of the parameter the similarity with the avoided grouping is much higher than with the non-avoided one, and for the values of β_N in which the two similarities come closer the quality of the result is very low and a small variation of the parameter produces an abrupt change in the quality values. Our intuition is that the cause of this behaviour has to do with the function which is minimised. With positive constraints, the function in (7) has its lower bound in zero, a value which, if obtained, would mean both that the clustering has good quality (NCut = 0) and that all the

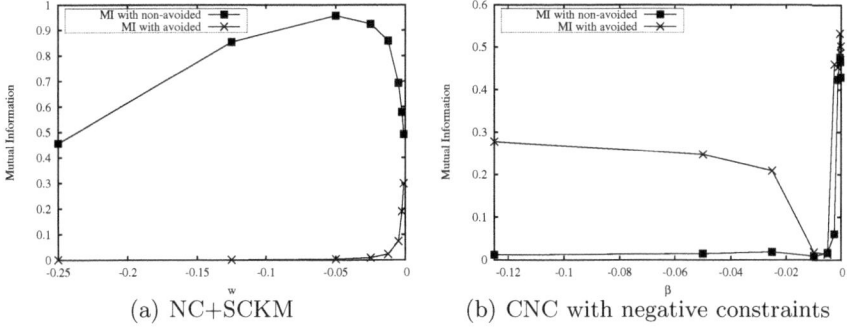

Fig. 1. Stability of the parameters of the two proposed algorithms in the training collection (Dataset (i), avoiding TOPIC)

constraints are respected ($||\beta U H||^2 = 0$). However, this is not what happens in the minimised function when negative constraints are involved (9). Here, a low value can be obtained if all the constraints are respected, regardless of the quality of the clustering, as one value is subtracted from the other. This makes tuning the value of β_N very hard, as a small change can alter dramatically the balance between those two factors.

5   Related Work

In the constrained clustering field [13], the problems of avoiding bias and finding alternative clusterings have gained popularity in the last years, with several authors looking into them and proposing different approaches. Bae and Bailey proposed in [14] a method similar to the one used in this paper, using negative constraints to try to steer the clustering away from the avoided grouping. They incorporate these constraints in a Average Link clustering algorithm, controlling with a parameter the compromise between obtaining a clustering of quality and honouring the constraints. However, they only report results in synthetic and numeric data collections with a very limited number of features.

Gondek and Hoffman introduced in [1] another strategy to find alternative clusters using Conditional Information Bottleneck clustering. Their approach tries to optimise an objective function which combines the objectives of yielding clusters of good quality and which should be different from the given clustering. To do so they need the complete distribution of each variable, which is one of the main drawbacks of the method.

In [15], Davidson and Qi present an approach to finding alternative clustering which also uses constraints, in this case to characterise the grouping to be avoided. A distance function matrix is learnt from these constraints, which is decomposed afterwards using Singular Value Decomposition (SVD). Finally, the matrices yielded by SVD are used to build an alternative distance function that is used to created transformed versions of the original data points, over which the clustering algorithm would be applied. Thus, this method has the advantage of being quite general, not being tied to any clustering clustering. Again, they tested their approach only in non-textual collections.

Cohn et al. introduced in [16] an algorithm to iteratively alter the grouping found by a clustering process according to the user feedback. They incorporate the user preferences altering the KL-divergence measure between the documents marked by the user, introducing a new factor to measure the importance of a term for distinguishing the documents. Even though they conduct their tests over textual documents, the collections are again very small.

Obviously, the avoiding bias method which is most related to the ones proposed in this paper is the one introduced by Ares et al. [3], which uses the soft constrained k-means algorithm, described in Sect. 2.3. It was used as baseline in our experiments (Sect. 4.5), and in one of the approaches proposed in this paper we have combined it with normalised cut 3.2. Another general constrained clustering algorithm which is also related to this paper is constrained normalised cut by Ji et al. [4], as it is the core of one of the Avoiding Bias methods proposed in this paper (Sect. 3.1). The unsuitability of that algorithm for the Avoiding Bias problem was discused in Sect. 4.5.

6   Conclusions

In this paper we have studied two approaches based on the use of negative constraints in conjunction with spectral clustering techniques to tackle the Avoiding Bias problem. While one of them, based in introducing the negative constraints in the core of constrained normalised clustering, did not yield good results, the second one, which combines normalised clustering and soft constrained clustering gave very good results in the experiments carried out, as it increased (in some cases dramatically) the quality of the clustering while maintaining a good avoidance of the known grouping. On a more general level, it should be noted that the possible fields of application of this approach are not limited to the Avoiding Bias problem on text. This algorithm can be applied in any general constrained clustering situation, where, opposed to constrained normalised cut (which would only allow the use of one kind of information), it lets the user use different kinds of knowledge (negative and positive, absolute and non absolute,...).

Acknowledgements. This work was co-funded by FEDER, Ministerio de Ciencia e Innovación, Xunta de Galicia and Ministerio de Educación under projects TIN2008-06566-C04-04 and 07SIN005206PR and FPU grant AP2007-02476.

References

1. Gondek, D., Hofmann, T.: Non-redundant data clustering. In: ICDM 2004: Proceedings of the Fourth IEEE International Conference on Data Mining, pp. 75–82. IEEE Computer Society, Los Alamitos (2004)
2. Davidson, I., Qi, Z.: Finding alternative clustering using constraints. In: ICDM 2008: Proceedings of the 2008 Eighth IEEE International Conference on Data Mining. IEEE Computer Society, Los Alamitos (2008)

3. Ares, M.E., Parapar, J., Barreiro, A.: Avoiding bias in text clustering using constrained k-means and may-not-links. In: Azzopardi, L., Kazai, G., Robertson, S., Rüger, S., Shokouhi, M., Song, D., Yilmaz, E. (eds.) ICTIR 2009. LNCS, vol. 5766, pp. 322–329. Springer, Heidelberg (2009)
4. Ji, X., Xu, W., Zhu, S.: Document clustering with prior knowledge. In: SIGIR 2006: Proceedings of the 29th Annual international ACM SIGIR conference on Research and development in information retrieval, pp. 405–412. ACM, New York (2006)
5. Shi, J., Malik, J.: Normalized cuts and image segmentation. IEEE Trans. Pattern Anal. Mach. Intell. 22(8), 888–905 (2000)
6. Ding, C.: A tutorial on spectral clustering. In: Tutorial presented at ICML 2004: 21st International Conference on Machine Learning (2004)
7. von Luxburg, U.: A tutorial on spectral clustering. Technical Report TR-149, Max Planck Institute for Biological Cybernetics (2006)
8. McQueen, J.: Some methods for classification and analysis of multivariate observations. In: Proceedings of the Fifth Berkeley Symposium on Mathematical Statistics and Probability, vol. 1, pp. 281–297 (1967)
9. Wagstaff, K., Cardie, C., Rogers, S., Schrödl, S.: Constrained k-means clustering with background knowledge. In: ICML 2001: Proceedings of the Eighteenth International Conference on Machine Learning, pp. 577–584, Morgan Kaufmann Publishers Inc., San Francisco (2001)
10. Pantel, P., Lin, D.: Document clustering with committees. In: SIGIR 2002: Proceedings of the 25th annual international ACM SIGIR conference on Research and development in information retrieval, pp. 199–206. ACM Press, New York (2002)
11. Rosell, M., Kann, V., Litton, J.E.: Comparing comparisons: Document clustering evaluation using two manual classifications. In: Proceedings of the International Conference on Natural Language Processing (2004)
12. Manning, C.D., Raghavan, P., Schtze, H.: Introduction to Information Retrieval. Cambridge University Press, New York (2008)
13. Basu, S., Davidson, I., Wagstaff, K.: Constrained Clustering: Advances in Algorithms, Theory, and Applications. Chapman & Hall/CRC, Boca Raton (2008)
14. Bae, E., Bailey, J.: COALA: A novel approach for the extraction of an alternate clustering of high quality and high dissimilarity. In: ICDM 2006: Proceedings of the Sixth International Conference on Data Mining, pp. 53–62. IEEE Computer Society, Los Alamitos (2006)
15. Davidson, I., Qi, Z.: Finding alternative clustering using constraints. In: ICDM 2008: Proceedings of the 2008 Eighth IEEE International Conference on Data Mining. IEEE Computer Society, Los Alamitos (2008)
16. Cohn, D., Caruana, R., McCallum, A.: Semi-supervised clustering with user feedback. Technical Report TR-2003-1892, Cornell University (2003)

An Efficient Similarity Join Algorithm with Cosine Similarity Predicate

Dongjoo Lee[1], Jaehui Park[1], Junho Shim[2], and Sang-goo Lee[1]

[1] School of Computer Science & Engineering,
Seoul National University, Seoul 151-742, Korea,
{therocks,jaehui,sglee}@europa.snu.ac.kr
[2] Dept of Computer Science, Sookmyung Women's University,
Seoul 140-742, Korea
jshim@sookmyung.ac.kr

Abstract. Given a large collection of objects, finding all pairs of similar objects, namely *similarity join*, is widely used to solve various problems in many application domains.Computation time of similarity join is critical issue, since similarity join requires computing similarity values for all possible pairs of objects. Several existing algorithms adopt *prefix filtering* to avoid unnecessary similarity computation; however, existing algorithms implementing the prefix filtering have inefficiency in filtering out object pairs, in particular, when aggregate weighted similarity function, such as *cosine similarity*, is used to quantify similarity values between objects. This is mostly caused by large prefixes the algorithms select. In this paper, we propose an alternative method to select small prefixes by exploiting the relationship between arithmetic mean and geometric mean of elements' weights. A new algorithm, *MMJoin*, implementing the proposed methods dramatically reduces the average size of prefixes without much overhead. Finally, it saves much computation time. We demonstrate that our algorithm outperforms a state-of-the-art one with empirical evaluation on large-scale real world datasets.

1 Introduction

Similarity join is an operation that finds all pairs of similar objects from given datasets. It is widely used to solve various problems in many application domains, such as data integration and cleansing [1,2], duplicate Web documents detection [3,4] and information retrieval [5].

More formally, similarity join can be defined as an operation that finds all pairs of objects whose similarity value quantified by the given *similarity function* is above the given *threshold* from the dataset. One issue of similarity join is how to quantify similarity values between objects. Various similarity functions, such as Jaccard-coefficient, cosine similarity, and edit similarity, are used to quantify similarity values between objects. In general, what similarity function to use depends on application domains and there is no best similarity function that works better than any other functions in all application domains. In [6],

P. García Bringas et al. (Eds.): DEXA 2010, Part II, LNCS 6262, pp. 422–436, 2010.

Chandel et al. grouped similarity functions into five classes, and showed accuracy and performance of similarity functions with various experimental analyses on textual data. From the results, aggregate weighted similarity function, such as cosine similarity showed comparatively good accuracy and performance in detecting errors from textual data. Also it was shown that cosine similarity produce high quality results across several domains [5,7,8,9]. In this paper, we focus on similarity join that uses cosine similarity to quantify similarity values between objects, especially when weights of elements need to be considered.

Another issue of similarity join is the computation time, since similarity join requires computing similarity values for all possible pairs of objects. Many of past researches used approximation techniques to reduce the running time of the operation, while undertaking some loss of expected answers; however, recent trend is to find all pairs of similar objects without any *false drop*. Many of recent works [4,10,11,12] in this trend adopt filtering techniques, such as *prefix filtering* and *positional filtering*, to avoid unnecessary similarity computation; however, positional filtering is not available when we should consider weights of elements. In addition, existing algorithms implementing the prefix filtering have inefficiency in filtering out object pairs, when weights of elements should be considered.

To our best knowledge, previously proposed *All-Pairs* algorithm, one of prefix filtering based methods, showed the best performance among algorithms applicable to our case [4]. From re-implementing the algorithm and analyzing experimental results with several datasets, we found out that prefix size strongly affects the running time of similarity join and All-Pairs has inefficiency in selecting prefixes. Therefore, we focused on how to select small prefixes, and finally, contrived an alternative prefix selection method by exploiting the relationship between arithmetic mean and geometric mean of elements' weights. A new algorithm, *MMJoin*, implementing the proposed prefix selection method reduces average prefix size without further overhead. Reduction of prefix size brings much more reduction of candidates, and finally saves much computation time. We demonstrate that MMJoin outperforms All-Pairs with empirical evaluation on large-scale real-world datasets.

The rest of the paper is organized as follows: Section 2 presents the problem definition with formal notations. Section 3 reviews an existing filtering-based approach. Section 4 describes our prefix selection method and similarity join algorithm. In Sect. 5, we demonstrate experimental results on large-scale real world datasets and give analyses about the results. Related work is covered in Sect. 6 and Sect. 7 concludes the paper.

2   Problem Statement

Given a set of objects $\mathcal{D}$, a similarity function $\text{sim}(x, y)$, and a similarity thresholds t, similarity join is defined as an operation that finds all pairs (x, y) such that $x, y \in \mathcal{D}$ and $\text{sim}(x, y) \geq t$, which is *similarity predicate*. We assume the similarity function is commutative. Thus, if the pair (x, y) satisfies the predicate, so does (y, x), and we need to include only one of them in the result.

According to the similarity function used to quantify similarities between objects, each object needs to be represented in a proper form. For example, *set* for overlap similarity, Jaccard coefficient and Dice coefficient, *vector* for cosine similarity and Tanimoto coefficient, and *sequence* for edit similarity. In this paper, we focus on aggregate weighted similarity function, particularly *cosine similarity*. Therefore, all objects are assumed to be weight vectors on pre-defined dimensions and denoted without right-pointing arrow in the rest of the paper. In addition, by default, sim(x, y) denotes cosine similarity between vectors x and y, unless otherwise stated. For simplicity, we assume all vectors have unit length. Then, given two vectors $x = \langle x[1], \ldots, x[m] \rangle$ and $y = \langle y[1], \ldots, y[m] \rangle$, cosine similarity is a dot product of two vectors as:

$$\text{sim}(x, y) = \frac{\sum_{i=1}^{m} x[i] \cdot y[i]}{\|x\| \|y\|} = \text{dot}(x, y) = \sum_{i=1}^{m} x[i] \cdot y[i], \quad (1)$$

where $x[i]$ and $y[i]$ are x's and y's weights on ith dimension respectively, and m is the total number of dimensions.

For many problem domains, especially those involving textual data, objects are *sparse* vectors where a vast majority of vector weights are 0. A sparse vector representation for a vector x is the set of all pairs $(i, x[i])$ such that $x[i] > 0$ over all $i = 1, \ldots, m$. Such pairs are called *features* of vector x. If there is a global ordering scheme $\mathcal{O}$ on dimensions $\mathcal{U}$, the sorted list of features is another representation of sparse vector. The *size* of a vector x, denoted by $|x|$, is the number of x's features. Vector size should not be confused with vector *length*, or *magnitude*, which is denoted by $\|x\|$.

For a given vector x, we denote the maximum value $x[i]$ over all i as maxw(x). For a given dimension i, we denote the maximum value $x[i]$ over all vectors x in the dataset $\mathcal{D}$ as maxw$_i(\mathcal{D})$. Let us consider an example dataset $\mathcal{X}$ shown in Fig. 1. $\mathcal{X}$ contains five weight vectors on dimensions $\mathcal{U} = \{A, \ldots, O\}$. In Fig. 1, we can see weights of five vectors (blanks mean zero weight), but also additional column and row for maxw$_i(\mathcal{X})$ and maxw(x) respectively.

Given a set of sparse vectors $\mathcal{D}$, an *inverted index* for the set consists of m lists $I_1, I_2, \ldots, I_m$ (one for each dimension), where list I_i, we simply refer *inverted list* for dimension i, includes pairs $(x, x[i])$ such that $x \in \mathcal{D}$ and $x[i] > 0$. For example, $I_A = \{(o_1, 0.46), (o_2, 0.46), (o_5, 0.15)\}$.

$\mathcal{U}$	A	B	C	D	E	F	G	H	I	J	K	L	M	N	O	maxw(x)
o_1	0.46	0.31			0.31			0.31		0.62	0.31				0.15	0.62
o_2	0.46	0.31		0.31	0.31			0.31		0.62					0.15	0.62
o_3		0.33		0.17		0.33	0.33		0.33	0.50	0.50		0.17			0.50
o_4		0.55	0.18	0.18	0.37	0.18		0.37	0.18		0.37		0.18		0.37	0.55
o_5	0.15		0.30		0.30			0.61				0.46		0.46		0.61
maxw$_i(\mathcal{X})$	0.46	0.55	0.30	0.31	0.37	0.33	0.33	0.61	0.33	0.62	0.50	0.46	0.18	0.46	0.37	

Fig. 1. Example dataset $\mathcal{X}$

For a given vector x, let $\dim(x)$ denote a set of all dimensions i such that $i \in \mathcal{U}$ and $x[i] > 0$. For example, $\dim(o_3) = \{B, D, F, G, I, J, K, M\}$. For given two vectors x and y, if $\dim(x) \cap \dim(y) \neq \emptyset$, then there exists at least one dimension i such that $x[i] > 0$ and $y[i] > 0$, which is equivalent to the predicate $\mathrm{dot}(x, y) > 0$. For example, $\dim(o_2) \cap \dim(o_3) = \{B, D, J\} \neq \emptyset$ and $\mathrm{dot}(o_2, o_3) = 0.465 > 0$, and $\dim(o_3) \cap \dim(o_5) = \emptyset$ and $\mathrm{dot}(o_3, o_5) = 0$.

For a given vector x, let the *prefix* of the vector be the first several features of x and denote it as x', and the *suffix* of the vector be the remaining features and denote it as x''. Accordingly, it is obvious that x' and x'' are also vectors, as $x' = \langle x[1], \ldots, x[p], 0, \ldots, 0 \rangle$ and $x'' = \langle 0, \ldots, 0, x[p+1], \ldots, x[m] \rangle$, where p is the last dimension on which x''s weight is nonzero. Prefix and suffix satisfy the followings;

- $x' + x'' = x$,
- $|x'| + |x''| = |x|$,
- $\|x'\|^2 + \|x''\|^2 = \|x\|^2$, and
- $\mathrm{dot}(x', y) + \mathrm{dot}(x'', y) = \mathrm{dot}(x, y) = \mathrm{sim}(x, y)$.

3   Filtering-Based Methods

A naïve approach to obtain similarity join result is to enumerate all possible pairs of vectors using *nested loops*, compute similarities of generated pairs, which we call *candidates*, and discard those whose similarity value is below the threshold. This approach generates total $\frac{n(n-1)}{2}$ candidates. Obviously, this approach is not feasible for large datasets due to the huge amount of comparisons.

An alternative approach may improve the performance of the similarity join by using *inverted index* used in IR community. We call this `InvertedIndexJoin` and its pseudo code is shown in Algorithm 1. While scanning each vector, InvertedIndexJoin dynamically constructs inverted index and accumulate similarity values in hash-based map by scanning the inverted index. This brings two benefits: this 1) guarantees that only one pair of (x, y) and (y, x) is considered, since each input vector is compared with vectors that had already been indexed in inverted lists and 2) reduces the overhead of scanning inverted lists, since size of inverted lists remains small in the early stage of the operation; however, still this approach is not feasible for large datasets, because this approach requires huge memory to keep the hash-based map for accumulating similarity values of candidates and yields much overhead to scan all inverted lists for each vector. Several existing algorithms improved the performance of InvertedIndexJoin by exploiting the threshold during matching and indexing.

For an input vector x, InvertedIndexJoin incrementally scans inverted lists from 1 to m such that $x[i] > 0$ (see line 6 - 8 of Algorithm 1.) Suppose that the operation is on dimension p and let x' and x'' denote the corresponding prefix and suffix for x, then $\mathrm{sim}(x, y) = \mathrm{dot}(x', y) + \mathrm{dot}(x'', y)$. [1] If $\mathrm{sim}(x, y) \geq t$ and $\mathrm{dot}(x'', y) < t$, then $\mathrm{dot}(x', y) > 0$, that is x' and y share at least one

---

[1] $x' = \langle x[1], \ldots, x[p], 0, \ldots, 0 \rangle$ and $x'' = \langle 0, \ldots, 0, x[p+1], \ldots, x[m] \rangle$.

Algorithm 1. InvertedIndexJoin$(\mathcal{D}, t)$

Input: a set of vectors $\mathcal{D}$, similarity threshold t
Output: $\{(x,y)|x,y \in \mathcal{D} \wedge \text{sim}(x,y) \geq t\}$
 1: $O \leftarrow \emptyset$
 2: $I_1, \ldots, I_m \leftarrow \emptyset$
 3: **for each** $x \in \mathcal{D}$ **do**
 4:     $C \leftarrow$ empty map from id to weight
 5:     **for** $i = 1$ **to** m such that $x[i] > 0$ **do**
 6:         **for each** $(y, y[i]) \in I_i$ **do**
 7:             $C[y] \leftarrow C[y] + x[i] \cdot y[i]$
 8:         **end for**
 9:         $I_i \leftarrow I_i \cup \{(x, x[i])\}$
10:     **end for**
11:     **for each** $y \in C$ **do**
12:         **if** $C[y] \geq t$ **then**
13:             $O \leftarrow O \cup \{(x,y)\}$
14:         **end if**
15:     **end for**
16: **end for**
17: **return** O

---

dimension. This is similar to the *prefix filtering principle* proposed in [2], which is based on the intuition that if two *canonicalized* objects are similar, some fragments of them should overlap with each other, otherwise the two objects cannot have enough overlap; however, the overlap-based prefix filtering principle does not cover aggregate weighted similarity functions, such as cosine similarity. Therefore, it needs to be extended to cover cosine similarity. Although Bayardo et al.[4] did not note explicitly that they used prefix filtering principle, their approach is on the similar intuition as the prefix filtering principle and we can extend the prefix filtering principle based on the notion used in [4]. An extended version of prefix filtering principle for aggregate weighted similarity functions is formalized in Lemma 1.

Lemma 1. (Aggregate Weighted Prefix Filtering Principle)
Consider two objects x and y, each of which is weight vector on dimension $\mathcal{U}$, which follows an ordering scheme $\mathcal{O}$. If $\text{dot}(x,y) \geq t$, then any x' and y', such that $\text{dot}(x'',y) < t$ and $\text{dot}(x,y'') < t$, share at least one dimension.

3.1   All-Pairs Algorithm

For a vector x, if we can determine the prefix x' such that $\text{dot}(x'',y) < t$ for all y in the dataset, we do not need to condier ys not observed until probing inverted lists I_i such that $i \in \dim(x')$. Also we only need to index x in inverted lists I_i such that $i \in \dim(x')$. Bayardo et al. used $\text{maxw}_i(\mathcal{D})$ to calculate the upper-bound of $\text{dot}(x'',y)$ for all y in the dataset as shown in (2).

$$\text{dot}(x'', y) = \sum_{i=p+1}^{m} x[i] \cdot y[i] \leq \sum_{i=p+1}^{m} x[i] \cdot \text{maxw}_i(\mathcal{D}) \tag{2}$$

Based on the `InvertedIndexJoin`, they devised an algorithm, `All-Pairs`, not only employing prefix filtering but also exploiting other factors affecting the performance. We rewrite the final version of All-Pairs as shown in Algorithm 2 and 3. Let us see briefly how they improved the performance of similarity join in addition to prefix filtering.

Algorithm 2. `All-Pairs`$(\mathcal{D}, t)$

Input: a set of vectors $\mathcal{D}$, similarity threshold t
Output: $\{(x, y) | x, y \in \mathcal{D} \wedge \text{sim}(x, y) \geq t\}$
1: Reorder the dimension $1 \ldots m$ such that dimension with the least non-zero entries in $\mathcal{D}$ appear first.
2: Denote the max. of $x[i]$ over all $x \in \mathcal{D}$ as $\text{maxw}_i(\mathcal{D})$.
3: Denote the max. of $x[i]$ for $1 \ldots m$ as $\text{maxw}(x)$.
4: $O \leftarrow \emptyset$
5: $I_1, I_2, \ldots, I_m \leftarrow \emptyset$
6: **for each** $x \in \mathcal{D}$ in decreasing order of $\text{maxw}(x)$ **do**
7:     $O \leftarrow O \cup \text{Find-Matches}(x, I_1, I_2, \ldots, I_m, t)$
8:     $b \leftarrow 0$
9:     **for** $i = m$ **to** 1 such that $x[i] > 0$ **do**
10:       $b \leftarrow b + x[i] \cdot \min(\text{maxw}(x), \text{maxw}_i(\mathcal{D}))$
11:       **if** $b \geq t$ **then**
12:         $I_i \leftarrow I_i \cup \{(x, x[i])\}$
13:       **end if**
14:     **end for**
15: **end for**
16: **return** O

Exploiting Specific Sort Order. Vectors are sequentially accessed in the algorithm. Suppose that vectors are sorted in decreasing order of $\text{maxw}(x)$ and accessed in that order. For a vector x, x is compared with indexed vectors before indexing it. After indexing x, vectors that are not accessed and indexed yet will be compared with x. Such vectors have smaller maximum weight $\text{maxw}(y)$ than x. Therefore, we can tighten the upper-bound of $\text{dot}(x'', y)$ for such ys that are not indexed yet as $\sum_{i=p+1}^{m} x[i] \cdot \min(\text{maxw}(x), \text{maxw}_i(\mathcal{D}))$ (line 11 of Algorithm 2.) We can determine prefix for x based on such upper-bound and the indexed amount for x can be reduced (line 9 - 15 of Algorithm 2.)

Size Filtering in Matching Phase. For two vectors x and y, if we know $|x|, |y|, \text{maxw}(x)$ and $\text{maxw}(y)$, we can obtain the upper-bound of $\text{dot}(x, y)$ as $\text{dot}(x, y) \leq \min(|x|, |y|) \cdot \text{maxw}(x) \cdot \text{maxw}(y) \leq |y| \cdot \text{maxw}(x)$. From this, we can obtain $|y| \cdot \text{maxw}(x) < t \leftrightarrow |y| < \frac{t}{\text{maxw}(x)} \rightarrow \text{dot}(x, y) < t$. Finally, when we probe inverted lists, we can remove y such that $|y| < \frac{t}{\text{maxw}(x)}$ from the inverted lists, since all remaining vectors to be compared with y have smaller or equal maximum weight over all dimensions than x (line 6 of Algorithm 3.)

Algorithm 3. Find-Matches$(x, I_1, I_2, \ldots, I_m, t)$

Input: a vector x, inverted lists $I_1, I_2, \ldots, I_m$, similarity threshold t
Output: $\{(x,y)|(y,y[i]) \in I_i \wedge \operatorname{sim}(x,y) \geq t\}$
1: $O \leftarrow \emptyset$
2: $C \leftarrow$ empty map from id to weight
3: $b \leftarrow \sum_{i=1}^{m} x[i] \cdot \operatorname{maxw}_i(\mathcal{D})$
4: $minsize \leftarrow \frac{t}{\operatorname{maxw}(x)}$
5: **for** $i = 1$ **to** m such that $x[i] > 0$ **do**
6:     Remove $(y, y[i])$ from I_i s.t. $|y| < minsize$
7:     **for each** $(y, y[i]) \in I_i$ **do**
8:         **if** $b \geq t$ **or** $C[y] > 0$ **then**
9:             $C[y] \leftarrow C[y] + x[i] \cdot y[i]$
10:         **end if**
11:     **end for**
12:     $b \leftarrow b - x[i] \cdot \operatorname{maxw}_i(\mathcal{D})$
13: **end for**
14: **for each** $y \in C$ **do**
15:     **if** $C[y] + \min(|x|, |y''|) \cdot \operatorname{maxw}(x) \cdot \operatorname{maxw}(y'') \geq t$ **then**
16:         **if** $C[y] + \operatorname{dot}(x, y'') \geq t$ **then**
17:             $O \leftarrow O \cup \{(x,y)\}$
18:         **end if**
19:     **end if**
20: **end for**
21: **return** O

Size Filtering in Verification Phase. Let x be input vector, y be one of indexed vectors and y' be indexed part, that is, prefix for y, then $\operatorname{dot}(x, y')$ is obtained in $C[y]$ after probing inverted lists (line 9 of Algorithm 3.) Then we can obtain $\operatorname{sim}(x, y)$ by adding $\operatorname{dot}(x, y'')$ to $C[y]$, where y'' is un-indexed part, that is, suffix for y. Before calculating $\operatorname{dot}(x, y'')$, we may avoid unnecessary computation for $\operatorname{dot}(x, y'')$ by calculating the upper-bound of $\operatorname{dot}(x, y'')$ as $\min(|x|, |y''|) \cdot \operatorname{maxw}(x) \cdot \operatorname{maxw}(y'')$ (line 15 of Algorithm 3.)

4   MMJoin

In this section, we describe what mostly affects the performance of All-Pairs and how we improved the performance of All-Pairs.

4.1   Effects of Filtering Techniques on Candidates Size

To see how each technique used in All-Pairs affects the performance of the algorithm, we implemented four versions of All-Pairs:

- **All-Pairs-0** implements basic prefix filtering method.
- **All-Pairs-1** exploits sort order of vectors based on All-Pairs-0.
- **All-Pairs-2** adopts size filtering in matching phase based on All-Pairs-1.
- **All-Pairs** adopts size filtering in verification phase based on All-Pairs-2.

We ran each version of All-Pairs over two datasets with varying the threshold (Details about the datasets are described in Sect. 5) and measured the total number of candidates each algorithm generates to see how much reduction of candidates is made by applying each technique.

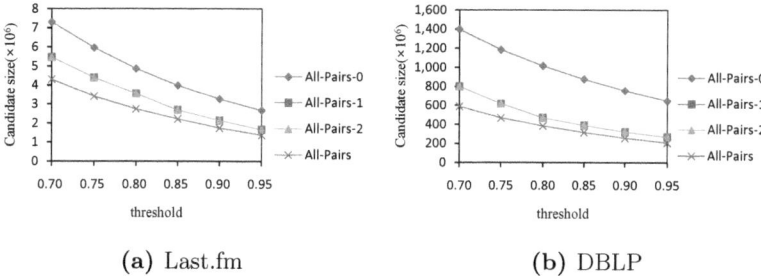

(a) Last.fm   (b) DBLP

Fig. 2. Effect of each technique on the number of candidates

In general, as shown in Fig. 2, exploiting the specific sort order of vectors brings more performance gain than others. Size filtering in matching phase has almost no effect on reducing candidate size. Although size filtering in verification phase reduces the number of candidates, we cannot except it brings as much reduction of time as candidates, since it needs overhead to compute upper-bound for all vectors remains until verification phases.

Besides, we measured average prefix size of All-Pairs-0 and could make an interesting observation about the relationship between average prefix size and the number of candidates All-Pairs-0 generates. In general, cube (or forth power) of average prefix size is almost proportional to the total number of candidates as shown in Fig. 3. This means that if we reduce average prefix size even *a little*, we will obtain *much* performance improvements. Therefore, we focused on reducing the size of prefixes.

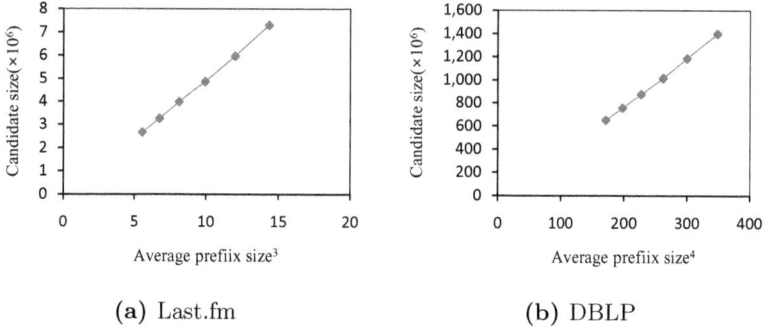

(a) Last.fm   (b) DBLP

Fig. 3. Effects of average prefix size on total number of candidates

4.2   Tightening Similarity Upper-Bound

If we do not consider weights of elements for quantifying similarity values between objects, prefix size is determined by only the given threshold based on the
overlap-based prefix filtering principle [2]; however, when we consider weights of
elements, prefix size varies according to how we calculate the upper-bound of
$\mathrm{dot}(x'', y)$ for all y following the Lemma 1.

In All-Pairs, for a vector x, $\mathrm{maxw}(x)$ and $\mathrm{maxw}_i(\mathcal{D})$ is used to calculate the
upper-bound of $\mathrm{dot}(x'', y)$ for all y to be compared with x. Let $M(x)$ be a vector
whose ith weight is $\min(\mathrm{maxw}(x), \mathrm{maxw}_i(\mathcal{D}))$. Accordingly, $\mathrm{dot}(x'', M(x)) =
\mathrm{dot}(x, M(x)) - \mathrm{dot}(x', M(x))$, and it can be thought that prefix for a vector is
determined by adding feature one by one until $\mathrm{dot}(x, M(x)) - \mathrm{dot}(x', M(x)) < t$.
From this, we can suppose that if $\mathrm{dot}(x, M(x))$ is big, many features need to be
included in prefix to fulfill the predicate. As a result, prefix becomes large. This
situation is easy to happen when the size of a vector is large. To overcome the
weakness of All-Pairs's prefix selection, we contrived an alternative method to
calculate the upper-bound of $\mathrm{dot}(x'', y)$ by exploiting the arithmetic mean and
geometric mean of elements' weights.

Once again, recall that $\mathrm{sim}(x, y) = \mathrm{dot}(x', y) + \mathrm{dot}(x'', y)$. Let y' and y'' be the
prefix and the suffix of y, each of which corresponds to x' and x'' respectively.
Then, obviously, $\mathrm{dot}(x'', y) = \mathrm{dot}(x'', y'')$. $\mathrm{dot}(x'', y'')$ can be rewritten as:

$$\mathrm{dot}(x'', y'') = x[p+1] \cdot y[p+1] + \ldots + x[m] \cdot y[m]. \tag{3}$$

By using the relationship between arithmetic mean and geometric mean of elements' weights, we can obtain the upper-bound of $\mathrm{dot}(x'', y'')$ as shown in (4).

$$
\begin{aligned}
& x[p+1] \cdot y[p+1] + \ldots + x[m] \cdot y[m] \\
\leq\ & \frac{x[p+1]^2 + y[p+1]^2}{2} + \ldots + \frac{x[m]^2 + y[m]^2}{2} \\
=\ & \frac{x[p+1]^2 + \ldots + x[m]^2}{2} + \frac{y[p+1]^2 + \ldots + y[m]^2}{2} \\
=\ & \frac{\|x''\|^2 + \|y''\|^2}{2}
\end{aligned} \tag{4}
$$

With $\|x\|^2 = \|x'\|^2 + \|x''\|^2 = 1$ and $\|x\|^2 \geq 0$ for all vectors, the upper-bound
of $\mathrm{dot}(x'', y'')$ can be calculated as (5).

$$
\begin{aligned}
\mathrm{dot}(x'', y'') \leq\ & \frac{\|x''\|^2 + \|y''\|^2}{2} = 1 - \frac{1}{2}\|x'\|^2 - \frac{1}{2}\|y'\|^2 \\
\leq\ & 1 - \frac{1}{2}\|x'\|^2
\end{aligned} \tag{5}
$$

Let $\mathrm{ubdot}(x'')$ denote upper-bound of $\mathrm{dot}(x'', y)$ for all y to be compared with
x. Then $\mathrm{ubdot}(x'') = \min(\mathrm{dot}(x, M(x)) - \mathrm{dot}(x', M(x)), 1 - \frac{1}{2}\|x'\|^2)$.

4.3   MMJoin Algorithm

Our algorithm is almost same with All-Pairs and still exploits its merits, because we only changed the way of selecting prefixes as shown in line 8-14 of Algorithm 4. Code for selecting prefix in matching phase of MMJoin-Find-Matches is almost same except that $\sum_{i=p+1}^{m} x[i] \cdot \text{maxw}_i(\mathcal{D})$ is used as $M(x)$ instead of $\sum_{i=p+1}^{m} x[i] \cdot \min(\text{maxw}(x), \text{maxw}_i(\mathcal{D}))$. Therefore we omit MMJoin-Find-Matches in this paper.

Algorithm 4. MMJoin($\mathcal{D}, t$)

Input: $\mathcal{D} = \{o_1, o_2, \ldots, o_n\}$, similarity threshold t
Output: $\{(x, y) | x, y \in \mathcal{D} \wedge \text{sim}(x, y) \geq t\}$

 1: Reorder the dimension $1 \ldots m$ such that dimension with the least non-zero entries in $\mathcal{D}$ appear first
 2: Denote the max. of $x[i]$ over all $x \in \mathcal{D}$ as $\text{maxw}_i(\mathcal{D})$
 3: Denote the max. of $x[i]$ for $1 \ldots m$ as $\text{maxw}(x)$
 4: $O \leftarrow \emptyset$
 5: $I_1, I_2, \ldots, I_m \leftarrow \emptyset$
 6: **for each** $x \in \mathcal{D}$ in decreasing order of $\text{maxw}(x)$ **do**
 7:    $O \leftarrow O \cup \text{MMJoin-Find-Matches}(x, I_1, I_2, \ldots, I_m, t)$
 8:    $b_1 \leftarrow \sum_{i=1}^{m} x[i] \cdot \min(\text{maxw}_i(\mathcal{D}), \text{maxw}(x))$
 9:    $b_2 \leftarrow 1$
10:    **for** $i = 1$ **to** m s.t. $x[i] > 0$ **while** $\min(b_1, b_2) \geq t$ **do**
11:        $b_1 \leftarrow b_1 - x[i] \cdot \min(\text{maxw}(x), \text{maxw}_i(\mathcal{D}))$
12:        $b_2 \leftarrow b_2 - \frac{1}{2}x[i]^2$
13:        $I_i \leftarrow I_i \cup \{(x, x[i])\}$
14:    **end for**
15: **end for**
16: **return** O

---

5   Experimental Evaluation

In this section, we compare the performance of MMJoin with All-Pairs. We do not compare with other algorithms, since All-Pairs shows the best performance among previous algorithms applicable to our cases[4,8].

5.1   Experimental Setup

We implemented all algorithms in Java 1.6 and used the standard java libraries to implement several data structures used in algorithms. All experiments were performed on a server with 2.83 GHz Intel Core2 Quad, 8 Gbytes of RAM and two 7200 RPM SATA II-IDE hard drives. The operating system is Windows Server 2003.

We ran two algorithms on five real world datasets to cover a wide spectrum of different characteristics. Some important statistics of datasets are summarized in Table 1.

Table 1. Statistics of Datasets

Dataset	n	avg_len	$|\mathcal{U}|$	avg_DF
DBLP	1,298,016	8.6	381,450	29.3
DBLP 4GRAM		23.9	135,204	224.5
LAST.FM	134,949	4.8	47,295	13.8
LAST.FM 4GRAM		11.2	44,272	34.3
TREC	348,566	77.1	298,302	90.1

DBLP is a snapshot of the bibliography records from the DBLP Web site[2]. It contains almost 1.3M records; each record is a concatenation of author name(s) and the title of a publication. We tokenized each record using white spaces and punctuations. The same DBLP dataset (with smaller size) was also used in previous studies [11,4,12,10,13].

TREC is from TREC-9 Filtering Track Collections[3]. It contains 0.35M references from the MEDLINE database. We extracted author, title, and abstract fields from records. Records are subsequently tokenized as in DBLP.

LAST.FM was gathered from last.fm web site[4]. It contains 0.13M randomly selected music tracks including artists and title. Each track is subsequently tokenized as in DBLP.

We made two additional datasets **DBLP 4GRAM** and **LAST.FM 4GRAM**, which are tokenized into 4-grams from DBLP and LAST.FM respectively. In particular, we extracted each 4-gram from tokens that had already been extracted with spaces and punctuations. After extracting tokens, we assigned weights on tokens based on tf-idf weighting scheme[14].

5.2   Experimental Results and Analysis

We ran All-Pairs and MMJoin over five datasets with varying thresholds from 0.70 to 0.95 by 0.05. We ran in-memory algorithm over LAST.FM, LAST.FM 4GRAM, and DBLP datasets; however, we could not use in-memory algorithm over DBLP 4GRAM and TREC in spite of we ran algorithms with excessive memory. Therefore, we ran disk-resident algorithm over DBLP 4GRAM and TREC. Disk resident version of All-Pairs and MMJoin were implemented in the same manner proposed in [4].

As discussed in Sect. 4.1, average prefix size affects the most on the candidate size. Also most time for similarity join is spent for calculating similarity values of candidates. Therefore, we focused on seeing these sequential effects by measuring

[2] Available at http://dblp.uni-trier.de/xml/
[3] Available at http://trec.nist.gov/data/t9_filtering.html
[4] http://www.last.fm/

average prefix size, total number of candidates, and total running time (see Fig. 4.) We can observe expected results in all datasets and parameters. Simple analyses about the experimental results is presented in following sub-sections.

Prefix Size. It is observed that the average prefix size increases when the threshold decreases(See Fig. 4a to 4c.) The average prefix size of MMJoin grows faster than that of All-Pairs when threshold decreases; however, the starting point of the MMJoin's average prefix size is much smaller than All-Pairs, especially when the average vector size is larer. In addition, MMJoin never generate bigger prefixes than All-Pairs as proved in Sect. 4.2.

Candidate Size and Time. Figure 4d to 4i shows the number of candidates generated by the algorithms and the time to complete the similarity join with varying the thresholds. We can make identical observations that had been made in previous work; the size of the join result grows modestly when the similarity threshold decreases, and all algorithms generate more candidate pairs with the decrease of the similarity threshold. Besides, time to complete the similarity

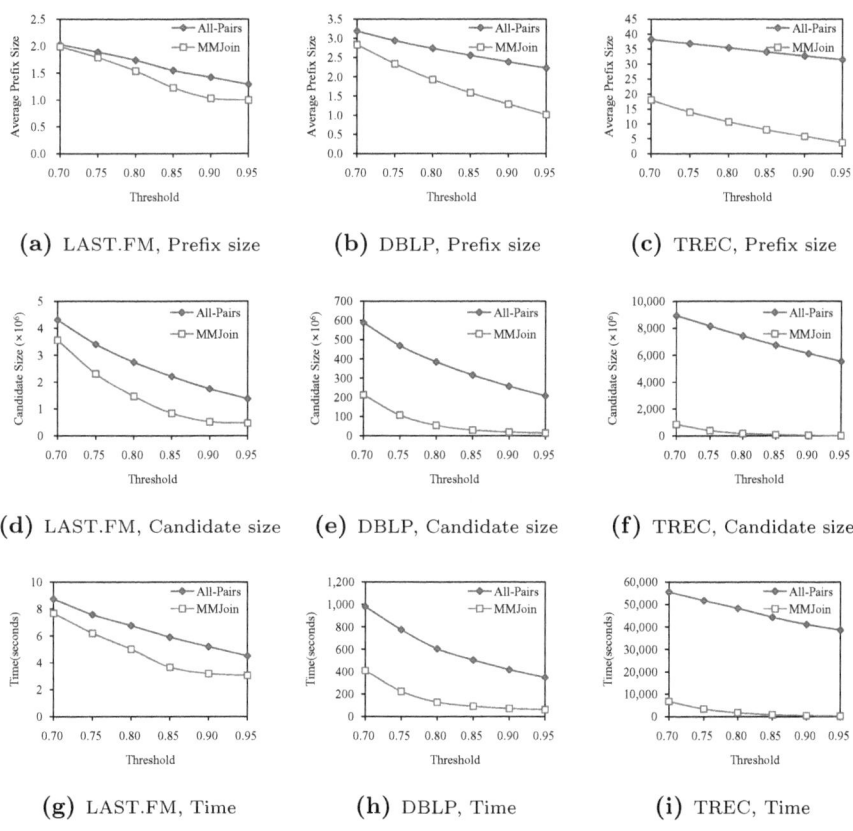

Fig. 4. Experimental Results

join mostly depends on the candidate size. This means that time to generate candidates do not occupy much portion of total running time in both algorithms. In all situations, MMJoin generates much smaller candidates than All-Pairs.

Performance Differences. MMJoin shows better performance than All-Pairs. It is much strongly observed when the average length of vectors is larger and the threshold is greater as shown in Fig. 5. Even in the case of TREC with threshold 0.95, speed-up of MMJoin is about 166x. This is exactly what we expected in Sect. 4.2.

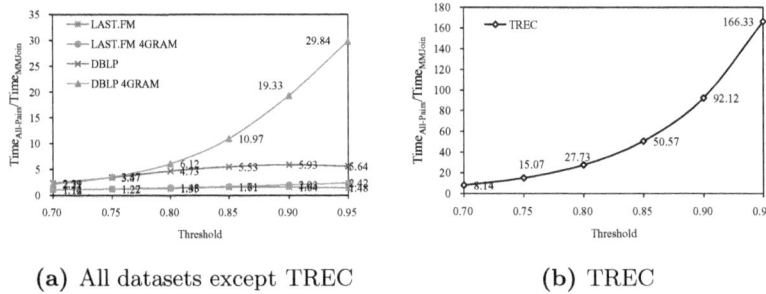

(a) All datasets except TREC   (b) TREC

Fig. 5. Performance differences between All-Pairs and MMJoin with varying threholds

6   Related Work

Early studies in similarity join were limited to binary similarity functions for sets including strict containment [15,16,17], equality [16], and non-zero overlap joins [16]. There are found several recent work that covers various partial overlap predicates using a variety of similarity functions including Jaccard coefficient, cosine similarity, edit distance, Hamming distance and their variants [4,12,10,13]. Similarity join in multi-dimensional non-binary space has also been studied[18,1]. [6] compared a large number of similarity functions experimentally with an evaluation on their performance and accuracy.

There is extensive related work in the IR community on designing efficient methods for indexing and compressing textual data [19] viewed as a set. Recent studies showed that this approach is effective to design efficient algorithms realizing similarity join [11,12,10,13]. Most of them use small part of object to reduce the number of objects that have to be fully compared with input object. Sarawagi et al. proposed a simple prefix filtering based algorithms with fully constructed inverted index[12]. Bayardo et al. improved the prefix filtering by dynamically constructing inverted index as well as considering other factors affecting the performance [4]. Recently proposed positional filtering shows remarkable performance improvements on similarity join with set-based similarity functions including Jaccard coefficient, overlap distance, and edit distance [10,13]; however positional filtering is not applicable to weighted cases, since they directly change positional information to measure similarities. To the best of our knowledge, All-Pairs is the best algorithm that is applicable to weighted similarity measures. We extended and improved the All-Pairs adapting a novel prefix selection method.

Our algorithm solves the similarity join problem with assuring that all similar pairs satisfying the given constraints are detected. Another line of work is to solve the similarity join problem with approximation. Locality Sensitive Hashing (LSH) [20] that is widely used in nearest neighbor search can be adapted to similarity join [11,4]. In [4], shingle-based technique was used to detect near duplicated web pages. Several alternatives are proposed to improve hash-based approaches [6].

7   Conclusion

Similarity join has been used in a wide range of application domains such as data integration and cleaning, pattern recognition, and information retrieval. Various similarity functions are used to define join conditions of similarity join. In this paper, we focused on an efficient algorithm for similarity join with cosine similarity predicate. We analyzed the previous algorithms and found out that the most critical process that affects the performance of similarity join is prefix selection. We contrived a novel prefix selection method that efficiently reduces the amount of indexed prefix size by exploiting the relationship between arithmetic mean and geometric mean of elements' weights. We proposed an algorithm, MMJoin, that implements our prefix selection method. Although we refined small part of previous algorithm, we obtained much performance gain through it. We demonstrated that the proposed algorithm outperforms state-of-the-art algorithm with empirical evaluation on large-scale real-world datasets.

Acknowledgments

This research was supported by the MKE(The Ministry of Knowledge Economy), Korea, under the ITRC(Information Technology Research Center) support program supervised by the NIPA(National IT Industry Promotion Agency). (grant number NIPA-2010-C1090-1031-0002).

References

1. Chaudhuri, S., Chen, B.C., Ganti, V., Kaushik, R.: Example-driven design of efficient record matching queries. In: VLDB (2007)
2. Chaudhuri, S., Ganti, V., Kaushik, R.: A primitive operator for similarity joins in data cleaning. In: ICDE (2006)
3. Henzinger, M.: Finding near-duplicate web pages: a large-scale evaluation of algorithms. In: SIGIR (2006)
4. Bayardo, R.J., Ma, Y., Srikant, R.: Scaling up all pairs similarity search. In: WWW (2007)

5. Chien, S., Immorlica, N.: Semantic similarity between search engine queries using temporal correlation. In: WWW (2005)
6. Chandel, A., Hassanzadeh, O., Koudas, N., Sadoghi, M., Srivastava, D.: Benchmarking declarative approximate selection predicates. In: SIGMOD (2007)
7. Chuang, S.L., Chien, L.F.: Taxonomy generation for text segments: A practical web-based approach. ACM Trans. Inf. Syst. 23(4), 363–396 (2005)
8. Sahami, M., Heilman, T.D.: A web-based kernel function for measuring the similarity of short text snippets. In: WWW (2006)
9. Spertus, E., Sahami, M., Buyukkokten, O.: Evaluating similarity measures: a large-scale study in the orkut social network. In: KDD (2005)
10. Xiao, C., Wang, W., Lin, X., Yu, J.X.: Efficient similarity joins for near duplicate detection. In: WWW (2008)
11. Arasu, A., Ganti, V., Kaushik, R.: Efficient exact set-similarity joins. In: VLDB (2006)
12. Sarawagi, S., Kirpal, A.: Efficient set joins on similarity predicates. In: SIGMOD (2004)
13. Xiao, C., Wang, W., Lin, X.: Ed-join: an efficient algorithm for similarity joins with edit distance constraints. In: VLDB (2008)
14. Jones, K.S.: A statistical interpretation of term specificity and its application in retrieval. Taylor Graham Series in Foundations of Information Science, pp. 132–142 (1988)
15. Helmer, S., Moerkotte, G.: Evaluation of main memory join algorithms for joins with set comparison join predicates. In: VLDB (1997)
16. Mamoulis, N.: Efficient processing of joins on set-valued attributes. In: SIGMOD (2003)
17. Ramasamy, K., Patel, J.M., Naughton, J.F., Kaushik, R.: Set containment joins: The good, the bad and the ugly. In: VLDB (2000)
18. Böhm, C., Braunmüller, B., Krebs, F., Kriegel, H.P.: Epsilon grid order: an algorithm for the similarity join on massive high-dimensional data. SIGMOD Rec. 30(2), 379–388 (2001)
19. Hersh, W.: Managing gigabytes—compressing and indexing documents and images (second edition). Inf. Retr. 4(1), 79–80 (2001)
20. Gionis, A., Indyk, P., Motwani, R.: Similarity search in high dimensions via hashing. In: VLDB (1999)

An Efficient Algorithm for Reverse Furthest Neighbors Query with Metric Index

Jianquan Liu[1], Hanxiong Chen[1], Kazutaka Furuse[1], and Hiroyuki Kitagawa[1,2]

[1] Department of Computer Science, Graduate School of SIE, University of Tsukuba,
1-1-1 Tennohdai, Tsukuba, Ibaraki 305-8573, Japan
[2] Center for Computational Sciences, University of Tsukuba,
1-1-1 Tennohdai, Tsukuba, Ibaraki 305-8573, Japan
ljq@dblab.is.tsukuba.ac.jp,
{chx,furuse,kitagawa}@cs.tsukuba.ac.jp

Abstract. The variants of similarity queries have been widely studied in recent decade, such as k-nearest neighbors (k-NN), range query, reverse nearest neighbors (RNN), an so on. Nowadays, the reverse furthest neighbor (RFN) query is attracting more attention because of its applicability. Given an object set O and a query object q, the RFN query retrieves the objects of O, which take q as their furthest neighbor. Yao et al. proposed R-tree based algorithms to handle the RFN query using Voronoi diagrams and the convex hull property of dataset. However, computing the convex hull and executing range query on R-tree are very expensive on the fly. In this paper, we propose an efficient algorithm for RFN query with metric index. We also adapt the convex hull property to enhance the efficiency, but its computation is not on the fly. We select external pivots to construct metric indexes, and employ the triangle inequality to do efficient pruning by using the metric indexes. Experimental evaluations on both synthetic and real datasets are performed to confirm the efficiency and scalability.

Keywords: Similarity search, Reverse furthest neighbors, Metric index, Convex hull.

1 Introduction

Similarity search has been well studied in the past decade, with its emerging application to the scientific researches and developments, such as pattern recognition [1,2], image retrieval [3], time-series matching [4] and the like. The variants of similarity queries include k-nearest neighbors (k-NN), range query, reverse nearest neighbors (RNN), and so on. Driven by the emerging requirement, a large number of techniques are developed for these query types to enhance the processing efficiency on the fly. They can be roughly divided into three categories, space-partitioning methods (e.g., grid-file [5]), data-indexing methods (e.g., R-tree [6]), and efficient methods for sequential scan (e.g., VA-file [7]). The authors usually perform their experimental evaluations, and naturally correspond to the opposite query type — "furthest" version as extension or future work. To the best of our knowledge, throughout the recent decade, the query type of reverse furthest neighbors (RFN) was out of concentration.

P. García Bringas et al. (Eds.): DEXA 2010, Part II, LNCS 6262, pp. 437–451, 2010.

Yao *et al.* fully identified the RFN query in [8]. For a large object set O and any random query object q (q may not belong to O), they formally retrieve the set of objects in O, which take q as their furthest neighbors among all objects in O, i.e., the reverse furthest neighbors (RFN). The variants of RFN queries are categorized as *monochromatic reverse furthest neighbors* (MRFN) query (i.e., RFN), and *bichromatic reverse furthest neighbors* (BRFN) query. The bichromatic version can be simply described as follows. Given a query set Q, a specified query $q \in Q$ and an object set O, the BRFN query is to retrieve the set of objects $o \in O$ that take q as their furthest neighbors, *comparing to the other objects in Q*. R-tree based algorithms are proposed to challenge the MRFN and BRFN problems in [8].

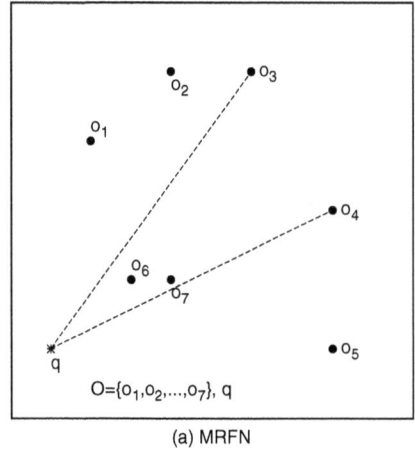

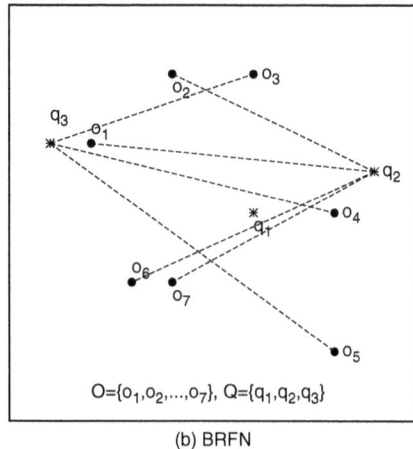

(a) MRFN

(b) BRFN

Fig. 1. Query examples of (a) MRFN, and (b) BRFN

In (a) of Figure 1, given the object set O and the query object q, then the answer to the RFN of q is $\{o_3, o_4\}$. Simply, the processing computes all the pairs of distances between q and o_i. As illustrated by the dotted lines, the objects o_3 and o_4 hold the furthest distance to q than other objects in O thus are included in the answer set.

The query example of BRFN is illustrated in (b) of Figure 1. The object set O is the same as in (a), but a different query object set $Q = \{q_1, q_2, q_3\}$ is given. Then the BRFN query returns the answer set corresponding to each $q \in Q$. For instance, pick up q_2, the the answer to its BRFN is $\{o_1, o_2, o_6, o_7\}$. In the figure, the dotted lines indicate the furthest distances from $o \in O$ to q_2 by comparing to the other $q \in Q$. Similarly, q_3's BRFN is $\{o_3, o_4, o_5\}$. However, the answer to q_1's BRFN becomes empty in this example, because all the object $o \in O$ does not take q_1 as their furthest neighbor w.r.t Q.

To give clear introduction to the RFN problems, besides the query examples reproduced above, we extend and emphasize the real applications as well, based on [8].

Application 1: Consider the application to tourism. The tourist would like to go shopping around their visiting sites. Usually, they firstly choose the nearest places unless for special reason. In this case, as the owner of a shop, he/she certainly hopes to be accessed by as many visitors as possible. Therefore, he/she should make more efforts to advertise his/her shop at the places (i.e., its reverse furthest neighbors), where the tourist most unlikely to visit his/her shop.

Application 2: In the approving authority of the government, e.g., the urban planning department, the officials have to approve the building applications from different industries. For instance, suppose that a building plan of constructing a chemical factory is submitted. The government must consider the hazardous influence to the local citizen. In this case, the potential locations are given as Q, and the residential location set O are known in hand. It is necessary to approve such a location $q \in Q$ that the citizen are far away from as many as possible. The decision comes to the location $q \in Q$ which has the maximum number of reverse furthest $o \in O$.

Application 3: The RFN query can be also applied to mine the correlations between customers and business companies. For example, the PC makers would make special marketing strategies to enlarge their profit. Given the customer set O who are the potential customer of their products, and the company set Q are the PC makers. Assume there is a quantized distance measure for the feedback from the customers to the makers. To sale more PCs, it is important for each maker to know potential customers who dislike their PC, and then carry out special strategy towards those customers. Here, the BRFN query helps to solve this problem for the PC makers.

Following Yao *et al.*'s concentration on the RFN problems, in this work, we motivate to enhance the processing efficiency against their algorithms. To make the improvement simple and understandable, we only focus on the MRFN query type in this paper. Indeed, the improved algorithm for MRFN can be easily applied to BRFN query. The main contributions of this work are as follows.

1. We analyze the expensive cost in the algorithms proposed in [8].
2. We summarize and extend the special properties of RFN query, and develop theoretical filtering.
3. Based on the properties, we design an efficient algorithm for RFN query, using the selected pivots with their metric indexes.
4. We perform extensive experiments on both synthetic and real data to evaluate the efficiency and I/O cost for the proposed algorithm.

The paper is organized as follows. Section 2 surveys the related work and analyze the expensive cost against the proposed algorithms for RFN query. Then we summarize the special properties of RFN query proposed in [8], and extend convex hull property for our approach in Section 3. Based on the properties, we propose a novel algorithm to enhance the processing efficiency for RFN query in Section 4. Experimental evaluations to confirm the efficiency and I/O cost are performed in Section 5. Section 6 comes to conclude this work and go into perspective of the future work.

2   Related Work

Techniques for similarity search have been extensively studied in the literature [7,9,10,11]. The similarity query types close to this work can be briefly classified in the following three categories: (k-)nearest neighbor(s) (NN, or k-NN), reverse (k-)nearest neighbors (RNN, or RkNN), and reverse furthest neighbors (RFN).

NN query. The early work proposed depth-first search based [12,13] and best-first search based [14] algorithms to answer a (k)NN query on a R-tree [6] indexed dataset. In these algorithms, the main idea is to determine whether the node should be visited according to the minimum distance between the query q and a R-tree node (MBR, or a data object). Due to the application value, various methods are still developed to speed up NN query in latest work like [15,16]. Athitsos et al. [15] proposed a space-indexing method using distance-based hashing to achieve efficient approximate nearest neighbor retrieval. Tao et al. [16] propose a new access method LSB-tree to enable fast high-dimensional nearest neighbor search with excellent quality.

RNN query. As an interesting type close to NN query, the RNN query is also attracted much attention [17,18,19,20,21,22,23,24]. The RNN query is firstly introduced by Korn and Muthukrishnan [17], who proposed the RNN-Tree to facilitate the query processing. Then, RdNN-Tree [18], TPL algorithm [21], MRkNNCop-Tree [22] are proposed to efficiently answer the RNN query. Many applications of RNN query are reported in [17,19,20,23]. Coming to the state of the art, Wu et al. [24] proposed the FINCH method to apply any RkNN algorithm to query processing, and evaluated RkNN queries on location data as well.

RFN query. The work most closely related to this paper is the recent work [8], which originally defined two RFN query types, MRFN and BRFN. The authors proposed the progressive furthest cell (PFC) algorithm and the convex hull furthest cell (CHFC) algorithm to handle RFN query. Both adopting R-tree index, and using the furthest Voronoi cell (fvc) to determine whether the points $o \in O$ are q's RFN. The $fvc(q, O)$ is to define a convex polygon w.r.t the query q in the given space of dataset O. To compute the $fvc(q, O)$, firstly draw the bisector line of each line segment oq ($o \in O$), then the space is separated into two subspaces, the $fvc(q, O)$ takes the intersection of all subspaces far away from the query q. The $fvc(q, O)$ strictly limits the answer set if and only if the point $o \in fvc(q, O)$. Straightforwardly, the authors proposed the PFC algorithm to compute $fvc(q, O)$ with the R-tree for each given query q on the fly. They also pointed out that the post-processing of PFC algorithm is expensive, so that they designed the faster one (CHFC) by deriving the important convex hull property of RFN query. The CHFC algorithm is represented in Figure 2.

Algorithm: CHFC(Query q; R-tree T)

---
1 Compute C_P with T using either the distance-priority or the depth-first algorithm;
2 if $q \subset C_P$ then return $\emptyset$;
3 else {
4   Compute C_{P*} using $C_P \cup \{q\}$;
5   Set $fvc(q, P^*)$ equal to $fvc(q, C_{P*})$;
6   Execute a range query using $fvc(q, P^*)$ on T;
7 }

Fig. 2. CHFC algorithm

In spite of the improvement against PFC, the CHFC algorithm is still expensive, because it has to compute the convex hull for each query q (line 1 and 4). The computation for executing a range query on R-tree (line 6) is expensive as well. These complaint can be confirmed by the illustration in Figure 6(a) based on our experiment result. Motivated by this, we propose a novel method 1) to avoid all the computation for convex hull, 2) to avoid executing range query on R-tree, 3) to reduce the distance computation on the fly as much as possible.

3 Theoretical RFN Filtering

In this section, we formally define the RFN query, summarize its special properties, and extend the convex hull property to explain our approach as well. As mentioned in Section 2, the furthest Voronoi cell ($fvc(q, O)$) is one of the properties that can tightly determine whether the points $o \in O$ are the answer to q's RFN. The other is the convex hull property of RFN query, which is explained by several lemmas in [8]. To re-use this property for our approach, we briefly conclude it as a basic Theorem 1. The complete proof of Theorem 1 can be confirmed in [8].

Definition 1. *Given an object set O and a query q, the answer to q's reverse furthest neighbors is defined by $RFN(q) = \{o | o \in O \land \forall p \in O, dist(o, q) > dist(o, p)\}$.*

Theorem 1. *Given an object set O, its convex hull C_O, and an arbitrary query q, the answer set to q's RFN is empty, if q is inside C_O.*

Theorem 1 essentially tells that *only if* q is on the boundary of, or outside the convex hull C_O, the answer to q's RFN maybe found. Along this property, we can perform powerful filtering to avoid range query using R-tree and hence reduce much distance computation. To achieve these purposes, we extend Theorem 1 to derive two filtering techniques as described in Lemma 1 and Lemma 2.

We assume that all vertices of the convex hull C_O are selected into a pivot set S_{piv}. By Theorem 1, only the objects $p \in S_{piv}$ have potential reverse furthest neighbors corresponding to O. In this case, p's RFN is the set of objects $o \in O$ such that o takes p as its furthest neighbor. The distance measure is on the Euclidean space, and $dist(\cdot, \cdot)$ denotes the distance function throughout this paper.

First of all, the filtering is based on the following baseline property:

$\forall o \in O$, if $dist(o, q) < max_{p_i \in S_{piv}}\{dist(o, p_i)\}$ then o is not included in the answer to q's RFN. On the other hand, if $dist(o, q) > max_{p_i \in S_{piv}}\{dist(o, p_i)\}$ then o is included in the answer to q's RFN.

The baseline property is easy to understand. For the former, because p_i is also a element of the dataset, $dist(o, q) < max_{p_i \in S_{piv}}\{dist(o, p_i)\}$ means that there is at least one object further than q from o. In other words, o will never take q as its furthest neighbour. The latter is similar.

Now, the problem is how to use this property to enhance the efficiency. It is meaningless if we have to calculate the $dist(o, q)$ and $max_{p_i \in S_{piv}}\{dist(o, p_i)\}$ whenever the query is issued. Rather all, we develop two techniques to avoid this calculation. To avoid the calculation of distances between o and p_i, we design a metric index which will be explained in the following section. To avoid the calculation of distance between o and q we develop the upper bound and lower bound as in the following two lemmas adopting the triangle inequality. Before formal description, a simple example is given

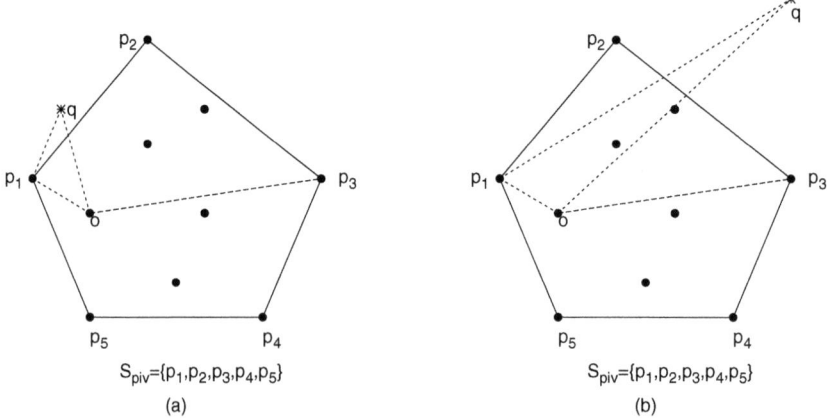

Fig. 3. Illustration for (a) Lemma 1 and (b) Lemma 2

in Figure 3. In (a), o can be safely excluded from q's RFN because the upper bound of $dist(o,q)$ (i.e.,$dist(p_1,q) + dist(o,p_1)$), is under $dist(o,p_3)$. In other words, at least o will take p_3 instead of q as its furthest neighbour. In (b), o is in the answer set because even the lower bound of $dist(o,q)$ (i.e., $|dist(p_1,q) - dist(o,p_1)|$), is larger than $dist(o,p_3)$. Since $dist(o,p_3)$ is the largest distance among $dist(o,p_i)$, we assure that no object o' exists such that $dist(o,o') > dist(o,p_3)$. In other words, for all object o', $dist(o,o') \leq dist(o,p_3) < dist(o,q)$ which means that q is the furthest neighbour of o.

Lemma 1. *Given an object set $O = \{o_1, \cdots, o_n\}$, suppose its convex hull is C_O and the pivot set $S_{piv} = \{p_1, \cdots, p_m\}$ represents the vertices of C_O. Given query q, for any object $o \in O$, o is not in the answer set of q's RFN if o satisfies the following Inequality 1*

$$min_{p_i \in S_{piv}}\{dist(o,p_i) + dist(p_i,q)\} < max_{p_i \in S_{piv}}\{dist(o,p_i)\} \qquad (1)$$

Proof. As illustrated in Figure 3(a), by the triangle inequality, we have

$$dist(o,q) \leq dist(o,p_i) + dist(p_i,q).$$

The exact distance $dist(o,q)$ is unknown and we want to give answer before calculating it. For all pairs of distance combinations $dist(o,p_i) + dist(p_i,q)$, the upper bound of the distance between o and q can be achieved by

$$U_{bound}(dist(o,q)) = min_{p_i \in S_{piv}}\{dist(o,p_i) + dist(p_i,q)\}.$$

On the other hand, the right hand side of Inequality 1 indicates that the object o takes such a pivot $p_i \in S_{piv}$ holding the maximum distance to o as o's furthest neighbor, i.e., o is p_i's reverse furthest neighbor. If Inequality 1 is satisfied, we have

$$U_{bound}(dist(o,q)) < max_{p_i \in S_{piv}}\{dist(o,p_i)\},$$

that means the distance from o to q is always smaller than the distance to the furthest pivot $p_i \in S_{piv}$. Consequently, the object o is impossible to become q's reverse furthest neighbor. This completes the proof.

Lemma 2. *Given an object set $O = \{o_1, \cdots, o_n\}$, select all vertices on its convex hull C_O as a pivot set $S_{piv} = \{p_1, \cdots, p_m\}$. When an arbitrary query q is given, for any object $o \in O$, if the next Inequality 2 is satisfied, the object o absolutely becomes q's reverse furthest neighbor.*

$$max_{p_i \in S_{piv}}\{|dist(o, p_i) - dist(p_i, q)|\} > max_{p_i \in S_{piv}}\{dist(o, p_i)\} \qquad (2)$$

Proof. Similar to Lemma 1, by the reverse triangle inequality, we have

$$dist(o, q) \geq |dist(o, p_i) - dist(p_i, q)| \,.$$

Similarly, for all pairs of distance differences $|dist(o, p_i) - dist(p_i, q)|$, the lower bound of the distance between o and q can be achieved by

$$L_{bound}(dist(o, q)) = max_{p_i \in S_{piv}}\{|dist(o, p_i) - dist(p_i, q)|\}.$$

In the same way, if Inequality 2 is satisfied, we have

$$L_{bound}(dist(o, q)) > max_{p_i \in S_{piv}}\{dist(o, p_i)\},$$

that indicates the distance from o to q is always bigger than the distance to the furthest pivot $p_i \in S_{piv}$. This results in that all the pivots on the convex hull C_O cannot be q's reverse furthest neighbor because $dist(o, q)$ is bigger. Therefore, the current object o safely becomes q's reverse furthest neighbor. The proof is completed.

Improvement for Computation

In the lemmas proofed above, the conditions for upper and lower bounds are very strict in terms of theory. However, for the computation in algorithm design, the inequality conditions can be improved as much as possible. Corresponding to Lemma 1, the Inequality 1 can be rewritten as

$$\exists p_i, \; dist(o, p_i) + dist(p_i, q) < max_{p_i \in S_{piv}}\{dist(o, p_i)\}. \qquad (3)$$

For the judgement to discard the false positive which is impossible to become q's reverse furthest neighbor, the computation can be simplified to the examination of whether there is such a pivot p_i satisfying the Inequality 3, instead of checking all the combinations for all pivots in S_{piv}. If such a pivot satisfying this condition exists, then we are sure that the real distance between o and q is smaller than the distance from o to its furthest pivot $p_i \in S_{piv}$. In other words, o has no chance to take into account q's reverse furthest neighbor, hence o can be safely discarded by this simplified condition.

Similarly, the Inequality 2 can be simplified to

$$\exists p_i, \; |dist(o, p_i) - dist(p_i, q)| > max_{p_i \in S_{piv}}\{dist(o, p_i)\}, \qquad (4)$$

which guarantees that at least one combination of the lower bound is bigger than the distance from o to its furthest pivot $p_i \in S_{piv}$. In this case, o can be safely included in the answer set of q's reverse furthest neighbor. To confirm this improvement for computation, we report the comparison result in the experimental evaluation.

4   Searching RFN with Metric Index

Based on the two Lemmas, we can speed up the RFN query on the fly, without using R-tree to execute range query. The cost for distance computation can be reduced as much as possible as well.

Regarding the assumption for Lemma 1 and 2, we pre-compute the convex hull C_O w.r.t the object set O, then use its vertices as the pivot set S_{piv} to prune the false positive and account the true positive. Naturally, we need to design metric indexes to store all the distances used by the lemmas. We design two hash tables as metric indexes in the following subsection. In addition, the distances between query q and all pivots $p \in S_{piv}$ can be easily computed on the fly.

4.1   Metric Indexes

For the $max_{p_i \in S_{piv}}\{dist(o, p_i)\}$, we use a hash table **MetricIndex-A** to store all the tuples $(o_{id}, max_{i=1}^{m}\{dist(o, p_i)\})$, where m is the number of pivots $p_i \in S_{piv}$. For the pairwise distance between each pivot $p_i \in S_{piv}$ and an object $o \in O$, we have to store a large matrix such as Figure 4. Instead of considering the storage for this large matrix, we map each value in the matrix into a hash table **MetricIndex-B** using the pair of (p_{id}, o_{id}) as the key. Due to these simple metric indexes using hash structure, all the distances can be fetched fast in cost $O(1)$. Since the memory ability is very large nowadays, we keep these two metric indexes in memory on the fly to reduce the I/O cost of reads from and writes to the disk. As the indexes are constructed by key-value hash tables, it is easy to store them to sequential record files. The overhead of this approach is the secondary storage space of $O(mN)$ times the size of an object ID.

Besides the construction, it is also very convenient to update the index tables when the dataset is updated. There are three situations when the update happens. The added/removed object is 1) inside the convex hull C_O (i.e., pivot set S_{piv}), or 2) on the boundary of the convex hull, or 3) outside the convex hull. For the first case, the pivot set does not change, therefore only one insertion/deletion to table **MetricIndex-A**, and only pairwise operations for each pivot can be processed on table **MetricIndex-B** within the complexity $O(m)$. It should be pointed out that the size of pivot set m is often much smaller than the size of the dataset N. For the last two cases, the convex hull has to be reconstructed. Although recomputation happens, the reconstruction only results in several insertions/deletions ($\ll m$), which is the same as the first case since $m \ll N$.

MetricIndex-A

o_{id}	$max_{p_i \in S_{piv}}\{dist(o, p_i)\}$
1	d_1
2	d_2
⋮	⋮
n	d_n

MetricIndex-B

$$\begin{array}{c} \\ p_1 \\ p_2 \\ \vdots \\ p_m \end{array} \begin{array}{cccc} o_1 & o_2 & \cdots & o_n \\ \left(\begin{array}{cccc} d_{11} & d_{12} & \cdots & d_{1n} \\ d_{21} & d_{22} & \cdots & d_{2n} \\ \vdots & \vdots & \ddots & \vdots \\ d_{m1} & d_{m2} & \cdots & d_{mn} \end{array} \right) \end{array}$$

Fig. 4. Metric indexes for storing distances

Moreover, the lemmas introduced above strictly follow the inequality property of metric space, this approach is applicable to all metric spaces.

4.2 Algorithm

Combining the Lemma 1 and 2, their improvements and the metric indexes mentioned above, we propose a complete algorithm (shortly named **PIV**) to handle RFN query efficiently. The Algorithm PIV is divided into two phases, the filtering processing (line 2-16), and the refinement processing (line 17-24). In the filtering phase, all the processing strictly conforms the theorem and lemmas mentioned before. Line 2 determines whether the query q is inside the convex hull by Theorem 1. O is scanned one pass from line 8 to 16. If Lemma 1 is satisfied, the current object o is safely discarded. Otherwise, Lemma 2 will be checked again, if satisfied, then object o should be added to the answer set *Answ*, otherwise it is inserted into the candidate set *Cand*. After the filtering processing, a few objects are remained in *Cand*. In the refinement phase, the real distance from each potential candidate to q should be computed to confirm if it can become q's RFN.

5   Experimental Evaluation

To confirm the efficiency of the proposed algorithm, we perform extensive experiments for evaluation. All the experiments are executed on an Intel-based computer and Linux OS. The CPU is Intel(R) Xeon (R) 2.83 GHz and the amount of main memory is 16.0GB. The programs are implemented in C++ language, using the open-source libraries: Spatial Index Library[1] and CGAL Library[2]. For taking into account IOs, the page size is set to 4KB.

In order to make the experimental comparison as fair as possible, we use the same datasets as that of the related work [8]. Two kinds of datasets are used for experiments: three synthetic datasets and a real dataset. One of the synthetic datasets is conforming Uniform distribution (UN), the other one is Random-Cluster distribution (RC), and the third is Correlated Bivariate (CB) (in Figure 5). The real dataset is obtained from the digital chart of the world server[3]. The real dataset contains 3 kinds of 2-D point data defining the road networks for California (CA) and its interest points, San Francisco (SF) and USA (US). As the same settings with [8], we also merge them into one dataset (named Map) after normalizing all the data points into the space $L = (0,0) \times (100000, 100000)$. Totally, the Map dataset contains 476,587 points.

For the performance measurement, we compare our proposed algorithm (PIV) to the other two algorithms (CHFC, and BFS — Brute-Force Search). The implementation of CHFC and BFS is also following their original descriptions in [8]. For all algorithms, we measure their CPU cost and the number of I/O cost. The final results are reported on the average for one query after issuing 100 queries. Because the area of the convex hull containing the dataset is almost touch the boundaries of the space L, we randomly

---

[1] http://research.att.com/~marioh/spatialindex/
[2] http://www.cgal.org/
[3] http://www.cs.fsu.edu/~lifeifei/SpatialDataset.htm

Algorithm: PIV(Query q, Pivots S_{piv}, Dataset O)
---
1 Initialize candidate set *Cand*← ∅, answer set *Answ*← ∅;
2 if (is_query_inside_convex_hull(q, S_{piv}) == TRUE) {
3   return ∅;
4 }
5 Compute each distance pair dist(q, p), $p \in S_{piv}$;
6 foreach $o \in O$ {
7   if ($\exists p_i$, $dist(o, p_i) + dist(p_i, q) < max_{p_i \in S_{piv}}\{dist(o, p_i)\}$) {
8   // Discard o by Lemma 1;
9   } else {
10   if ($\exists p_i$, $|dist(o, p_i) - dist(p_i, q)| > max_{p_i \in S_{piv}}\{dist(o, p_i)\}$) {
11   Insert o into *Answ*; //o is one of the answer by Lemma 2;
12   } else {
13   Insert o into *Cand*; // objects need to be checked further;
14   } //end if
15   } //end if
16 } //end foreach
17 foreach $o \in Cand$ {
18   Compute real distance $dist(o, q)$;
19   if ($dist(o, q) < max_{p_i \in S_{piv}}\{dist(o, p_i)\}$) {
20   Discard o;
21   } else {
22   Insert o into *Answ*;
23   } //end if
24 } //end foreach
25 return *Answ*;
---

generate the queries from the space L', the area of which is 2 times of L. This setting guarantees the selection of queries is fair, which makes the possibility of having answers or no answers to RFN query equal to each other. For the extensive comparison, we also randomly generate different sizes of sub datasets: 10K, 50K, 100K, 200K, 300K, and 400K, as input to the programs.

5.1 Cost Analysis

Responding to the complaint mentioned in Section 2, we perform experiments to compare the partial cost in CHFC algorithm with our PIV algorithm. As shown in Figure 6(a), it is clear to confirm that the "hull" and "range query" phases in CHFC algorithm occupy very high CPU cost. However, our PIV algorithm much outperforms the CHFC. Moreover, it is exact to confirm that any one of the two most expensive phases in CHFC is over the total CPU cost of PIV.

Meanwhile, for confirming the improvement for computation mentioned in Section 3, we illustrates the CPU cost comparison between the naive and improved PIV algorithms. In the naive PIV algorithm, we use Inequality 1 and 2 different from improved PIV algorithm (line 7 and 10). As shown in Figure 6(b), the result confirms that the improved pruning conditions (Inequality 3 and 4) are more efficient than the naive but tight conditions.

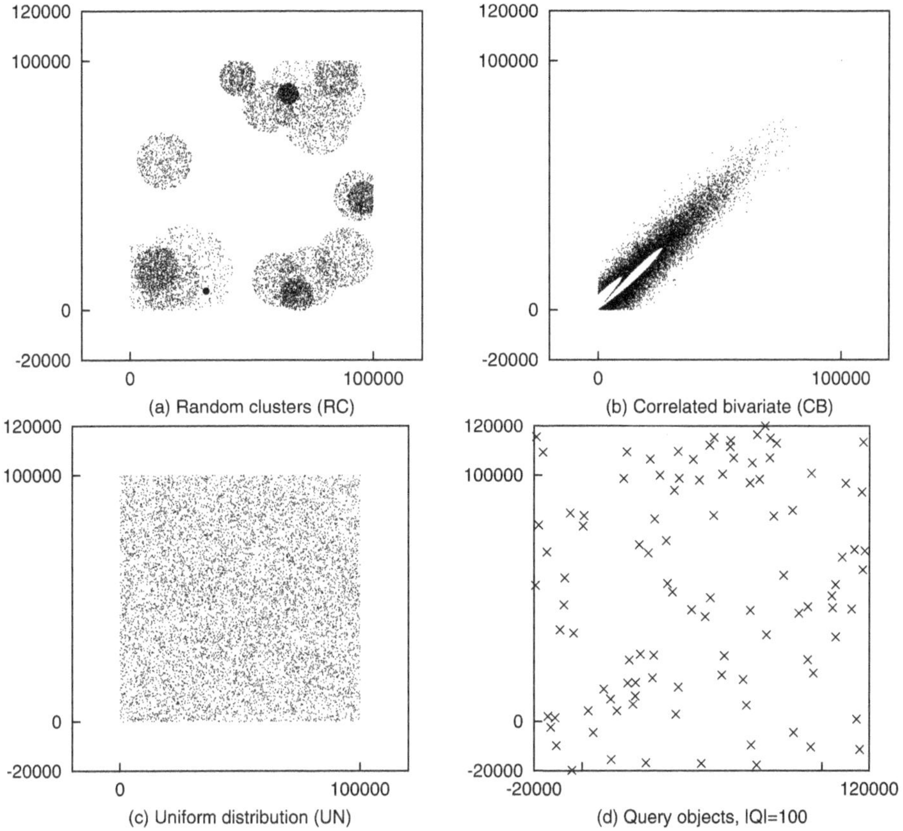

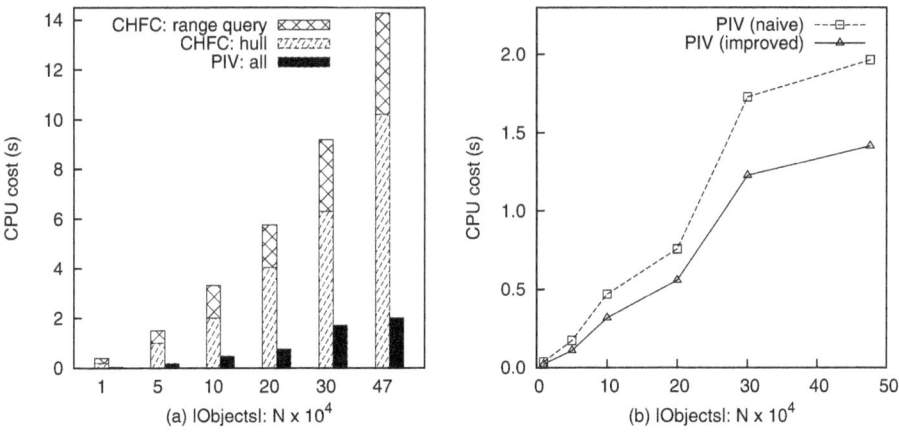

Fig. 5. Synthetic datasets: (a) Random-Cluster (RC), (b) Correlated Bivariate (CB), (c) Uniform distribution (UN), (d) Query objects

Fig. 6. (a) CHFC vs. PIV. on CPU cost; (b) naive vs. improved PIV on CPU cost.

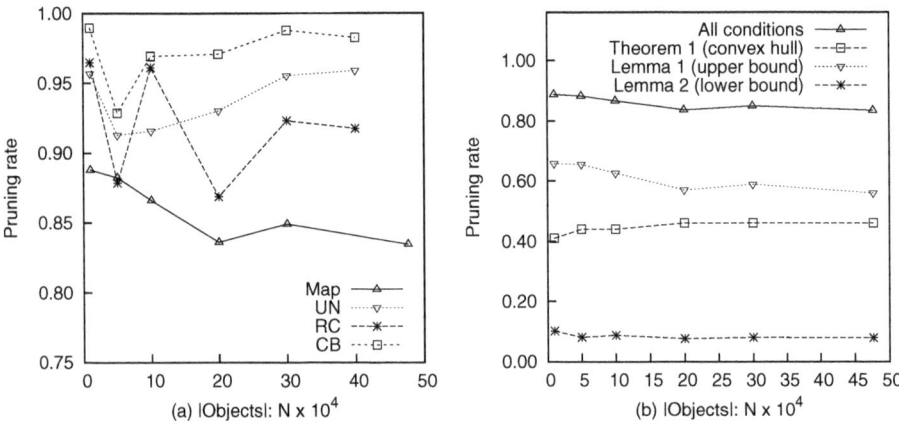

Fig. 7. (a) Pruning power of filtering phase in PIV algorithm; (b) Pruning power comparison with different conditions in PIV algorithm

On the other hand, the pruning power of the filtering phase mostly influences the cost of PIV algorithm. Thus we examine the full pruning power for the filtering phase by different datasets, and report the different pruning rates by real dataset when different filter conditions are employed. In Figure 7(a), it is confident that the pruning power reaches very high percentage. For the synthetic datasets (UN,RC,CB), the pruning power is over 90%, even close to 100%. The real dataset (MAP) also performs the high pruning power about 85%. As shown in Figure 7(b), the result indicates that the pruning power of Lemma 1 employing the upper bound is greater than Theorem 1 and Lemma 2. However, Lemma 2 performs the worst with extremely low pruning rate. The reason for such low rate is because the searching space is very limited (only $2L$) where the queries are generated from. Suppose if the space is unlimited, when a query q is very far from the dataset, Lemma 2 will work well, safely including almost all the objects into the answer set of q's RFN. Oppositely, Lemma 1 will become the worst and hardly discard fewer objects in this case. The result in Figure 7(b) still implies that the total power to prune the false positives is not equal the sum of the other three independent pruning rates. It is because there is overlapping effect between Theorem 1 and Lemma 1 when a query q is inside the convex hull C_O. In this case, all the objects will be discarded only by Theorem 1. Nevertheless, only using Lemma 1 without Theorem 1 is still possible to discard many objects that are false positives.

After all, according to analysis derived from the extensive experiments above, it is clear to understand the cost of proposed PIV algorithm and its efficiency.

5.2 Comparison Results

Finally, we examine the efficiency and scalability of the proposed algorithm PIV, comparing with CHFC and BFS. As illustrated in Figure 8 and 9, the experiment results are reported for the comparison on real data (MAP) and synthetic data (RC). It is obvious that no matter the execution on real data or synthetic data, the PIV algorithm outperforms the other two algorithms in terms of either CPU cost or I/O cost. To the exact

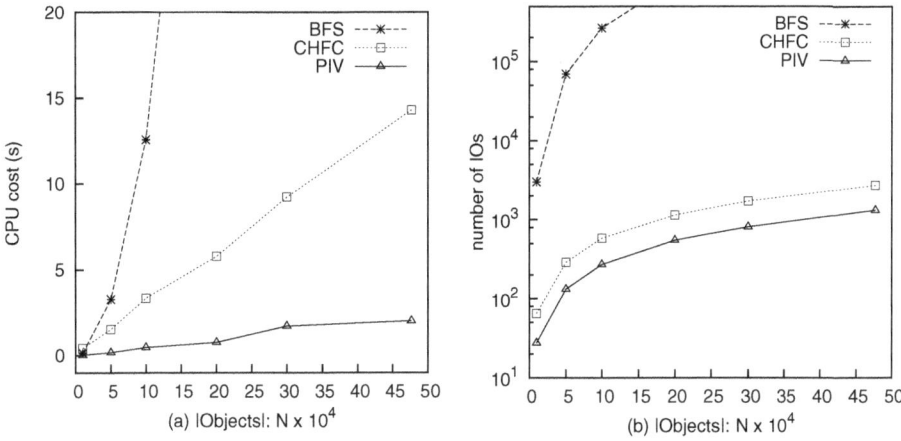

Fig. 8. Comparison on real data (MAP): (a) CPU cost, and (b) number of IOs

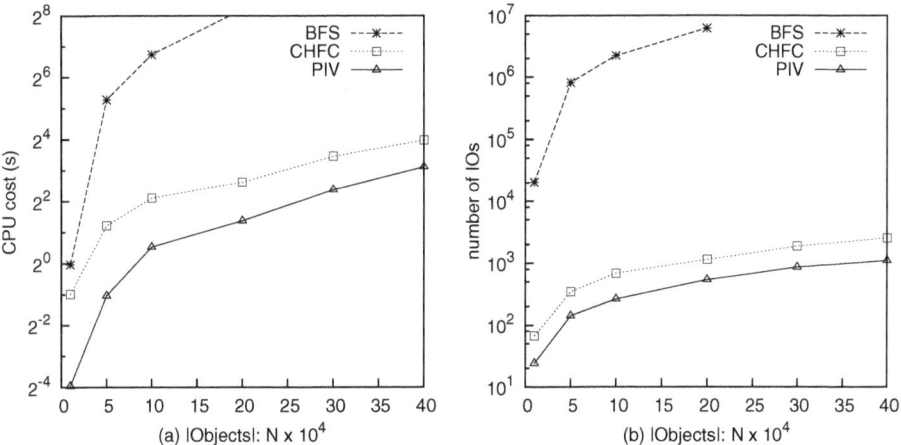

Fig. 9. Comparison on synthetic data (RC): (a) CPU cost, and (b) number of IOs

quantity, it is easy to find that the PIV algorithm is about 10 times faster than the CHFC algorithm on CPU cost, and can save about 5 times of I/O cost than the CHFC algorithm. Beside, the cost of the PIV algorithm is without quick increase against the CHFC and BFS algorithms, which verifies the stable scalability of the PIV algorithm.

6   Conclusion and Future Work

This work analyzes the expensive cost in the related algorithm CHFC, and quantizes the internal cost for each phase of CHFC by experiment. The special property — convex hull of RFN problem is summarized and extended to derive two lemmas. Based on the lemmas, a novel algorithm for RFN query is proposed. The extensive experiments confirm its efficiency and scalability scientifically. As the perspective of future work, we are planning to extend this work to answer RFN query for high-dimensional data, and put efforts to apply this work to the real applications such as company marketing strategy mentioned in the introduction. Finally, the lemmas derived in this work still can make much stricter to enhance the pruning power for the PIV algorithm.

Acknowledgment

This research has been supported in part by the Grant-in-Aid for Scientific Research from MEXT (#21013004).

References

1. Belongie, S., Malik, J., Puzicha, J.: Shape matching and object recognition using shape contexts. IEEE Trans. Pattern Anal. Mach. Intell. 24(4), 509–522 (2002)
2. Grauman, K., Darrell, T.: Fast contour matching using approximate earth mover's distance. In: CVPR, (1), pp. 220–227 (2004)
3. Datta, R., Joshi, D., Li, J., Wang, J.Z.: Image retrieval: Ideas, influences, and trends of the new age. ACM Comput. Surv. 40(2) (2008)
4. Agrawal, R., Faloutsos, C., Swami, A.N.: Efficient similarity search in sequence databases. In: Lomet, D.B. (ed.) FODO 1993. LNCS, vol. 730, pp. 69–84. Springer, Heidelberg (1993)
5. Nievergelt, J., Hinterberger, H., Sevcik, K.C.: The grid file: An adaptable, symmetric multi-key file structure. ACM Trans. Database Syst. 9(1), 38–71 (1984)
6. Guttman, A.: R-trees: A dynamic index structure for spatial searching. In: SIGMOD Conference, pp. 47–57 (1984)
7. Weber, R., Schek, H.J., Blott, S.: A quantitative analysis and performance study for similarity-search methods in high-dimensional spaces. In: VLDB, pp. 194–205 (1998)
8. Yao, B., Li, F., Kumar, P.: Reverse furthest neighbors in spatial databases. In: ICDE, pp. 664–675 (2009)
9. Lian, X., 0002, L.C.: Similarity search in arbitrary subspaces under lp-norm. In: ICDE, pp. 317–326 (2008)
10. Skopal, T., Bustos, B.: On index-free similarity search in metric spaces. In: Bhowmick, S.S., Küng, J., Wagner, R. (eds.) DEXA 2009. LNCS, vol. 5690, pp. 516–531. Springer, Heidelberg (2009)
11. Zhang, Z., Ooi, B.C., Parthasarathy, S., Tung, A.K.H.: Similarity search on bregman divergence: Towards non-metric indexing. PVLDB 2(1), 13–24 (2009)
12. Roussopoulos, N., Kelley, S., Vincent, F.: Nearest neighbor queries. In: SIGMOD Conference, pp. 71–79 (1995)
13. Cheung, K.L., Fu, A.W.C.: Enhanced nearest neighbour search on the r-tree. SIGMOD Record 27(3), 16–21 (1998)
14. Hjaltason, G.R., Samet, H.: Distance browsing in spatial databases. ACM Trans. Database Syst. 24(2), 265–318 (1999)
15. Athitsos, V., Potamias, M., Papapetrou, P., Kollios, G.: Nearest neighbor retrieval using distance-based hashing. In: ICDE, pp. 327–336 (2008)
16. Tao, Y., Yi, K., Sheng, C., Kalnis, P.: Quality and efficiency in high dimensional nearest neighbor search. In: SIGMOD Conference, pp. 563–576 (2009)
17. Korn, F., Muthukrishnan, S.: Influence sets based on reverse nearest neighbor queries. In: SIGMOD Conference, pp. 201–212 (2000)

18. Yang, C., Lin, K.I.: An index structure for efficient reverse nearest neighbor queries. In: ICDE, pp. 485–492 (2001)
19. Stanoi, I., Riedewald, M., Agrawal, D., Abbadi, A.E.: Discovery of influence sets in frequently updated databases. In: VLDB, pp. 99–108 (2001)
20. Singh, A., Ferhatosmanoglu, H., Tosun, A.S.: High dimensional reverse nearest neighbor queries. In: CIKM, pp. 91–98 (2003)
21. Tao, Y., Papadias, D., Lian, X.: Reverse knn search in arbitrary dimensionality. In: VLDB, pp. 744–755 (2004)
22. Achtert, E., Böhm, C., Kröger, P., Kunath, P., Pryakhin, A., Renz, M.: Efficient reverse k-nearest neighbor search in arbitrary metric spaces. In: SIGMOD Conference, pp. 515–526 (2006)
23. Tao, Y., Papadias, D., Lian, X., Xiao, X.: Multidimensional reverse nn search. VLDB J. 16(3), 293–316 (2007)
24. Wu, W., Yang, F., Chan, C.Y., Tan, K.L.: Finch: evaluating reverse k-nearest-neighbor queries on location data. PVLDB 1(1), 1056–1067 (2008)

A Scalable and Self-adapting Notification Framework

Anthony Okorodudu, Leonidas Fegaras, and David Levine

University of Texas at Arlington, CSE, 416 Yates Street, Arlington, TX 76019
aokorodudu@uta.edu, fegaras@cse.uta.edu, levine@cse.uta.edu

Abstract. There has been a great interest in publish/subscribe systems in recent years. This interest, coupled with the pervasiveness of light-weight electronic devices, such as cellular phones and PDAs, has opened a new arena in publish/subscribe networks. Currently, many broker overlay networks are static and rarely change in structure. Often, a network overlay structure is predefined or manually modified. We present a dynamic broker network for disseminating XML data. Our work builds upon previous network optimization research on ad-hoc publish/subscribe networks. Our framework utilizes user-defined cost functions to satisfy quality of service (QoS) constraints. We reduce the broker network optimization problem to an incremental search problem to generate low cost network configurations with respect to the provide cost functions. We also address certain reliability issues by providing a scheduling algorithm to selectively retransmit information and handle broker connectivity failures.

Keywords: Publish/Subscribe, Broker Overlay Network.

1 Introduction

Many stream-oriented applications have emerged, such as financial quote tickers, and sensor monitors, which require the data to be delivered to a large number of clients continuously. The traditional method of accomplishing this is to use the pull-based dissemination paradigm. In this paradigm, each individual client sends a request to the server to retrieve the required data. The server then sends a response to each client separately. Each client request has a measurable cost in both bandwidth and CPU processing. The server uses unnecessary CPU and IO resources by processing each client request individually instead of taking advantage of the similarities across client requests. Publish subscribe systems are push-based alternatives for disseminating data.

In publish subscribe systems, subscribers express the desired content by using a high-level specification language such as XPath. A content-based publish/subscribe system is generally more effective than a pull-based system with respect to network bandwidth usage and client filtering costs for massive data dissemination. Publisher/subscribe technology is widely used in message-oriented middleware systems [2].

In the most basic and common form of publish subscribe systems, a single machine can be used to filter data from the publisher and subsequently route it to interested clients. This is a viable solution for trivial scenarios where there is little resource demand on the broker. However, this solution does not scale well for larger number of clients and profiles. To accommodate large number of diverse profiles, several machines can cooperate to distribute the filtering and routing tasks. This network of

P. García Bringas et al. (Eds.): DEXA 2010, Part II, LNCS 6262, pp. 452–461, 2010.
© Springer-Verlag Berlin Heidelberg 2010

computers is called a broker network [1]. In a broker network, each individual machine is a broker. A broker's primary tasks are to filter data from publishers and route them to interested parties. The interested parties are either subscribers or other intermediary brokers in the network.

Current systems predominantly utilize a static network topology. Some systems allow for manual modifications of the network. In this paper, we introduce a framework for a self-organizing topology management system (TMS) that disseminates XML data. Our proposed broker network system extends our previous work on XML fragmentation to disseminate XML [3]. The unit of broadcast in our system is an XML fragment.

The broker network is constructed in an ad hoc fashion by adding brokers randomly to the network. A new broker joins the network by designating an existing broker to server as its parent. These simple rules lead to the formation of a broker network on top of a tree overlay. Subscribers connect with a random known broker in the network. Over time, the network degrades to a higher cost state determined by relevant QoS requirements such as bandwidth usage. To combat this, we employ various transformation operations to small localized parts of the network. We intend to show that over time, these small transformations will improve the overall fitness of the broker network. We also differentiate our work putting forth a mechanism to address broker failures in the self organizing network.

2 Related Work

This paper draws from prior works on publish/subscribe systems. Publish/Subscribe systems route data from publishers to subscribers based on profiles. Earlier works were limited to subject-based routing. In recent years, more work has been done on content-based routing, which allows for richer client profiles in terms of expressiveness. ONYX [1] and SemCast [5] introduced content-based publish/subscribe using XML data and XPath profiles. These works are purely event-based and transfer the message filtering overhead and routing documents to clients. Many earlier systems do not support self-organization [10]. XPORT tackles incremental optimization of profile driven distributed systems [4]. However, XPORT does not address broker failures which we propose a solution to. A crash resilient topology was put forth in [11] by maintaining a list of all neighboring brokers in the event of a failure. Our approach does not maintain such an extensive list. Instead, each broker knows only about its parent and grandparent. There is also a lookup service for locating random brokers. [15] Puts forth a heuristic based approach to optimizing the network using a training and reconfiguration phase. Our work differs allowing for more flexibility in configuring the heuristic functions. [7] proves that finding an optimal configuration for the broker network is NP-hard. Therefore, the goal of our framework is to incrementally increase the fitness of the network. Our findings show improvements over a static broker network.

Peer-to-peer (P2P) systems have received much attention lately [6]. They allow the distribution of resources among many peers. This allows systems to scale fairly well since the load is better balanced. A P2P system is also not sensitive to poor network connectivity of peers, which is desirable in some publish/subscribe environments. Our work follows a different network topology. We put forth a tree based topology in our framework.

3 Preliminaries

The scheduling of repeating fragments to be multicast is addressed by using algorithms from packet fair queueing [8]. Each fragment is treated as a data item. As the number of fragments increase, the expected delay also increases. To decrease this delay, multiple schedules can be utilized to decrease the overall expected delay of fragments. Table 1 shows the description of the symbols used throughout this section.

Table 1. Descriptiojn of symbols

Description	Symbol
Total number of clients	N
A particular client/subscriber	C_i
An XPath query belonging to a client's (C_i) profile. P_i is the set of all XPath queries belonging to a client (C_i).	P_{ij}
A particular fragment specification (fragment spec). A fragment spec is specified as an XPath indicating the fragmentation point. Fragment specs are ideally derived from client profiles.	F_i
The weight of a given fragment specification (F_i)	W_i
Timestamp for a particular query within a client's (C_i) profile (P_{ij})	T_{ij}
Computes the dampening factor for a particular timestamp. Queries with older timestamps have less impact than newer queries.	$D(T_{ij})$
Computes the relevance of P_{ij} to F_k	$R(P_{ij}, F_k)$
Initial weight of all fragments	K

A fragment is the unit of broadcast in our system [3]. Given an XML fragment F_i, to compute its weight W_i, we need to determine how relevant this fragment is to all the profiles. Fragments with a higher relevance with respect to the profiles are disseminated to the clients more frequently than those with a lower relevance. The formula for computing a fragment's (F_k) weight (W_k) is:

$$W_k = K + \left(\sum_{i=1}^{N} \left(\sum_{j=1}^{|P_i|} R(P_{ij}, F_k) \right) / N \right)$$

$$R(P_{ij}, F_k) = \left(|P_{ij} \cap F_k| / |F_k| \right) * D(T_{ij})$$

(1)

That is, a fragment's weight is the sum of all the profiles it satisfies. All profiles have an equal effect on a matching fragment so that a single client cannot boost the weight of a particular fragment by registering a profile with multiple queries targeted at the same fragment. Fragments that do not match any profiles are given a small fixed weight so they will be included in the schedule.

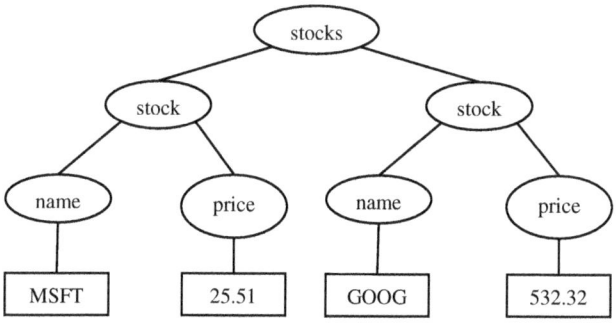

Fig. 1. A sample XML document

Below, we show the computation of fragment weights. Figure 1 shows a simple XML document. The document is broken up into five fragments as shown in Table 2. The fragmentation points are expressed using XPath.

Table 2. Fragment and associated computed weights

ID	Fragment Spec	# Covered profiles	Weight
1	/stocks	0	1
2	/stocks/stock[name="MSFT"]	0	1
3	/stocks/stock[name="MSFT"]/price	1	2
4	/stocks/stock[name="GOOG"]	1	2
5	/stocks/stock[name="GOOG"]/price	2	3

4   The Broker Network

In this section, we put forth the various QoS and privacy constraints that must be addressed in a broker network. We classify the costs into structural or behavioral. Structural costs are based on the profile distribution which, impacts the route table construction. Behavioral costs are influenced by the way the system interacts with its environment. This paper will focus on the structural network characteristics.

4.1   Network Construction

The broker network is first created in an ad hoc fashion. Initially, there is only a single broker. This broker will serve as the root of the entire network overlay tree. Additional brokers are added to the network by specifying the connection endpoint of an arbitrary existing broker in the network to serve as their parent. A broker lookup service can be used to locate an existing broker. Following these simple rules, the network forms a tree topology, since a broker registers with a single parent. Brokers are added randomly to the network without regard for optimization. Over time, the network will be optimized incrementally using the heuristic functions discussed in detail in the following sections.

Subscribers register their profiles with any known broker using the broker lookup service. No initial effort is made to find the optimal broker for a given subscriber. The system relies upon incremental optimization to eventually migrate the client to a broker who best suits its specified interests.

4.2 Privacy and Security

Special consideration must be taken into account for both privacy and security. Transport Layer Security (TLS) is used in bilateral mode to maintain the integrity and confidentiality of data in the transport layer. TLS allows clients and servers to communicate in such a way as to prevent eavesdropping, tampering, and message forgery [13]. Bilateral mode ensures that both broker and clients cannot misrepresent themselves. The TLS protocol encrypts data across the network. This thwarts network sniffer attacks. Phishing is prevented by utilizing TLS in bilateral mode and verifying the certificates of both communicating parties [14].

5 Centralized Optimization Strategy

Given sufficient processing power, bandwidth, and memory, a centralized optimization strategy can be used to optimize the broker network. The coordinator must have knowledge of all the brokers' route tables and statistics. Each broker must transmit the required information to the coordinator, who must then use a brute force algorithm to iterate every possible network configuration and select the best one for a given set of weighted cost functions. The routing permutations problem on trees has been shown to be NP-hard [9]. Upon selecting the optimal configuration, the coordinator transmits the changes that each broker must perform to its route table to transform the overall network configuration. Thus, for this centralized approach, the process of reconfiguring the network requires a lot of inter-process communication (IPC) and is very CPU intensive. The amount of time to compute the optimal configuration is prohibitively expensive in a dynamic environment.

6 Distributed Optimization Strategy

In broker networks, where brokers have limitations in processing capabilities and memory, an alternative optimization strategy is necessary. Our framework builds on related work [4] by using a distributed incremental optimization strategy to converge the network structure to a low cost state for a given load. We further supplement [4] by increasing the search depth per transformation. In order to narrow the search space, the unit of transformation is limited to three levels within a tree: 1) a local coordinator, 2) its children, and 3) its grandchildren. Any broker with at least two descendants has an opportunity to become a coordinator. A coordinator is responsible for initiating a local transformation. The transformations are termed local since transformations to the local unit have no impact on connectivity outside of the transformation unit.

Once a broker has been validated to become a coordinator, it immediately notifies all of its children that it is the coordinator of the transformation unit in order to lock them into the unit. The coordinator's children in turn notify their children to lock them into the transformation unit. The coordinator evaluates alternative network configurations based on the finite transformation functions discussed below. Each possible configuration is given a cost based upon a cost function or group of cost functions. The configuration with the smallest cost above a specified delta threshold is chosen and the coordinator informs the rest of the unit how to reconfigure itself.

To avoid falling into a local minimum, the cost function may include an element of randomness. This randomness makes our transformation similar in many ways to the mutation operator in genetic algorithms [12]. We use three primary transformations: upgrading, downgrading, and shifting transformations. Figure 2 through 4 illustrate the various transformations.

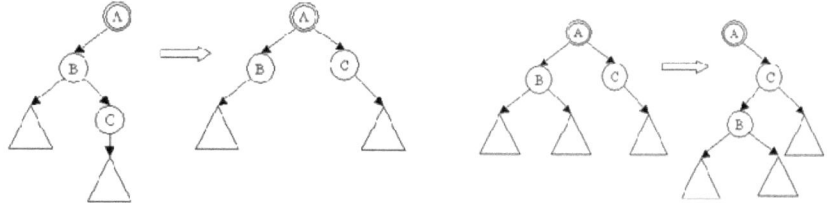

Fig. 2. Upgrade transformation   **Fig. 3.** Downgrad transformation

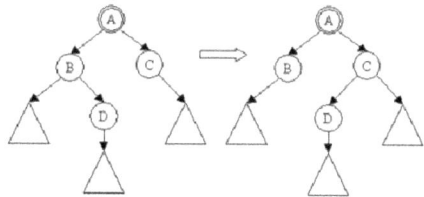

Fig. 4. Shift transformation

7 Broker Failures

It is imperative that the problem of broker failure be addressed for many systems. Current research in this area does not fully address this issue. A lot of research pertaining to the publish/subscribe broker networks use a tree overlay to organize the brokers. The disadvantage of this topology is that trees contain single-points of failures. The loss of a critical broker can bring down the entire network. We can address this issue by introducing a protocol for orphan brokers to rejoin the network.

The first issue to address is how to detect broker failures. Since our framework does not mandate an established connection amongst connected brokers, we cannot rely on detecting a closed socket connection. Polling can be used as a viable solution to detect broker failures. However, polling brokers to determine if they are alive uses unnecessary bandwidth which we strive to conserve. It takes at least one complete round trip message to determine if a broker is alive.

We put forth using a push-based mechanism for determining if a broker is alive. We accomplish this by using heartbeats. Brokers periodically send heartbeats to the brokers they are connected to. If a broker does not send a heartbeat within a certain timeout period, it is assumed to be down and necessary actions take place to maintain the integrity of the network.

In the event that a broker goes down, all its children are regarded as orphans. Each orphaned broker reestablishes connectivity with the rest of the network by requesting a non-descendant broker to adopt it. An orphan pushes up its profiles to its new found parent using the broker registration process. If the root broker fails, one of the orphans will serve as the new root for the network overlay.

8 Evaluating a Network

A particular network configuration is evaluated based on cost functions. There are several costs that a network may choose to minimize. In many situations, these costs have conflicting requirements. For example, minimizing message hops leads to decreasing the overall depth of the broker network. However, this conflicts with the goal of distributing workload. Distributing the workload based on profiles promote having similar size route tables amongst the brokers.

In real world scenarios, we have to balance these conflicting costs based on the application requirements. Our framework provides a mechanism for combing cost functions. Each individual cost function is assigned a weight which determines its overall impact on the overall cost. To evaluate the cost of a particular structure, the framework contains several built-in evaluation operations. Statistics are maintained by the brokers in order to evaluate the behavioral dependent costs.

We uniformly spread share profiles among brokers by employing the *Distribute-WorkloadCost* function (*DWC*). This cost function takes a transformation unit as its input and computes the standard deviation among the route table lengths. The size of the route table |R| is equivalent to the number of profiles registered at the broker. The cost is calculated as follows.

$$DWC = \sqrt{\frac{1}{|R|}\sum_{i=1}^{|R|}\left(|R_i| - \overline{|R|}\right)^2} \tag{2}$$

For some requirements, we need to minimize the cumulative size of the route tables. This will push the broker network to group similar profiles together. The framework can accomplish this with the following *MinimizeRouteTableCost* function (*MRTC*).

$$MRTC = \frac{1}{|R|}\sum_{i=1}^{|R|}|R_i| \tag{3}$$

Multiple cost functions, h, are chained together by assigning a weight to each individual cost function and summing them together. The weights should sum to 1. The assigned weight impacts the influence of the cost function on the overall cost.

$$h = w_1 \times DWC + w_2 \times MRTC \tag{4}$$

9 Experiments

We implemented our framework simulation using Java. We created several random network topologies varying in size to determine the impact of the network size on the transformations. In particular, we wanted to determine how many transformations it took to converge to an optimal configuration. The factors we consider are the broker network size, client population size, and query population. We used an average of three profiles per client. Brokers are selected randomly to act as local coordinators of an optimization unit. The configuration with the lowest cost is chosen and the others in the search space are discarded. Over the course of these local transformations, the broker network decreases its cost. We present our findings below. The results show how many transformations occurred before the system eventually stabilized to a lower cost state. The number of transformations required to reach this state is directly proportional to the number of brokers and profiles. It is worth noting that performing optimizations in parallel decreases the overall time to converging on optimal solution.

Table 3. Effects of network size on optimization

# Brokers	# Clients	Average Transformations
100	1000	54
200	2000	115
300	3000	147
400	4000	196

We evaluated various broadcast schedules to evaluate the mean inter-arrival time of relevant fragments. The fragment specification probabilities follow a ZipF distribution. The client interests on the fragments follow the same distribution. The document is broken up into 1,000 fragments. We use a schedule length of 10,000 data items and vary the number of clients. The experiment results are shown in table 3. We are primarily interested in the mean inter-arrival metric. This metric shows how long a client has to wait for a fragment that it is interested in. The smaller the inter-arrival time, the more frequently the client receives updates to satisfy its queries.

We tested three schedulers. The first scheduler creates a random schedule for the fragments. The second scheduler orders the fragments in descending order according to the probability of the fragment. The final scheduler uses the single channel fragment broadcast algorithm based on [8]. The results show that our single channel fragment broadcast algorithm performs the best in terms of mean inter-arrival time. The mean inter-arrival time of the other schedules is roughly equal to the number of fragments in the system.

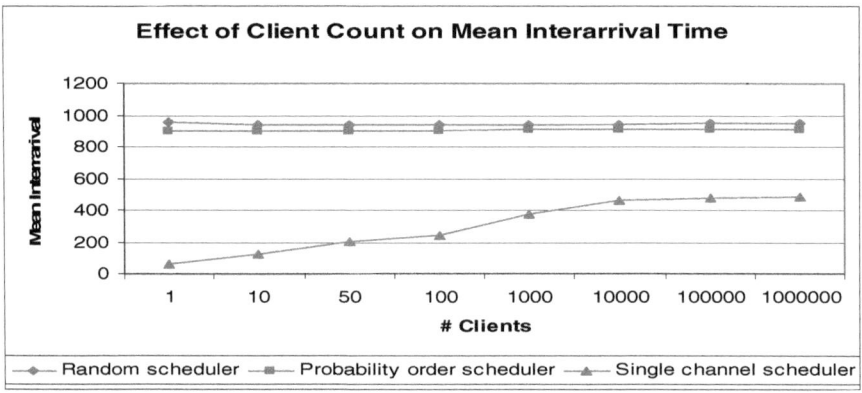

Fig. 5. The effect of client count on mean inter-arrival time per time unit (milliseconds)

10 Conclusion

In this paper, we present a generalized framework for incrementally optimizing an ad hoc publish/subscribe network. Our framework utilizes user-defined heuristic functions to achieve QoS constraints. Our experiments demonstrated that the system eventually converges to a significantly lower cost state over time in comparison to a static broker network. The integrity of the system is also maintained, which is not the case in a comparable tree-based static topology. The experiments also illustrate the advantages of using a single channel scheduler for retransmitting repeating fragments.

References

1. Diao, Y., et al.: Towards an Internet-Scale XML Dissemination Service. In: VLDB (2004)
2. Tian, F., DeWitt, D., Pirahesh, H., Reinwald, B., Mayr, T., Myllymaki, J.: Implementing a Scalable XML Publish/Subscribe System Using a Relational Database System. In: Proc. of SIGMOD 2004 (2004)
3. Fegaras, L., Levine, D., Bose, S., Chaluvadi, V.: Query Processing of Streamed XML Data. ACM, New York (2002)
4. Papaemmanouil, O., Ahmad, Y., Cetintemel, U., Jannotti, J., Yildirim, Y.: Extensible Optimization in Overlay Dissemination Trees. SIGMOD (2006)
5. Papaemmanouil, O., Cetintemel, U.: SemCast: Semantic Multicast for Content-Based Stream Dissemination. In: ICDE (2005)
6. Terpstra, W., Behnel, S., Fiege, L., Zeidler, A., Buchmann, A.: A Peer-to-Peer Approach to Content-Based Publish Subscribe. ACM, New York (2003)
7. Jaeger, M., Parzyjegla, H., Muhl, G., Herrmann, K.: Self Organizing Broker Topologies for Publish/Subscribe Systems. ACM, New York (2007)
8. Hameed, S., Vaidya, N.: Log-time Algorithms for Scheduling Single and Multiple Channel Data Broadcast, pp. 90–99. ACM, New York (1997)
9. Barth, B., Corteel, S., Denise, A., Gardy, D., Valencia-Pabon, M.: On the Complexity of Routing Permutations on Trees by Arc-Disjoint Paths. Theoretical Informatics, 308–317 (2000)

10. Carzaniga, A., Rosenblum, D., Wolf, A.: Design and Evaluation of a Wide-Area Notification Service. ACM Transactions on Computer Systems, 332–338 (2001)
11. Baldoni, R., Beraldi, R., Querzoni, L., Virgillitom, A.: A self-organizing crash-resilient topology management system for content-based publish/subscribe. In: 3rd International Workshop on Distributed Event-Based Systems (DEBS). IEEE, Los Alamitos (2004)
12. Mei-yi, L., Zi-xing, C., Guo-yun, S.: An Adaptive Genetic Algorithm with Diversity-Guided Mutation and its Global Convergence Property. Journal of Central South University of Technology, 323–327 (2004)
13. Dierks, T., Allen, C.: The TLS Protocol Version 1.0, RFC Editor (1999)
14. Oppliger, R., Gajek, S.: Effective protection against phishing and web spoofing. In: Dittmann, J., Katzenbeisser, S., Uhl, A. (eds.) CMS 2005. LNCS, vol. 3677, pp. 32–42. Springer, Heidelberg (2005)
15. Nitto, E., Dubois, D., Mirandola, R.: Overlay self-organization for traffic reduction in multi-broker publish-subscribe systems. In: International Conference on Autonomic Computing (2009)

Enrichment of Raw Sensor Data to Enable High-Level Queries*

Kenneth Conroy[1] and Mark Roantree[2]

[1] CLARITY: Centre for Sensor Web Technologies, School of Computing,
Dublin City University
{kconroy,mark}@computing.dcu.ie
[2] Interoperable Systems Group, School of Computing, Dublin City University

Abstract. Sensor networks are increasingly used across various application domains. Their usage has the advantage of automated, often continuous, monitoring of activities and events. Ubiquitous sensor networks detect location of people and objects and their movement. In our research, we employ a ubiquitous sensor network to track the movement of players in a tennis match. By doing so, our goal is to create a detailed analysis of how the match progressed, recording points scored, games and sets, and in doing so, greatly reduce the effort of coaches and players who are required to study matches afterwards. The sensor network is highly efficient as it eliminates the need for manual recording of the match. However, it generates raw data that is unusable by domain experts as it contains no frame of reference or context and cannot be analyzed or queried. In this work, we present the UbiQuSE system of data transformers which bridges the gap between raw sensor data and the high-level requirements of domain specialists such as the tennis coach.

1 Introduction

Many new applications employ sensors or networks of sensors to automatically monitor and generate reports and analysis across domains. Increasingly, elite sports men and women are monitored to determine the effects of various exercises on their bodies. In almost every case, sports coaches will watch video recordings of previous matches to look for player faults and determine strengths and weaknesses. The problem with this effort is that it is extremely time-consuming as the coach must search for key moments in the match. In the case of tennis, this problem is exacerbated as matches can be up to five hours in length, and coaches will often have many players in their charge. What these coaches require is an automatic analysis of tournament and practice matches with the possibility to query to retrieve key segments. In more advanced scenarios, they require data mining functionality to analyze matches based on duration of points and games, analysis of defensive against attacking play; and details on games where the player had service. Our research involves a collaboration with tennis coaches

---

* This work is supported by Science Foundation Ireland under grant 07/CE/I1147.

P. García Bringas et al. (Eds.): DEXA 2010, Part II, LNCS 6262, pp. 462–469, 2010.

in Ireland to determine if it was possible to capture game analysis automatically and provide some form of query interface for coaches, to facilitate extracting the type of information described above. In a generic sense, we present a framework and methodology for automated processing of wireless sensor data so that it can be queried using a standard query language. The major benefit of our research is to provide queryable information to knowledge or domain workers using the sensed data. Our contribution is in the development of a framework and data management layer, with algorithms, to automate the analysis of a ubiquitous sensing environment.

In a ubiquitous computing environment we have a space and a selection of basic sensor data. Our space is equipped with a *Ubisense* [3] setup, where sensors are fitted on all sides of the space. Portable *ubitags* are then held by a participant and the sensors track its movement through the space. The raw data output is primitive, consisting of only three distinct properties: Ubitag ID, timestamp and 3D location (x,y,z). There is a significant gap between the raw sensor data generated by the ubiquitous environment and the query and analysis needs of the coach/domain specialist. To bridge this gap, a system of both generic and domain-specific layers is required to provide meaning to the data and solve complex queries. This system uses generic functions to first allow basic queries in XPath/XQuery[12], the query languages for XML, and then allow an interaction between structurally enriched sensor data and a domain specific context database. The system consists of three layers, the sensing layer which provides the raw data, the process layer which applies the data processing; and the storage layer where context is provided and enriched files are stored. To test our system in a real-world scenario, we collaborated with Tennis Ireland [2], the governing body of tennis in Ireland, deploying Ubisense on an indoor tennis court, and worked with a national coach to meet their analytical requirements after tennis matches. The sport coaches need to query the data for certain events, such as a players serve, a point being scored or the end of a particular game. The system developed allows the coaches to request a breakdown of the entire match, in terms of its different states (games, points, serves).

This setup illustrates a real-world application where there is a substantial gap between the raw data sensed and the query requirements of the coaches. At present, event detection techniques in both tennis and other sports is primarily tackled by video and audio recognition techniques. These are generally computationally expensive and inaccuracies can develop when players leave the line-of-sight of the camera, as well as poor picture quality and the inclusion of graphics, replays and advertisements on broadcast media. The Ubisense system uses ubitag IDs to ensure no mix up in identification, and the static nature of the environment provides an ideal training environment where complex actions can be detected without noisy data.

2   The UbiQuSE Architecture

In this section we present the *Ubiquitous Query Sensing Environment (UbiQuSE)* architecture which was designed to facilitate both live and offline queries, for both

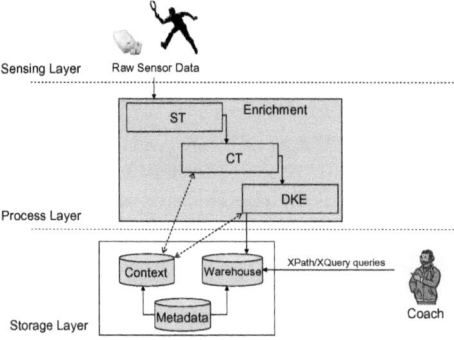

Fig. 1. UbiQuSE in the Tennis Domain

tracking and context purposes. The architecture consists of three broad layers: Sensing, Process and Storage. However, it is the processors of the Process Layer that merit detailed discussion. Figure 1 illustrates the UbiQuSE architecture in the tennis domain.

The sensing layer contains the devices which generate the raw data in a sensing environment. The hardware platform for our current domain application is the Ubisense network. Ubisense generates data based on a participants movement, specifically tracking position in space using triangulation of ultra-wide band (UWB) radio waves. When a participant carrying a ubitag moves, Ubisense detects the movement and the timestamp and location of the ubitag, in (x,y,z) coordinates, are transmitted. Essentially, this is the hardware layer for UbiQuSE and provides the raw sensor data.

The process layer contains the main processing algorithms and reduces the gap between user query requirements and raw sensor data. There are three processors which when used together will transform the data into an XML format, which users can query using XQuery expressions. The *Structural Transformation (ST)* and *Contextual Transformation (CT)* processors are generic and remain unchanged across application scenarios. A separate *Domain Knowledge Enrichment (DKE)* processor is required as UbiQuSE is used in different environments.

ST Processor: Structural Transformation. The role of the Structural Transformation processor is to convert the sensor data from an unstructured stream of data to a structurally enriched XML format. The basic conversion wraps the Ubisense sensor data into `<x>`,`<y>`, `<z>`, `<player>`, and `<timestamp>` tags which are generic to any Ubisense environment. Very basic queries, limited to the exact position of a player in space can be determined using XQuery following enrichment.

CT Processor: Contextual Transformation. Our vision for ubiquitous computing is that all sensed data is associated with a specific *Zone* (as part of a Smart Space). Each activity may have a series of *States* and in the case of a tennis match, these states will be Game Number, Set Number etc. The combination of these Zones and States provide a powerful enrichment to basic sensed data as it permits us to make certain assumptions with varying degrees of confidence. Furthermore, with our metamodel approach, the application of Zone and State information is performed in a generic fashion, standard to any activity that required Zones and/or States.

The Contextual Transformation processor provides contextual enrichment for sensed data. This processor is not domain specific but the UbiQuSE framework can still add Zone and State information using the metamodel approach illustrated in Fig. 2. Both the Zone and State elements have templates to suit application scenarios so that the system remains unchanged as it moves from domain to domain. The zones are contained within newly created `<zone>` tags, which appear as children to each `<location>` element in the sensor data file. In our tennis case study, the context specific zonal tags include `<special_zone>` and `<side>`. A further generic operation - `addStates` - is carried out during this stage. This function adds domain tags relevant to future enrichment of the data. The exact tags created depends on the domain, in the case of tennis `<game>`, `<serve>`, `<point>`, `<hit>`, `<receive>`, `<duration>`, `<change_side>` are added.

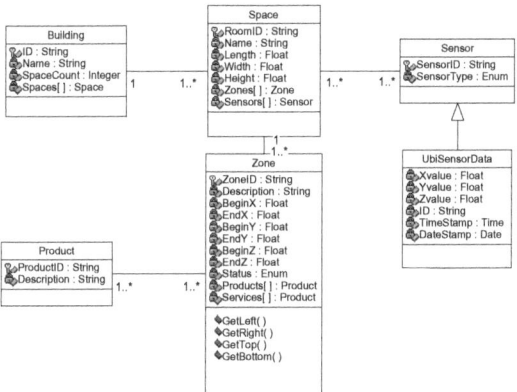

Fig. 2. SmartSpace Metamodel

DKE Processor: Domain Knowledge Enrichment. The role of the Domain Knowledge Enrichment processor is to apply the rules of the particular domain (or sport) to the sensor data to allow the user the ability to return the required query results. The input to this processor is the output from the CT processor. The output is a fully enriched XML file on which XPath/XQuery expressions can be used to detect complex domain specific states and events. It provides the final step in providing a complete break down of the domain into its constituent states. As this processor changes from domain to domain, we focus a detailed discussion of the tennis match case study in the following section.

The storage layer contains both the context repository and the final XML output following enrichment, available for querying. The context repository is accessed throughout processing in order to identify the space being used, its corresponding zonal information and the domain specific rules or expectancies of the scenario being recorded.

3   Tennis Match Case Study

In tennis, the zones originate from the actual structure of the tennis court, as illustrated in previous works [4] [5]. The game itself has a rigid structure: when a player is serving, the receiving player is usually beyond the baseline on the opposite side of the court. However in practice, the receiver can be closer to the net, or within the baseline, which requires flexibility in our zoning rules. To improve accuracy, the boundaries of the special zones(`<special_zone>`) were altered and a new receiver zone was defined. By zoning the court, we can use the rules of tennis to define a rule set to deduce probable actions during the match.

We applied the rules of tennis [1] to our system and as required by the domain specialists, the complete breakdown of a tennis match was automatically computed based solely on the Ubisense data. Serves, points and games are the key aspects of tennis that were identified by our system. The functions used are `detectServes`, `applyPointBoundaries` and `applyGameBoundaries`, the logic for identifying each of these is explained below.

1. `detectServes` - We know that if we can detect when a new player is serving, then the match has changed state (a new game or possibly set has commenced). Accuracy in detecting this change in serve is crucial to providing correct results. Thus, the special zones mentioned above were created. The algorithm consists of examining the position of both players for a specific time period. A lack of consecutive serve event detections by the same player indicates a false positive which is then dropped.
2. `applyPointBoundaries` - Point scoring is identified by checking multiple instances of serves by the same player. As stated in the rules of tennis [1], players who serve from one side of the baseline must serve from the other side following a point being scored (by either player), providing the point does not also result in the end of a game - in which case players switch service. The duration of a point and the time between each point can be calculated by examining the timestamps of the current and preceding points.
3. `applyGameBoundaries` - The game boundaries are based on which side of the court a player stands as well as when each player switches serve. The rules state that every odd-numbered game (1,3,5...) of a set is followed by a change of side and all games are followed by a change of service (from one player to the other).

4   Experimental Data

After the sensor data is processed by the UbiQuSE processors, data in the XML database is sufficiently transformed to enable querying using the XQuery language. An example of some of the queries are expressed in XQuery and presented

Table 1. XQuery implementations of required queries

Query
1 let $c := collection('db/ubisense/trainingFeb10') return $c//player[@id=1] /UbiQuSE/state[game=1]/hit
2 let $c := collection('db/ubisense/trainingFeb10') return $c//state[game=1]/point
3 let $c := collection('db/ubisense/trainingFeb10') return $c//player[@id=1] /UbiQuSE/location[@timestamp=8000]/zone

in Table 1. More complex queries are presented in the Technical Report [6]. In fact, the enrichment of the XML requires simple query expressions to return previously complex requirements. In this section, we report on the accuracy of our results as our algorithms make many assumptions based on player location and movement, combined with the semantics of the sport of tennis. In order to identify the accuracy of the Ubisense system, we took a sample set and examined player positions during a number of games. In our setup, inaccuracy of Ubisense is confined to a maximum error of 15cm with the greater inaccuracies occur on the periphery of the Ubisense space. Other researchers have shown inaccuracies of up to 48cm on the periphery of their setup [8]. We have taken Ubisense inaccuracy into account when devising our algorithms.

The focus of our experiments was on determining the accuracy of the UbiQuSE algorithms in the Domain Knowledge Enrichment processor which detects the key events in a tennis match. In particular, we examine how the system performs in detecting player serves, points scored, and game boundaries. Our experimental data consisted of a number of best-of-5-set training matches between two elite tennis players playing in a competitive manner. We randomly selected one of the sets for experimental evaluation, used UbiQuSE to enrich the raw data, and visually inspected the video recording to time the exact occurrence for all serves, points and game boundaries. These manual records and then compared with events identified by UbiQuSE, the results of which are summarised below.

- Out of a total of 72 serves made in the set, 70 were correctly detected by our system, representing 97% of the total. Apart from the 2 undetected serves, we had 9 false positives where we believed serves to be taking place. End-users reported a preference for serve detection over the low number of false positives as these can easily be ignored.
- Regarding point detection, a total of 44 points were scored in the set we examined. All but one of these were detected by `applyPointBoundaries`. As a result 98% of points are correctly identified by the system.
- Game detection is built on the previous algorithms for point and serve detection and can be affected by poor accuracy. In our experiments, UbiQuSE determined all six games in the set were correctly identified by our system.

5 Related Research

A general model to represent semantics of a smart space based on lower-level location contexts is provided by [7]. Several sensor types, including Ubisense, generate sensor data which is combined with a time factor and a conditional confidence value factor to identify valid contexts for that situation. Context is described using an RDF triple. Like UbiQuSE, space is limited to where the sensors (Ubisense) can get a signal. However, we have built a model where we can change zonal boundaries depending on a changing domain, or apply to another Ubisense setup. A formal structured format for representing contextual information of a smart space environment is outlined by [9]. They use RDF to define context ontologies, and provide a generic querying mechanism to infer high-level contexts based on rules. While this approach does define an infrastructure for smart spaces and represents contexts as easily interpreted semantic markups from which higher-level contexts can be inferred, the context is generally provided by external sensing sources. This is in contrast to our approach where we take previously meaningless raw data from one source (Ubisense) and transform it structurally and semantically in order to allow the detection of complex actions and events.

The identification of events in sports, and in tennis in particular are mainly based on video analysis [10][11]. Approaches make use of audiovisual data to extract relevant features of a video clip. These features when combined with the semantic concepts and game structure, can identify the required events. In [10] ace detection has an accuracy of 93%, however, a net approach in the same system has an accuracy of only 50%. Our system has an accuracy of 100% when detecting net approaches. Video recognition can suffer from poor camera angles, poor conditions and is usually very expensive computationally to perform.

6   Conclusions

In this paper, we presented the UbiQuSE system which provides a data management layer for sensor networks. It contains two generic processors, the Structural Transformation and Contextual Transformation processors, that operate on any sensor network that captures location and movement within a smart space. These processors take the raw sensor data and provide both structure and basic semantic enrichment as they place each player movement within some context. The third processor, the Domain Knowledge Enrichment processor, differs across domains, and is used to populate generic data structures and provide higher levels of enrichment. Our goal in creating UbiQuSE was to provide the missing data management layer that bridges the gap between raw data and the high-level query expressions of specialist users. To demonstrate the effectiveness of UbiQuSE, we took a query set specified by the domain specialist (in this case a tennis coach) and expressed them using XQuery, the standard XML language. In our experiment, these expressions generated the required results after the sensor data was passed through UbiQuSE. Our experiments focused on the accuracy of our algorithms in detecting key events in the sensed data and results are reported and discussed in this paper. Future work is focused both on improving the accuracy of our event detection algorithms and widening the scope of the coaches requirements to include doubles matches and the complexities of the tie-break in tennis.

References

1. Rules of Tennis 2009 (2009),
   http://www.tennisireland.ie/userfiles/File/Database/
   IO_38417_original.pdf
2. Tennis Ireland, http://www.tennisireland.ie
3. Ubisense English Site, http://www.ubisense.net/
4. Shaeib, A., Conroy, K., Roantree, M.: Extracting tennis statistics from wireless
   sensing environments. In: Proceedings of the Sixth international Workshop on Data
   Management For Sensor Networks, DMSN (2009)
5. Technical Report,
   http://www.computing.dcu.ie/˜isg/publications/ISG-09-02.pdf
6. UbiQuSE Technical Report,
   http://www.computing.dcu.ie/˜isg/publications/ISG-10-04.pdf
7. Ye, J., McKeever, S., Coyle, L., Neely, S., Dobson, S.: Resolving uncertainty in
   context integration and abstraction: context integration and abstraction. In: Proceedings of the 5th international conference on Pervasive services, ICPS 2008 (2008)
8. Coyle, L., Ye, J., Loureiro, E., Knox, S., Dobson, S., Nixon, P.: A Proposed Approach to Evaluate the Accuracy of Tag-based Location Systems. In: Workshop on
   Ubiquitous Systems Evaluation, USE 2007 (2007)
9. Wang, X., Dong, J.S., Chin, C., Hettiarachchi, S., Zhang, D.: Semantic Space: An
   Infrastructure for Smart Spaces. IEEE Pervasive Computing 3(3), 32–39 (2004)
10. Tien, M., Lin, Y., Wu, J.: Sports wizard: sports video browsing based on semantic
    concepts and game structure. In: Proceedings of the Seventeen ACM international
    Conference on Multimedia, MM 2009 (2009)
11. O'Connaire, C., Kelly, P., Connaghan, D., O'Connor, N.: TennisSense: A Platform
    for Extracting Semantic Information from Multi-camera Tennis Data. In: Proceed-
    ings of the 16th International Conference on Digital Signal Processing, DSP 2009
    (2009)
12. http://www.w3.org/TR

Transductive Learning from Textual Data with Relevant Example Selection

Michelangelo Ceci

Dipartimento di Informatica, Università degli Studi di Bari
via Orabona, 4 - 70126 Bari, Italy
ceci@di.uniba.it

Abstract. In many textual repositories, documents are organized in a hierarchy of categories to support a thematic search by browsing topics of interests. In this paper we present a novel approach for automatic classification of documents into a hierarchy of categories that works in the transductive setting and exploits relevant example selection. While resorting to the transductive learning setting permits to classify repositories where only few examples are labelled by exploiting information potentially conveyed by unlabelled data, relevant example selection permits to tame the complexity of the task and increase the rate of learning by focusing only on informative examples. Results on real world datasets show the effectiveness of the proposed solutions.

1 Introduction

Transductive learning is an inference mechanism adopted from several classification algorithms capable of exploiting, as in *semi-supervised learning*, information potentially conveyed by unlabelled data to better estimate the data distribution when making predictions. However, transductive learning differs from *semi-supervised learning* since, instead of learning a function to be used to make predictions on any possible example, it is only possible to make predictions for the given set of unlabeled data. This means that transductive learning needs no general hypothesis and appears to be an easier problem than both semi-supervised learning and classical inductive learning.

Several transductive learning methods have been proposed in the literature for classification tasks. They exploit SVMs ([1] [9] [12] [5]), k-NN classifiers ([13]) and even general classifiers ([14]). However, a common problem in this learning setting comes from the high dimensionality of unlabeled data and labeled data that have to be simultaneously analyzed during learning. In order to face this problem, two orthogonal directions can be exploited: the first direction aims at simplifying the classification process by considering that categories can be organized hierarchically. The second direction aims at simplifying the classification process by considering only a subset of relevant examples for learning (relevant examples selection).

Indeed, both directions can be profitably pursued in the context of document categorization [22] that we consider in this paper. In fact, Hierarchical text categorization, that is, the process of automatically assigning one or more predefined

P. García Bringas et al. (Eds.): DEXA 2010, Part II, LNCS 6262, pp. 470–484, 2010.

categories to text documents where the pre-defined categories are organized in a tree-like structure, has received increasing attention in the last years [17,18,7,21]. From an information retrieval viewpoint, this hierarchical arrangement is essential when the number of categories is high, since thematic search is made easier by browsing topics of interests. Yahoo, Google Directory, Medical Subject Headings (MeSH), Open Directory Project and Reuters Corpus Volume I provides typical examples of organization of documents in topic hierarchies.

The hierarchical structure of categories may help to simplify the classification process: while in flat classification a given example is assigned to a category on the basis of the output of one or a set of classifiers, in hierarchical classification the assignment of a document to a category can be done on the basis of the output of multiple sets of classifiers, which are associated to different levels of the hierarchy and distribute example among categories in a top-down way. The advantage of this hierarchical view of the classification process is that the problem is partitioned into smaller subproblems, each of which can be effectively and efficiently managed [4].

As for relevant example selection, in [2], the authors observed that there are at least three reasons for selecting examples to be used during the learning process: *i)* the learning process is computationally intensive, *ii)* the cost of manual labelling is high, *iii)* it is necessary to increase the rate of learning by focusing only on informative examples. In this context, all these motivations make relevant example selection particularly suited. Surprisingly, there is not much research on relevant example selection for text classification. The issue is mostly addressed either with the traditional statistical approach of sampling [27] or by more elaborate, but sometimes heuristic, approaches. For Instance, in [26] the problem is addressed using a distance measure. In essence instances that are "closer" to each other tend to bear overlapping information; therefore, some of them can be discarded.

In this paper, we investigate the use of transductive learning by exploiting both hierarchical classification and relevant example selection. At this aim, we exploit a modified version of an inductive hierarchical learning framework that permits to classify examples (documents) in internal and leaf nodes of a hierarchy of categories. The learner is asked to take into account only a subset of the original documents. This way it is possible to speed up learning times without loosing in accuracy. Transductive learning exploits the Spectral Graph Transducer (SGT) [13], in the context of a hierarchical classification framework. Experimental results on real world datasets are reported.

This paper is organized as follows. The problem to solve and the background work are introduced in the next section. The proposed solution is described in Sections 3 and 4. Experimental results on real world datasets are reported in Section 5 while conclusions are drawn in Section 6.

2   Preliminaries

The problem we intend to solve can be formalized as follows:

Let D be a set of documents and $\Psi : D \to Y$ be an unknown target function, whose range is a finite set $Y = \{C_1, C_2, \ldots, C_L\}$ where $\{C_1, C_2, \ldots, C_L\}$ are categories organized according to a tree-like structure such that $\forall i = 2, \ldots, L \quad \exists | j = 1, \ldots, L,\ i \neq j$ such that C_i is a subcategory of C_j (C_1 is the root category). Then, the transductive classification problem can be defined as follows:

   Given:

   – a training set TS of pairs (d_i, y_i) where d_i represents a document and $y_i \in Y$ represents the class (label)
   – a working set WS of unlabelled documents;

Find: a prediction of the class value of each document in the working set WS which is as accurate as possible.

The learner receives full information (including labels) on the documents in TS and partial information (without labels) on the documents in WS and is required to predict the class values only of the examples in WS.

The hierarchical organization of categories adds additional sources of complexity to the transductive learning problem. First, documents can either be associated to the leaves of the hierarchy or to internal nodes. Second, the set of features selected to build a classifier can either be category specific or the same for all categories (corpus-based). Third, the training set associated to each category may or may not include training documents of subcategories. Fourth, the classifier may or may not take into account the hierarchical relation between categories. Fifth, a stopping criterion is required for hierarchical classification of new documents in non-leaf categories. Sixth, performance evaluation criteria should take into account the hierarchy when considering classification errors.

We face such complexity by resorting to solutions investigated in a previous work done on hierarchical classification in the classical inductive setting [4]. Those solutions have been implemented in the system WebClass. In WebClass, the search proceeds top-down from the root to the leaves according to a greedy strategy. When the document reaches an internal category C, it is represented on the basis of the feature set associated to C. The classifier of category C returns a score for each direct subcategory. Score thresholds, which are automatically determined for all categories, are used to filter out the set of candidate subcategories. If the set is empty, then search is stopped, otherwise the subcategory corresponding to the highest score is selected and the (greedy) search recursively proceeds with that subcategory (if not leaf). The last crossed node in the hierarchy is returned as the candidate category for document classification (*single-category classification*). If the search stops at the root, then the document is considered *unclassified*. An example is illustrated in Figure 1.

Each document is represented at decreasing levels of abstraction by considering features selected according to the $maxTF \times DF^2 \times ICF$ [4]. According to the definition of such measure, features tend to be more specific for lower level categories. These different representations of a document make the classification scores incomparable across different nodes in the hierarchy and prevent

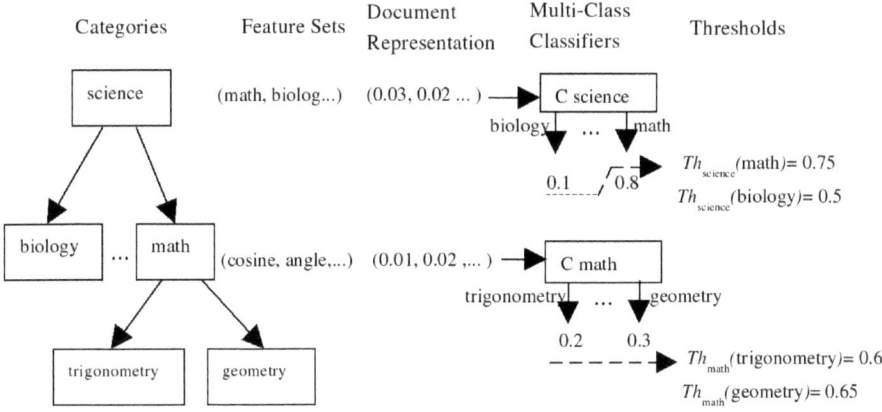

Fig. 1. Classification of a new document. On the basis of the scores returned by the first classifier (associated to the category *science*) the document is passed down to *math*. The scores returned by the second classifier (associated to the category *math*), are not high enough to pass down the document to either *trigonometry* or *geometry*. Therefore, the document is classified in the *math* category.

the correct application of an exhaustive search strategy instead of the proposed greedy strategy.

During learning, for each internal category C of the hierarchy and for each document in WS to be classified, the decision on which category C' among the direct subcategories of C is the most appropriate to receive the document has to be taken. In general, however, a document should not be necessarily passed down to a subcategory of C. This makes sense in the case that the document to be classified deals with a general rather than a specific topic, or in the case that the document to be classified belongs to a specific category that is not present in the hierarchy and it makes more sense to classify the document in the "general category" rather than in a wrong category.

To support the classification of documents also in the internal categories of the hierarchy, it is necessary to compute the thresholds that represent the "minimal score" (returned by the classifier), such that a document can be considered to belong to a direct subcategory. More formally, let $\gamma_{C \to C'}(d)$ denote the score returned by the classifier associated to the internal category C when the decision of classifying the document d in the subcategory C' is made. Thresholds are used to decide if a new testing document is characterized by a score that justifies the assignment of such a document to C'. Formally, a new document $d \in WS$ temporary assigned to a category C will be passed down to a category C' if $\gamma_{C \to C'}(d) > Th_C(C')$, where $Th_C(C')$ is the score threshold.

The algorithm for the automated determination of thresholds $Th_C(C')$ is based on a bottom-up strategy and minimizes a measure based on a *tree distance*[4].

3   Relevant Example Selection

When working in the transductive setting, we do not distinguish between learning and classification steps. However, the hierarchical organization of categories
requires a preliminary step during which thresholds are automatically identified. Later on, in a second stage, the transductive classification is performed.
Indeed, the two phases are not completely independent each other since the algorithm for automatic threshold identification estimates thresholds on the basis
of a simulation of the classification step on the training set. Relevant example
selection is performed both in the automatic threshold determination and in the
transductive classification task. In particular, while in automatic threshold determination relevant examples are determined from the training set TS for each
internal category C of the hierarchy, in the classification case, both examples in
TS and examples in WS are analyzed for each internal category C of the hierarchy. Transductive classification of relevant examples in WS is then extended
to other examples in WS by means of a K-NN label propagation by means of
the classical K-NN classifier.

For relevant example selection, we consider two different approaches: the first
approach reduces the number of documents by exploiting clustering algorithms,
while the second approach identifies and keeps only documents that are at the
boundary of the class. Before describing how relevant documents are selected,
we present details on their representation.

3.1   Document Representation

Document representation depends on a preprocessing step which aims at:

1. Removing *stopwords*, such as articles, adverbs, prepositions and other frequent words.
2. Determining equivalent stems (*stemming*) by means of Porter's algorithm
   for English texts [20].

After these preprocessing steps, documents are represented by means of a feature
set which is determined on the basis of some statistics whose formalization is
reported below. Let

- C be an internal node in the hierarchy of categories,
- C' a direct subcategory of C,
- d a training document from C',
- w a token of a stemmed (non-stop)word in d,
- $TF_d(w)$ the *relative* frequency of w in d,
- $Training(C) \subseteq TS$ the set of documents in C and its subcategories,
- $TF_{C'}(w) = max_{d \in Training(C')} TF_d(w)$ the maximum value of $TF_d(w)$ on all
  training documents d of category C',
- $DF_{C'}(w) = \frac{|\{d \in Training(C')| \quad w \; occurs \; in \; d\}|}{|Training(C')|}$ the percentage of documents of
  category C' in which w occurs,
- $CF_C(w)$ the number of subcategories $C'' \in DirectSubCategories(C)$ such
  that w occurs in a document $d \in Training(C'')$.

Then the following measure: $v_i = TF_{c'}(w_i) \times DF_{c'}^2(w_i) \times \frac{1}{CF_c(w_i)}$ is used to select relevant tokens for the representation of documents in C.

Tokens that maximize v_i ($maxTF \times DF^2 \times ICF$ criterion) are those commonly used in documents of category C' but not in its sibling categories. The *category dictionary* of C', $Dict_{C'}$, is the set of the best n_{dict} terms with respect to v_i, where n_{dict} is a user defined parameter.

For each learning task, the following feature set is used:

$$FeatSet_C = \bigcup_{C' \in DirectSubCategories(C)} Dict_{C'} \qquad (1)$$

and documents are represented according to the classical $TF \times idf$ measure [22].

3.2 Clustering-Based Relevant Example Selection

This approach follows the main idea of cluster sampling where the goal is to sample a set S documents into l subsets $N_1, N_2, \ldots, N_l$ respectively. These subsets (called strata) are non-overlapping, and together they comprise the whole of the data set (i.e., $\cup_{i=1..l} N_i = N$). When the strata have been determined, a sample is drawn from each stratum. Drawings are performed independently in different strata. Cluster sampling is often used in some applications where we wish to divide a heterogeneous data set into subsets, each of which is internally homogeneous [15].

For relevant examples selection, we consider the simple *k-means* [16] clustering algorithm for the identification of strata.

Once the strata have been identified, each cluster is represented by means of its surrogate. In our approach, as in the case of the Rocchio classifier [22], the surrogate of the cluster is its centroid $d'(i)$:

$$d'(i) = \sum_{d_j \in N_i} \frac{d_j}{|N_i|} \qquad (2)$$

We also evaluate the opportunity of considering, in alternative to the centroid a representative example, that is, the example in N_i that appears to be closer to the cluster centroid. Formally:

$$d''(i) = \arg\min_{d_j \in N_i} d_1(d_j, d'(i)) \qquad (3)$$

where $d_1(\cdot, \cdot)$ is the euclidean distance measure between document vectors.

For example reduction purposes, for each internal category C of the hierarchy, both documents in TS and in WS are represented according to features in $FeatSet_C$ and according to the $TF \times idf$ measure.

3.3 Class Border Identification for Relevant Example Selection

In this alternative approach to example reduction, the main idea is that of exploiting support vectors extracted by support vector machines [24] in order to identify the class border.

Indeed, in this case, the class is associated to the learning task that permits to establish whether an example should be passed down from a category C to its descendant category C' or not. This means that we extract a set of relevant examples by considering as positive examples the documents that belong to $Training(C')$ and as negative examples the documents that belong to $Training(C) - Training(C')$.

As in the case of clustering-based relevant example selection, this approach permits to reduce examples both in TS and in WS.

Let $(\{\mathbf{x}_1, y_1\}, (\mathbf{x}_2, y_2), \ldots, (\mathbf{x}_N, y_N)\}$ be the set of training documents in $Training(C)$ such that $\mathbf{x}_i \in \mathbb{R}^{|FeatSet_C|}$ ($\mathbf{x}_i$ is a document vector) and $y_i = +1$ if $\mathbf{x}_i \in Training(C')$ and $y_i = -1$ if $\mathbf{x}_i \in Training(C) - Training(C')$. An SVM identifies the hyperplane in $\mathbb{R}^{|FeatSet_C|}$ that linearly separates positive and negative examples with the maximum margin (*optimal separating hyperplane*). In general, the hyperplane can be constructed as the linear combination of all training examples, however, only some examples, called *support vectors*, do actually contribute to the optimal separating hyperplane which can be represented as:

$$f(x) = \sum_{i=1}^{N} y_i \alpha_i \mathbf{x}_i \cdot \mathbf{x} + b \tag{4}$$

Indeed, we are only interested in identifying support vectors, that is, vectors for which $\alpha_i \neq 0$ (see Figure 2a).

The coefficients of the linear combination α_i and b are determined by solving a large-scale quadratic programming (QP) problem, for which efficient algorithms that find the global optimum exist.

The linear separability appears to be a strong limitation, however, as experimentally observed by [11], most text categorization problems are linearly separable.

SVMs are based on the *Structural Risk Minimization* principle: a function that can classify training data accurately and which belongs to a set of functions

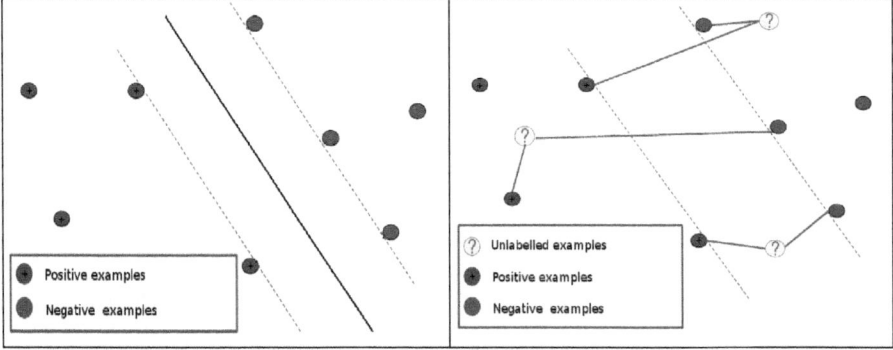

Fig. 2. a) Support vectors (examples on the dashed lines). b) Relevant examples selection from WS.

with the lowest capacity (particularly in the VC-dimension) [24] will generalize best, regardless of the dimensionality of the feature space $|FeatSet_C|$. Therefore, SVMs can generalize well even in large feature space, such as those used in text categorization. In the case of the separating hyperplane, minimizing the VC-dimension corresponds to maximizing the margin.

The SVM we use in the identification of support vectors is a modified version of the Sequential Minimal Optimization classifier (SMO) [19] with linear kernels. SMO is very fast and is based on the idea of breaking a large QP problem down into a series of smaller QP problems that can be solved analytically. This allows us to directly identifying document vectors x_i for which $\alpha_i \neq 0$.

Let $D(C, C') = \{x_i | x_i \in Training(C), \alpha_i\}$ be the set of support vectors that have been identified, they are used in order to identify the subset of documents in WS to be considered in the classification phase. In particular, we are only interested in keeping the most discriminative $p\%$ documents from WS according to the score function $score : WS \to \mathbb{R}$ defined as follows:

$$score(d) = \min_{x_i \in Training(C');\ x_j \in Training(C)-Training(C')} d_1(d, x_i) + d_1(d, x_j) \quad (5)$$

Intuitively, we only consider examples in WS that are close to both positive and negative class margins.

4   SGT Hierarchical Classifier

In this section, the proposed learning method is described in detail. In particular, we first introduce the hierarchical transductive classification, and, then, we detail the application of SGT.

4.1   Hierarchical Transductive Classification

We assume that a classifier returns a numerical score $\gamma_{C \to C'}(d)$ that expresses a "belief" that a document d belonging to C also belongs to a direct subcategory C'. The document d is passed down if $\gamma_{C \to C'}(d)$ is greater than a threshold, which is automatically determined for each class by an algorithm that minimizes, on the training set, a tree distance. This distance measures the number of edges in the hierarchy of categories between the actual class of a document and the class returned by the hierarchical classifier [8].

As in [4], the computation proceeds bottom-up, from leaves to the root. The difference is that in this work we learn, for each internal category C, m two-class classifiers, one for each subcategory C' and compare the scores. This is quite different from what proposed in [4], where a 1-of-m classifier is learnt for each internal node. This would permit us to exploit two-class classifiers and avoid computational problems coming from a subsequent pair-wise coupling classification [10].

The classifier used to classify examples belonging to internal nodes of the hierarchy is based on the Spectral Graph Transducer algorithm (SGT) proposed in [13] that works in the transductive setting. Although, in its final formulation SGT returns hard class assignments, we use the SGT algorithm in order to compute the scores $\gamma_{C \to C'}(d)$. This way, the algorithm can be used both to compute thresholds and to classify examples in the working set. The problem solved by each application of SGT can be formalized as follows:

Given:

- An internal category C;
- A direct subcategory C' of C;
- A set of l relevant labeled examples (documents) belonging to C and its descendants (identified as specified in the previous section). Positive examples (labeled with +1) refer to documents in $Training(C')$ and all its descendands, while negative examples (labeled with -1) refer to all other examples in categories descendants of C (in $Training(C) - Training(C')$);
- A set of relevant unlabeled examples (possibly) belonging to C and its descendants;

the task of the transductive algorithm is to compute the score $\gamma_{C \to C'}(d)$ for each relevant document d in the training or in the working set such that error is minimized.

4.2   Application of SGT Algorithm

The algorithm builds a nearest neighbor graph $G = (N, E)$, with labeled and unlabeled examples as vertexes, and dissimilarity measure $(d_2(d_i, d_j))$ between the neighboring examples as edge weights. SGT assigns labels to unlabeled examples by cutting G into two subgraphs G^- and G^+, and tags all examples corresponding to vertexes in G^- (G^+) with -1 (+1). To give a good prediction of labels for unlabeled examples, SGT chooses the cut of G that maximizes the normalized cut cost.

$$\max_{y} \frac{cut(G^+, G^-)}{|\{i|y_i = +1\}||\{i|y_i = -1\}|} \qquad (6)$$

where $y = [y_i]_{\{i=1,..,n\}}$ is the prediction vector (where n is the number of both labeled and unlabeled relevant examples), and $cut(G^+, G^-)$ is the sum of the weights of all edges that cross the cut (i.e., edges with one end in G^- and the other in G^+). The optimization is subjected to the following constraints: (i) $y_i \in \{-1, +1\}$ and (ii) labels for labeled training examples must be correct, i.e., vertexes corresponding to positive (negative) labeled relevant training examples must lie in G^+ ($G-$). As this optimization is NP-hard, SGT performs approximate optimization by means of a spectral graph method which solves the following problem [6]:

$$\min_{Z} \quad Z^T L Z + c(Z - y)^T C(Z - y) \qquad (7)$$

$$\text{such that } Z^T 1 = 0 \text{ and } Z^T Z = n$$

where

- Z is the transformed prediction vector with comparable scores,
- L is computed as the Laplacian matrix $L = (B - A)$ in the case of RATIO CUT or, alternatively, as the normalized Laplacian matrix obtained as $L = B^{-1}(B - A)$ in the case of NORMALIZED CUT [23];
- $A = [a_{i,j}]_{\{i,j=1,..,n\}} = [a'_{i,j} + a'_{j,i}]_{\{i,j=1,..,n\}}$ where $a'_{i,j} = d_2(d_i, d_j)$;
- $B = [b_{i,j}]_{\{i,j=1,..,n\}}$ is the diagonal matrix such that $b_{i,i} = \sum_j a_{i,j}$;
- c is a user-defined parameter;
- $C = [c_{i,j}]_{\{i,j=1,..,n\}}$ is a diagonal cost matrix with $c_{i,i} = l/(2l+)$ for positive relevant examples, $c_{i,i} = l/(2l-)$ for negative and $c_{i,i} = 0$ for unlabelled relevant examples;
- $l+$ $(l-)$is the number of positive (negative) relevant labeled examples and $l \leq n$ is the number of relevant labelled examples;
- $\gamma = [\gamma_i]_{\{i=1,..,n\}}$ is a vector with $\gamma_i = \sqrt{l - /l+}$ for positive examples, $\gamma_i = \sqrt{l + /l-}$ for negative examples and $\gamma_i = 0$ for unlabelled examples.

This minimization problem leads to compute

$$Z^* = V(M - \lambda^* I)^{-1} b \tag{8}$$

where V is the matrix with all eigenvectors of L except the smaller; $b = CV^T C\gamma$; $M = (D + cV^T I)$; D is the diagonal matrix with the square of all eigenvalues of L except the smaller; λ^* is the smaller eigenvalue of $\begin{bmatrix} M & -I \\ \frac{-1}{n}bb^T & M \end{bmatrix}$.

The vector $Z^* = [z_i^*]_{\{i=1,..,n\}}$ is then used to compute the score $\gamma_{C \to C'}(d_i)$. In particular:

$$\gamma_{C \to C'}(d_i) = (z_i^* - \min_j z_j^*)/(\max_j z_j^* - \min_j z_j^*) \tag{9}$$

The dissimilarity measure $d_2(\cdot, \cdot)$ used in this work is the cosine dissimilarity computed as follows:

$$d_2(d_i, d_j) = 1 - \frac{\boldsymbol{d_i} \cdot \boldsymbol{d_j}}{\|\boldsymbol{d_i}\|_2 \|\boldsymbol{d_j}\|_2} \tag{10}$$

where $\boldsymbol{d_i}$ $(\boldsymbol{d_j})$ represents the $TF \times idf$ representation of d_i (d_j).

5   Experiments

To evaluate the applicability of the proposed approach, we performed experiments on distinct experimental settings involving two distinct datasets. As baseline we considered the Hierarchical SGT transductive classifier that do not exploit relevant example selection [3].

Results are obtained with the following parameters: $c = 10^4$ as proposed in [13]; $n_{dict}=100$; $\alpha = 0.4$; K used in the K-NN label propagation is set to the highest odd integer such that $K \leq \sqrt{n}$ (according to [25]) where n is the total number of both labelled and unlabelled examples that (possibly) belong to the processed category C; $n_s = 5\% \times n$; $p\% = 5\%$. Values of n_{dict} and α are estimated after an empirical evaluation. Values of n_s and $p\%$ are set in order to make the comparison between relevant example selection algorithms fair.

Results obtained with the different experimental settings aim at comparing the trasductive algorithm without example reduction (that we indicate in these experiments as HSGT- Hierarchical SGT) with the trasductive algorithm with relevant example selection based on k-means clustering and $d'(\cdot)$ cluster representation (KmSGT), the trasductive algorithm with relevant example selection based on k-means clustering and $d''(\cdot)$ cluster representation (SelectSGT) and the trasductive algorithm with relevant example selection based on support vectors (SVSGT).

5.1 Datasets

Reuters Corpus Volume I (RCV1). RCV1 is a benchmark dataset widely used in text categorization and in document retrieval[1]. It consists of over 800,000 newswire stories, collected by the Reuters news and information agency. The stories have been manually coded using three orthogonal hierarchical category sets. In our study, similarly to other authors [28], we use topic codes for categorization. Topics hierarchy consists of a set of 104 categories organized in 4-levels.

We pre-processed documents as proposed by Lewis et al. and, in addition, we considered only documents associated to a single category. This selection is due to the fact that in this study we are interested in investigating single category assignment [4]. We separated the training set and the testing set using the same split adopted by Lewis et al. In particular, documents published from August 20, 1996 to August 31, 1996 (document IDs 2286 to 26150) were included in the training set, while documents published from September 1, 1996 to August 19, 1997 (document IDs 26151 to 810596) were included in the working set. The result was a split of the 804,414 documents into 23,149 training documents and 781,265 working documents. After multiple-label document removal, we had 150,765 documents, (4,517 training documents and 146,248 testing documents).

In our experiments we analyze three large subsets of RCV1: A subset rooted in the category "*C3*" (1,647 training documents, 50,345 working documents); a subset rooted in the category "*C18*" (1,438 training documents, 44,148 working documents); a subset rooted in the category "*MCAT*" (10,715 training documents, 163,592 working documents).

Dmoz dataset. Dmoz data is obtained from the documents referenced by the Open Directory Project (ODP) (www.dmoz.org)[2]. We extracted all actual Web documents referenced at the top five levels of the Web directory rooted in the branch "*Health\ Conditions_and_Diseases*". Empty documents, documents containing only scripts, and documents whose size is less than 3Kb are removed. At the end, the dataset contains 3,668 documents organized in 203 categories.

---

[1] The dataset cannot be made available on-line without maintainers authorization.
[2] The dataset is available at http://www.di.uniba.it/%7ececi/micFiles/ dmoz_health_conditions_and_diseases_docs.zip.

The dataset is analyzed by means of a 3-fold cross-validation (CONDITION). Two subset of this dataset rooted in the category *"Cancer"* and in the category *"Cardiovascular disorders"* respectively are also analyzed by means of a 3-fold cross-validation. It is noteworthy that, differently from usual, in this paper the t-fold cross-validation uses in turn one fold for training and the remaining $t-1$ folds as working set. This is coherent with principles motivating the transductive approach where the working set is generally larger than the training set.

5.2 Results

Accuracy results[3] are reported in Tables 1 and 2. In particular, results in Table 1 show that relevant example selection permits to obtain classification accuracies that are generally comparable with those obtained with *HSGT*. By analyzing results in Table 2 it is possible to see that when considering only relevant

Table 1. Average accuracies obtained with HSGT, KmSGT, SelectSGT and SVSGT. Thresholds are obtained on the whole set of training examples.

DATASET	cut	HSGT	KmSGT	SelectSGT	SVSGT
CANCER	RATIO	64%	62%	55%	29%
	NORMALIZED	60%	56%	53%	32%
CARDIOVASCULAR	RATIO	63%	61%	53%	31%
	NORMALIZED	60%	55%	47%	42%
CONDITION	RATIO	37%	29%	26%	12%
	NORMALIZED	34%	23%	26%	12%
C3	RATIO	–	30%	32%	32%
	NORMALIZED	–	29%	7%	29%
C18	RATIO	–	65%	68%	49%
MCAT	RATIO	–	49%	50%	%5

Table 2. Average accuracies obtained with HSGT, KmSGT, SelectSGT and SVSGT. Thresholds are obtained on the set of relevant training examples.

DATASET	cut	HSGT	KmSGT	SelectSGT	SVSGT
CANCER	RATIO	64%	65%	59%	32%
	NORMALIZED	60%	60%	54%	28%
CARDIOVASCULAR	RATIO	63%	64%	57%	33%
	NORMALIZED	60%	60%	51%	34%
CONDITION	RATIO	37%	39%	36%	13%
	NORMALIZED	34%	37%	32%	14%
C3	RATIO	–	31%	28%	18%
	NORMALIZED	–	31%	12%	29%
C18	RATIO	–	69%	77%	70%
MCAT	RATIO	–	49%	50%	5%

---

[3] Due to space complexity problems, it was not possible to run HSGT on the datasets C3, C18 and MCAT.

examples in automatic threshold determination, accuracy significantly increases. In fact, in most of cases $KmSGT$ outperforms $HSGT$ even if it works on smaller set of examples. A possible reason can be found in the fact that, in this way, the classification performed by SGT is coherent with the automatic threshold determination phase.

By comparing results obtained with relevant example selection, we can see that the clustering algorithm permits to identify a good representative set of examples to be used during the learning phase. We cannot draw the same conclusion for $SVSGT$ that, as SMO [19], suffers from the high imbalanced distribution of examples. In fact, for C3 and C18, where categories are almost uniformly distributed $SVSGT$ provides interesting results.

It is also noteworthy that the RATIO CUT outperforms the NORMALIZED CUT both in terms of accuracy and efficiency (for this reason we do not report NORMALIZED CUT results for C18 and MCAT). This means that the use of a normalized cut in transductive learning is not as beneficial as in the case of image processing [23].

Finally, results reported in Table 3 give a clear perspective of the learning time reduction obtained with relevant example selection. As expected, this advantage is more clear when the automatic threshold determination algorithm works only on relevant examples.

Table 3. (Average) classification times with Ratio cut (in secs.). NS refers to thresholds obtained on the whole set of training examples. S refers to thresholds obtained on the set of relevant training examples.

DATASET	HSGT	KmSGT		SelectSGT		SVSGT	
		NS	S	NS	S	NS	S
CANCER	64	56	50	53	49	45	42
CARDIOVASCULAR	63	51	43	41	40	41	42
CONDITION	58803	19939	10482	16200	6971	11685	7800
C3	–	35991	16541	12581	16881	12354	30762
C18	–	25801	35000	15440	16671	4853	3203
MCAT	–	413548	253410	168962	175239	234431	62072

6   Conclusions

In this paper, we present a novel approach for automatic classification of documents into a hierarchy of categories that exploits relevant example selection and works in the transductive setting. The proposed approach is based on a framework that exploits the SGT classifier in internal nodes of the hierarchy. This way, it can pass down examples to more specific categories on the basis of scores returned by the classifier. Documents can also be classified in internal nodes of the hierarchy according to some automatically learned thresholds. The SGT algorithm is used both for learning thresholds and for classifying examples.

Relevant example selection is performed according to two different approaches: the first approach reduces the number of documents by exploiting clustering algorithms, while the second approach identifies and keeps only documents that are at the boundary of the class. Results empirically prove that relevant example selection based on clustering algorithms permits to tame the computational complexity and, at the same time, permits to increase predictive capabilities of the learning algorithm.

For future work, we intend to exploit the proposed transductive learning in a multi-label classifier that permits to classify documents by considering more than one classification dimension.

Acknowledgment

This work is in partial fulfillment of the research objective of the project "DM19410 - The Molecular Biodiversity LABoratory Initiative".

References

1. Bennett, K.P.: Combining support vector and mathematical programming methods for classification. In: Advances in kernel methods: support vector learning, pp. 307–326 (1999)
2. Blum, A.L., Langley, P.: Selection of relevant features and examples in machine learning. Artificial Intelligence 97(1-2), 245–271 (1997)
3. Ceci, M.: Hierarchical text categorization in a transductive setting. In: ICDM Workshops, pp. 184–191. IEEE Computer Society, Los Alamitos (2008)
4. Ceci, M., Malerba, D.: Classifying web documents in a hierarchy of categories: a comprehensive study. J. Intell. Inf. Syst. 28(1), 37–78 (2007)
5. Chen, Y., Wang, G., Dong, S.: Learning with progressive transductive support vector machines. Pattern Recognition Letters 24, 1845–1855 (2003)
6. Dhillon, I.S.: Co-clustering documents and words using bipartite spectral graph partitioning. In: KDD 2001: Proceedings of the seventh ACM SIGKDD international conference on Knowledge discovery and data mining, pp. 269–274. ACM, New York (2001)
7. Dumais, S., Chen, H.: Hierarchical classification of web content. In: Proceedings of the 23rd annual international ACM SIGIR conference on Research and development in information retrieval, pp. 256–263. ACM Press, New York (2000)
8. Esposito, F., Malerba, D., Tamma, V., Bock, H.: Analysis of Symbolic Data. Exploratory methods for extracting statistical information from complex data. In: Classical resemblance measures. Studies in Classification, Data Analysis, and Knowledge Organization, vol. 15, pp. 139–152. Springer, Heidelberg (2000)
9. Gammerman, A., Azoury, K., Vapnik, V.: Learning by transduction. In: Proc. of the 14th Annual Conference on Uncertainty in Artificial Intelligence, UAI 1998, pp. 148–155. Morgan Kaufmann, San Francisco (1998)
10. Hastie, T., Tibshirani, R.: Classification by pairwise coupling. In: NIPS 1997: Proceedings of the 1997 conference on Advances in neural information processing systems, vol. 10, pp. 507–513. MIT Press, Cambridge (1998)

11. Joachims, T.: Text categorization with support vector machines: Learning with many relevant features. In: Nédellec, C., Rouveirol, C. (eds.) ECML 1998. LNCS, vol. 1398, pp. 137–142. Springer, Heidelberg (1998)
12. Joachims, T.: Transductive inference for text classification using support vector machines. In: Proc. of the 16th International Conference on Machine Learning, ICML 1999, pp. 200–209 (1999)
13. Joachims, T.: Transductive learning via spectral graph partitioning. In: Proc. of the 20th International Conference on Machine Learning, ICML 2003 (2003)
14. Kukar, M., Kononenko, I.: Reliable classifications with machine learning. In: Elomaa, T., Mannila, H., Toivonen, H. (eds.) ECML 2002. LNCS (LNAI), vol. 2430, pp. 219–231. Springer, Heidelberg (2002)
15. Liu, H., Motoda, H.: On issues of instance selection. Data Min. Knowl. Discov. 6(2), 115–130 (2002)
16. MacQueen, J.B.: Some methods for classification and analysis of multivariate observations. In: Cam, L.M.L., Neyman, J. (eds.) Proc. of the fifth Berkeley Symposium on Mathematical Statistics and Probability, vol. 1, pp. 281–297. University of California Press, Berkeley (1967)
17. McCallum, A., Rosenfeld, R., Mitchell, T.M., Ng, A.Y.: Improving text classification by shrinkage in a hierarchy of classes. In: Proceedings of the Fifteenth International Conference on Machine Learning, pp. 359–367. Morgan Kaufmann Publishers Inc., San Francisco (1998)
18. Mladenić, D.: Machine learning on non-homogeneus, distribuited text data. PhD thesis, University of Ljubjana, Ljubjana, Slovenia (1998)
19. Platt, J.: Fast training of support vector machines using sequential minimal optimization. In: Advances in kernel methods - support vector learning (1998)
20. Porter, M.F.: An algorithm for suffix stripping. In: Readings in information retrieval, pp. 313–316 (1997)
21. Ruiz, M.E., Srinivasan, P.: Hierarchical text categorization using neural networks. Inf. Retr. 5(1), 87–118 (2002)
22. Sebastiani, F.: Machine learning in automated text categorization. ACM Computing Surveys 34(1), 1–47 (2002)
23. Shi, J., Malik, J.: Normalized cuts and image segmentation. IEEE Trans. Pattern Anal. Mach. Intell. 22(8), 888–905 (2000)
24. Vapnik, V.: The Nature of Statistical Learning Theory. Springer, Heidelberg (1995)
25. Wettschereck, D.: A study of Distance-Based Machine Learning Algorithms. PhD thesis, Oregon State University (1994)
26. Wilson, D.R., Martinez, T.R.: Instance pruning techniques. In: ICML 1997: Proceedings of the Fourteenth International Conference on Machine Learning, pp. 403–411. Morgan Kaufmann Publishers Inc., San Francisco (1997)
27. Yang, Y.: Sampling strategies and learning efficiency in text categorization. In: AAAI Spring Symposium on Machine Learning in Information Access, pp. 88–95 (1996)
28. Zhang, J., Jin, R., Yang, Y., Hauptmann, A.G.: Modified logistic regression: An approximation to svm and its applications in large-scale text categorization. In: Proceedings of the 20th International Conference on Machine Learning (2003)

A Discretization Algorithm for Uncertain Data

Jiaqi Ge[1,*], Yuni Xia[1], and Yicheng Tu [2]

[1] Department of Computer and Information Science,
Indiana University – Purdue University, Indianapolis, USA
{jiaqge,yxia}@cs.iupui.edu
[2] Computer Science and Engineering
University of South Florida
ytu@cse.usf.edu

Abstract. This paper proposes a new discretization algorithm for uncertain data. Uncertainty is widely spread in real-world data. Numerous factors lead to data uncertainty including data acquisition device error, approximate measurement, sampling fault, transmission latency, data integration error and so on. In many cases, estimating and modeling the uncertainty for underlying data is available and many classical data mining algorithms have been redesigned or extended to process uncertain data. It is extremely important to consider data uncertainty in the discretization methods as well. In this paper, we propose a new discretization algorithm called UCAIM (Uncertain Class-Attribute Interdependency Maximization). Uncertainty can be modeled as either a formula based or sample based probability distribution function (*pdf*). We use probability cardinality to build the quanta matrix of these uncertain attributes, which is then used to evaluate class-attribute interdependency by adopting the redesigned *ucaim* criterion. The algorithm selects the optimal discretization scheme with the highest *ucaim* value. Experiments show that the usage of uncertain information helps UCAIM perform well on uncertain data. It significantly outperforms the traditional CAIM algorithm, especially when the uncertainty is high.

Keywords: Discretization, Uncertain data.

1 Introduction

Data discretization is a commonly used technique in data mining. Data discretization reduces the number of values for a given continuous attribute by dividing the range of the attribute into intervals. Interval labels are then used to replace actual data values. Replacing numerous values of a continuous attribute by a small number of interval labels thereby simplifies the original data. This leads to a concise, easy-to-use, knowledge-level representation of mining results [32]. Discretization is often performed

---

* Please note that the LNCS Editorial assumes that all authors have used the western naming convention, with given names preceding surnames. This determines the structure of the names in the running heads and the author index.

P. García Bringas et al. (Eds.): DEXA 2010, Part II, LNCS 6262, pp. 485–499, 2010.

prior to the learning process and has played an important role in data mining and knowledge discovering. For example, many classification algorithms as AQ [1], CLIP [2], and CN2 [3] are only designed for category data, therefore, numerical data are usually first discretized before being processed by these classification algorithms. Assume A is one of the continuous attributes of a dataset, A can be discretized into n intervals as $D = \{[d_0, d_1), [d_1, d_2),\ldots, [d_{n-1}, d_n]\}$, where d_i is the value of the endpoints in each interval. Then D is called as a discretization scheme on attribute A. A good discretization algorithm not only produces a concise view of continuous attributes so that experts and users can have a better understanding of the data, but also helps machine learning and data mining applications to be more effective and efficient [4]. A number of discretization algorithms have been proposed in literature, most of them focus on certain data. However, data tends to be uncertain in many applications [9], [10], [11], [12], [13]. Uncertainty can originate from diverse sources such as data collection error, measurement precision limitation, data sampling error, obsolete source, and transmission error. The uncertainty can degrade the performance of various data mining algorithms if it is not well handled. In previous work, uncertainty in data is commonly treated as a random variable with probability distribution. Thus, uncertain attribute value is often represented as an interval with a probability distribution function over the interval [14], [15].

In this paper, we propose a data discretization technique called Uncertain Class-Attribute Interdependency Maximization (UCAIM) for uncertain data. It is based on the CAIM discretization algorithm and we extend it with a new mechanism to process uncertainty. Probability distribution function (*pdf*) is commonly used to model data uncertainty and *pdf* can be represented as either formulas or samples. We adopt the concept of probability cardinality to build the quanta matrix for uncertain data. Based on the quanta matrix, we define a new criterion value *ucaim* to measure the interdependency between uncertain attributes and uncertain class memberships. The optimal discretization scheme is determined by searching the one with the largest *ucaim* value. In the experiments, we applied the discretization algorithm as the preprocessing step of an uncertain naïve Bayesian classifier [16], and measured the discretization quality by its classification accuracy. Results illustrated that the application of the UCAIM algorithm as a front-end discretization algorithm significantly improve the classification performance.

The paper is organized as following. In section 2, we discuss related work. Section 3 introduces the model of uncertain data. In section 4, we present the ucaim algorithm in detail. The experiments results are shown in section 5, and section 6 concludes the paper.

2   Related Work

Discretization algorithms can be divided into top-down and bottom-up methods according to how the algorithms generate discrete schemes [6]. Both top-down and bottom-up discretization algorithms can be further subdivided into unsupervised and supervised methods [17]. Equal Width and Equal Frequency [5] are well-known unsupervised top-down algorithms, while the supervised top-down algorithms include MDLP [7], CADD (class-attribute dependent discretize algorithm) [18], Information

Entropy Maximization [19], CAIM (class-attribute interdependent maximization algorithm) [8] and FCAIM (fast class-attribute interdependent maximization algorithm) [20]. Since CAIM selects the optimal discretization algorithm that has the highest interdependence between target class and discretized attributes, it is proven to be superior to other top-down discretization algorithms in helping the classifiers to achieve high classification accuracy [8]. FCAIM extends CAIM by using a different strategy to select fewer boundary points during the initialization, which speeds up the process of finding the optimal discretization scheme.

In the bottom-up category, there are widely used algorithms such as ChiMerge [21], Chi2 [22], Modified Chi2 [23], and Extended Chi2 [24]. Bottom-up method starts with the complete list of all continuous value of the attribute as cut-points, so its computational complexity is usually higher than the top-down method [29]. Algorithms like ChiMerge require users to provide some parameters such as significant level and minimal/ maximal interval numbers during the discretization process. [25] illustrates that all these different supervised discretization algorithms can be viewed as assigning different parameters to a unified goodness function, which can be used to evaluate the quality of discretization algorithms. There also exist some dynamic discretization algorithms [26] which are designed for particular machine learning algorithms such as decision tree and naïve Bayesian classifier.

All the algorithms mentioned above are based on certain datasets. To the best of our knowledge, no discretization algorithm has been proposed for uncertain data that are represented as probability distribution functions. In the recent years, there have been growing interests in uncertain data mining. For example, a number of classification algorithms have been extended to process uncertain datasets, as uncertain support vector machine [27], uncertain decision tree [28], uncertain naïve Bayesian classifier. It is extremely important that data preprocessing techniques like discretization properly handle this kind of uncertainty as well. In this paper, we propose a new discretization algorithm for uncertain data.

3 Data Uncertainty Model

When the value of a numerical type attribute A is uncertain, the attribute is called an uncertain attribute (UNA), denoted by A^{u_n} [29]. In uncertain dataset D, each tuple t_i is associated with a feature vector $Vi = (f_{i,1}, f_{i,2}, \ldots, f_{i,k})$ to model its uncertain attributes. $f_{i,j}$ is a probability distribution function (pdf) representing the uncertainty of attribute $A_{ij}^{u_n}$ in tuple t_i. Meanwhile, a probability distribution c_i is assigned to the t_i's uncertain class label C_i as class membership.

In practice, uncertainties are usually modeled in forms of Gaussian distributions, and parameters such as mean μ and standard variance σ are used to describe the Gaussian distributed uncertainty. In this case, uncertain attribute $A_{ij}^{u_n}$ has a formula based probability representation over the interval $[A_{ij}^{u_n}.l, A_{ij}^{u_n}.r]$ as

$$p_{ij} = \int_{A_{ij}^{u_n}.l}^{A_{ij}^{u_n}.r} \frac{1}{\sqrt{2\pi}\sigma} e^{\frac{(x-\mu)^2}{\sigma^2}} dx .$$

Here p_{ij} is the probability distribution of uncertain attribute A_{ij}^{un} which can be seen as a random variable.

In case that the uncertainty cannot be modeled by any mathematical formula expression, a sample based method is often used to model the probability distribution:

$$p_{ij} = \{A_{ij}^{un} | (x_1 : p_1), (x_2 : p_2), \dots (x_i : p_i), \dots, (x_n : p_n)\}.$$

Where, $X = \{x_1, x_2 \dots x_i \dots x_n\}$ is the set of all possible values for attribute A_{ij}^{un}, and p_i is the probability that $A_{ij}^{un} = x_i$.

Not only can the attributes be uncertain, class labels may also contain uncertainty. Instead of having the accurate class label, a class membership may be a probability distribution as following:

$$C_i = \{c | (c_1 : p_1), (c_2 : p_2), \dots, (c_n : p_n)\}$$

Here, $\{c_1, c_2, \dots, c_n\}$ is the set containing all possible class labels, and p_i is the probability that this instance t_i belongs to class c_i.

Table 1 shows an example of an uncertain database. Both attributes and class labels of the dataset are uncertain. Their precise values are unavailable and we only have knowledge of the probability distribution. For attribute 1, its uncertainty is represented as a Gaussian distribution with parameters (μ, σ) to model the *pdf*. For attribute 2, it lists all possible values with their corresponding probabilities for each instance. Note that the uncertainty of class label is always represented in the sample format as the values are discrete.

Table 1. An example of uncertain dataset

ID	Class Type	Attribute 1	Attribute 2
1	T: 0.3, F :0.7	(105, 5)	(100: 0.3, 104: 0.6, 110:0.1)
2	T: 0.4, F:0.6	(110,10)	(102:0.2, 109: 0.8)
3	T: 0.1, F:0.9	(70,10)	(66: 0.4, 72:0.4, 88:0.2)

4   UCAIM Discretization Algorithm

4.1   Cardinality Count for Uncertain Data

According to the uncertainty models, an uncertain attribute A_{ij}^{un} is associated with a *pdf* either in a formula based or sample based format. The probability that the value of A_{ij}^{un} falls in a partition [*left, right*] is:

For formula based *pdf*:

$$P_{A_{ij}^{un}} = \int_{left}^{right} A_{ij}^{un} \cdot f(x) \, dx \tag{1}$$

Where, $A_{ij}^{un} \cdot f(x)$ is the probability density distribution function of A_{ij}^{un}.

For sample based *pdf*:

$$p_{A_{ij}^{u_n}} = \sum_{A_{ij}^{u_n}.x_k \in [left, right]} A_{ij}^{u_n} \cdot p_k \qquad (2)$$

Where, $A_{ij}^{u_n}.x_k$ is the possible value of $A_{ij}^{u_n}$, and $A_{ij}^{u_n}.p_k$ is the probability that $A_{ij}^{u_n} = x_k$.

We assume that class uncertainty is independent to the probability distributions of attribute values. Thus, for a tuple t_i belonging to class C, the probability that its attribute value $A_{ij}^{u_n}$ falls in the interval $[left, right]$ is:

$$P\left(A_{ij}^{u_n} \in [left, right], c_i = C\right) = p_{A_{ij}^{u_n}} * p(c_i = C) \qquad (3)$$

$p_{A_{ij}^{u_n}}$ is defined in formula (1) and (2) and $p(c_i = C)$ is the probability that t_i belongs to class C.

For each class C, we compute the sum of the probabilities that an uncertain attribute A_{ij}^{u} falls in partition $[left, right]$ for all the tuples in dataset D. This summation is called *probabilistic cardinality*. For example, the probability cardinality of partition $P = [a,b)$ for class C is calculated as:

$$P_C(p) = \sum_{i=1}^{n} P\left(A_{ij}^{u_n} \in [a, b)\right) * P(c_i = C) \qquad (4)$$

Probability cardinalities provide us valuable insight during the discretization process and it used to build the quanta matrix for uncertain data, as shown in the next section.

4.2 Quanta Matrix for Uncertain Data

The discretization algorithm aims to find the minimal number of discrete intervals while minimizing the loss of class-attribute interdependency. Suppose F is a continuous numeric attribute, and there exists a discretization scheme D on F, which divides the whole continuous domain of attribute F into n discrete intervals bounded by the endpoints as:

$$D: \{[d_0, d_1), [d_1, d_2), [d_2, d_3), ..., [d_{n-1}, d_n] \} \qquad (5)$$

where d_0 is the minimal value and d_n is the maximal value of attribute F; $d_1, d_2, ..., d_{n-1}$ are cutting points arranged in ascending order.

For certain dataset, every value of attribute F is precise; therefore it will fall into only one of the n intervals defined in (5). However, the value of an uncertain attribute can be an interval or a series of values with associated probability distribution. Therefore, it could fall into multiple intervals. The class membership for a specific interval in (5) varies with different discretization scheme D.

The class variable and the discretization variable of attribute F are treated as two random variables defining a two-dimensional quanta matrix (also known as the contingency table). Table 2 is an example of quanta matrix.

In Table 2, q_{ir} is the probability cardinality of the uncertain attribute $A_F^{u_n}$ which belongs to the ith class and has its value within the interval $[d_{r-1}, d_r]$. Thus, according to formula (4), q_{ir} can be calculated as:

$$q_{ir} = P_c(c = Ci, A_F^{u_n} \in [d_{r-1}, d_r]) \qquad (6)$$

Table 2. Quanta matrix for uncertain attribute A_F^{un} and discretization scheme D

class	Interval					Class Total
	$[d_0,d_1)$	...	$[d_{r-1},d_r)$	...	$[d_{n-1},d_n]$	
C_1	q_{11}	...	q_{1r}	...	q_{1n}	M_{1+}
:	...	...	...	...	...	
C_i	q_{i1}	...	q_{ir}	...	q_{in}	M_{i+}
:	...	...	...	...	...	
C_s	q_{s1}	...	q_{sr}	...	q_{sn}	M_{s+}
Interval Total	M_{+1}		M_{+r}		M_{+n}	M

M_{i+} is the sum of the probability cardinality for objects belonging to the ith class, and M_{+r} is the total probability cardinality of uncertain attribute A_F^{un} that are within the interval $[d_{r-1}, d_r]$, for i = 1, 2... S, and r= 1, 2... n.

The estimated joint probability that uncertain attribute values A_F^{un} is within the interval $D_r= [d_{r-1}, d_r]$, and belong to class C_i can be calculated as:

$$p_{ir} = p(C_i, D_r | A_F^{un}) = \frac{q_{ir}}{M} \tag{7}$$

4.3 Uncertain Class-Attribute Interdependent Discretization

We first introduce the Class-Attribute Interdependency Maximization (CAIM) discretization approach. CAIM is one of the classical discretization algorithms. It generates the optimal discretization scheme by quantifying the interdependence between classes and discretized attribute, and its criterion is defined as following:

$$CAIM(C, D | A_F^{un}) = \frac{\sum_{r=1}^{n} \frac{max_r^2}{M_{+r}}}{n} \tag{8}$$

Where n is the number of intervals, r iterates through all intervals, i.e. $r=1, 2,..., n$, max_r is the maximum value among all q_{ir} values (maximum value within the rth column of the quanta matrix), $i = 1,2,...,S$, M_{+r} is the total probability of continues values of attribute F that are within the interval $D_r= [d_{r-1}, d_r]$.

From the definition, we can see that caim value increases when the values of max_i grow, which corresponds to the increase of the interdependence between the class labels and the discrete intervals. Thus CAIM algorithm finds the optimal discretization scheme by searching the scheme with the highest *caim* value. Since the maximal value max_r is the most significant part in the definition of CAIM criterion, the class which max_r corresponds to is called main class and the larger max_r the more interdependent between this main class and the interval $D_r= [d_{r-1}, d_r]$.

Although CAIM performances well on certain datasets, it encounters new challenges in uncertain case. For each interval, CAIM algorithm only takes the main class into account, but does not consider the distribution over all other classes, which leads to problems when dealing with uncertain data. In an uncertain dataset, each instance no longer has a deterministic class label, but may have a probability distribution over all possible classes and this reduces the interdependency between attribute and class. We use the probability cardinality to build the quanta matrix for uncertain attributes, and we observe that the original *caim* criterion causes problems when handling uncertain quanta matrix. Below we give one such example. Suppose a simple uncertain dataset containing 5 instances is shown in Table 3. Its corresponding quanta matrix is shown in Table 4.

Table 3. An example of uncertain dataset

Attribute (x: p_x)	Class (label: probability)
(0.1:0.3), (0.9: 0.7)	0: 0.9, 1: 0.1
(0.1:0.2), (0.9: 0.8)	0: 0.9, 1: 0.1
(0.9: 1.0)	0: 1.0, 1: 0.0
(0.2:0.7), (0.8: 0.3)	0: 0.1, 1: 0.9
(0.1:0.7), (0.8: 0.2), (0.9:0.1)	0: 0.1, 1: 0.9

From table 3, we can calculate the probability distribution of attribute values x each class as following:

$P(x=0.1, C=0) = 0.3*0.9 + 0.2*0.9 + 0.7*0.1 = 0.52$
$P(x=0.1, C=1) = 0.3*0.1 + 0.2*0.1 + 0.7*0.9 = 0.68$
$P(x=0.2, C=0) = 0.7*0.1 = 0.07$
$P(x=0.2, C=1) = 0.7*0.9 = 0.63$
$P(x=0.8, C=0) = 0.3*0.1 + 0.2*0.1 = 0.05$
$P(x=0.8, C=1) = 0.3*0.9 + 0.2*0.9 = 0.45$
$P(x=0.9, C=0) = 0.7*0.9 + 0.8*0.9 + 0.1*0.1 +1.0*1.0= 2.36$
$P(x=0.9, C=1) = 0.7*0.1 + 0.8*0.1 + 0.1*0.9 = 0.24$

Table 4. Quanta Matrix for the uncertain dataset

class	Interval
	[0, 1]
0	3
1	2

According to formula (8), the *caim* value for the quanta matrix in table 4 is: $caim = 3^2/(3+2) = 1.8$. From the distribution of attribute values in each class, we can see the attribute values of instances in class 0 have a high probability around $x = 0.9$; and those for instances in class 1 are mainly located in the small end around x=0.1 and 0.2. Obviously, $x = 0.5$ is a reasonable cutting point to generate the discretization scheme {[0, 0.5) [0.5, 1]}. After the splitting, the quanta matrix is shown in table 5, whose corresponding *caim* value is:

$$caim = \left(\frac{\frac{1.31^2}{1.31+0.59} + \frac{2.41^2}{2.41+0.69}}{2} \right) = 1.38$$

Table 5. Quanta Matrix after splitting at $x = 0.5$

class	Interval	
	[0, 0.5)	[0.5, 1]
0	0.59	2.41
1	1.31	0.69

The goal of the CAM algorithm is to find the discretization scheme with highest caim value, so {[0, 0.5) [0.5, 1]} will not be accepted as a better discretization scheme, because the *caim* value decreases from 1.8 to 1.38 after splitting at $x = 0.5$.

Data uncertainty obscures the interdependence between classes and attribute values by flatting the probability distributions. Therefore, when the original CAIM criterion is applied to uncertain data, it results in two problems. First, it usually does not create enough intervals in the discretization scheme or it stops splitting too early, which causes the loss of much class-attribute interdependence. Second, in order to increase the caim value, it is possible that the algorithm generates intervals with very small probability cardinalities, which reduces the robustness of the algorithm.

For uncertain data, the attribute-class interdependence is in form of a probability distribution. The original caim definition as in formula (8) ignores this distribution, and only considers the main class. Therefore, we need to revise the original definition to handle uncertain data. Now that uncertainty blurs the attribute-class interdependence and reduces the difference between the main class and the rest of the classes, we try to make the CAIM value more sensitive to change of values in quanta matrix. We propose the uncertain CAIM criterion UCAIM, which is defined as follows:

$$UCAIM \left\langle C, D \middle| A_F^{Un} \right\rangle = \frac{\sum_{r=1}^{n} \frac{\max_v^2 \times Offset_r}{M_{+r}}}{n} \qquad (9)$$

Where

$$Offset_r = \frac{\sum_{i=1, q_{ir} \neq \max_r}^{s} (\max_r - q_{ir})}{s-1} \qquad (10)$$

In formula (9), max_r is the maximum value among all q_{ir} values (maximum value within the rth column of the quanta matrix), $i = 1, 2,..., S$, M_{+r} is the total probability of continues values of attribute F that are within the interval $D_r = [d_{r-1}, d_r]$. $Offset_r$ defined in (10) is the average of the offsets or differences for all other q_{ir} values to max_r.

Because the larger the attribute-class interdependence, the larger the value max_r/M_{+r}, CAIM therefore uses it to identify splitting points in formula (8). In the UCAIM definition we proposed, $Offset_r$ shows how significant the main class is, compared to other classes. When $Offset_r$ is large, it means that within interval r, the probability an instance belongs to the main class is much higher than the other classes, so the interdependence between interval r and the main class becomes is also high. Therefore, we propose the ucaim definition in formula (9) for the following reasons:

1) Compared with max_r/M_{+r}, we multiply it with the factor $Offset_r$ to make the value $Offset_r*max_r/M_{+r}$ more sensitive to interdependence changes, which are usually less significant for uncertain data.

2) The value max_r/M_{+r} may be large merely because M_{+r} is small, which happens when there are not many instances falling into interval r. However, $Offset_r$ does not have such this problem, because it measures the relative relationship between main class and other classes.

Now we apply the new definition to the sample uncertain dataset in Table 3. For the original quanta matrix as in Table 4, the $ucaim$ value is

$$ucaim = \frac{3^2 \times (3-2)}{5} = 1.8$$

For the quanta matrix after splitting as in Table 5, we have

$$S1 = 1.31\text{-}0.59 = 0.72;\ S2 = 2.41\text{-}0.69 = 1.72$$

$$ucaim = \left(\frac{\dfrac{1.31^2}{1.31+0.59} \times 0.72 + \dfrac{2.41^2}{2.41+0.69} \times 1.72}{2} \right) = 1.98$$

Since ucaim value increases after the splitting, the cutting point x = 0.5 will be accepted in the discretization scheme. We can see in this example that ucaim is more effective in finding the interdependence between attribute and class, compared to the original approach.

Table 6. UCAIM discretization algorithm

Algorithm
1. Find the maximal and minimal possible values of the uncertain attribute A_F^{un}, recorded as d_0, d_n. 2. Create a set B of all potential boundary endpoints. For uncertain attribute modelled in sample based *pdf*, simply sort all distinct possible values and use them as the set; for uncertain data modelled as formula based pdf, we use the mean of each distribution to build the set. 3. Set the initial discretization scheme as D:$\{[d_0, d_n]\}$, set GlobalUCAIM = 0 4. initialize k=1; 5. tentatively add an inner boundary, which is not already in D, from B and calculate corresponding UCAIM value 6. after all the tentative additions have been tested, accept the one with the highest value of UCAIM 7. if UCAIM > GlobalUCAIM or k<S, update D with the accepted boundary and set GlobalUCAIM = UCAIM, else terminate 8. set k=k+1 and go to 5 Output: D

4.4 Uncertain Discretization Algorithm

The Uncertain Discretization algorithm is shown in table 6. It consists of two steps: (1) initialization of the candidate interval boundaries and the initial discretization scheme; (2) iterative additions of new splitting points to achieve the highest value of the UCAIM criterion.

The time complexity of ucaim algorithm is similar to caim algorithm. For a single attribute, in the worst case, the running time of caim is $O(Mlog(M))$ [8], and M is the number of distinct values of the discretization attributes. In ucaim algorithm, the additional computation is to calculate S_r, whose time complexity is $O(C*M)$. C is the number of classes, and usually very small comparing with M. Therefore, the addition in time complexity is $O(M)$, and the final running cost of ucaim is still $O(Mlog(M))$. Please note that this algorithm works on certain data as well since certain data can be viewed as a special case of uncertain data.

5 Experiments

In this section, we present the experimental results of *ucaim* discretization algorithm on eight datasets. We compare our technique with the traditional CAIM discretization algorithm, to show the effectiveness of UCAIM algorithm on uncertain data.

5.1 Experiment Setup

The datasets selected to test the ucaim algorithm are: Iris Plants dataset (iris), Johns Hopkins University Ionosphere dataset (ionosphere), Pima Indians Diabetes dataset (pima), Glass Identification dataset (glass), Wine dataset (wine), Breast Cancer Wisconsin Original dataset (breast), Vehicle Silhouettes dataset (vehicle), Statlog Heart dataset (heart). All these datasets were obtained from UCI ML repository [30], and their detailed information is shown in table 7.

Table 7. Properties experimental datasets

Datasets	# of class	# of instance	# of attribute	# of continues attribute
iris	3	1	150	4
ionosphere	2	351	34	34
pima	2	768	8	8
glass	7	214	10	10
wine	3	178	13	13
breast	2	699	10	10
vehicle	4	846	18	18

These datasets are made uncertain by adding a Gaussian distributed noise as in [31][9][14]. For each attribute, we add a Gaussian noise with a zero mean, and a standard variance drawn from the unification distribution [0, 2*f*Sigma]. Here, Sigma is the standard variance of the attribute values, and f is an integer parameter used to define different uncertain level. The value of f is selected from the set {1, 2, 3}. For class label uncertainty, we assume the original class for each instance is the main class, and assign it a probability p_{mc}, and there is a uniform distribution over all other classes. As a comparison, assume the real data does not center in the original position, but sit in the noised value, and the noises are in the same distribution as those described above.

We use the accuracy of uncertain naïve Bayesian classifier to evaluate the quality of discretization algorithms. As the purpose of our experiment is to compare discretization algorithms, when we build the classifier, we ignore nominal attributes. In the experiments, we first compare our UCAIM algorithm for uncertain data with the original algorithm CAIM-O which does not take the uncertainty into account. We also compare the UCAIM with a discretization algorithm named CAIM-M which simply applies CAIM-O algorithm on uncertain quanta matrix (without using the Offset).

5.2 Experiment Results

The accuracy of uncertain naïve Bayesian classifier on these 8 dataset is shown in table 8. Table 9 shows the average classification accuracy under different uncertain level for all three discretization algorithms. Figure 1 shows detailed performance comparison of these algorithms at each uncertain level.

Table 8. Accuracies of the uncertain Naïve Bayesian classifier with different discretization algorithms

dataset	uncertain level	UCAIM	CAIM-M	CAIM-O
iris	f=1, p_{mc}=0.9	88.67%	81.67%	80.58%
	f=2, p_{mc}=0.8	76.67%	73.33%	69.56%
	f=3, p_{mc}=0.7	72.66%	71.33%	63.85%
wine	f=1, p_{mc}=0.9	96.07%	94.38%	85.39%
	f=2, p_{mc}=0.8	93.09%	89.32%	85.39%
	f=3, p_{mc}=0.7	88.44%	73.59%	77.53%
glass	f=1, p_{mc}=0.9	61.07%	57.94%	47.66%
	f=2, p_{mc}=0.8	57.94%	53.27%	37.07%
	f=3, p_{mc}=0.7	50.93%	43.92%	35.98%
ionosphere	f=1, p_{mc}=0.9	74.09%#	81.26%	76.31%
	f=2, p_{mc}=0.8	78.34%	77.13%	72.17%
	f=3, p_{mc}=0.7	77.20%	75.88%	69.66%
pima	f=1, p_{mc}=0.9	77.13%	75.74%	71.35%
	f=2, p_{mc}=0.8	72.32%	70.89%	63.97%
	f=3, p_{mc}=0.7	70.45%	68.66%	62.33%
breast	f=1, p_{mc}=0.9	95.42%	94.27%	93.36%
	f=2, p_{mc}=0.8	90.70%	87.83%	87.14%
	f=3, p_{mc}=0.7	87.83%	83.12%	80.68%
vehicle	f=1, p_{mc}=0.9	61.22%	55.39%	50.13%
	f=2, p_{mc}=0.8	57.44%	52.12%	44.72%
	f=3, p_{mc}=0.7	53.19%	43.61%	37.87%
Heart	f=1, p_{mc}=0.9	82.59%	78.88%	75.33%
	f=2, p_{mc}=0.8	78.19%	72.16%	70.15%
	f=3, p_{mc}=0.7	73.63%	69.95%	67.76%

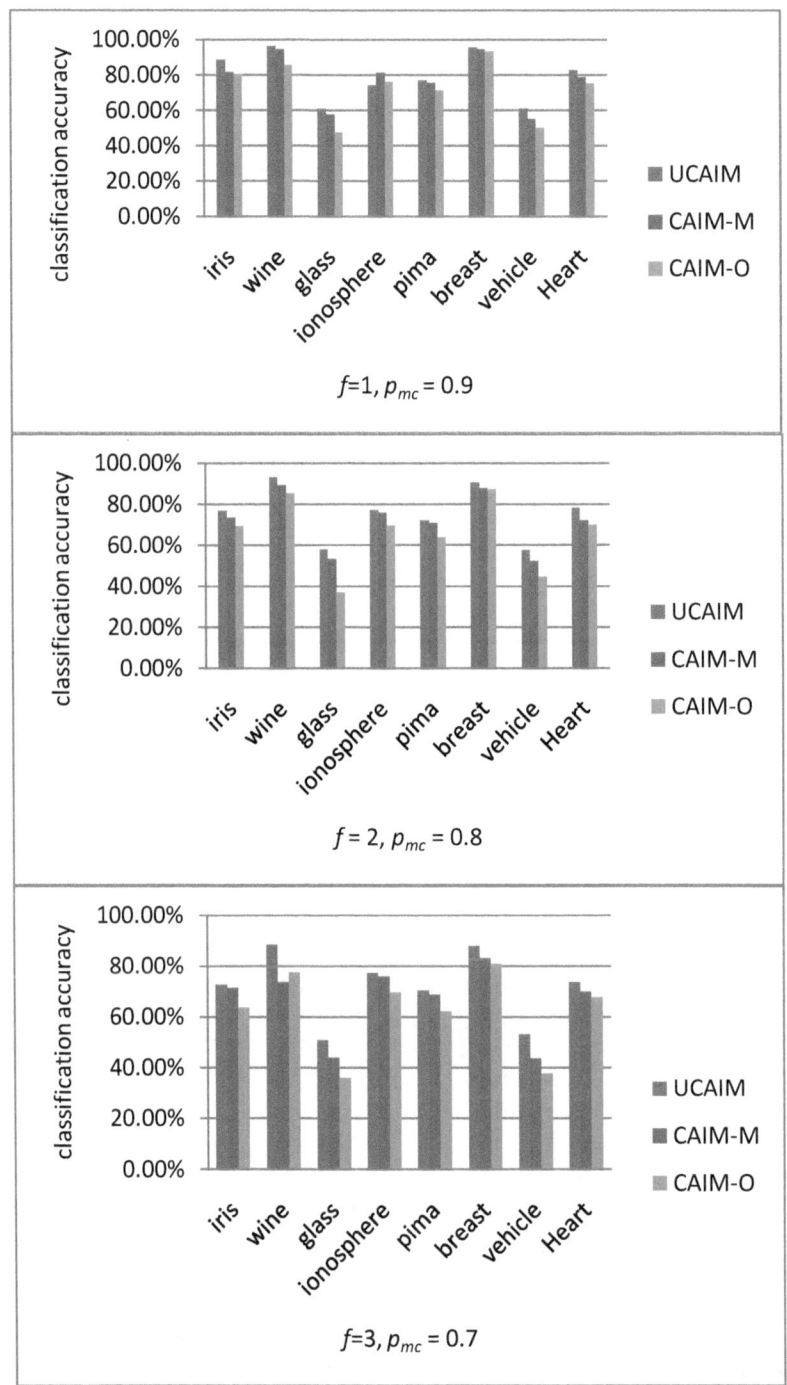

Fig. 1. Classification accuracies with different discretization methods under different uncertain level

Table 9. Average classification accuracies with different discretization methods under different uncertain level

Uncertain level	UCAIM	CAIM-M	CAIM-O
$f=1, p_{mc}=0.9$	79.53%	77.44%	72.51%
$f=2, p_{mc}=0.8$	78.19%	72.16%	70.15%
$f=3, p_{mc}=0.7$	71.79%	66.26%	61.96%

From table 8, table 9 and figure 1, we can see that UCAIM outperforms the other two algorithms in most cases. Particularly, UCAIM has a more significant performance improvement for datasets with higher uncertainty. That is because UCAIM utilizes extra information such as probability distribution of uncertain data, and employs the new criterion to retrieve the class-attribute interdependency which is not obvious when data is uncertain. Therefore, the discretization process of UCAIM is more sophisticated and comprehensive, and the discretized data can help data mining algorithms such as Naïve Bayesian classifier to gain a higher accuracy.

6 Conclusion

In this paper, we propose a new discretization algorithm for uncertain data. We employ both the formula based and sample based probability distribution function to model data uncertainty. We use probability cardinality to build the uncertain quanta matrix, which is then used to calculate *ucaim* to find the optimal discretization scheme with highest class-attribute interdependency. Experiments show that our algorithm can help the naïve Bayesian classifier to reach higher classification accuracy. We also observe that the proper use of data uncertainty information can significantly improve the quality of data miming results and we plan to explore more data mining approaches for various uncertain models in the future.

References

1. Kaufman, K.A., Michalski, R.S.: Learning from inconsistent and noisy data: the AQ18 approach. In: Proceeding of 11th International Symposium on Methodologies for Intelligent Systems (1999)
2. Cios, K.J., et al.: Hybrid inductive machine learning: an overview of clip algorithm. In: Jain, L.C., Kacprzyk, J. (eds.) New Learning Paradigms in Soft Computing, pp. 276–322. Springer, Heidelberg (2001)
3. Clark, P., Niblett, T.: The CN2 Algorithm. Machine Learning 3(4), 261–283 (1989)
4. Catlett, J.: On Changing Continues Attributes into Ordered Discrete Attributes. In: Kodratoff, Y. (ed.) EWSL 1991. LNCS, vol. 482, pp. 164–178. Springer, Heidelberg (1991)
5. Liu, H., Hussain, F., Tan, C.L., Dash, M.: Discretization: An Enable Technique. Data Mining and Knowledge Discovery 6, 393–423 (2002)

6. Fayyad, U.M., Irani, K.B.: Multi-Interval Discretization of Continues- Valued Attributes for Classification Learning. In: Proceedings of the 13th Joint Conference on Artificial Intelligence, pp. 1022–1029 (1993)
7. Hanse, M.H., Yu, B.: Model Selection and the Principle of Minimum Description Length. Journal of the American Statistical Association (2001)
8. Kurgan, L.A.: CAIM Discretization Algorithm. In: IEEE Transactions on Knowledge and Data Engineering, p. 145 (2004)
9. Aggarwal, C.C., Yu, P.: A framework for clustering uncertain data streams. In: IEEE International Conference on Data Engineering, ICDE (2008)
10. Cormode, G., McGregor, A.: Approximation algorithms for clustering uncertain data. In: Principle of Data base System, PODS (2008)
11. Kriegel, H., Pfeifle, M.: Density-based clustering of uncertain data. In: ACM SIGKDD Conference on Knowledge Discovery and Data Mining (KDD), pp. 672–677 (2005)
12. Singh, S., Mayfield, C., Prabhakar, S., Shah, R., Hambrusch, S.: Indexing categorical data with uncertainty. In: IEEE International Conference on Data Engineering (ICDE), pp. 616–625 (2007)
13. Kriegel, H., Pfeifle, M.: Hierarchical density-based clustering of uncertain data. In: IEEE International Conference on Data Mining (ICDM), pp. 689–692 (2005)
14. Aggarwal, C.C.: On Density Based Transforms for uncertain Data Mining. In: IEEE International Conference on Data Engineering, ICDE (2007)
15. Aggarwal, C.C.: A Survey of Uncertain Data Algorithms and Applications. IEEE Transactions on Knowledge and Data Engineering 21(5) (2009)
16. Ren, J., et al.: Naïve Bayes Classification of Uncertain Data. In: IEEE International Conference on Data Mining (2009)
17. Dougherty, J., Kohavi, R., Sahavi, M.: Supervised and Unsupervised Discretization of Continues Attributes. In: Proceedings of the 12th International Conference on Machine Learning, pp. 194–202 (1995)
18. Linde, Y., Buzo, A., Gray, R.M.: An Algorithm for Vector Quantizer Design. IEEE Transactions on Communications 28, 84–95 (1980)
19. Wong, A.K.C., Chiu, D.K.Y.: Synthesizing Statistical Knowledge from Incomplete Mixed-Mode Data. IEEE Transactions on Pattern Analysis and Machine Intelligence 9, 796–805 (1987)
20. Kurgan, L., Cios, K.J.: Fast Class-Attribute Interdependence Maximization (CAIM) Discretization Algorithm. In: Proceeding of International Conference on Machine Learning and Applications, pp. 30–36 (2003)
21. Kerber, R.: ChiMerge: discretization of numeric attributes. In: Proceeding of 9th International Conference on Artificial Intelligence, pp. 123–128 (1992)
22. Liu, H., Setiono, R.: Feature Selection via discretization. IEEE Transactions on knowledge and Data Engineering 9(4), 642–645 (1997)
23. Tray, F., Shen, L.: A modified Chi2 algorithm for discretization. IEEE Transactions on Knowledge and Data Engineering 14(3), 666–670 (2002)
24. Su, C.T., Hsu, J.H.: An extended Chi2 algorithm for discretization of real value attributes. IEEE Transactions on Knowledge and Data Engineering 17(3), 437–441 (2005)
25. Jing, R., Breitbart, Y.: Data Discretization Unification. In: IEEE International Conference on Data Mining, p. 183 (2007)
26. Berzal, F., et al.: Building Multi-way decision Trees with Numerical Attributes. Information Sciences 165, 73–90 (2004)
27. Bi, J., Zhang, T.: Support Vector Machines with Input Data Uncertainty. In: Proc. Advances in Neural Information Processing Systems (2004)

28. Qin, B., Xia, Y., Li, F.: DTU: A Decision Tree for Classifying Uncertain Data. In: Theera-munkong, T., Kijsirikul, B., Cercone, N., Ho, T.-B. (eds.) PAKDD 2009. LNCS, vol. 5476, pp. 4–15. Springer, Heidelberg (2009)
29. Cheng, R., Kalashnikov, D., Prabhakar, S.: Evaluating Probabilistic Queries over Impre-cise Data. In: Proceedings of the ACM SIGMOD, pp. 551–562 (2003)
30. Asuncion, A., Newman, D.: UCI machine learning repository (2007), http://www.ics.uci.edu/mlearn/MLRepository.html
31. Aggarwal, C.C., Yu, P.S.: Outlier Detection with Uncertain Data. In: SIAM International Conference on Data Mining (2009)
32. Han, J., Kamber, M.: Data Mining: Concepts and Techniques, 2nd edn. Morgan Kaufmann, San Francisco (2006)

Author Index

GPSR Compliance

*The European Union's (EU) General Product Safety Regulation (GPSR)
is a set of rules that requires consumer products to be safe and our
obligations to ensure this.*

*If you have any concerns about our products, you can contact us on
ProductSafety@springernature.com*

In case Publisher is established outside the EU, the EU authorized
representative is:

Springer Nature Customer Service Center GmbH
Europaplatz 3
69115 Heidelberg, Germany

Batch number: 09473985

Printed by Printforce, the Netherlands